Electrical Brain Stimulation: Mechanisms and Modulation of Neuronal Activity

Zhouyan Feng

Electrical Brain Stimulation: Mechanisms and Modulation of Neuronal Activity

Springer

Zhouyan Feng
College of Biomedical Engineering
and Instrument Science
Zhejiang University
Hangzhou, Zhejiang, China

ISBN 978-981-95-4144-7 ISBN 978-981-95-4145-4 (eBook)
https://doi.org/10.1007/978-981-95-4145-4

This Springer imprint is published by the registered company Springer Nature Singapore Pte Ltd.
The registered company address is: 152 Beach Road, #21-01/04 Gateway East, Singapore 189721, Singapore

Preface

The brain is the most complex organ in human body, playing a crucial role in regulating physiological functions and maintaining normal life activity. It is responsible for essential functions such as memory, thinking, and body control. Brain disorders can severely impact human health. Understanding, protecting, and creating the brain are the three major goals of the Human Brain Project. Neural electrical stimulation is one of the most important technologies to achieve these goals, playing crucial roles in exploring the brain, treating brain diseases, and developing brain-like intelligence. It includes deep brain stimulation (DBS), a cutting-edge technique for treating brain disorders. DBS uses implanted electrodes to deliver electrical stimulations to specific brain targets, controlling symptoms of certain brain diseases. Compared to techniques that permanently alter diseased brain regions, such as surgical tissue removal or ablation, DBS offers several advantages. It is minimally invasive, has fewer side effects, provides controllable and reversible effects, and results in fewer complications. DBS has shown promise for treating neurological disorders that resist conventional medication. Unlike drug therapy, electrical stimulation can target specific brain regions, and its "dose" can be adjusted in real time through parameter programming.

DBS began its clinical application in the 1980s. The groundbreaking work of neurosurgeon Alim Louis Benabid and his colleagues at the University of Grenoble, France, is regarded as a milestone in developing DBS. Currently, DBS has been successfully used in treating Parkinson's disease, movement disorders, and refractory epilepsy. Studies have suggested that it also has potentials for treating other neurological and psychiatric disorders, including refractory obsessive-compulsive disorder, depression, anorexia, drug addiction, and Alzheimer's disease. However, the action mechanisms of DBS remain inconclusive and its stimulation paradigms are limited, which restricts its applications for effectively treating more brain diseases.

Brain functions occur through the generation and transmission of diverse neural electrical signals. In theory, external electrical stimulations delivered into the brain can control symptoms of brain diseases by modulating or replacing abnormal neural activity, thus achieving therapeutic efficacies. However, current DBS paradigms are limited. Treatments for various conditions typically rely only on high-frequency pulse stimulation with constant parameters—a strategy that works well for Parkinson's disease treatment but lacks the variety of stimulation options needed for treating other conditions. Clinical experience has preceded mechanistic understanding in DBS applications. Therefore, in-depth research into the action mechanisms of electrical brain stimulations and development of new stimulation paradigms are required to advance and expand DBS applications. This book summarizes our research in this field, covering basic theory, experimental methods, our findings, and mechanism analyses in brain stimulations.

DBS uses thin, needle-like electrodes to minimize tissue damage during implantation. The electrical current delivered through the tiny tips (contacts) of these electrodes must elicit sufficient neural responses to achieve therapeutic effects. Due to its shortest chronaxie, the axon is the most sensitive neuronal structure to the narrow pulses typically used in DBS. Additionally, the action potentials produced in the axons can propagate widely through neural networks, expanding the stimulation effects. As a result, axonal responses play a crucial role in DBS. This book reveals how axonal stimulations affect activity of neurons and their networks.

Neurons and their networks exhibit highly nonlinear characteristics. The classic Hodgkin-Huxley model has revealed the foundation of the nonlinear property in neuronal membrane. In fact, from membrane ion channels at the molecular level, to neuronal structures—such as dendrites, soma, axons, and synapses—and further to the neural networks, all are nonlinear in their fundamental mechanisms. In a nonlinear system, different time-varying stimulations can produce numerous neuromodulation effects, creating extensive possibilities for designing diverse stimulation patterns. Like how different drugs treat diseases, stimulation patterns with different "efficacies" and "dosages" can provide more options for clinical applications, thus expanding the DBS scope. Developing the new stimulation patterns requires in-depth preclinical investigations, including animal experiments and theoretical analyses. These are essential for understanding how neurons and their networks respond to various time-varying stimulations, providing guidance for clinical applications.

With this goal in mind, we conducted a series of stimulation experiments in rat brain in-vivo by utilizing the unique structure of hippocampal region. We verified that narrow-pulse high-frequency stimulations can generate intermittent axon block, demonstrated various neuromodulation effects of different pulse sequences, and revealed their possible underlying mechanisms. Our research shows that the intermittent block induced by high-frequency stimulation can make the axon behave like a taut "string". In this situation, small parameter changes in the pulse sequences can lead to substantially different responses in neurons and their networks, potentially offering options to meet various neuromodulation needs.

This book consists of two parts. Part I (including four chapters) introduces the basics of neuro-electrophysiology, electrical stimulations, and experimental methods. It covers the fundamental structures of neurons and nervous systems, neural circuits in the hippocampus, rat experimental techniques, and methods for recording and processing neural signals. Part II (including five chapters) describes various neuromodulation effects we found through in-vivo rat experiments using different axonal stimulations, along with discussions on their possible mechanisms. These stimulations include pulse sequences with constant and time-varying parameters, alternating polarities, and so on. This part also explores the effects and mechanisms of high-frequency stimulations in controlling epileptiform discharges. This book, featuring vivid illustrations and practical methods, can serve as a valuable reference for researchers, students, and enthusiasts in the fields of neurophysiology, neural signal analysis, neural engineering, neuromodulation, brain stimulation, and brain-computer interfaces.

I am profoundly grateful for the guidance from my supervisors and help from collaborators. I especially thank Prof. Jiguang Ge, my undergraduate thesis advisor and master's supervisor. During my time in his lab, I laid a solid foundation in hardware and software design and instrument fabrication. I am grateful to my doctoral supervisor, Prof. Xiaoxiang Zheng, who encouraged me to freely explore my research interests, enabling my entry into the field of neuro-electrophysiology. I appreciate my postdoctoral advisor, Prof. Dominique M. Durand, who sponsored my two-year work at the Neural Engineering Center at Case Western Reserve University and continued our collaboration for years. This experience significantly advanced my research in neurophysiology and neural engineering. Finally, I thank Prof. Xuefeng Wei from the College of New Jersey, whose expertise in mathematical modeling and simulation has been instrumental in my research.

I am deeply grateful to my graduate students. The establishment of our laboratory from scratch, the completion of research projects, and the publication of academic papers all reflect their diligent efforts and invaluable contributions. They are Zhaoxiang Wang, Lvpiao Zheng, Zheshan Guo, Yue Yuan, Yifan Hu, Ying Yu, Yang Wang, Jing Wang, Xiaojing Zheng, Hanhan Hu, Jiayue Cao, Ziyan Cai, Wenjie Zhou, Weijian Ma, Gangsheng Yang, Yipeng Xu, Xiangyu Ye, Lu Huang, Yufang Zhu, Cong Tian, Na Hu, Dan Chen, Chen Qiu, Zhenhua Hu, Pengju Yang, Yunyun Wu, Chuchu Lu, Xiaoqian Chen, Qianjiang Xiao, Haoyu Xing, Bailu Chen, and Lei Guang, among others. Many undergraduate students have also made significant contributions to our research, particularly those who completed their undergraduate theses in our laboratory. I extend my gratitude to all of them! The major research work described in this book was conducted at the College of Biomedical Engineering and Instrument Science, Zhejiang University. I am profoundly thankful for the support and assistance from the college leadership and my colleagues. Lastly, I am eternally grateful for the consistent support and help from my family over the years. I especially thank my life partner, Prof. Zheng Xu, whose encouragement, motivation, and support have strengthened my resolve to complete this book.

I dedicate this book to my alma mater, Zhejiang University, where I have studied and worked all my life.

Zhouyan Feng
Professor, College of Biomedical
Engineering and Instrument Science
Zhejiang University
Hangzhou, Zhejiang, China
fengzhouyan@zju.edu.cn

Competing Interests The author has no competing interests to declare that are relevant to the content of this manuscript.

Contents

Abbreviations

$[K^+]_o$	Extracellular K^+ concentration
$[Na^+]_i$	Intracellular Na^+ concentration
4-AP	4-aminopyridine
ACSF	Artificial cerebrospinal fluid
AD	After-discharge
ADC	Analog to digital converter
A-HFS	Antidromic high-frequency stimulation
AIS	Axon initial segment
ANOVA	Analysis of variance
AP	Action potential
APS	Antidromically-evoked population spike
ASE	Antidromic stimulation electrode
ATS	Antidromic test stimulation
CA	Cornu ammonis
CA1	Cornu ammonis field 1 of the hippocampal formation
CA3	Cornu ammonis field 3 of the hippocampal formation
CAP	Compound action potential
CF	Commissural fiber
CNS	Central nervous system
CSD	Current source density
DAC	Digital to analog converter
DAQ	Data acquisition
DBS	Deep brain stimulation
DG	Dentate gyrus
ECF	Extracellular fluid
ECoG	Electrocorticogram
EEG	Electroencephalogram
EGTA	Ethylene glycol tetraacetic acid
EPSP	Excitatory postsynaptic potential
fEPSP	Field excitatory postsynaptic potential
FES	Functional electrical stimulation

FFT	Fast Fourier Transform
FR	Firing rate
GABA	γ-aminobutyric acid
gcl	Granule cell layer of the dentate gyrus
*h*_Na	Inactivation variable of Na^+ channel
hf	Hippocampal fissure
HFS	High-frequency stimulation
I/O	Input-output port
IN	Interneuron
I-O	Input-output test
IPG	Inter-phase gap
IPI	inter-pulse-interval
IQR	Interquartile range
ISI	Inter-spike-interval
JPSSD	Joint peri-stimulus scatter diagram
JPSTH	Joint peri-stimulus time histogram
L	Latency
LFP	Local field potential
LTD	Long-term depression
LTP	Long-term potentiation
ml	Molecular layer of the dentate gyrus
MUA	Multiple unit activity
NI	National Instruments
Node	Node of Ranvier
O-HFS	Orthodromic high-frequency stimulation
OPS	Orthodromically-evoked population spike
OSE	Orthodromic stimulation electrode
OTS	Orthodromic test stimulation
pc	Principal component
pcl	Pyramidal cell layer
pl	Polymorphic layer of the dentate gyrus
PPD	Paired-pulse depression
PPF	Paired-pulse facilitation
PPS	Paired-pulse stimulation
PS	Population spike
PSD	Power spectral density
PST	Post-stimulation time
PSTH	Post-stimulus time histogram or peri-stimulus time histogram
PTX	Picrotoxin
Pyr	Pyramidal neuron
RE	Recording electrode
RMS	Root mean square
RNS	Responsive neurostimulation
RP	Refractory period
SCS	Spinal cord stimulation

SD	Spreading depression or standard deviation
sm	Stratum moleculare
SNS	Sacral nerve stimulation
so	Stratum oriens
SPW	Small positive-going waveform
sr	Stratum radiatum
STFT	Short-time Fourier transform
SUA	Single unit activity
TEA	Tetraethyl ammonium
TNS	Trigeminal nerve stimulation
TTX	Tetrotoxin
US	Unit spike
VHC	Ventral hippocampal commissure
VNS	Vagus nerve stimulation
WT	Wavelet transform

Part I
Fundamentals and Experimental Methods

Chapter 1
Fundamentals of Neuro-electrophysiology

Abstract This chapter describes the fundamentals of neuro-electrophysiology. The electrical properties of neuronal membranes can be simulated by equivalent circuits composed of resistance and capacitance. Ion concentration differences across the membrane, combined with varying membrane permeability to ions, form the basis of resting and action potentials. The chapter introduces quantitative membrane indices (including time constant, space constant, rheobase and chronaxie), membrane cable equations, and Hodgkin-Huxley computational model. It also covers electrical stimulation approaches for brain neurons, including intracellular versus extracellular modes, current versus voltage modes, and unipolar versus bipolar modes. Finally, it addresses stimulation safety rules to prevent tissue damage and electrode corrosion.

Keywords Neuro-electrophysiology · Membrane properties · Time constant · Space constant · Action potential · Hodgkin-Huxley model · Neuron stimulation · Stimulation safety

The nervous system plays a crucial role in regulating and controlling the activities of humans and other organisms. Its basic structural and functional unit is neuron. Electrical activity generated by neuronal membranes is essential for nervous system functions. The early understanding of bioelectricity and subsequent developments in related theories and technologies have laid the foundations for neuro-electrophysiology.

1.1 Brief History of Bioelectricity

The discovery of bioelectricity in fishes can date back to ancient times (Wickens 2015). Egyptian murals depicted Nile catfish capable of producing powerful electrical currents. People used these fish-generated currents to punish prisoners and treat ailments like headaches. However, without understanding the true nature of electricity

Z. Feng, *Electrical Brain Stimulation: Mechanisms and Modulation of Neuronal Activity*, https://doi.org/10.1007/978-981-95-4145-4_1

and bioelectricity, they simply attributed these mysterious phenomena to ghostly forces.

In 1791, Luigi Galvani (1737–1798), an anatomist and physiologist at the University of Bologna in Italy, published a paper on muscle contractions caused by touching metal arches to frog leg muscles or nerves (Piccolino 1998). Galvani pointed out that the muscle contractions were caused by electrical energy released from muscle cells. This discovery marked a significant milestone in electrophysiology. His finding generated immediate excitement in the academic community and caught the attention of prominent Italian physicist Alessandro Volta (1745–1827). However, Volta offered a different explanation. He argued that the current causing muscle contractions came not from the muscles themselves, but from chemical reactions between the different metals in the arches and the liquid on the muscles. These reactions created a potential difference to generate current through the muscles. This disagreement sparked a famous debate. Both Galvani and Volta pursued further research in two distinct directions, and eventually became pioneers in electrophysiology and electrical engineering (with Volta inventing chemical batteries), respectively. Their achievements were closely intertwined with the advances in electrical physics during that period. For more details about this scientific debate, refer to Sect. 12.2 History in the textbook "Introduction to Biomedical Engineering" (Enderle and Bronzino 2012).

Since Galvani's discovery over 200 years ago, electrophysiology has made significant progress, driven by the advances in electrical science and the inventions of measurement technologies and instruments. The field development has been marked by several key milestones. When Galvani first discovered bioelectricity, no instruments existed for measuring it directly. In 1827, the invention of ammeters enabled scientists to measure potential differences between injured and intact muscle tissues. By 1849, German physiologist Emil du Bois Reymond became another pioneer in electrophysiology by successfully recording resting and action potentials in nerves.

It took over 100 years for scientists to understand the mechanism of bioelectricity after its discovery. In 1902, German scientist Julius Bernstein proposed the cell membrane theory of bioelectricity. He suggested that a potential difference exists across cell membrane, related to the membrane permeability to potassium ions. This theory led to significant advances in electrophysiology. In the 1920s, American scientists Herbert Spencer Gasser and Joseph Erlanger made a breakthrough by using a cathode ray tube to observe and study the potential conduction on nerve fibers, earning them the 1944 Nobel Prize. In the 1940s and 1950s, British scientists Alan Lloyd Hodgkin and Andrew Fielding Huxley studied the conductivity changes of potassium and sodium ion channels in cell membranes by using voltage clamp technology with glass microelectrodes. They created a computational model for these ion channels. Additionally, Australian scientist John Carew Eccles discovered excitatory and inhibitory synaptic potentials. In 1963, Eccles, Hodgkin, and Huxley jointly received the Nobel Prize in Physiology or Medicine. Another major advancement came in 1976 when German scientists Erwin Neher and Bert Sakmann invented the patch clamp technique—a crucial method for studying membrane ion channels—leading to their

1991 Nobel Prize. These milestones showcase how science and technology mutually enhance each other to continuously build theoretical foundations and technical supports in the electrophysiology field. Today, extensive researches have deepened our understanding of cell membrane composition, structure, and electrical properties.

1.2 Electrical Properties and Membrane Potentials in Neurons

1.2.1 Electrical Properties of Neuronal Membrane

The neuronal membrane consists mainly of a lipid bilayer containing membrane proteins. It exhibits electrical properties of both resistance and capacitance.

1. **Membrane Resistance**

The electrical resistivity of membrane lipid bilayer is extremely high, ranging from 10^{13} to 10^{15} Ω·cm. In contrast, the resistivity of the fluids on either side of the membrane—the intracellular fluid (cytosol) and extracellular fluid—is only 60–80 Ω·cm. This means that the membrane resistivity is more than ten orders of magnitude higher than the surrounding fluid. Additionally, the membrane resistance is not fixed, but varies with multiple factors, including changes in transmembrane potential and in intracellular and extracellular environments. When neurons shift between resting and excited states, their membrane resistance can span several orders of magnitude.

Similar to the definition of resistance in physics, membrane resistance (R_m) is defined as

$$R_m = \rho_m \frac{l}{s} \tag{1.1}$$

Its unit is Ω. In this formula, ρ_m is electrical resistivity—the resistance per unit length across unit cross-sectional area—measured in Ω·cm. l is length and s is cross-sectional area. When calculating membrane resistance (or transmembrane resistance) for current flowing between the inside and outside of a neuron, the l represents membrane thickness while s represents membrane area. For example, in a slender neuronal axon, the l represents the thickness of axonal membrane and the s represents the membrane area of the axonal surrounding cylindrical surface. Beyond membrane resistance, other neuronal components also have resistance, calculated using the same formula. When calculating the axoplasmic resistance as current flows along the neuronal axon, the l represents the axon length and the s represents its cross-sectional area. Note that the definitions of length and cross-sectional area change based on the current path when calculating resistance in different neuronal components.

The reciprocal of membrane resistance R_m is the membrane conductivity G_m (or g_m), measured in Siemens (S). Membrane permeability to charged ions is typically expressed by the ionic conductivity.

2. **Membrane Capacitance**

The lipid bilayer in neuronal membrane has very low conductivity, making the membrane act like an insulator—similar to the dielectric layer in a capacitor. The ion-rich intracellular and extracellular fluids on both sides of the lipid bilayer act like the parallel plates of a capacitor. This structure of conductive fluids separated by the thin insulating membrane creates the membrane capacitance.

The capacitance of a parallel-plate capacitor is directly proportional to both the dielectric constant of insulating layer and the plate area, while inversely proportional to the distance between the plates. Membrane capacitance follows the same principle and is typically represented as capacitance per unit area, known as specific capacitance ($\hat{C}_m$), which is defined as

$$\hat{C}_m = \frac{\varepsilon_m}{l} \tag{1.2}$$

Here ε_m is the dielectric constant of membrane, and l is the thickness of membrane lipid bilayer. The ε_m of neuronal membrane is remarkably great at approximately 3–5, even greater than that of polypropylene plastic film used in common capacitors, which is only 2.25. The thickness of neuronal membranes is only 5–10 nm, resulting in a great membrane capacitance ($\hat{C}_m$) of about 1 $\mu F/cm^2$. This $\hat{C}_m$ is stable, varying hardly between resting and excited membrane states. Additionally, the $\hat{C}_m$ values are similar across different neuron types.

Changes in transmembrane potential can generate charging or discharging currents across membrane capacitance. This capacitive current can interfere with the measurements of ionic currents flowing through membrane channels. Using voltage clamp to fix the membrane potential and eliminate the capacitive current can allow accurate measurements of the ionic currents. By using this technique, Hodgkin and Huxley created the famous mathematical membrane model describing ion channel dynamics, as detailed in Sect. 1.3 (note that transmembrane potential is simplified as membrane potential).

Membrane capacitance prevents any sudden changes in membrane potential. When subjected to an electrical stimulation with a sudden change in potential or current (e.g., a step waveform stimulation), the membrane potential always undergoes a transient state before reaching a steady state. The duration of this transient state can be described by a time constant.

3. **Time Constant**

Neuronal membrane at rest can be represented by an equivalent RC circuit with a resistor (R_m) and a capacitor (C_m) in parallel (Fig. 1.1A). As illustrated in Fig. 1.1B, after a glass pipette stimulation-electrode is inserted into the cell membrane, a square current is applied between this microelectrode and a grounded electrode outside

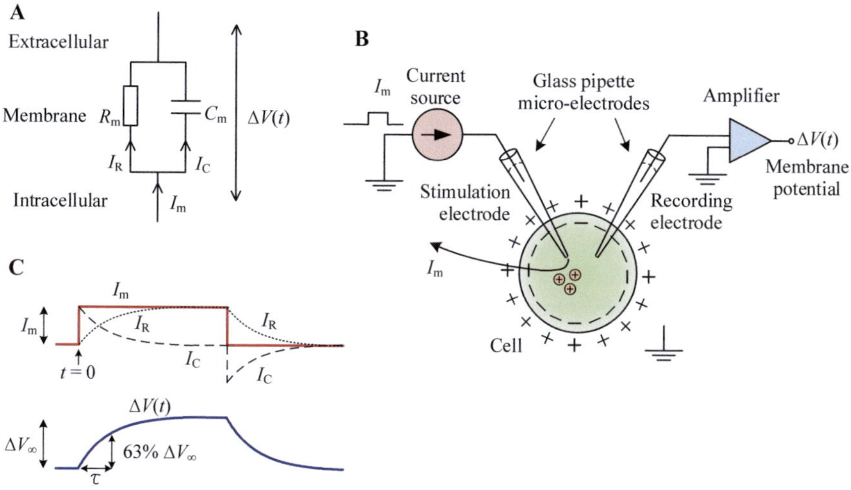

Fig. 1.1 Equivalent circuit for quantitative analysis of membrane potential changes. **A** Equivalent RC circuit. **B** Schematic diagram showing membrane potential measurement during a square current injection into a cell. **C** Change in membrane potential $\Delta V(t)$ caused by step changes in membrane current I_m

the cell. When the duration of square current exceeds the charging time of the membrane capacitor, the step current changes in the rising and falling edges can cause the membrane potential to change exponentially over time toward a steady state (Fig. 1.1C), rather than a step potential change. This transient period can be characterized by a time constant.

When a small-amplitude step current (I_m) is applied across the membrane at time $t = 0$ (a subthreshold stimulus that cannot induce an action potential), the membrane potential begins to deviate from its resting value by $\Delta V(t)$. The applied I_m divides into two components as it flows through the membrane: a capacitive current (I_C) and a resistive current (I_R) (Fig. 1.1A, C). According to the Kirchhoff's current law, we have

$$I_m = I_C + I_R = C_m \frac{d\Delta V(t)}{dt} + \frac{\Delta V(t)}{R_m} \tag{1.3}$$

The solution to this differential equation is

$$\Delta V(t) = \Delta V_\infty (1 - e^{-t/\tau}) \tag{1.4}$$

Here $\Delta V_\infty = I_m R_m$ is the steady-state value of $\Delta V(t)$, representing the maximum change in membrane potential induced by the stimulus I_m. The time constant τ equals $R_m C_m$. When $t = \tau$, $\Delta V(t) = \Delta V_\infty(1 - e^{-1}) = 0.63\Delta V_\infty$. This means that after applying a step current, the change of membrane potential can reach 63% of its maximum value within one time constant. Conversely, by measuring how long it takes for the membrane potential to reach $63\%\Delta V_{\infty}$, we can determine the membrane time constant τ. Obviously, the larger the membrane resistance and capacitance, the longer the time constant τ becomes, resulting in a longer transient period of membrane potential. In contrast, the membrane potential can approach its steady-state level more quickly. After a duration of 5τ, the membrane potential reaches over 99% of its steady-state value, indicating that the transient period approximately ends and the potential enters its steady-state. When the applied current is removed, the membrane potential undergoes a transient change in the opposite direction, immediately following the falling edge of the square current. The time constant τ remains the same.

Let $V_{\mathrm{m}}(t)$ represent the absolute value of membrane potential and V_{rest} be the resting potential, then

$$V_{\mathrm{m}}(t) = \Delta V_{\infty}(1 - \mathrm{e}^{-t/\tau}) + V_{\mathrm{rest}} \tag{1.5}$$

The direction of membrane potential change—whether depolarization or hyperpolarization—depends on the direction of transmembrane current (simplified as membrane current). When the current flows from inside to outside, it causes membrane depolarization (as shown in Fig. 1.1B), raising the negative resting potential. Conversely, current flowing from outside to inside results in hyperpolarization.

4. **Rheobase and Chronaxie**

The RC circuit in Fig. 1.1A can only describe the membrane response to subthreshold stimulation when membrane resistance remains constant approximately. Once the change of membrane potential exceeds the activation threshold, an action potential (AP) is initiated through a series of ion channel opening and closing processes. During the processes, the membrane resistance changes significantly (refer to Sect. 1.3). Let's denote the AP threshold as V_{th} and the maximum potential change caused by the applied current I_{m} as $\Delta V_{\infty} = I_{\mathrm{m}}R_{\mathrm{m}}$. Then, the minimum I_{m} required to raise the membrane potential to the threshold V_{th} is known as rheobase I_{rh}:

$$I_{\mathrm{rh}} = \frac{V_{\mathrm{th}}}{R_{\mathrm{m}}} \tag{1.6}$$

This current I_{rh} must be maintained indefinitely (i.e. $t = \infty$) to raise the membrane potential to threshold V_{th}. When the capacitor C_{m} in the RC circuit is fully charged, its branch becomes open, and the applied current I_{rh} flows entirely through the resistor R_{m}. TheR_{m} correlates with cell size. For a spherical cell, its membrane resistance can be simplified as $R_{\mathrm{m}} = c/r^2$, where c is a constant and r is the cell radius. Based on Eq. (1.6), the rheobase I_{rh} is directly proportional to the square of cell radius. Consequently, a larger cell requires a greater rheobase I_{rh} to reach threshold potential due to its smaller resistance R_{m}. This principle extends to synaptic transmission—larger postsynaptic neurons require greater synaptic currents (integrated from more synaptic inputs) for activation. This is because current spreads across the entire membrane area. The current intensity per unit area, rather than total current, determines whether the membrane can be excited sufficiently to generate an AP. Therefore, a larger membrane area needs a proportionally larger current.

For a current applied as a brief pulse with width T rather than lasting infinitely long, based on Eq. (1.4), to reach the threshold V_{th} at the pulse end (i.e. $t = T$), the current intensity must be at least

$$I_{\mathrm{s,th}} = \frac{V_{\mathrm{th}}}{R_{\mathrm{m}}(1 - e^{-T/\tau})} = \frac{I_{rh}}{1 - e^{-T/\tau}} \tag{1.7}$$

As shown in Fig. 1.2, this equation describes how the threshold current $I_{\mathrm{s,th}}$ changes with pulse width T, forming what is called the strength-duration curve, or I-T curve (Horch and Dhillon 2004). According to Eq. (1.7), when the current $I_{\mathrm{s,th}}$ is twice the rheobase I_{rh} (i.e., $I_{\mathrm{s,th}} = 2I_{\mathrm{rh}}$), the minimum pulse width (denoted as T_{chr}) required for the membrane potential to reach the threshold V_{th} is

$$T_{\mathrm{chr}} = \tau \ln 2 \approx 0.69\tau \tag{1.8}$$

This T_{chr}, known as the chronaxie, is approximately 2/3 of the membrane time constant. For spherical cells, the membrane capacitance can be simplified as $C_{\mathrm{m}} = kr^2$, where k is a constant and r is the cell radius. The time constant $\tau = R_{\mathrm{m}}C_{\mathrm{m}} = (c/r^2)\,(kr^2) = ck$. The membrane resistance R_{m} is inversely proportional to the membrane area ($\propto 1/r^2$), while the membrane capacitance C_{m} is directly proportional to the membrane area ($\propto r^2$). Both the time constant τ and the chronaxie T_{chr} appear independent of cell size. However, for non-spherical neurons and other cells, τ and T_{chr} do show some size dependence, though their relations to cell size are much weaker than that of the rheobase I_{rh} (Horch and Dhillon 2004). Additionally, c and k

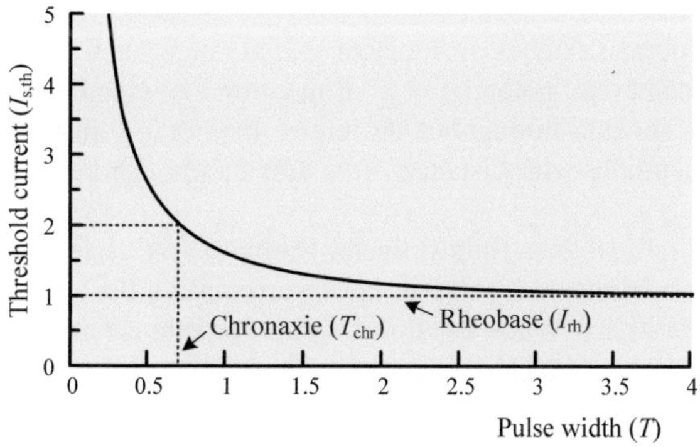

Fig. 1.2 Strength-duration curve of square current stimulation applied to a cell. Rheobase is defined as the minimum activation threshold for an infinite pulse duration. Chronaxie is the minimum pulse duration needed to activate a cell with a current twice the rheobase. The vertical axis ($I_{\mathrm{s,th}}$) is normalized to the rheobase I_{rh}, while the horizontal axis (T) is normalized to the time constant τ. In short, let $I_{\mathrm{rh}} = 1$ and $\tau = 1$

are not true constants—larger membrane areas typically have greater time constants. Thus, larger cells usually have greater τ values, while smaller cells have smaller ones.

Different neural tissues and neuronal structures (such as dendrites, cell bodies, and axons) have different membrane morphologies and features, resulting in distinct strength-duration curves and chronaxies. By using electrical pulses with an appropriate width based on various chronaxies, stimulations can selectively activate specific neural tissues or structures. For instance, the chronaxie of C-type axonal fibers in peripheral nerves exceeds 1 ms (Schneider et al. 2023), while neuronal axons in the central nervous system have chronaxies ranging only from 30 to 300 μs. Thicker myelinated axons have shorter chronaxies than thinner unmyelinated axons, making them more responsive to narrow pulses. Neuronal cell bodies (somata) can have a chronaxie as long as 1 ms (Ranck 1975; McIntyre and Grill 1999; Udupa and Chen 2015), and dendrites have a chronaxie about 10 times longer than axons (Stern et al. 2015). Therefore, narrow pulses ($\leq$ 100 μs), commonly used in deep brain stimulation (DBS), can activate axons first due to their short chronaxies (Nowak and Bullier 1998; DiLorenzo and Bronzino 2008). Beyond neural tissues, other excitable tissues also have diverse chronaxies. For example, ordinary myocardial cells have a chronaxie of approximately 2 ms, while specialized myocardial cells with autonomy (His bundle) have a chronaxie of about 0.5 ms (Jastrzębski et al. 2019). Some smooth muscles have chronaxies reaching up to 10 ms. This explains why neural stimulations typically require pulse widths shorter than 2 ms, while muscle tissue stimulations may need pulses as long as 10 ms.

When the intensity of applied pulses is only slightly greater than the threshold, small changes in tissue excitability can prevent its activation. Therefore, in practical applications, stimulation intensity is typically set at least twice the threshold to ensure reliable effects.

5. **Space Constant and Cable Equation**

When a cell membrane extends over a large spatial range, such as in a slender axon, the change of membrane potential at a stimulation site cannot maintain constant magnitude as it spreads throughout the entire membrane. Instead, the potential decreases exponentially with distance. This attenuation can be characterized by a space constant.

A slender cylindrical axon (unmyelinated) behaves like a cable (Fig. 1.3A), with its membrane serving as an insulating layer surrounding the axoplasm which has a much lower resistivity. When the tip of a glass pipette electrode is inserted into the axonal membrane and locally injects stimulation current into the axoplasm, part of the current can flow along the axoplasm bidirectionally toward both ends of the axon, known as axial current. Simultaneously, the current continuously penetrates the axonal membrane along its path, creating transmembrane radial current (indicated by brown arrows in Fig. 1.3A). Finally, the current converges at the extracellular electrode (e.g., a grounded electrode) to complete the current loop. As the current passes through the membrane at different sites along the axon, it alters the membrane potential there. The change magnitude of membrane potential varies with its distance from the stimulation site. The potential change is largest at the injection site and gradually decreases with distance, exhibiting an exponential decay (Fig. 1.3A, bottom curve). This pattern of potential distribution, formed by current diffusion within the axoplasm and across the membrane, is known as electrotonic potential.

Assume that a slender axon is evenly divided into identical small cylindrical segments with a length of Δx (Fig. 1.3B). The membrane of each segment is represented as an equivalent circuit with a resistor (R_m) and a capacitor (C_m) in parallel. The axoplasm resistance between two adjacent segments is R_i. The resistance of extracellular fluid is ignored due to its negligible value compared to the membrane and axoplasm resistances. Under these conditions, no potential gradient exists in the extracellular space, and the extracellular potential is set to 0. This axon model is known as core conductor model. From this model, the so called axon cable equation can be derived as below.

From the resistance formula Eq. (1.1), the axoplasm resistance R_i of cylindrical segment is

$$R_i = \rho_i \frac{\Delta x}{\pi (d/2)^2} = r_i \Delta x \tag{1.9}$$

Here, ρ_i is the axoplasm resistivity, typically ranging from 50 to 250 Ω·cm (Koch 1999). Δx is the length of cylindrical segment, d is its diameter, and r_i is the axoplasm resistance per unit length (measured in Ω/cm).

The membrane resistance R_m of cylindrical segment is

$$R_m = \frac{\rho_m l}{\pi d \Delta x} = \frac{\hat{R}_m}{\pi d \Delta x} = \frac{r_m}{\Delta x} \tag{1.10}$$

Here, ρ_m is the membrane resistivity, l is the membrane thickness, $\hat{R}_m$ is the membrane resistance per unit area (in Ω·cm^2), and r_m is the membrane resistance per unit length (in Ω·cm).

The membrane capacitance C_m of cylindrical segment is

$$C_m = \frac{\varepsilon_m \pi d \Delta x}{l} = \hat{C}_m \pi d \Delta x = c_m \Delta x \tag{1.11}$$

Here, ε_m is the membrane dielectric constant, $\hat{C}_m$ is the membrane capacitance per unit area (in μF/cm^2), and c_m is the membrane capacitance per unit length (in μF/cm).

As shown in Fig. 1.3B, set the extracellular potential $V_e = 0$. x is the axial distance from the stimulation site. $V_i(x, t)$ is the intracellular potential (i.e., membrane potential) and $i_m(x, t)$ is the membrane current, at the position x and time t. Then, we have

$$V_i(x + \Delta x, t) - V_i(x, t) = r_i \Delta x \cdot i_i(x + \Delta x, t) \tag{1.12}$$

Rearranging this equation yields

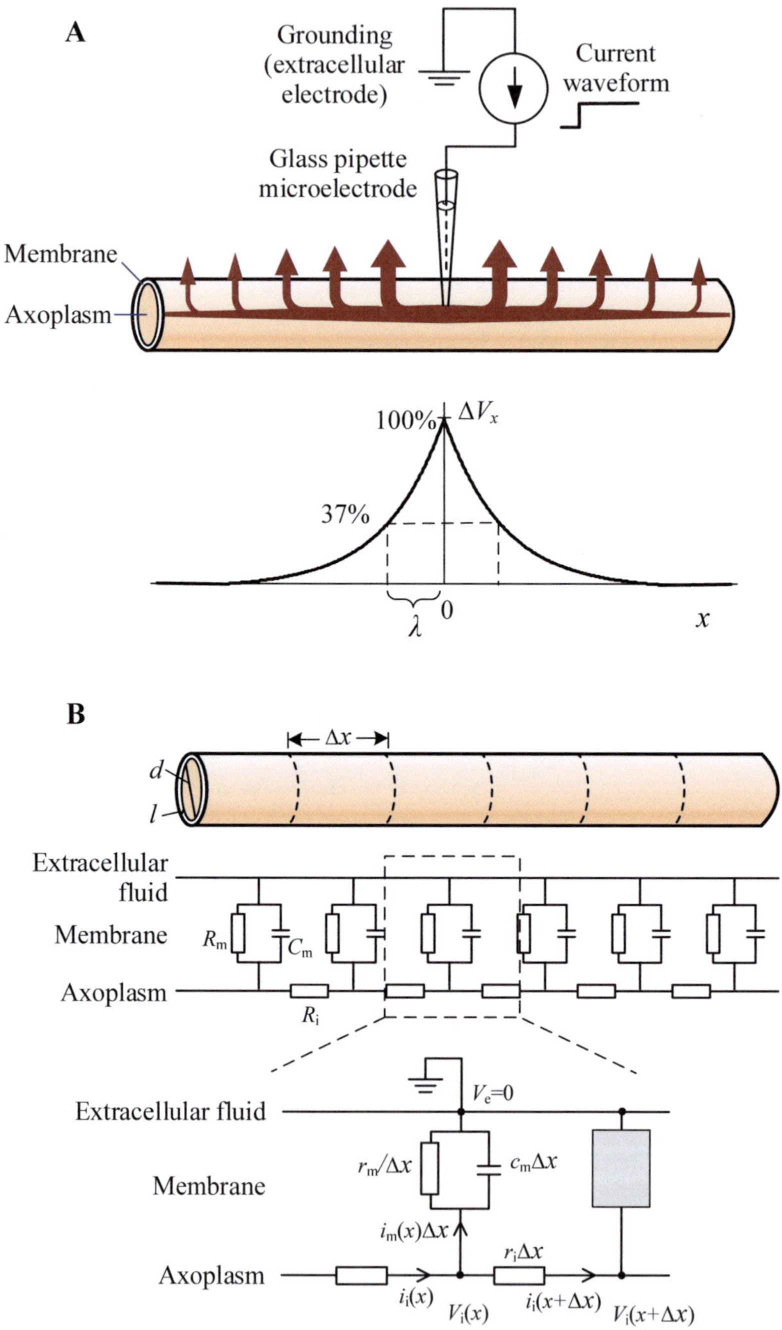

Fig. 1.3 Creation of electrotonic potential along an axon and definition of space constant (**A**) and axonal equivalent circuit (**B**). For simplicity, the time variable t is omitted from the current and potential variables in (**B**)

$$i_{\mathrm{i}}(x+\Delta x, t)=\frac{1}{r_{\mathrm{i}}} \cdot \frac{V_{\mathrm{i}}(x+\Delta x, t)-V_{\mathrm{i}}(x, t)}{\Delta x} \tag{1.13}$$

On the other hand, from the law of current conservation, we have

$$i_{\mathrm{m}}(x, t)=\frac{i_{\mathrm{i}}(x+\Delta x, t)-i_{\mathrm{i}}(x, t)}{\Delta x} \tag{1.14}$$

The membrane current is the sum of currents flowing through both the resistance and the capacitance:

$$i_{\mathrm{m}}(x, t) \Delta x=\frac{V_{\mathrm{i}}(x, t)}{r_{\mathrm{m}} / \Delta x}+c_{\mathrm{m}} \Delta x \frac{\partial V_{\mathrm{i}}(x, t)}{\partial t} \tag{1.15}$$

Omitting Δx in the items, we have

$$i_{\mathrm{m}}(x, t)=\frac{V_{\mathrm{i}}(x, t)}{r_{\mathrm{m}}}+c_{\mathrm{m}} \frac{\partial V_{\mathrm{i}}(x, t)}{\partial t} \tag{1.16}$$

By substituting Eqs. (1.13) and (1.14) into Eq. (1.16) and let $\Delta x \rightarrow 0$, we have the partial differential equation

$$\frac{r_{\mathrm{m}}}{r_{\mathrm{i}}} \frac{\partial^2 V_{\mathrm{i}}(x, t)}{\partial x^2}=V_{\mathrm{i}}(x, t)+r_{\mathrm{m}} c_{\mathrm{m}} \frac{\partial V_{\mathrm{i}}(x, t)}{\partial t} \tag{1.17}$$

It is the linear cable equation, which has analytical solutions in two special cases.

The first case occurs when the membrane potential along the axon reaches a steady state with no further changes over time. For example, as shown in Fig. 1.3A, when a step current is applied at $x=0$ and maintained constant long enough for the membrane potential to reach steady state, the potential at each site along the axon remains unchanged with time. As a result, the second term on the right side of Eq. (1.17) becomes 0, yielding the following second-order ordinary differential equation

$$\frac{r_{\mathrm{m}}}{r_{\mathrm{i}}} \frac{\mathrm{d}^2 V_{\mathrm{i}}(x)}{\mathrm{d} x^2}=V_{\mathrm{i}}(x) \tag{1.18}$$

Given the boundary conditions $V_{\mathrm{i}}(x)=\mathrm{V}_0$ at $x=0$ and $V_{\mathrm{i}}(x)=0$ at $x \rightarrow \infty$, solving Eq. (1.18) yields

$$V_{\mathrm{i}}(x)=\mathrm{V}_0 \mathrm{e}^{-|x| / \lambda} \tag{1.19}$$

where λ is

$$\lambda=\sqrt{\frac{r_{\mathrm{m}}}{r_{\mathrm{i}}}}=\sqrt{\frac{\hat{R}_{\mathrm{m}}}{\rho_{\mathrm{i}}} \cdot \frac{d}{4}} \tag{1.20}$$

The λ is known as space constant, representing the distance ($x = \lambda$) at which the membrane potential decreases to e^{-1}, or 37% of the original potential V_0 at the stimulation site ($x = 0$). Equation (1.20) shows that the λ is directly proportional to the square root of the axon diameter d. For an axon with diameter $d = 4$ μm, membrane resistance per unit area $\hat{R}_m = 20{,}000\ \Omega\cdot\text{cm}^2$ and axoplasm resistivity $\rho_i = 200\ \Omega\cdot\text{cm}$, its space constant is $\lambda = 1$ mm. When the axon diameter decreases to 1 μm with other parameters remaining the same, the λ decreases to 0.5 mm. In describing membrane potential propagation through space, spatial distances are often expressed as dimensionless multiples of λ.

The second case for the analytical solution of Eq. (1.17) occurs when there is no potential difference within the entire axon. For example, inserting a bare metal wire axially through an axon interior can create a spatial clamp, equalizing the potential throughout the axoplasm. In this case, the left term of Eq. (1.17) becomes 0, resulting the following first-order ordinary differential equation

$$r_m c_m \frac{dV_i(t)}{dt} = -V_i(t) \tag{1.21}$$

Given the initial condition $V_i(t) = V_0$ at $t = 0$, solving Eq. (1.21) yields

$$V_i(t) = V_0 e^{-\frac{t}{\tau}} \tag{1.22}$$

where τ is the time constant

$$\tau = r_m c_m = \hat{R}_m \hat{C}_m \tag{1.23}$$

It is consistent with the time constant of a spherical cell membrane shown in Fig. 1.1B. Time variables can be converted to dimensionless values by dividing them by τ, yielding multiples of τ. The x-axis in Fig. 1.2, labeled "pulse width (T)", shows this dimensionless scale in τ units.

By substituting Eqs. (1.20) and (1.23) into Eq. (1.17), the linear cable equation becomes

$$\lambda^2 \frac{\partial^2 V_i(x,t)}{\partial x^2} = V_i(x,t) + \tau \frac{\partial V_i(x,t)}{\partial t} \tag{1.24}$$

Using the dimensionless spatial distance $X = x/\lambda$ and time $T = t/\tau$, we have

$$\frac{\partial^2 V_i(X,T)}{\partial X^2} = V_i(X,T) + \frac{\partial V_i(X,T)}{\partial T} \tag{1.25}$$

If the applied current is a waveform varying with time and space, the equation changes accordingly to

$$\frac{\partial^2 V_i(X,T)}{\partial X^2} = V_i(X,T) + \frac{\partial V_i(X,T)}{\partial T} + \frac{I_m(X,T)}{\lambda\tau} \tag{1.26}$$

where $I_m(X, T) = \lambda\tau I_m(x, t)$ is the applied current waveform with spatial and temporal coordinates normalized by λ and τ, respectively. Here, $I_m(x, t)$ is the actual current waveform in original spatial and temporal coordinates (Koch 1999).

The above describes the basic electrical properties of cell membranes in subthreshold situations without inducing any action potential. They are known as passive properties that can be simulated using an equivalent RC circuit with constant parameters. Among them, the time constant describes the speed of membrane potential changing over time, which is determined by membrane resistance and capacitance. The space constant describes the attenuation of membrane potential over spatial distance, mainly determined by the membrane resistance and intracellular plasma resistance. Although the cable equation can describe passive potential propagation along axons, axonal fibers are essentially different from electric cables that are made of metal conductors and insulation layers for power transmission and information communication. Beyond passive properties, axons also possess active properties created by ion channels distributed in their membranes. When the membrane is stimulated to reach its activation threshold and triggered an action potential, the resulting changes in ion channels and membrane potential must be described using a time-varying equivalent circuit (introduced in Sect. 1.3) with so called active electrical properties. Before discussing those properties, let's first review the resting potential and action potential, along with their generation mechanisms.

1.2.2 Resting Potential and Action Potential

1. **Resting Potential**

When an excitable cell is at rest and unstimulated, its intracellular potential is typically tens of millivolts lower than its extracellular potential, resulting in a resting potential. Different species of cells have distinct resting potentials. For example, brain neurons have a resting potential of around − 70 mV, while visual sensory cells (such as rod and cone cells) in the retina have a resting potential of approximately − 40 mV.

The resting potential is formed by ion concentration differences between the inner and outer fluids. For instance, the intracellular potassium (K^+) concentration (~ 100 mM) is much higher than its extracellular concentration (~ 5 mM), resulting in an approximate 20-fold difference. In contrast, the extracellular sodium (Na^+) concentration (~ 150 mM) is higher than its intracellular concentration (~ 15 mM), with a tenfold difference. The distribution of chloride ions is similar to that of sodium ions (Bear et al. 2016). Additionally, various membrane channels have different permeabilities to ions. At rest, the permeability of K^+ channels in the neuronal membrane is much higher than that of other ion channels. Thus, driven by the great K^+ concentration gradient across the membrane, K^+ ions diffuse outward, carrying positive charges out. This results in accumulations of positive charges outside and negative charges inside the membrane. These opposite charges attract each other across the membrane, establishing a potential difference across the membrane. The force on K^+ ions from this potential difference opposes the force from the concentration gradient. When these opposing forces reach equilibrium, K^+ inflow equals outflow, resulting in no net movement of K^+ ions across the membrane. At this equilibrium state, the potential difference that exactly balances an ionic concentration gradient is known as the transmembrane equilibrium potential for the ion, or simply equilibrium potential, termed as E_{ion}, e.g., E_K for K^+.

The E_{ion} of a specific ion, determined by its concentration gradient across the membrane, can be calculated using the Nernst equation

$$E_{ion} = \frac{RT}{Z_{ion}F} \ln \frac{[ion]_o}{[ion]_i} = 2.303 \frac{RT}{Z_{ion}F} \lg \frac{[ion]_o}{[ion]_i} \tag{1.27}$$

Here R is the gas constant of 8.314 J/(mol·K), T is the absolute temperature which equals Celsius temperature plus 273, F is the Faraday constant of 9.6485×10^4 C/mol, Z_{ion} is the number of charges carried by the ion, $[ion]_o$ is the extracellular ionic concentration, and $[ion]_i$ is the intracellular ion concentration. The equilibrium potential E_{ion} is measured in volts (V).

According to the Nernst equation, the equilibrium potential of K^+ at 37 °C is $E_K = -80$ mV. The resting potential of neurons measured in actual experiments is slightly higher, around − 70 mV. This difference occurs because at rest, while K^+ channels have high permeability, the membrane also has some permeability to Na^+, chloride (Cl^-) and other ions. The final equilibrium potential resulting from multiple ions permeating simultaneously can be calculated by the Goldman equation, also known as the GHK equation

$$E_{ion} = 2.303 \frac{RT}{F} \lg \frac{P_K[K^+]_o + P_{Na}[Na^+]_o + P_{Cl}[Cl^-]_i}{P_K[K^+]_i + P_{Na}[Na^+]_i + P_{Cl}[Cl^-]_o} \tag{1.28}$$

Here P_K, P_{Na} and P_{Cl} are the permeability coefficients of K^+, Na^+ and Cl^- ions, respectively. $[K^+]_o$, $[Na^+]_o$ and $[Cl^-]_o$ are the extracellular concentrations of these three ions, while $[K^+]_i$, $[Na^+]_i$ and $[Cl^-]_i$ are their intracellular concentrations.

At rest, the permeability ratio of K^+, Na^+ and Cl^- is $P_K{:}P_{Na}{:}P_{Cl} = 1{:}0.02{:}0.45$. Therefore, the final equilibrium potential of the membrane at 37 °C is $E_{ion} = -67$ mV. The resting potential can vary with the permeability ratio of ion channels, with a typical resting potential around − 70 mV. If the $P_K{:}P_{Na}$ ratio decreases, the resting potential becomes less negative; conversely, if the ratio increases, the resting potential becomes more negative, tending toward -80 mV—the K^+ equilibrium potential (E_K).

During an AP, the opening and closing of Na^+, K^+ and other ion channels change dynamically, causing corresponding changes in the permeability ratio of these ions. For instance, when the membrane potential reaches the AP peak, the opening of Na^+ channels leads to $P_K{:}P_{Na}{:}P_{Cl} = 1{:}20{:}0.45$. At this moment, the equilibrium potential at 37°C is approximately 50 mV, which is close to the Na^+ equilibrium potential (E_{Na}).

Either at rest or during an AP, ion channels allow K^+ and Na^+ to flow along concentration gradients across the membrane to some extent. Won't these concentration gradients eventually disappear? Actually not! Sodium–potassium pumps in the membrane maintain these ion concentration gradients thereby maintaining the resting potential. Sodium–potassium pumps are integrated proteins embedded in the membrane that transport K^+ and Na^+ ions across the membrane against their concentration gradients by consuming energy from adenosine triphosphate (ATP). Each pump exchanges three internal Na^+ for two external K^+. The sodium–potassium pump protein has a molecular weight of approximately 2.75×10^5 Daltons (1 g = 6×10^{23} Da) and a size of approximately 6×8 nm. Neuronal membrane contains roughly 100–200 pumps/μm^2, which means that a single neuron can have up to one million pumps. At peak speed, each pump can transport approximately 200 Na^+ and 130 K^+ ions per second.

In summary, ion concentrations are different between the inside and outside of a neuron. At rest, the neuronal membrane is highly permeable to K^+ through K^+ channels, resulting in K^+ efflux that creates an equilibrium potential across the membrane. And, energy-consuming sodium–potassium pumps maintain the ion concentration gradients. Together, these mechanisms form the resting potential. The K^+ concentration gradient plays a crucial role in generating the resting potential, which approximates the E_K. The resting potential serves as the foundation for cell activity and bioelectricity generation. Significant changes in resting potential can seriously disrupt normal physiological functions and potentially cause fatal damage. To maintain stable resting potential, the extracellular K^+ concentration must remain constant. In the brain, astrocyte glial cells create an effective K^+ buffering system. These cells can rapidly absorb excess K^+ nearby, then transport, diffuse and release the K^+ ions to distant areas through extensive astrocytic networks. Once local neural activity causes an abnormal increase in $[K^+]_o$, the astrocytic buffer system can quickly adjust K^+ distribution to protect neurons (Bear et al. 2016).

2. **Action Potential**

When a neuron receives excitatory synaptic inputs or is activated directly by an external stimulus to depolarize its membrane up to the activation threshold, an AP can be triggered. The AP generation follows this process: the initial depolarization causes voltage-gated Na^+ channels to open, resulting in Na^+ influx driven by the great Na^+ concentration gradient across the membrane. This Na^+ influx further accelerates membrane depolarization and opens more Na^+ channels. Through this positive feedback, the regenerative Na^+ influx rapidly raises the intracellular potential from negative to positive until it reaches a peak. This forms the AP rising phase, where the inside positive potential is called overshoot. The Na^+ channels remain open briefly before entering an inactive state, causing a rapid decrease in Na^+ influx. In the meantime, the membrane depolarization activates K^+ channels, leading to a K^+ efflux to form the AP falling phase—the repolarization phase—until the membrane potential approaches the equilibrium potential of K^+ (E_K). The end of falling phase can become more negative than the resting potential, creating a so called undershoot or after-hyperpolarization (AHP). The AHP then gradually disappears as the membrane

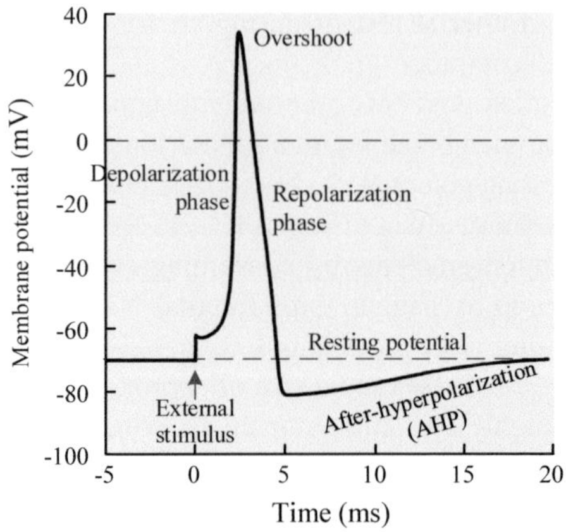

Fig. 1.4 Typical waveform of a neuronal action potential with its defined phases

potential returns to its resting potential (Fig. 1.4). During the repolarization, Na^+ channels recover from their inactive state to closed state. Thus, Na^+ channels cycle through three states: closed, open and inactive. During the inactive state, although the channels appear "open" due to membrane depolarization, their altered protein structure prevent Na^+ from flowing through. Once membrane repolarization occurs, the Na^+ channels are relieved of their inactivation and return to closed state.

Neuronal action potentials typically exhibit an "all-or-none" feature. A subthreshold stimulus is too weak to trigger an AP, but once the stimulus exceeds the activation threshold, both the waveform and amplitude of the triggered AP are independent of stimulus intensity. Through Na^+ channel opening regeneration, the membrane automatically reaches peak potential before returning to resting potential. A sustained suprathreshold stimulus with a higher intensity can only increase the AP occurrence rate without significantly altering their waveforms. However, AP waveforms may change under certain special circumstances. Additionally, different neuron species have distinct AP waveforms, as do different parts of neuronal structures, such as the soma and axon (Bean 2007).

AP generation produces a refractory period consisting of two phases: an absolute refractory period followed by a relative one. The refractory period occurs because inactive Na^+ channels prevent Na^+ from flowing into the cell. During the absolute refractory period, no new AP can be triggered, regardless of stimulus strength. As the membrane gradually repolarizes, Na^+ channels begin to recover. This recovery phase is called the relative refractory period, during which an AP can be triggered, but only by a stimulus significantly stronger than its normal threshold.

The above provides only a qualitative description of AP, while the classic Hodgkin-Huxley model (HH model), named after its creators, offers a quantitative description as introduced below.

1.3 Modeling Membrane Ion Channels (HH Model)

In the 1950s, British scientists Hodgkin and Huxley created a computational model to describe the electrical properties of K^+ and Na^+ channels in axon membranes, long before the physical and chemical structures of these ion channels were understood. Using a voltage clamp technique, they measured membrane currents flowing through ion channels at different membrane potentials. Their electrophysiological experiments used giant squid axons with a diameter about 0.5 mm (Hodgkin et al. 1952). Through studying conductance changes in ion channels during AP generation, they created the classic HH model. This groundbreaking work earned them the 1963 Nobel Prize in Physiology and Medicine. This section will present their experimental data and mathematical equations.

1.3.1 Voltage-Clamp Experiments on Squid Giant Axons

During an AP, the voltage-gated Na^+ and K^+ channels in the membrane rapidly change their states, causing changes in ion channel conductance. As Na^+ and K^+ ions flow through these channels, they result in a rapid change in membrane potential, which always generates a charging or discharging current through membrane capacitor. Consequently, the measured membrane current includes currents flowing through both ion channels and the membrane capacitor. Measuring pure ion channel currents is challenging under this situation.

Figure 1.5 illustrates an equivalent circuit describing the active electrical properties of membrane during AP period. Compared to the RC circuit that describes the passive electrical properties during subthreshold stimulation (Fig. 1.1A), this circuit adds two conductance variables for Na^+ and K^+ channels (g_{Na} and g_K), along with three ion equilibrium potentials (E_{Na}, E_K and E_L). Now, the total current (I_m) flowing through the membrane is

$$I_m = C_m \frac{dV_m}{dt} + I_{Na} + I_K + I_L \tag{1.29}$$

Here V_m is the membrane potential, I_{Na}, I_K and I_L are the currents through Na^+, K^+ and other ion channels, respectively, and C_m is the membrane capacitance.

Using a voltage-clamp technique, the membrane potential can be clamped at a preset level and kept constant for a sufficient duration, resulting in $\frac{dV_m}{dt} = 0$. Under this condition, $I_m = I_{Na} + I_K + I_L$, the pure current through ion channels can be measured after eliminating the capacitive current $C_m \frac{dV_m}{dt}$. Since ion channels are purely resistive, we have $I_{Na} = g_{Na}(V_m - E_{Na})$, $I_K = g_K(V_m - E_K)$ and $I_L = g_L(V_m - E_L)$. By clamping the membrane potential (V_m) at different depolarization levels respectively and measuring the correspondingI_{Na} and I_K, the conductances of Na^+ and K^+ channels (g_{Na} and g_K) can be determined. These conductances are

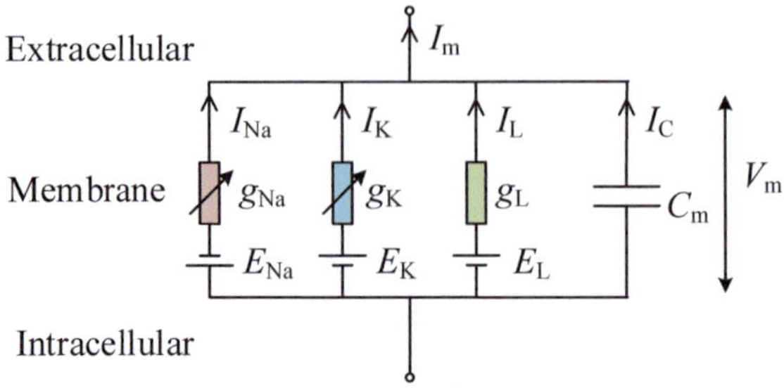

Fig. 1.5 Simplified equivalent circuit for the active properties of squid giant axon membrane. Variables g_{Na}, g_K and g_L represent the conductance of Na^+, K^+ and other ion channels (the leakage conductance), respectively. E_{Na}, E_K and E_L represent the equilibrium potentials of Na^+, K^+ and other ions, respectively

functions of time (t) and membrane potential (V_m). The other parameters in the equivalent circuit (Fig. 1.5), including the ionic equilibrium potentials E_{Na}, E_K and E_L, the membrane capacitance C_m, and the leakage conductance g_L, can be treated as constants. After obtaining the experiment g_{Na} and g_K data at different t and V_m values, mathematical equations can be developed for simulating membrane dynamics. To collect the necessary data, Hodgkin and Huxley conducted experiments on squid giant axons.

Figure 1.6 illustrates their voltage-clamp experiment setup, which included both intracellular space-clamp and transmembrane-clamp. The space-clamp was achieved by inserting a thin bare metal wire axially inside the axon to keep an isoelectric potential along its entire length. The transmembrane-clamp was achieved using a feedback circuit formed as following: the tip of a glass microelectrode was inserted inside the axon to measure the membrane potential with another electrode placed outside. The measured potential was amplified by a high-gain operational amplifier to obtain the membrane potential V_m. Then, through a feedback amplifier, V_m was compared with a preset clamping voltage—a step potential from a signal generator. When V_m differed from the clamping voltage, the feedback amplifier injected a positive or negative current into the axon to drive V_m toward the clamping voltage and maintained it. The output current of feedback amplifier was measured as the membrane current I_m. In this way, under the situation of both space-clamp and transmembrane-clamp, the potential within the entire axon was clamped at different preset voltages one by one. When the membrane potential remained at constant for a sufficient duration, the current through the membrane capacitor became zero ($I_C = 0$). Thus, the measured I_m represented the total current flowing through the ion channels (Eq. (1.29) and Fig. 1.5).

Define the direction of membrane current from inside to outside as positive. As shown in Fig. 1.7, when the applied clamping voltage was small and caused only a subthreshold reaction on the membrane, an initial I_C appeared before the membrane current I_m settled at a constant low level—the leakage current I_L (Fig. 1.7A). When

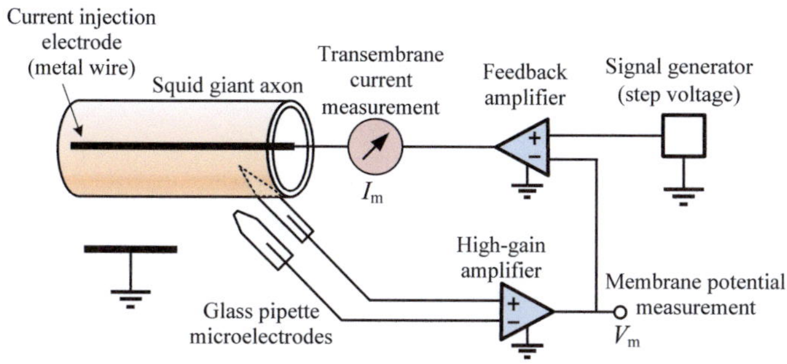

Fig. 1.6 Schematic diagram of the voltage-clamp experiment on a squid giant axon

the applied clamping voltage was large enough to cause a super-threshold depolarization, I_m changed with an initial I_C followed by a rapid inward Na^+ current (I_{Na}) and then reversed to an outward K^+ current (I_K) (Fig. 1.7B). Today, channel blockers, such as tetraethylamine (TEA) for K^+ channels and tetrodotoxin (TTX) for Na^+ channels, can readily separate the two ion currents, I_K andI_{Na}. However, in the 1950s, the nature of ion channels was unknown, let along channel blockers. Hodgkin and Huxley devised an elegant method to remove Na^+ from the external solution, thereby eliminating I_{Na} from the measured I_m. ThisI_m then contained only I_K and I_L. By subtracting this I_m from the I_m measured with normal external solution containing Na^+, they estimated I_{Na} (Hodgkin and Huxley 1952a). With I_L obtained from subthreshold measurement (Fig. 1.7A), they then determined I_K.

1.3.2 Ion Channel Model

1. **Equations for Potassium Channel Conductance**

Using the determined I_K data and clamping voltage V_m values, the experimental result of K^+ channel conductance (g_K) can be determined by the formula $g_K = \frac{I_K}{V_m - E_K}$. Figure 1.8 shows the g_K data at three different clamping voltages—depolarization potentials (denoted on the top). The depolarization potential was the difference between the clamped V_m and the resting potential, denoted by V. When the V_m was clamped at a specific depolarization level, a series of g_K measurements was determined over time (t), shown by small circles in the plots. The original paper by the two scientists included more depolarization levels than shown in Fig. 1.8. The smooth curves were simulation data calculated by the HH model, closely matching the experimental data.

The HH model equations were developed by fitting curves to the experimental data. The conductance equation for K^+ channels was assumed to be

$$g_K = \bar{g}_K n^4 \tag{1.30}$$

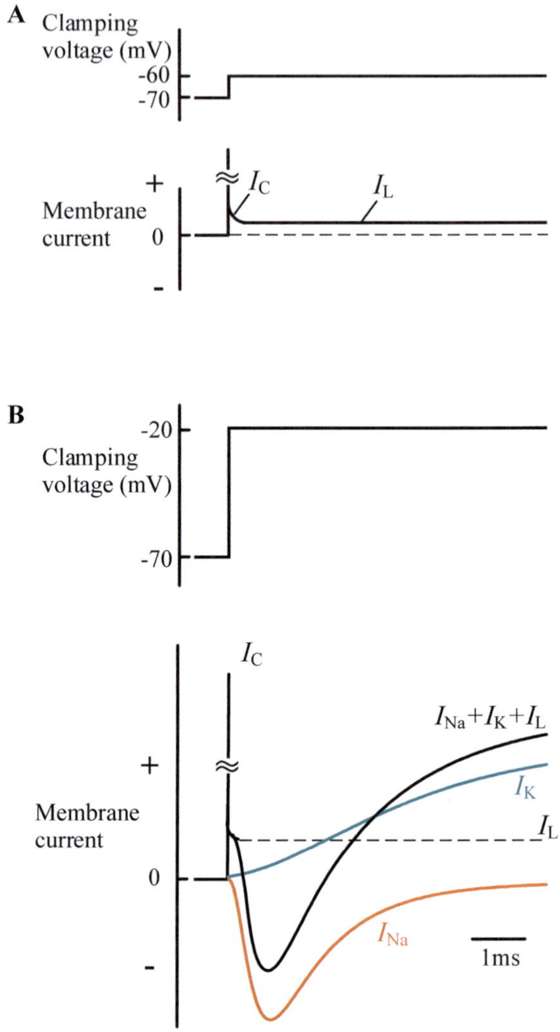

Fig. 1.7 Membrane currents measured during voltage-clamp at subthreshold (**A**) and super-threshold (**B**) depolarization levels

with

$$\frac{\mathrm{d}n}{\mathrm{d}t} = \alpha_n(1 - n) - \beta_n n \tag{1.31}$$

Here $\overline{g}_K$ is a constant representing the maximum value of K^+ conductance (g_K). n represents the probability of K^+ channel opening ($0 \leq n \leq 1$), also known as K^+ channel gating variable. The variables α_n and β_n are rate constants (measured in s^{-1}) describing the ion channel transitions between open and closed states at different membrane potentials:

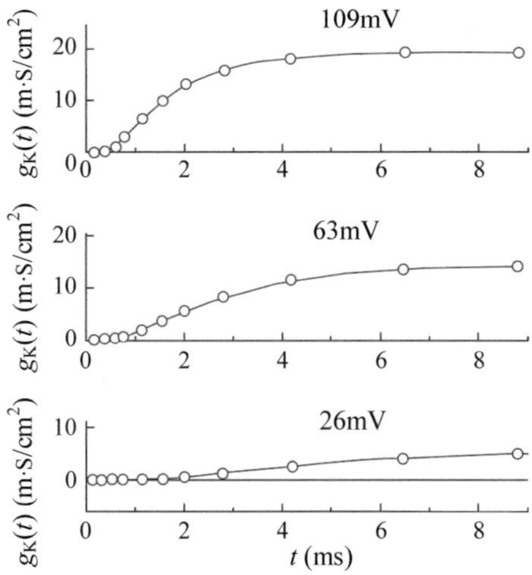

Fig. 1.8 Changes in K^+ channel conductance (g_K) over time (t) at three different levels of membrane depolarization. The small circles represent experimental data, while the smooth curves are the simulation results of HH model

$$\text{(Open)} \quad n \underset{\alpha_n}{\overset{\beta_n}{\rightleftharpoons}} 1 - n \quad \text{(Closed)}$$

Thus, Eq. (1.31) is the first-order kinetic equation describing how n varies with both time and membrane potential. At rest ($V = 0$), n has a resting value of $n_0 = \frac{\alpha_{n0}}{\alpha_{n0}+\beta_{n0}}$. With the boundary condition of $n = n_0$ at$t = 0$, the solution of Eq. (1.31) is

$$n = n_\infty - (n_\infty - n_0)\mathrm{e}^{-t/\tau_n} \tag{1.32}$$

where the steady-state value n_∞ and the time constant τ_n are

$$n_\infty = \frac{\alpha_n}{\alpha_n + \beta_n} \tag{1.33}$$

and

$$\tau_n = \frac{1}{\alpha_n + \beta_n} \tag{1.34}$$

The rate constants α_n and β_n vary with V. They are determined using the experimental data as follows. From Eq. (1.30), we have

$$n = \left[\frac{g_K}{\bar{g}_K}\right]^{1/4} \tag{1.35}$$

Substituting it into Eq. (1.32), we obtain

$$g_K = \left\{(g_{K\infty})^{1/4} - \left[(g_{K\infty})^{1/4} - (g_{K0})^{1/4}\right]\mathrm{e}^{-t/\tau_n}\right\}^4 \tag{1.36}$$

where g_{K0} is the initial conductance value at $t = 0$. $g_{K\infty}$ is the steady-state conductance value that finally stabilizes after sufficient time. The values of g_{K0}, $g_{K\infty}$ and $\bar{g}_K$ can be read from the experimental data at different depolarization levels (Fig. 1.8). Choose an appropriate value for τ_n to optimize the fit between the g_K curve calculated by Eq. (1.36) and the experimental data. Then, use the following three formulas to calculate series of n_∞, α_n and β_n values at different depolarization levels (shown by the small circles in Fig. 1.9):

$$n_\infty = \left[\frac{g_{K\infty}}{\bar{g}_K}\right]^{1/4} \tag{1.37}$$

$$\alpha_n = \frac{n_\infty}{\tau_n} \tag{1.38}$$

$$\beta_n = (1 - n_\infty)/\tau_n \tag{1.39}$$

Using these calculated α_n and β_n values, the following equations can be determined by mathematical fitting:

$$\alpha_n = \frac{0.01(10 - V)}{e^{\frac{10-V}{10}} - 1} \tag{1.40}$$

$$\beta_n = 0.125e^{\frac{-V}{80}} \tag{1.41}$$

The smooth curves in Fig. 1.9 were calculated using Eqs. (1.33), (1.34), (1.40) and (1.41). They show how the variables n_∞, τ_n, α_n and β_n in the conductance model of K^+ channels vary with the depolarization potential V.

One may wonder why the variable n was raised to the fourth power in the conductance equation $g_K = \bar{g}_K n^4$. Actually, Hodgkin and Huxley tried different powers in their fitting process. They found that although higher powers yielded better fits between simulation results and experimental data, the fourth power provided sufficient accuracy. Using higher powers would have only added unnecessary computational complexity.

Totally, the K^+ channel model consists of the following four equations

$$g_K = \bar{g}_K n^4 \tag{1.30}$$

$$\frac{dn}{dt} = \alpha_n(1 - n) - \beta_n n \tag{1.31}$$

$$\alpha_n = \frac{0.01(10 - V)}{e^{\frac{10-V}{10}} - 1} \tag{1.40}$$

$$\beta_n = 0.125e^{\frac{-V}{80}} \tag{1.41}$$

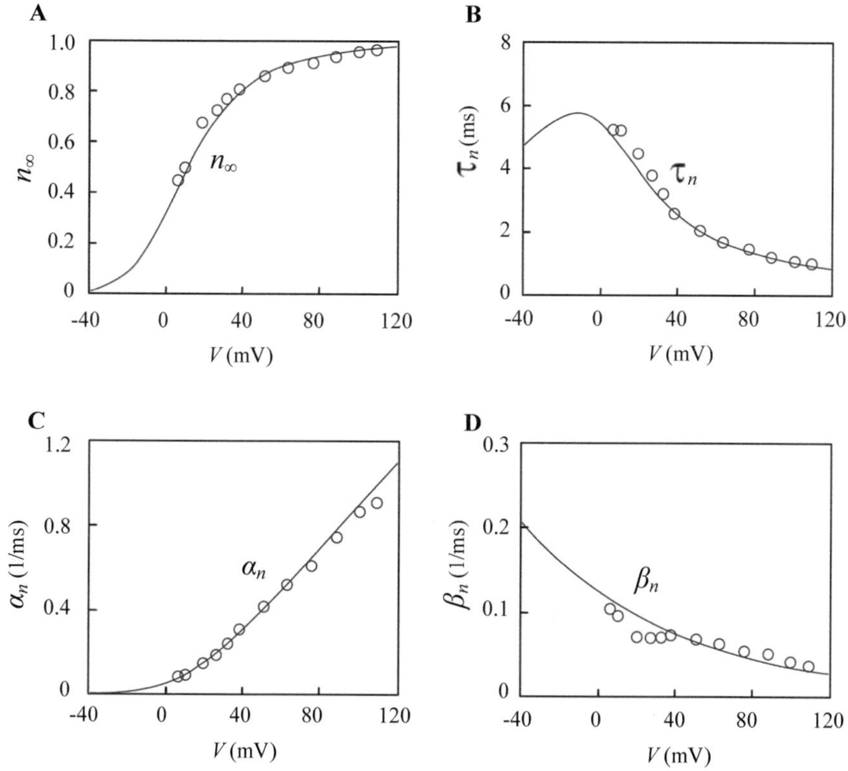

Fig. 1.9 Parameters of K^+ channel model versus membrane potential, including the steady-state value of gating variable n_∞ (**A**), time constant τ_n (**B**), and rate constants α_n (**C**) and β_n (**D**). In each plot, the abscissa V is the depolarization potential relative to resting potential. A positive V value indicates depolarization, while a negative V value indicates hyperpolarization. The small circles are the values determined from experiment measurements, while the smooth curves are the simulation values from the HH model

2. Equations for Sodium Channel Conductance

Similarly, the Na^+ channel model can be determined. First, Na^+ currents (I_{Na}) were measured at different depolarization potentials during voltage-clamp experiments. Then, using the applied clamping voltages and the measured I_{Na} data, the experimental data of Na^+ conductance $\left(g_{Na} = \frac{I_{Na}}{V_m - E_{Na}}\right)$ were determined, as shown by the small circles in Fig. 1.10.

Unlike K^+ channels which have only two states (open and closed), Na^+ channels have an additional inactive state. Although voltage-gated Na^+ channels can be activated by a superthreshold depolarization, they can remain open briefly before automatically entering an inactive state. The inactive Na^+ channels can return to their closed state—becoming ready for reactivation—only after membrane repolarization occurs. To describe how Na^+ channel conductance changes during activation and

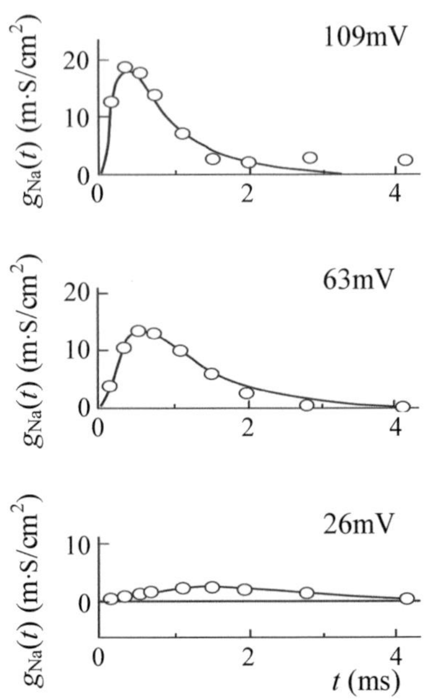

Fig. 1.10 Changes in Na^+ channel conductance (g_{Na}) over time at three different levels of membrane depolarization. The small circles represent experimental data, while the smooth curves are the simulation results of HH model

inactivation, two gating variables, m and h, were used in the conductance equation

$$g_{Na} = \overline{g}_{Na} m^3 h \tag{1.42}$$

with

$$\frac{dm}{dt} = \alpha_m (1 - m) - \beta_m m \tag{1.43}$$

$$\frac{dh}{dt} = \alpha_h (1 - h) - \beta_h h \tag{1.44}$$

Here $\overline{g}_{Na}$ is a constant representing the maximum value of Na^+ conductance (g_{Na}). The two pairs of rate constants α and β are functions of membrane potential, since the Na^+ channel is voltage-gated.

The two scientists proposed Eq. (1.42) based on the following hypothesis: Na^+ channel activation is determined by three independent activation particles (M) and one inactivation particle (H). A Na^+ channel opens only when all three M particles and one H particle locate in specific positions on the membrane. The probability of the channel opening is then represented by m^3h, where the variable m represents the probability of an M particle being in the open position ($0 \leq m \leq 1$), and the variable h represents the probability of an H particle being in the open position ($0 \leq h \leq 1$). Thus, $1 - m$ and $1 - h$ represent the probabilities of the two particles being in the closed position, respectively. Similar to the K^+ channel model, variables α_m, β_m and

α_h, β_h represent the rate constants describing the particle transitions between open and closed states:

$$\text{(Open)} \quad m \underset{\alpha_m}{\overset{\beta_m}{\rightleftharpoons}} 1 - m \quad \text{(Closed)}$$

$$\text{(Open)} \quad h \underset{\alpha_h}{\overset{\beta_h}{\rightleftharpoons}} 1 - h \quad \text{(Closed)}$$

Equations (1.43) and (1.44) describe the first-order kinetics of m and h over time (t) and membrane potential (V). With the boundary conditions $m = m_0$ and $h = h_0$ at $t = 0$, the solutions of these equations are

$$m = m_\infty - (m_\infty - m_0)\mathrm{e}^{-t/\tau_m} \tag{1.45}$$

$$h = h_\infty - (h_\infty - h_0)\mathrm{e}^{-t/\tau_h} \tag{1.46}$$

where

$$\begin{aligned} m_\infty &= \frac{\alpha_m}{\alpha_m + \beta_m}, \ \tau_m = \frac{1}{\alpha_m + \beta_m}; \\ h_\infty &= \frac{\alpha_h}{\alpha_h + \beta_h}, \ \tau_h = \frac{1}{\alpha_h + \beta_h} \end{aligned} \tag{1.47}$$

Here we need to determine the functions of α_m, β_m and α_h, β_h varying with the membrane potential. Since the Na^+ channel conductance at rest is very small, the m_0 becomes negligible compared to the m_∞ under large depolarization. In addition, the inactivation of Na^+ channel is approximately complete. Thus, h_∞ is negligible compared to h_0. Under these conditions, substituting Eqs. (1.45) and (1.46) into Eq. (1.42) yields

$$g_{\mathrm{Na}} = g'_{\mathrm{Na}}(1 - \mathrm{e}^{-t/\tau_m})^3\mathrm{e}^{-t/\tau_h} \tag{1.48}$$

where $g'_{\mathrm{Na}} = \bar{g}_{\mathrm{Na}} m_\infty^3 h_0$. Choose appropriate values for τ_m and τ_h to reach the g_{Na} curve that best fits the experimental data at different depolarization levels. Then, use the following equations to calculate series of α and β values at different depolarization potentials

$$\alpha_m = \frac{m_\infty}{\tau_m}, \quad \beta_m = (1 - m_\infty)/\tau_m \tag{1.49}$$

$$\alpha_h = \frac{h_\infty}{\tau_h}, \quad \beta_h = (1 - h_\infty)/\tau_h \tag{1.50}$$

Finally, using these calculated α and β values, the following equations for the four rate constants can be determined by mathematical fitting:

$$\alpha_m = \frac{0.1(25 - V)}{e^{\frac{25-V}{10}} - 1} \tag{1.51}$$

$$\beta_m = 4e^{\frac{-V}{18}} \tag{1.52}$$

$$\alpha_h = 0.07e^{\frac{-V}{20}} \tag{1.53}$$

$$\beta_h = \frac{1}{e^{\frac{30-V}{10}} + 1} \tag{1.54}$$

Equations (1.51) to (1.54) and (1.42) to (1.44) together form the Na^+ channel model.

3. **HH Model Equations**

The complete HH model includes the following equations:

$$I_m = C_m\frac{dV}{dt} + \overline{g}_K n^4(V - V_K) + \overline{g}_{Na} m^3 h(V - V_{Na}) + \overline{g}_L(V - V_L) \tag{1.55}$$

$$\frac{dn}{dt} = \alpha_n(1 - n) - \beta_n n \tag{1.12}$$

$$\frac{dm}{dt} = \alpha_m(1 - m) - \beta_m m \tag{1.43}$$

$$\frac{dh}{dt} = \alpha_h(1 - h) - \beta_h h \tag{1.44}$$

$$\alpha_n = \frac{0.01(10 - V)}{e^{\frac{10-V}{10}} - 1} \tag{1.40}$$

$$\beta_n = 0.125e^{\frac{-V}{80}} \tag{1.41}$$

$$\alpha_m = \frac{0.1(25 - V)}{e^{\frac{25-V}{10}} - 1} \tag{1.51}$$

$$\beta_m = 4e^{\frac{-V}{18}} \tag{1.52}$$

$$\alpha_h = 0.07e^{\frac{-V}{20}} \tag{1.53}$$

$$\beta_h = \frac{1}{e^{\frac{30-V}{10}} + 1} \tag{1.54}$$

Here V is the displacement of membrane potential from its resting potential. V_{Na}, V_K and V_L are the displacements in equilibrium potentials of Na^+, K^+ and other ions from the resting potential. All these potentials are measured in mV. The membrane current I_m is measured in μA/cm^2, conductance in mS/cm^2, capacitance in μF/cm^2, time in ms, and all rate constants α and β in ms^{-1}. In addition, the α and β values are temperature-dependent. The equations above were derived at 6.3°C. At a temperatures of T °C, the α and β values should be adjusted by a multiplication factor $\varphi = 3^{\frac{T-6.3}{10}}$ (Hodgkin and Huxley 1952a).

1.3.3 Example Simulations of the HH Model

The HH model has no analytical solution but numerical ones. Using a computer program, we can solve the model equations and perform simple simulations. For example, using the MATLAB program provided in the appendix at the end of this book, we can calculate the curves of the three pairs of α and β over V values (Fig. 1.11A, B), and time constants and steady-state values of the gating variables (m, h and n) for Na$^+$ and K$^+$ channels against V values (Fig. 1.11C, D). The abscissa (V) in these plots is the difference between the membrane potential and its resting potential, where positive values represent depolarization and negative values represent hyperpolarization.

Among the time constants of three gating variables (Fig. 1.11C), τ_m of Na$^+$ channel activation is the smallest, much smaller than both τ_h of Na$^+$ channel inactivation and τ_n of K$^+$ channel activation. This indicates that Na$^+$ channels open very rapidly. Additionally, as membrane depolarization increases, both activation variables m_∞ and n_∞ increase monotonically (Fig. 1.11D). However, m_∞ rises more rapidly than n_∞, indicating that Na$^+$ channels can be activated by smaller depolarization displacements. Nevertheless, with increased membrane depolarization, the inactivation variable h_∞ of Na$^+$ channels decreases, thereby reducing Na$^+$ channel opening. Under excessive depolarization, complete inactivation occurs with a very smallh_∞, resulting in a depolarization block that prevents action potential generation. For example, applying a prolonged positive step current into a neuron can cause sustained depolarization leading to this block. On the other hand, when the membrane is excessively hyperpolarized, the activation variable m_∞ can decrease to a very small value (Fig. 1.11D), causing a hyperpolarization block. This block occurs through deep closure of Na$^+$ channels, distinct from depolarization block caused by Na$^+$ channel inactivation (Bhunia et al. 2015).

Assume no axial current within the axon, as shown in the space-clamp experiment in Fig. 1.6. Under this condition, the net current across the membrane remains zero unless an external current is applied. In other words, the I_m in Eq. (1.55) represents an applied stimulation current. Using the MATLAB program in the appendix, we can simulate neuronal responses to step (or square) current stimulations (I_m) with various intensities and directions. The I_m is represented by the variable Iapp in the program. As shown in Fig. 1.12, with $t = 0$ marking the start of stimulation, the changes in membrane potential are shown in the left column, while the changes in gating variables n, m and h of Na$^+$ and K$^+$ channels are shown in the right column.

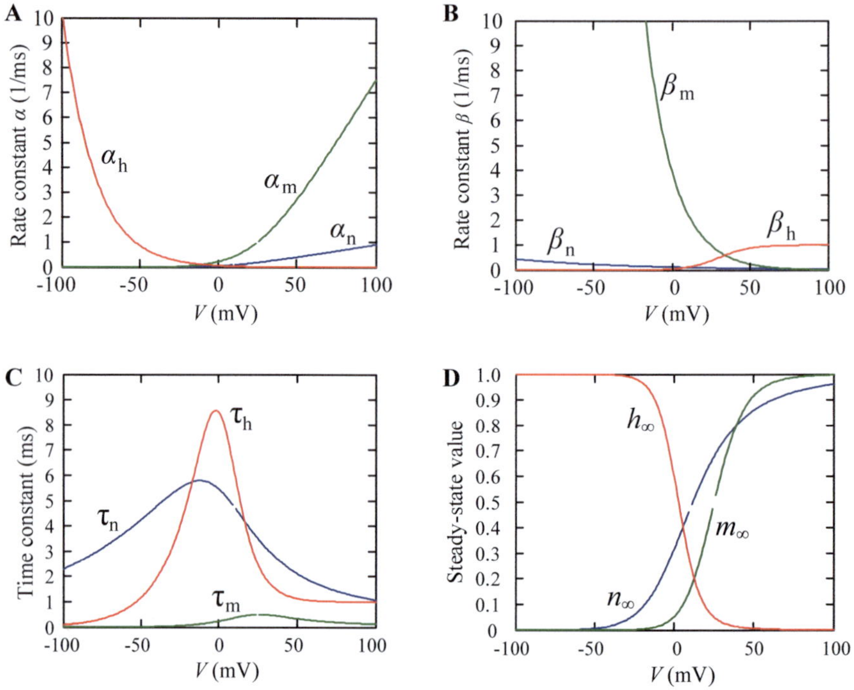

Fig. 1.11 Simulated curves of rate constants (**A** and **B**), time constants (**C**) and steady-state values (**D**) for Na^+ and K^+ channel gating variables in the HH model. The four plots are the outputs of the MATLAB script "constants_variables.m" provided in the appendix. The β_m curve is truncated in (**B**) to better display the β_n and β_h curves. Gaps in the curves occur where corresponding equations have zero denominators. The abscissa V represents the depolarization potential relative to resting potential

These changes induced by various I_m stimulations can be explained based on the properties of both ion channels shown in Fig. 1.11.

The resting potential is set to −70 mV. Figure 1.12A–E show the simulation results with a positive step currentI_m (i.e., an outward current) starting at $t = 0$, with an intensity ranging from 2 to 150 μA/cm^2. At a subthreshold small current of $I_m =$ 2 μA/cm^2 (Fig. 1.12A), only small membrane potential fluctuations between − 70 and − 65 mV are induced in the initial period. Then, the membrane potential stabilizes at − 68.5 mV, maintaining a depolarization of about 1.5 mV without triggering an action potential (AP). The changes in the gating variables of ion channels are also very small. The activation variable m of Na^+ channels increases slightly from 0.053 to 0.063, while its inactivation variable h decreases from 0.60 to 0.54. The activation variable n of K^+ channels increases slightly from 0.32 to 0.34. These changes follow the curves shown in Fig. 1.11D, slightly increasing the conductance of both types of channels. However, the inward Na^+ flow cannot exceed the outward K^+ flow. The

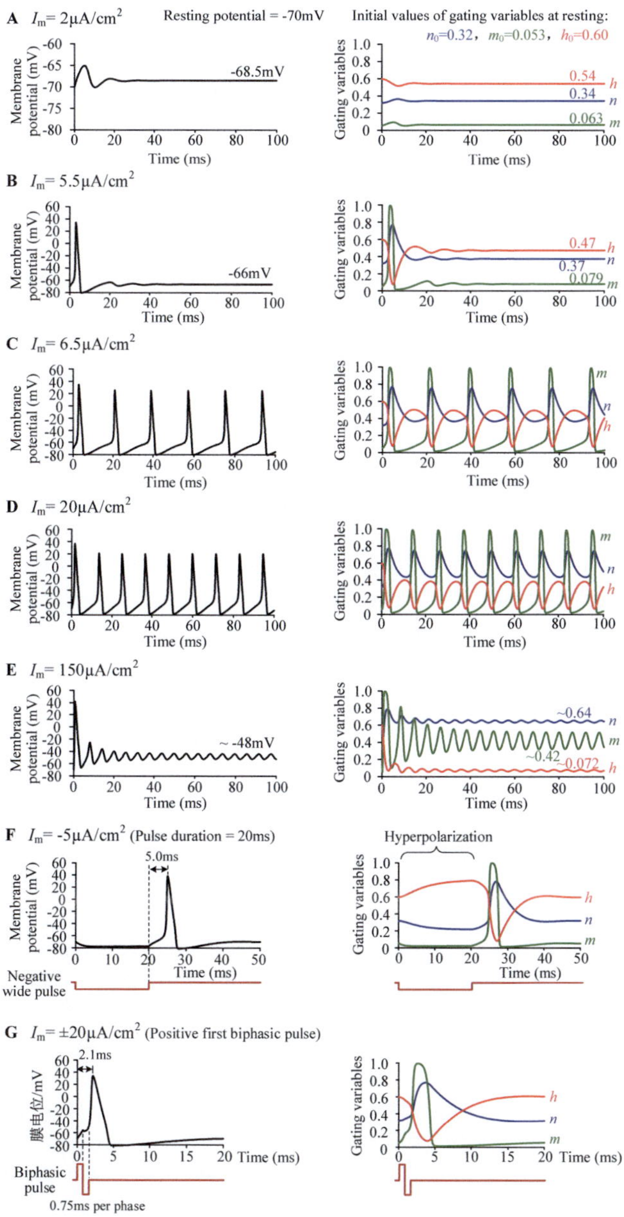

Fig. 1.12 Changes in the membrane potential (left column) and gating variables (right column) induced by different current stimulations—positive step currents with different intensities starting at time zero (**A–E**), a negative wide pulse (**F**) and a narrow biphasic pulse (**G**). These curves were generated by the MATLAB script "HH_simulation.m" provided in the appendix. You can produce different simulation results by adjusting the waveform and amplitude of the applied current I_m (represented by the variable Iapp in the script). The Iapp settings are detailed in the script comments

two opposite ionic flows reach an equilibrium at the new membrane potential of −68.5 mV, slightly above the original resting potential of −70 mV.

Increasing the step current to I_m = 5.5 μA/cm^2 initially induces an AP (Fig. 1.12B). Afterwards, the applied current generates only small sustained depolarization at about − 66 mV. This depolarization reduces the inactivation variable h of Na^+ channels. Although conductance increases in both types of channels, the Na^+ flow still cannot overcome the K^+ flow, preventing continuous AP generation. When the applied current increases further to I_m = 6.5 μA/cm^2 (Fig. 1.12C), the activation variable m of Na^+ channels periodically reaches its peak (~ 1.0), causing a sharp increase in Na^+ conductance. The Na^+ flow exceeds the K^+ flow, inducing periodic APs. Increasing the applied current to I_m = 20 μA/cm^2 enhances the AP occurrence rate (Fig. 1.12D). However, when the current is set to I_m = 150 μA/cm^2 (Fig. 1.12E), APs cannot occur continuously, because the membrane cannot repolarize sufficiently to allow inactive Na^+ channels to recover after the first AP. The inactivation variable h remains too low to generate new AP. Eventually, the membrane potential only oscillates slightly around − 48 mV, resulting in a sustained depolarization of ~ 22 mV. This depolarization keeps the h at a low level while enhancing the activation variables m and n significantly higher than their resting values (see Fig. 1.11D).

When the applied current changes direction from outward to inward, the stimulation can hyperpolarize membrane. Although hyperpolarization typically cannot induce an AP directly, it may trigger one when the stimulation is withdrawn, provided the current has sufficient intensity and duration (Fig. 1.12F). This occurs because Na^+ channels at rest are partially inactivated (e.g., $h_0 = 0.60 < 1$). Hyperpolarization decreases the activation variable m, but simultaneously increases the inactivation variable h (Fig. 1.12F, right), thereby reducing inactivation. Once the hyperpolarization ends and membrane potential rises, the$m$ increase is much faster than the h decrease due to their different time constants (see Fig. 1.11C). Thus, the increased m combined with the still-high h leads to a rapid increase in Na^+ channel conductance ($\propto m^3h$). Meanwhile, the recovery of K^+ channel activation (n), which has decreased during hyperpolarization, is slow, resulting in a slower increase in K^+ channel conductance ($\propto n^4$) than that of Na^+ channels. As a result, the increased Na^+ inflow generates an AP. This phenomenon is known as "anodic break excitation" or "rebound excitation" (Durand 2000). The AP latency in this case is long, as shown in the left of Fig. 1.12F. The latency between the AP peak and the end of negative current stimulation (20 ms width) is about 5.0 ms. In comparison, with positive stimulations (Fig. 1.12B–E), the latencies between the first AP peak and the current start point (t = 0) are only 3.0, 2.7, 1.5 and 0.6 ms. The greater the positive current, the shorter the latency becomes.

When a biphasic current pulse is applied with a first positive phase followed by a second negative phase (Fig. 1.12G), the negative phase cannot counteract the effect of positive phase to prevent AP generation. When the positive phase has sufficient intensity and duration, an AP can still generate even if the negative phase occurs before its full generation. In the example shown in Fig. 1.12G, the negative phase spans from t = 0.75 to 1.5 ms, occurring earlier than the AP peak at 2.1 ms. However, the threshold current intensity required for a biphasic pulse is higher than that of a single-phase positive pulse.

These simulation results show that membrane depolarization can generate APs. However, depolarization is not always beneficial for AP generation. In some cases, it can actually prevent APs. When membrane maintains a certain level of depolarization, even at a subthreshold level, the depolarization can increase the threshold of AP initiation due to increased Na^+ channel inactivation (see Fig. 1.11D). Continuous excessive depolarization can prevent AP generations (see Fig. 1.12E). In contrast, membrane hyperpolarization can reduce Na^+ channel inactivation and lower the threshold of AP initiation. Once this established hyperpolarization is removed, an AP can occur as the membrane potential rises toward its resting potential (Fig. 1.12F).

Note that the description above refers to intracellular stimulation, where the stimulation current flowing from inside to outside is defined as positive (Fig. 1.1B). Under this definition, a positive current depolarizes the cell membrane, while a negative current hyperpolarizes it. This differs from extracellular stimulation, where a negative pulse depolarizes the cell membrane. Please refer to Sect. 1.4.1 for detail differences between intracellular and extracellular stimulations.

The simulation results shown in Fig. 1.12 were obtained by solving the HH model equations using the MATLAB program in the appendix. The parameter settings were: $C_m = 1\ \mu F/cm^2$, $\overline{g}_K = 36\ mS/cm^2$, $\overline{g}_{Na} = 120\ mS/cm^2$, $\overline{g}_L = 0.3\ mS/cm^2$, $V_K = -12\ mV$, $V_{Na} = 115\ mV$, and $V_L = 10.6\ mV$. With a resting potential of $-70\ mV$, all potentials (V, V_K, V_{Na} and V_L) represented the differences of corresponding membrane potentials from the resting potential. The membrane potentials displayed in the left column of Fig. 1.12 were calculated as ($V - 70\ mV$). The initial gating variables (resting values) were $n_0 = 0.3177$, $m_0 = 0.0529$ and $h_0 = 0.5961$.

The HH model enables simulation and exploration of neuronal responses to various external stimulations. While the nonlinear characteristics of ion channels in neuronal membranes make it difficult to intuitively predict these responses, computational simulations can demonstrate complex responses and reveal their underlying mechanisms. In addition to the HH model, other computational models have emerged to simulate the behaviors of various neurons and neural networks. Some of them simplify calculations to address the high computation demands of the HH model, particularly for complex neural networks containing many neurons. However, the HH model remains a classic and irreplaceable foundation for other models. Nevertheless, the HH model was originally based on squid giant axons. Different neuron types possess distinct ion channels in their membranes, including various subtypes of Na^+ and K^+ channels with different properties. Additional ion channels, such as calcium (Ca^{2+}) channels, may also exist on membranes. Therefore, when applying the model, parameters and equations must be adjusted according to specific simulation objects to achieve results that better align with actual physiological situations.

The original literature on the HH model consists of five consecutive papers published in the Journal of Physiology in 1952 (Hodgkin et al. 1952; Hodgkin and Huxley 1952a, b, c, d). This journal has since published commemorative articles every decade (Schwiening 2012; Catterall 2012; Catacuzzeno and Franciolini 2022). Many books and publications cover the HH model introduction and applications, including teaching materials to help beginners (Hopper et al. 2022).

1.4 Electrical Stimulation of Neurons

Neurons can respond to various external stimuli to generate APs or to alter their firing activity. While other stimulations—such as heat, mechanical force and magnetic field—can activate neurons by depolarizing their membranes to reach the AP threshold, electrical stimulation is the most common one. It is widely used in both research and therapeutic applications. In electrophysiological research, for example, it can help revel AP mechanisms, investigate synaptic excitations and inhibitions, measure nerve fiber conduction velocity, as well as explore pathways, structures, and functions of nervous systems. In clinical applications, electrical stimulation has been used in diagnosing neuromuscular junction disorders, treating sensory deficits and motor paralysis, controlling pain and epileptic seizures, and modulating brain disorders. Both intracellular (Fig. 1.1B) and extracellular electrical stimulation methods are used for research purposes. However, clinical applications requiring simultaneous activation of many neurons must use extracellular stimulation. It was also the method used in the in-vivo rat experiments described in Part II of this book. Additionally, electrical stimulation can be delivered through various forms, including voltage or current waveforms and monopolar or bipolar modes. Stimulation safety is a crucial prerequisite for both research and clinical applications. These aspects are discussed in the following sections.

1.4.1 Intracellular and Extracellular Stimulations

To activate neurons, external electrical stimulation can be applied either intracellularly or extracellularly. In the squid axon experiment shown in Fig. 1.6 and the model simulations in Fig. 1.12, intracellular stimulations were used. By delivering an outward positive current through a stimulation electrode (anode) placed inside a cell, we can depolarize the cell membrane to generate APs. Additionally, through rebound excitation, an inward negative current can also induce an AP (Fig. 1.12F). However, negative current is rarely used for intracellular stimulation because its activation efficiency is much lower, requiring longer duration. A key limitation of intracellular stimulation is that each electrode can only act on one cell at a time, making it challenging to activate multiple neurons simultaneously. Moreover, maintaining long-term stimulation of even a single cell is difficult. Current technology restricts intracellular manipulations to microelectrodes made of thin glass pipettes, which cannot maintain stability for long periods. This limitation is particularly pronounced in in-vivo neural electrical stimulation. As a result, intracellular neuronal stimulation is confined to scientific research rather than clinical applications. This technique is used primarily in isolated neural tissues (such as brain slices) or cultured neurons, and less frequently in live animal experiments.

In contrast, extracellular electrical stimulation can activate many neurons around the electrode simultaneously. Its clinical applications include deep brain stimulation (DBS), vagus nerve stimulation (VNS), spinal cord stimulation (SCS), cochlear implants (CI), and visual prostheses, among many others. Unlike intracellular stimulation, extracellular stimulation achieves higher activation efficiency when its working electrode functions as a cathode delivering negative current. Figure 1.13A illustrates how an axon is activated extracellularly by the exposed tip of a wire electrode. The tip, also called a contact, acts as a point source to deliver stimulation current. When a negative pulse is applied, it produces an outward current on the axonal membrane directly beneath the point source, causing membrane depolarization. (Let's call this membrane area the central site.) If the pulse is strong enough to make the depolarization reach the activation threshold, an AP occurs at the central site. Simultaneously, this pulse produces inward current on the membranes on both flank sides as the current completes its electrical loop. These inward currents hyperpolarize the flank membranes. The AP initiated at the central site must overcome the flank hyperpolarization before propagating along the axon bilaterally. The propagation toward the axon terminals is called orthodromic propagation, aligning with the normal physiological direction, while the propagation toward the soma is called antidromic.

The magnitude of membrane depolarization and hyperpolarization produced by a negative pulse depends on the distance between the stimulation point source and the membranes—the shorter this distance, the greater the change in membrane potential. As shown in the cross-sectional view of axon fibers in Fig. 1.13B, where a point source is surrounded by axons perpendicular to the figure plane, three distinct zones emerge. In "Zone I", closest to the point source, a sufficiently strong negative pulse can produce substantial hyperpolarization on the flank membranes, preventing the outward propagation of centrally initiated AP (Durand 2000; van de Steene et al. 2020). As distance increases, in "Zone II" (the activatable zone), the flank hyperpolarization weakens, allowing APs to propagate outward successfully. In "Zone III", the most distant zone, the depolarization produced by the negative pulse cannot reach the activation threshold, thus no APs generate. Consequently, for axons with similar characteristics, effective activation that can propagate outward occurs only on axons with an appropriate distance from the stimulation site. Axons too close to or too far from the stimulation site may not generate effective activation. Even at the same distance, different types of axons may respond differently to stimulation.

With sufficient intensity, a positive pulse can also activate axons extracellularly. Figure 1.13C shows that while a positive pulse generates hyperpolarization at the axonal membrane directly beneath the point source, depolarizations occur on both flank sides under the action of current loop. The depolarizations can produce APs that propagate outward. Since there is no hyperpolarization barrier in their propagation pathway, axons in the zone close to the point source can be activated, while axons outside this zone cannot, due to subthreshold depolarization at a distance (Fig. 1.13D). With a same pulse intensity, the activatable zone of a positive pulse is much smaller than that of a negative pulse. Therefore, in extracellular stimulation, negative pulses are typically much more efficient than positive pulses. However,

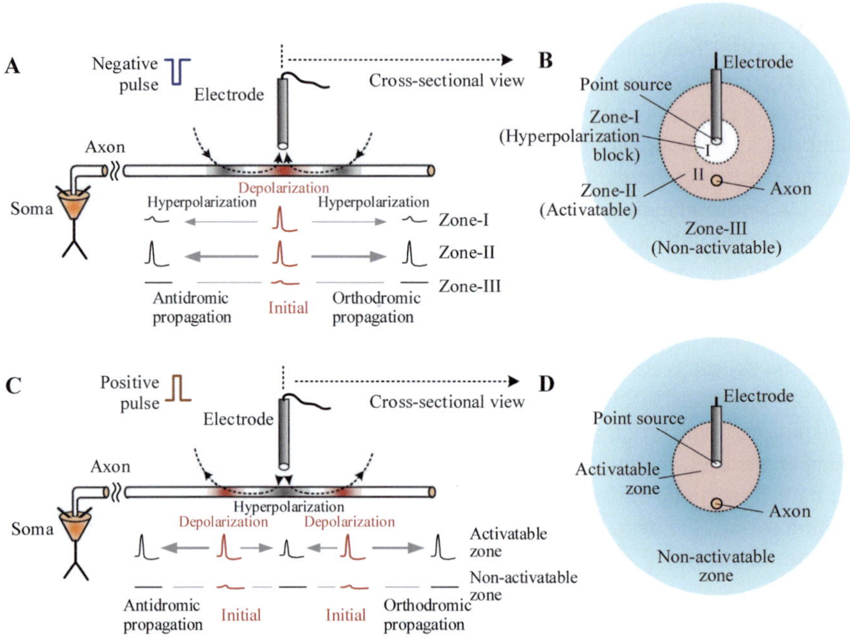

Fig. 1.13 Schematic diagrams illustrating axonal responses to extracellular pulse stimulations. **A** Depolarization and hyperpolarization sites on the axonal membrane under a negative pulse stimulation. **B** Three distinct response zones (within axon fibers) at different distances from the stimulation point source, shown in the cross-sectional view through the point source. **C** and **D** Corresponding diagrams for a positive pulse stimulation

under special circumstances, such as during sustained high-frequency pulse stimulation, the axonal membrane can change, altering the activation efficiency of positive and negative pulses. For more details, please refer to Chap. 7 in this book.

For extracellular pulse stimulation, it is crucial to select an appropriate pulse width based on the chronaxie of target cell membrane (refer to Fig. 1.2). For example, narrow pulses with about 100 μs width can effectively activate axonal fibers of brain neurons, especially myelinated ones, but cannot activate smooth muscle cells with a chronaxie of several milliseconds (Horch and Dhillon 2004). Different neuronal structures—axons, dendrites, and samata (cell bodies)—have distinct membrane properties. Among them, axonal membranes have the shortest chronaxie and are most readily activated by narrow pulses (Ranck 1975; Nowak and Bullier 1998; Buzsáki 2006; Brocker and Grill 2013). Dendrites contain numerous synapses that receive excitatory and inhibitory inputs to generate postsynaptic potentials. Their membranes generally have few voltage-gated Na^+ channel capable of generating APs. Therefore, dendrite potentials spread primarily as passive electrotonic potential, decaying with distance (see Fig. 1.3A). In contrast, soma and axon membranes are rich in voltage-gated Na^+ channels. The regenerative response of Na^+ channels can generate an "all-or-none" AP that propagates along the axon without attenuation.

Due to its shorter chronaxie, an axon is more sensitive to narrow pulses commonly used in electrical neuromodulations than a soma. Therefore, narrow pulse stimulation in the brain can typically first induce APs in axons, which then propagate bilaterally outward (see Fig. 1.13). In myelinated axons, the multi-layer myelin sheath increases resistance and reduces capacitance. Their space constant is typically twice the internode length between Ranvier's nodes. These allow APs to jump between bare nodes, significantly accelerating AP propagation speed.

Extracellular stimulation of peripheral nerves has long shown that thicker axons require lower current intensity for activation than thinner ones (Gilbert et al. 2023). This occurs because thick axons have lower axial resistance and larger axial current proportional to the square of their diameter. Their larger diameters also result in longer space constants. In myelinated axon fibers, thicker axons feature longer internode lengths and space constants, which enhance excitation conduction. Through graded stimulation intensities, axons of different thicknesses can be selectively activated—lower intensities activate thicker axons, while higher intensities gradually recruit thinner ones. Axon activation also depends on other factors, such as the distance from stimulation site, as well as the relationship between axon direction and stimulating current flow.

The in-vivo rat experiments described in this book employed extracellular electrical stimulation. Unless noted otherwise, the following contents focuse specifically on extracellular stimulation.

1.4.2 Voltage and Current Stimulations

In extracellular stimulation, both voltage and current waveforms can be used. During pulse stimulation, voltage waveforms maintain constant voltage while current waveforms maintain constant current. The efficiency of electrical stimulation on excitable tissues mainly depends on the current flowing through the target tissue. For brain tissue specifically, neuronal responses to stimuli depend on the current flowing through the neuronal membrane—the amount of injected charges per unit time (Montgomery 2014).

When applying electrical stimulation, the stimulation current flows through several parts in the circuit loop: instruments (including a stimulator) that generate stimulation waveforms, electrodes that deliver stimulation, interfaces between electrodes and biological tissues, and the stimulated object—the biological tissues. In voltage stimulation, despite constant output voltage from the stimulator, impedance changes in any parts of the circuit can alter the current flowing through the stimulated object (such as neurons), thereby affecting the stimulation efficiency. These impedance changes can arise from electrode polarization, inflammation and glial proliferation around electrodes, and other factors during sustained stimulations (Lempka et al. 2010; Cheung et al. 2013). In current stimulation, however, the stimulator automatically adjusts its output voltage to compensate for impedance changes.

This adjustment maintains the output current at the preset level, ensuring the stimulated neurons receive a steady current. However, if abnormal changes occur in the stimulation circuit, the stimulator may fail to maintain its preset constant current.

Nevertheless, voltage stimulation is commonly used when the impedance of stimulation circuit remains stable. To monitor impedance changes during stimulation, a resistor (like 10 kΩ) can be connected in series with the stimulation electrode. Measuring the voltage across the resistor allows detection of current and impedance fluctuations in the circuit (Kim et al. 2012).

In animal experiments, current stimulation is typically used to avoid the influence of circuit impedance changes. Clinical applications, however, present a more complex situation. The complexity of current-mode stimulators makes them difficult to manufacture as implantable devices. Additionally, voltage stimulation is generally safer than current stimulation, since the latter may apply excessive voltage to the human body during impedance accidents. For these reasons, early implantable products for neural stimulation therapies mainly used voltage-mode. Over time, these products evolved to offer both stimulation modes. Some of the most recent products have even provided current stimulation exclusively (Kern et al. 2020; Gilbert et al. 2023). Theoretically, current stimulation should improve efficiency during unstable impedance periods, such as the postoperative adaptation period. However, its clinical applications still require further verification.

1.4.3 Unipolar and Bipolar Stimulations

Based on the arrangement of cathodic and anodic electrodes, extracellular stimulations can be applied in two modes: unipolar or bipolar.

In unipolar stimulation, the working electrode is placed in the target area, while the other electrode—called independent or ground electrode—is placed far away and serves only to complete the current loop without contributing to stimulation effect. Figure 1.13 illustrates an ideal unipolar stimulation—point source stimulation. To depolarize the cell membrane to its activation threshold, the stimulation must produce a sufficient current density (current per unit area). With a same applied current intensity, an electrode contact with a larger surface area produces a lower current density. Therefore, in neural stimulation, the exposed tip (the contact) of working electrode is typically small to generate sufficient current density. This electrode usually delivers a negative voltage or current to depolarize surrounding cell membranes. Its structural design must also consider both potential tissue damage during implantation and the desired shape and size of its action area. The independent electrode usually has a much larger surface area to prevent depolarizing nearby cells.

The effect of unipolar stimulation decays rapidly with distance, focusing its action near the working electrode. However, the broader spread of stimulation current between the two distant electrodes may activate cells in non-target areas and cause side effects. Nevertheless, its broad action range offers an advantage: therapeutic effects can be achieved even with less precise electrode positioning.

Unipolar stimulation is widely used in clinical applications. For example, for the implant DBS electrodes with multiple contacts, it can be achieved by designating one contact as the working electrode while using the stimulator outer shell (buried under the clavicle) as the independent electrode (Wang et al. 2021). Unipolar stimulation is also common in non-invasive treatments, such as using skin surface electrodes to activate superficial nerves and muscle tissues.

In bipolar stimulation, both electrodes are placed in or near the target area. The two electrode contacts are small and positioned close to each other. This stimulation can be created by using two thin wires with exposed tips only about 0.5 mm apart, or by using the inner and outer poles of a concentric bipolar electrode. The contacts can also be separated by several millimeters. Which contact serves as the working electrode depends on factors such as its position relative to the target, the polarity of the stimulator output it is connected to, and its contact size. For optimal efficiency, the working electrode should be positioned closer to the target area, connected to the stimulator anodic output, and have a smaller contact surface area. Bipolar stimulation offers a key advantage: it can confine the stimulation action to the target area and immediate surroundings, thereby achieving higher efficiency than unipolar stimulation. Additionally, during simultaneous neural signal recording, bipolar stimulation produces smaller stimulation artifacts (see Sect. 4.5 for details). However, it has a disadvantage: when cathodic and anodic contacts are very close, their respective depolarization and hyperpolarization effects may interfere with each other, reducing stimulation efficiency.

1.4.4 Electrical Stimulation Safety

The safety of electrical stimulation involves two key aspects: preventing tissue damage and avoiding electrode corrosion. In clinical applications, both implanted electrodes and applied voltages/currents must meet these safety requirements. Long-term safety must also be maintained during continuous use. Similarly, safety is crucial in animal experiments to prevent tissue damage and electrode corrosion from affecting experimental results. Even when metal electrodes show no corrosion in in-vitro tests, sustained current application through implanted electrodes in a living body may still damage both biological tissues and electrodes (Merrill et al. 2005).

Electrical neural stimulation in clinical applications typically uses metal electrodes. When implanted, the electrodes form an electrode–electrolyte interface with the surrounding tissue fluid—extracellular fluid (ECF). This ECF is an electrolyte solution rich in various ions, including Na^+, K^+, Cl^-, Ca^{2+} and etc. In metal electrodes and their connected electrical circuits, electrons carry charges to form current. In ECF, however, ions carry charges to form current. These two types of currents convert to each other at the electrode–electrolyte interface. This conversion of charge carriers—from electrons in the metal electrode to ions in the electrolyte—occurs mainly through two mechanisms: non-Faradaic reaction and Faradaic reaction (Merrill et al. 2005; Gilbert et al. 2023). The non-Faradaic reaction is capacitive and reversible. It does not involve real electron transfer from electrode to ECF. Instead, it only redistributes charges at the interface—much like charging a capacitor. As electrons accumulate on the electrode side, they attract positively charged ions from the ECF side.

In contrast, Faradaic reaction involves electrochemical processes, transferring electrode electrons to ECF ions to form reactants. This reaction can be reversible providing the electrode delivers opposite charges (reversing the current direction) before the reactants disperse. However, if the reactants diffuse away and cannot be recovered, the electrode gradually corrodes and damages the surrounding tissue. At the anode, the electrochemical reaction is oxidation, while at the cathode, it is reduction. For example, water can be reduced to form hydroxide ions and hydrogen gas, or oxidized to produce hydrogen ions and oxygen gas. Thus, during electrical stimulation, tissue and electrode damages may manifest as bubble formation in the surrounding tissues. Thus, an appearance of bubbles around stimulation electrodes during an experiment may indicate damages and require immediate inspection.

In extracellular stimulation, negative pulses activate nerve tissue most efficiently. However, sustained applying negative pulses, especially at high-frequencies, can cause local charge accumulations, leading to irreversible electrochemical reactions and tissue damage. To prevent this, clinical applications typically use biphasic pulses—a negative phase immediately followed by a charge-balanced positive phase. The positive charges of subsequent phase quickly neutralize the negative charges of preceding phase, preventing irreversible reactions. Once the negative phase depolarizes the cell membrane and activate Na^+ channels, their regenerative response continues despite the following positive phase. Therefore, the positive phase cannot fully counteract the effect of negative phase, and an action potential can still be triggered. However, the positive phase may weaken the activation of negative phase. To minimize this impact, the positive phase can be slightly delayed or delivered with reduced amplitude and increased width. Various waveforms for the balanced positive phase have been proposed (Merrill et al. 2005).

Additionally, limiting the stimulation current below the Faradaic reaction threshold can ensure safety by restricting the stimulation process to non-Faradaic reactions only. Previous studies have shown that when the charge density delivered by the electrode is kept sufficiently low, oxidation–reduction reactions become negligible. The recommended charge density limitation ranges from 15 to 65 $\mu C/cm^2$ per pulse. This serves only as a guideline, as it can vary with electrode material, shape, size, and stimulation waveform, etc (Shannon 1992; Cogan et al. 2016).

For current pulse stimulation, the charge Q conveyed by each pulse equals the integral of the pulse current $I(t)$, i.e., the product of the constant current I and the pulse width PW:

$$Q = \int I(t)\mathrm{d}t = I \times \mathrm{PW} \tag{1.56}$$

The charge density D is the charge Q per unit area of the electrode contact surface, calculated by dividing Q by the surface area A:

$$D = \frac{Q}{A} \tag{1.57}$$

Shannon (1992) proposed using the following k value to determine the current limitation delivered by an electrode:

$$k = \lg D + \lg Q \tag{1.58}$$

where D is measured in μC/cm^2/pulse-phase, and Q in μC/pulse-phase. A stimulus is considered safe when k falls below 1.5–2.0. Higher values may cause tissue damage (Shannon 1992; Cogan et al. 2016; Gilbert et al. 2023). This guideline serves as a reference only and should not be treated as an absolute rule.

1.5 Summary

This chapter introduces characteristics of neuronal membranes. The membrane electrical properties can be modeled using an equivalent circuit composed of resistance and capacitance. During subthreshold changes, both membrane resistance and capacitance remain constant. The time constant can describe how fast the membrane potential changes after applying a step current stimulation. The rheobase is the minimum current required to trigger an AP, while the chronaxie is the minimum pulse duration required to trigger an AP at twice the rheobase current. The space constant and cable equations describe how the subthreshold changes in membrane potential spread along a neuron.

Ion concentration differences between the inside and outside of neurons, along with varying membrane permeabilities to different ions, create the foundations of resting potential and AP. During an AP, voltage-gated Na^+ and K^+ channels manifest nonlinear changes in their conductivity against time and membrane potential. These changes can be simulated by Hodgkin-Huxley (HH) model, which consists of differential equations and nonlinear algebraic equations that require computer programs to solve. This foundational model can simulate responses to stimulations in both individual neurons and neural networks.

Electrical stimulation can activate neurons through both intracellular and extracellular methods. Extracellular stimulation—widely used in clinical applications and in-vivo animal experiments—typically employs square pulse waveforms with either constant current or constant voltage. The pulse width is determined by the characteristics of target tissue. Among neuronal structures, the axonal membrane has the shortest chronaxie, which makes it most responsive to narrow pulses. Extracellular stimulation can be applied in unipolar or bipolar manners, depending on specific requirements. Safety is a prerequisite in electrical stimulation applications—excessive current density must be avoided to prevent tissue damage and electrode corrosion.

References

Bean BP (2007) The action potential in mammalian central neurons. Nat Rev Neurosci 8(6):451–465
Bear MF, Connors BW, Paradiso MA (2016) Neuroscience: exploring the brain (4th ed.). Wolters Kluwer
Bhunia S, Majerus S, Sawan M (2015) Implantable biomedical microsystems: design principles and applications. Elsevier, Amsterdam
Brocker DT, Grill WM (2013) Principles of electrical stimulation of neural tissue. In: Handbook of clinical neurology, vol 116. Elsevier, Amsterdam, pp 3–18
Buzsáki G (2006) Rhythms of the brain. Oxford University Press, Oxford
Catacuzzeno L, Franciolini F (2022) The 70-year search for the voltage-sensing mechanism of ion channels. J Physiol 600(14):3227–3247
Catterall WA (2012) Voltage-gated sodium channels at 60: structure, function and pathophysiology. J Physiol 590(11):2577–2589
Cheung T, Nuño M, Hoffman M et al (2013) Longitudinal impedance variability in patients with chronically implanted DBS devices. Brain Stimul 6(5):746–751
Cogan SF, Ludwig KA, Welle CG et al (2016) Tissue damage thresholds during therapeutic electrical stimulation. J Neural Eng 13(2):021001
DiLorenzo DJ, Bronzino JD (2008) Neuroengineering. CRC Press, Taylor & Francis Group, Boca Raton
Durand DM. 2000. Electrical stimulation of excitable tissue. In: Bronzino JD (ed) The biomedical engineering handbook. Springer, Berlin
Enderle JD, Bronzino JD. 2012. Introduction to biomedical engineering (3rd ed.). Elsevier, Amsterdam
Gilbert Z, Mason X, Sebastian R et al (2023) A review of neurophysiological effects and efficiency of waveform parameters in deep brain stimulation. Clin Neurophysiol 152:93–111
Hodgkin AL, Huxley AF, Katz B (1952) Measurement of current-voltage relations in the membrane of the giant axon of Loligo. J Physiol 116(4):424–448
Hodgkin AL, Huxley AF (1952a) A quantitative description of membrane current and its application to conduction and excitation in nerve. J Physiol 117(4):500–544
Hodgkin AL, Huxley AF (1952b) Currents carried by sodium and potassium ions through the membrane of the giant axon of Loligo. J Physiol 116(4):449–472
Hodgkin AL, Huxley AF (1952c) The components of membrane conductance in the giant axon of Loligo. J Physiol 116(4):473–496
Hodgkin AL, Huxley AF (1952d) The dual effect of membrane potential on sodium conductance in the giant axon of Loligo. J Physiol 116(4):497–506
Hopper AJ, Beswick-Jones H, Brown AM (2022) A color-coded graphical guide to the Hodgkin and Huxley papers. Adv Physiol Educ 46(4):580–592
Horch KW, Dhillon GS. 2004. Neuroprosthetics: theory and practice. World Scientific Publishing Company.
Jastrzębski M, Moskal P, Bednarek A et al (2019) His bundle has a shorter chronaxie than does the adjacent ventricular myocardium: implications for pacemaker programming. Heart Rhythm 16(12):1808–1816
Kern DS, Fasano A, Thompson JA et al (2020) Constant current versus constant voltage: clinical evidence supporting a fundamental difference in the modalities. Stereotact Funct Neurosurg 99(2):171–175
Kim E, Owen B, Holmes WR et al (2012) Decreased afferent excitability contributes to synaptic depression during high-frequency stimulation in hippocampal area CA1. J Neurophysiol 108(7):1965–1976
Koch C (1999) Biophysics of computation: Information processing in single neurons. Oxford University Press, UK

Lempka SF, Johnson MD, Miocinovic S et al (2010) Current-controlled deep brain stimulation reduces in vivo voltage fluctuations observed during voltage-controlled stimulation. Clin Neurophysiol 121(12):2128–2133

McIntyre CC, Grill WM (1999) Excitation of central nervous system neurons by nonuniform electric fields. Biophys J 76(2):878–888

Merrill DR, Bikson M, Jefferys JGR (2005) Electrical stimulation of excitable tissue: design of efficacious and safe protocols. J Neurosci Methods 141(2):171–198

Montgomery EB (2014) Intraoperative neurophysiological monitoring for deep brain stimulation: principles, practice, and cases. Oxford University Press

Nowak LG, Bullier J (1998) Axons, but not cell bodies, are activated by electrical stimulation in cortical gray matter. Exp Brain Res 118(4):489–500

Piccolino M (1998) Animal electricity and the birth of electrophysiology: the legacy of Luigi Galvani. Brain Res Bull 46(5):381–407

Ranck JB (1975) Which elements are excited in electrical stimulation of mammalian central nervous system: a review. Brain Res 98(3):417–440

Schneider T, Filip J, Soares S et al (2023) Optimized electrical stimulation of c-nociceptors in humans based on the chronaxie of porcine C-fibers. J Pain 24(6):957–969

Schwiening CJ (2012) A brief historical perspective: Hodgkin and Huxley. J Physiol 590(11):2571–2575

Shannon RV (1992) A model of safe levels for electrical stimulation. IEEE Trans Biomed Eng 39(4):424–426

Stern S, Agudelo-Toro A, Rotem A et al (2015) Chronaxie measurements in patterned neuronal cultures from rat hippocampus. PLoS ONE 10(7):e0132577

Udupa K, Chen R (2015) The mechanisms of action of deep brain stimulation and ideas for the future development. Prog Neurobiol 133:27–49

van de Steene T, Tanghe E, Tarnaud T et al (2020) Sensitivity study of neuronal excitation and cathodal blocking thresholds of myelinated axons for percutaneous auricular vagus nerve stimulation. IEEE Trans Biomed Eng 67(12):3276–3287

Wang Z, Feng Z, Yang G et al (2021) Advances in the development of new stimulation paradigms of deep brain stimulation. Prog Biochem Biophys 48(3):263–274 (in Chinese)

Wickens AP (2015) A history of the brain: from stone age surgery to modern neuroscience. Psychology Press

Chapter 2
Neural Circuits and Evoked Potentials

Abstract The animal experiments described in this book focus on the rat hippocampus. This chapter introduces hippocampal structures and neural circuits. It explains the generations of various evoked potentials in neuronal population responses to pulse stimulations, including orthodromically and antidromically evoked potentials. The chapter also explores the effects of feedforward and feedback inhibitory circuits through paired-pulse stimulations in the hippocampal CA1 and CA3 regions.

Keywords Hippocampus · Evoked potential · Local inhibitory circuit · Orthodromic activation · Antidromic activation · Feedback inhibition · Paired-pulse stimulation

2.1 Overview of Nervous System

2.1.1 Neuron and Nervous System

The mammalian nervous system consists of two main parts: the central nervous system (CNS) and the peripheral nervous system (PNS). CNS includes the brain and spinal cord, while PNS includes cranial, spinal, and visceral nerves. The nervous system contains two main cell types: neurons and glial cells. Neurons are the basic structural and functional units that receive input signals, generate action potentials, and transmit them. Glial cells support neurons by providing protection, repair, and nutrition. They also form myelin sheaths and transport metabolic substances.

A neuron typically consists of three main structures: soma, dendrites, and axon (Fig. 2.1). The soma, varying in size and shape, serves as both the signal integration center and the metabolic and nutritional hub. It comprises membrane, nucleus, and cytoplasm. Cytoplasm contains both common organelles and specialized structures such as Nissl bodies and neurofibrils. Dendrites, typically shorter than axons, form tree-like structures that branch repeatedly and become progressively thinner. They receive input signals and propagate them to soma. Each neuron typically has one

Z. Feng, *Electrical Brain Stimulation: Mechanisms and Modulation of Neuronal Activity*, https://doi.org/10.1007/978-981-95-4145-4_2

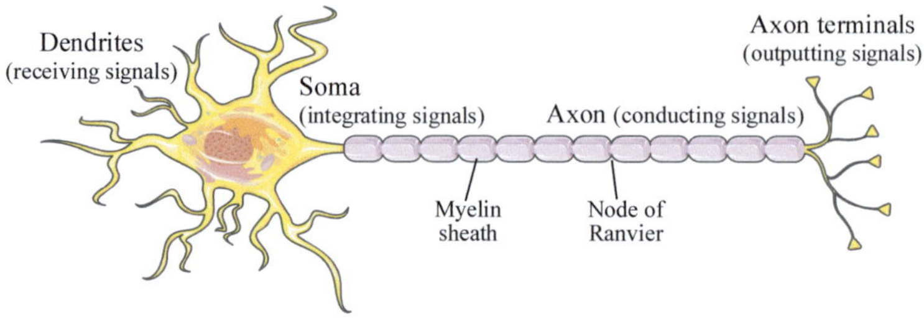

Fig. 2.1 Schematic diagram of neuronal structures and their functions

axon, ranging in length from a few micrometers to over a meter. Axons carry action potentials (also known as neural impulses) and transmit them to other neurons or effectors through synapses at axonal terminals. An axon maintains relatively uniform thickness throughout its length, beginning at the axon hillock where it emerges from the soma. The axon initial segment (AIS) is slightly thicker and lacks a myelin sheath. Axons with myelin sheaths are called myelinated nerve fibers, while those without are called unmyelinated nerve fibers. Axons typically end by splitting into multiple branches, each with swollen spines that form synapses with other neurons or muscle cells.

2.1.2 Synapses and Synaptic Potentials

Neurons communicate with each other and with effector cells through specialized junctions called synapses. These synapses can be classified based on their locations, such as axon-dendrite, axon-soma, or axon-axon synapses. They can also be categorized into chemical and electrical synapses. Chemical synapses use neurotransmitters to transmit signals, converting electrical signals to chemical ones and back again. Electrical synapses, in contrast, conduct signals directly through localized currents.

Electrical synapses, also called gap junctions, are protein channels connecting the adjacent membranes of two neurons. These channels work like hollow rivets, allowing ions and small molecules to flow through. Unlike chemical synapses, electrical synapses have no distinct presynaptic and postsynaptic membranes. This allows them to transmit signals quickly in both directions with almost no delay. Though their exact functions are not fully understood, electrical synapses likely play a key role in promoting synchronous neuronal activity.

Chemical synapses, often simply called synapses, are the primary means of signal transmission in the nervous system. A typical synapse consists of three components: presynaptic membrane, synaptic gap, and postsynaptic membrane. The synaptic gap is about 20–40 nm wide, while both the presynaptic and postsynaptic membranes are approximately 7.5 nm—slightly thicker than typical neuronal membranes. Inside the presynaptic membrane, there exist numerous synaptic vesicles that store neurotransmitters. When neural impulses (action potentials) reach and depolarize the presynaptic membrane to threshold level, voltage-gated calcium ion (Ca^{2+}) channels open, allowing extracellular Ca^{2+} to flow in. This influx causes a rapid increase in Ca^{2+} concentration in the axoplasm ($[Ca^{2+}]_i$) inside presynaptic membrane. The increased $[Ca^{2+}]_i$ triggers synaptic vesicles to fuse with the presynaptic membrane, releasing neurotransmitters into the synaptic gap. The amount of released neurotransmitters is directly proportional to the $[Ca^{2+}]_i$ level. Once in the synaptic gap, the neurotransmitters diffuse to the postsynaptic membrane, bind to specific receptors, and open specific ion channels. This process leads to either depolarization or hyperpolarization in the postsynaptic membrane, creating synaptic potentials.

Synaptic potentials have two types: excitatory postsynaptic potential (EPSP) and inhibitory postsynaptic potential (IPSP). These potentials arise from the opening of ligand-gated Na^+, Cl^-, and K^+ channels, which depends on the types of neurotransmitters and their receptors. In the CNS, glutamate and aspartate are the main excitatory amino acid neurotransmitters. Their receptors include α-amino-3-hydroxy-5-methyl-4-isoxazole propionic acid (AMPA), kainic acid (KA) and N-methyl-D-aspartate (NMDA). AMPA and KA receptors—together termed non-NMDA receptors—respond quickly to glutamate and mainly increase membrane permeability to Na^+ and K^+. In contrast, NMDA receptors respond slowly to glutamate and, when activated, become permeable to Na^+, K^+ and Ca^{2+}. Unlike non-NMDA receptors, which are purely ligand-gated, NMDA receptors are both ligand-gated and voltage-gated. They contain binding sites for both glutamate and magnesium ions (Mg^{2+}). Mg^{2+} ions block NMDA receptor channels in a voltage-dependent manner. When bound with glutamates, the post-synaptic membrane must also reach a specific level of depolarization before Mg^{2+} ions move away from their blocking sites and let the NMDA receptor channels open. The dual requirements explain why NMDA receptors respond slowly to glutamate. Most glutamate target neurons contain both NMDA and AMPA receptors, allowing the rapid membrane depolarization caused by AMPA receptors to trigger the opening of NMDA receptor channels.

In the CNS, the main inhibitory amino acid neurotransmitters include γ-aminobutyric acid (GABA) and glycine (Gly). Postsynaptic membranes contain two types of GABA receptors: $GABA_A$ and $GABA_B$. $GABA_A$ is an ionotropic receptor with Cl^- channels that, when activated, increases Cl^- influx. $GABA_B$ is a G-protein-coupled metabotropic receptor that, when activated, inhibits adenylate cyclase, activates K^+ channels, and increases K^+ efflux. Both Cl^- influx and K^+ efflux can hyperpolarize the postsynaptic membrane, generating IPSPs. Inhibitory synapses are prevalent in local neuronal circuits of the brain. They play a crucial role in controlling neuronal excitability, modulating information integration and action potential generation, regulating synaptic plasticity, and promoting synchronous neuronal activity (Cobb et al. 1995; Maccaferri and Dingledine 2002; Staff and Spruston 2003).

Synaptic transmission exhibits plasticity—the ability to undergo long-lasting changes in efficiency through repeated activations. Long-term potentiation (LTP) and long-term depression (LTD) are the two main forms of synaptic plasticity. They are considered as the foundations of high-level brain functions such as learning and memory. The mechanisms of LTP and LTD formations are outlined below.

1. **Long-Term Potentiation**

LTP refers to an enhancement in EPSPs that occurs in postsynaptic neurons after presynaptic neurons are activated by a brief, high-frequency pulse train. This enhancement can persist for a long period. For example, applying 50–100 pulses at 100 Hz can produce LTP in the hippocampal CA1 pyramidal neurons, lasting for hours, days or even months. LTP generates when high-frequency stimulation increases EPSP through superimposition, which promotes the opening of NMDA receptor channels gated by both ligands and membrane potential. This results in increased Ca^{2+} influx and $[Ca^{2+}]_i$ inside the postsynaptic membrane. The increased $[Ca^{2+}]_i$ activates Ca^{2+}-CaM dependent protein kinase IIs, which phosphorylate AMPA receptor-coupled channels and increase their conductance. It also causes more AMPA receptors to bind to the postsynaptic membrane, increasing receptor density. As a result, when presynaptic membrane releases neurotransmitters, there is increased opportunity and efficiency for neurotransmitters binding to AMPA receptors on the postsynaptic membrane. This enhances EPSP, resulting in LTP.

2. **Long-Term Depression**

LTD refers to a decrease in synaptic transmission efficiency which can be also produced by repeated pulse stimulation but at a low frequency. For example, LTD can generate by applying 1–5 Hz pulses for a few minutes to the Schaffer collaterals—afferent fibers of the hippocampal CA1 region. While its generation follows a similar route as LTP, LTD produces the opposite outcome. The low-frequency pulses open only a small number of NMDA receptor channels, resulting in a slight increase in $[Ca^{2+}]_i$ inside the postsynaptic membrane. This leads to dephosphorylation, rather than phosphorylation, of Ca^{2+}-CaM dependent protein kinase II. The enzymes then dephosphorylate AMPA receptors, causing a decrease in AMPA receptor density on the postsynaptic membrane, thereby reducing EPSP. Both LTP and LTD generations involve NMDA receptor channels, Ca^{2+} influx, and changes in AMPA receptor efficiency on the postsynaptic membrane, though with different outcomes.

2.1.3 Information Integration in Neurons

1. **Synaptic Integration**

Each neuron's dendrites contain numerous synapses—typically tens of thousands in total. These synapses are either excitatory or inhibitory, producing EPSPs or IPSPs, respectively. Spatiotemporal integration of these postsynaptic potentials can trigger an action potential (AP) when the membrane becomes sufficiently depolarized to reach the activation threshold. The AP typically starts at the AIS due to its high density of voltage-gated Na^+ channels and small diameter, which requires fewer charges to achieve membrane depolarization during capacitance charging (Stuart et al. 1997). Once triggered, the AP propagates along the axon to its terminals to complete signal conduction. The AP generation depends on multiple factors, including EPSP magnitude, the spatiotemporal integration of EPSPs, and the activity of inhibitory synapses near soma (Bear et al. 2016).

First, larger EPSPs contribute more in triggering an AP. The postsynaptic membrane of each synapse contains many neurotransmitter-gated channels that activate in response to neurotransmitters released from the presynaptic membrane. The neurotransmitter release is controlled by Ca^{2+} influx into the presynaptic membrane, which varies with the strength and frequency of incoming APs. Because each released synaptic vesicle contains a fixed number of neurotransmitter molecules, the EPSP magnitude at postsynaptic membrane changes in discrete steps rather than continuously.

Second, the integration of multiple EPSPs over space and time can produce a large depolarization potential, increasing the likelihood to trigger an AP. This integration depends on the electrical properties of dendrites. Synaptic current gradually decays as it spreads toward the soma. The synapses' contribution to soma depolarization decreases with increasing distance, governed by both the intracellular and membrane impedances of the dendrites. The intracellular impedance is determined by the dendrite diameter and the cytoplasm feature.

Finally, inhibitory synapses in peri-somatic region can exert bypass effects. These synapses, primarily of the axon-soma type, locate close to the soma. When they are activated during dendritic EPSPs, their IPSPs (produced by opening Cl^- channels near the soma) can bypass the spreading currents from the EPSPs, significantly reducing the EPSPs' contribution to soma depolarization and preventing AP generation.

Moreover, the EPSP integration from a few excitatory synapses alone is usually insufficient to trigger an AP. A single EPSP typically produces only about 1 mV of soma depolarization, while its activation threshold is about 20–30 mV. This means that even without opposing IPSPs, a postsynaptic neuron needs 20–50 EPSPs occurring almost simultaneously to trigger an AP.

The traditional view of neuronal signal conduction is unidirectional: dendrite synapses receive input signals, which trigger the soma (or AIS) to generate an AP that then travels along the axon to terminal synapses, as shown in Fig. 2.1. This model casts dendrites and axons as simple conductors—dendrites for input and axons for output. The soma acts as an integrator, like a threshold calculator, combining dendrite signals to produce an output signal. However, research has challenged this basic model, revealing that dendrites also perform integrations and show complex hierarchical information processing. Additionally, APs can travel backward from soma to dendrites. These perspectives are outlined below.

2. **Hierarchical Information Integration in Neurons**

With advanced techniques like multi-electrode patch clamps and two-photon microscopy imaging, the structure and function of tiny dendrites have been revealed. Studies have shown that dendrites themselves can perform complex signal processing. When they receive inputs from numerous synapses, dendrites can carry out both linear and nonlinear signal processing, including integration, amplification, reduction, and detection of synchronous inputs (London and Häusser 2005; Sidiropoulou et al. 2006).

Synapses typically distribute on dendritic spines—tiny protrusions on dendrite branches. The pyramidal neurons in cerebral cortex and the Purkinje cells in cerebellar cortex have high densities of dendritic spines, with thousands per neuron. These spines not only greatly increase the surface area available for receiving input signals but also perform information integration (Spruston 2008).

Dendritic spines contain high densities of voltage-gated Na^+ and Ca^{2+} channels and NMDA receptors, giving them active electrical properties. Like the ion channels on soma and axon that have a threshold mechanism for APs, conductance in spine channels can also change and result in varying gains to synaptic inputs. This mechanism enables dendritic spines to integrate input signals nonlinearly. Strong inputs can trigger localized dendritic spikes. These spikes, driven by voltage-gated ion channels, are regenerative and have distinct waveforms—occurring only when input signals reach a specific threshold. Unlike APs, these spikes have limited propagation and are confined to tiny dendritic branches, exhibiting independence and locality. However, they can help amplify input signals at neighboring dendritic spines.

As shown in Fig. 2.2, when dendritic spines receive dense inputs simultaneously, their nonlinear integrations—a threshold mechanism (denoted by S-shape symbols in the figure)—can enhance dendritic signals locally. In contrast, sparse inputs can decay quickly due to the filtering effect of spine membrane and spine narrow neck structure (Yuste and Urban 2004). The numerous dendritic spines on dendrites act as independent computational subunits (Poirazi et al. 2003). They first integrate their respective synaptic inputs by a threshold mechanism. Then, the outputs of these subunits are further integrated and threshold-computed on the dendritic trunk and soma to produce the final output of the entire neuron. This creates a two-level integration mechanism.

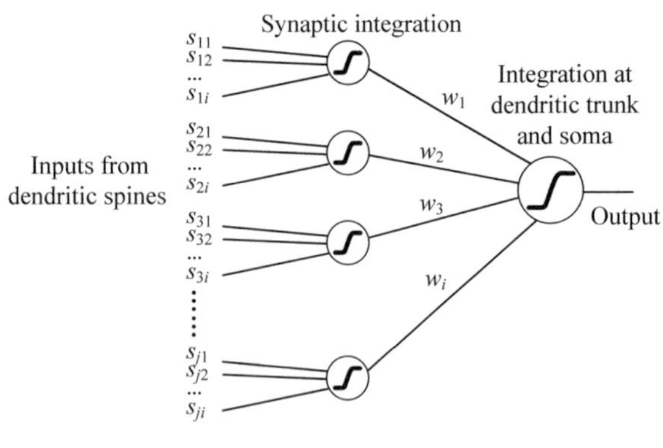

Fig. 2.2 Schematic diagram of the two-level integration model in pyramidal neurons. The S-shape symbols in the circles represent threshold computations. w_i represents subunit weight

For a pyramidal neuron with extensive apical dendrites (as shown in Fig. 2.3), Häusser and Mel (2003) proposed a three-level integration model for its information processing. The three levels are: first, the integration of dendritic spines at the distal apical dendrites; second, the integration of fine branches at the proximal apical dendrites; and third, the output integration at soma based on the first two levels. Experimental data have supported this model, demonstrating areas that can produce secondary spikes at the main bifurcation of apical dendrites. The branches at distal apical dendrites function as independent integration units with their own spike generation mechanism—the first-level integration. When membrane potential exceeds a threshold, it triggers a Ca^{2+}-dominated dendritic spike, producing significant dendrite depolarization that can substantially contribute to the somatic membrane potential. Additionally, a so-called "coupling zone" at the proximal apical dendrites forms the second-level integration, where input signals can modulate the interaction between somatic/axonal spikes and distal dendritic spikes.

The multi-level integration model reveals the important role of apical dendrites in pyramidal neurons in brain regions such as the cerebral cortex and hippocampus. These neurons have extensive apical dendrites with numerous synaptic inputs located far from their soma. Due to passive potential attenuation in the dendritic membrane, it is challenging to explain how signals from these distant apical dendrites contribute to the somatic potentials without some compensatory mechanism. The persistence of these distal dendrites throughout biological evolution demonstrates their essential physiological functions.

Incoming signals from dendritic synapses do not always directly affect a neuron to generate action potentials. Rather, these signals undergo screening and processing

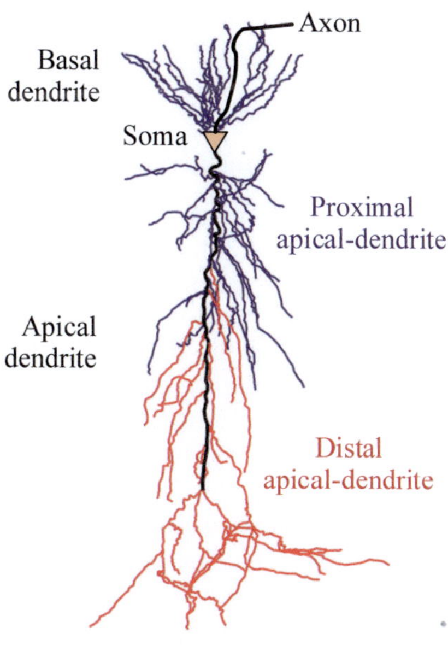

Fig. 2.3 Schematic diagram of CA1 pyramidal neuron morphology

before final integration at soma, where some are filtered out while others are enhanced. The dendrites' ability to integrate information depends on complex factors, including their morphological structures, the timing and location of synaptic inputs, the interaction between excitatory and inhibitory inputs, and the distributions of voltage-gated channels on dendritic membranes.

The above multi-level models only describe the unidirectional propagation of neural signals—the traditional flow from dendrites to soma and to axon. However, research has shown that action potentials can also travel backward from the soma to dendrites (Stuart et al. 1997). This reveals that a neuron's information processing is not simply "open-loop", but involves feedback mechanisms. These mechanisms play important roles in dendritic function and synaptic plasticity (London and Häusser 2005). For instance, backpropagation of action potentials can activate slow voltage-gated ion channels on dendrites. The resulting channel currents can change the potentials at the soma and axon initial segment, triggering new action potentials. This soma-dendrite interaction can lead neurons to produce burst firing, a characteristic of certain principal neurons like hippocampal pyramidal neurons. Burst firing can enhance the reliability of neural information transmission (Buzsáki 2006; Koch 1999).

In summary, a neuron itself has sophisticated mechanisms for information processing. These include not only soma integration but also threshold processing through active properties in dendritic spines and main dendritic branches, as well as backpropagation of somatic action potentials. Each neuron acts as a collection of smaller units working together to integrate information. When a group of neurons connect through synapses to form neural networks (or neural circuits), they achieve even more complex information integration. The next section will examine integration at the neural circuit level and the responses of neuronal populations to external stimulations.

2.2 Hippocampal Structure and Evoked Potentials

The hippocampus, an essential part of the brain limbic system, is named for its resemblance to seahorse. The information processing in its neural circuits is closely related to brain functions such as learning and memory, as well as emotion and sleep. Though crucial for memory formation, the hippocampus does not permanently store memories. People with hippocampal damage can retain old memories but struggle to form new ones. In the 1950s, an epilepsy patient named Henry Molaison (H.M.) became an anterograde amnesiac after undergoing brain tissue removal involving hippocampus. This famous case, among others, confirms the vital role of hippocampus in learning and memory. Later, in the 1970s, Bliss and colleagues made a groundbreaking discovery—long-term potentiation (LTP) in hippocampal synaptic transmission (Bliss and Lømo 1973; Bliss and Gardner-Medwin 1973). This finding revealed that synapses have long-term plasticity—possibly the biological basis for brain learning and memory functions. In that same decade, the discovery of place cells in the hippocampus also indicated its crucial role in spatial cognition and memory (O'Keefe and Dostrovsky 1971). Additionally, the hippocampus is a common focus region of diseases such as epilepsy and Alzheimer's disease. Furthermore, its clear layered structure and well-defined neural circuits make it fascinating to researchers across various neuroscience fields, including neurophysiology, psychology, pathology, neural engineering, and computational neurobiology.

2.2.1 Hippocampal Structure and Major Excitatory Pathways

The hippocampus has a unique anatomical structure, with neuron types and fiber pathways fundamentally similar across various mammalian species (Andersen et al. 2007). As shown in Fig. 2.4A, the rat hippocampus lies at the bottom of lateral ventricle in the temporal lobe, forming part of the medial surface of cerebral hemisphere. It curves along its long axis into a crescent shape, extending first from anterior to posterolateral, then anteromedially. The hippocampi in the left and right cerebral hemispheres resemble a pair of sheep horns. The anterior portions of the two hippocampi connect through axonal fibers of the ventral hippocampal commissure (VHC), which pass beneath the corpus callosum. The hippocampal lateral surface forms the medial wall of lateral ventricle, while its inferomedial surface wraps around the brainstem. Its posterior edge contains the dentate gyrus, which connects to the entorhinal cortex through the subiculum.

The principal neurons in the hippocampal region are arranged densely and orderly, forming distinct neural circuit structures. Figure 2.4B shows a sagittal section at 2.4 mm mediolateral to bregma from the rat brain atlas (Paxinos and Watson 2007), where stained somata appear blue-purple. In the hippocampal region, neuronal somata form two dark curves: one in the cornu ammonis (CA) and another in the dentate gyrus (DG). These curves stand out against surrounding areas due to the dense soma arrangement. The hippocampus well-defined laminar structure is visible to the naked eye even in unstained brain slices. Figure 2.4C shows a sagittal view of a rat brain, captured with a Canon camera (model EOS 6D) equipped with a macro lens (EF 100 mm f/2.8 USM). This photo displays the brain after paraformaldehyde immersion, sectioned at approximately 2.7 mm mediolateral to bregma. In the center of the photo, the hippocampus and its structure are visible, standing distinctly apart from surrounding brain tissues.

Figure 2.4D illustrates the major excitatory pathways in the two hippocampal regions—the CA and the DG. The CA forms a C-shape and is divided into four subregions (CA1, CA2, CA3 and CA4) based on differences in internal development and fiber formation. Pyramidal neurons (or pyramidal cells) are CA principal neurons. The DG, with granule cells as its principal neurons, curves into a V- or U-shape (depending on the septotemporal position) and nestles within the CA's C-shape, mostly surrounded by the CA.

The excitatory synaptic connections of hippocampus form a "tri-synaptic circuit": (1) The perforant path (PP), formed by the axons of entorhinal cortex neurons, serves

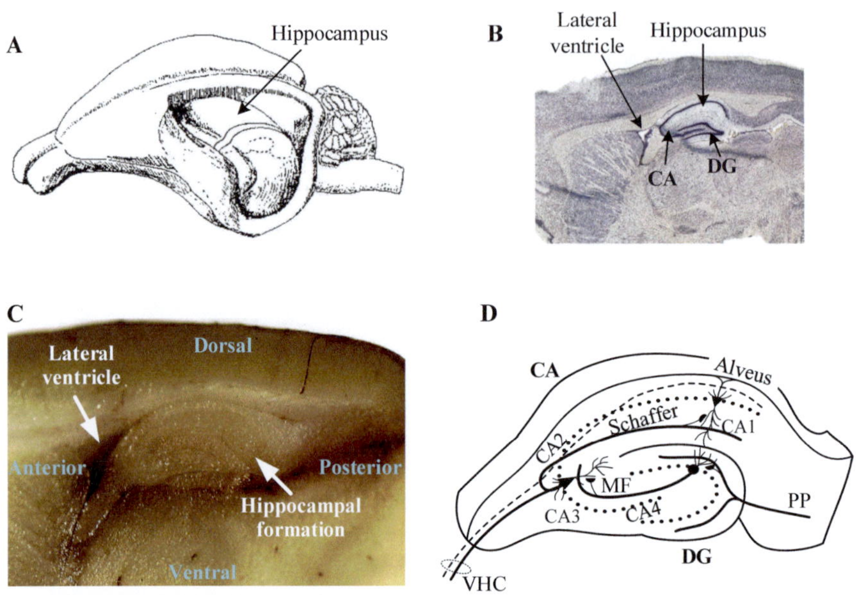

Fig. 2.4 Anatomical location, sagittal section views, and major excitatory pathways of the rat hippocampus. **A** Schematic diagram showing the hippocampus location in the rat brain. **B** Sagittal section at 2.4 mm mediolateral to bregma from the rat brain stereotaxic atlas (Paxinos and Watson 2007). **C** A photo showing a sagittal section view of the rat hippocampus. **D** Schematic diagram of the "tri-synaptic circuit" showing excitatory pathways in the cornu ammonis (CA) and dentate gyrus (DG)

as the afferent fibers to the hippocampus and forms synapses with the dendrites of DG granule cells. (2) The axons of these granule cells form mossy fibers (MF), which connect to and form synapses with the dendrites of CA3 pyramidal neurons. (3) The axons of CA3 pyramidal neurons form the Schaffer collaterals that then make synapses with the dendrites of CA1 pyramidal neurons. Finally, the myelinated axons of CA1 pyramidal neurons form the hippocampal white matter, known as alveus—the efferent fiber of the hippocampus. The alveus creates a thin white sheet covering the dorsal surface of hippocampus, giving it a silvery appearance. When the overlying cerebral cortex is removed, the exposed hippocampus can be easily identified by this silvery color. Section 3.3 will describe an experiment approach with exposed hippocampus.

In addition to the ipsilateral excitatory connections, the axonal branches of rodent CA3 pyramidal neurons connect to neurons in the contralateral hippocampal region through VHC (Fig. 2.4D). Furthermore, inhibitory GABAergic interneurons exist throughout the hippocampal regions. They have short axons that form inhibitory synapses with nearby principal neurons and other interneurons, releasing GABA neurotransmitters to create local inhibitory effects. Although interneurons comprise only about 10% of the total neuron population—far fewer than principal neurons (i.e., CA pyramidal neurons and DG granule cells) (Andersen et al. 2007), they exhibit diverse morphologies, chemical properties and functions (Klausberger and Somogyi 2008; Kepecs and Fishell 2014; Pelkey et al. 2017). The axon terminals of each interneuron connect with multiple neurons to form inhibitory synapses. These synapses serve different functions depending on their locations: those on neuronal somata and axon initial segments can effectively regulate action potential generation, while those on dendrites can affect synaptic potential integrations. The effects of local inhibitory circuits can be tested using electrical pulse stimulations, as detailed in the subsequent sections.

2.2.2 Hippocampal Evoked Potentials by Pulse Stimulations

The rat hippocampus has a tightly packed cell body layer that contains only 3–5 layers of somata. The somata are densely arranged within a thin layer. The dendrites align directionally, forming a distinct layered structure. When the neurons are activated synchronously, this arrangement can result in unique potential waveforms in each layer (Richardson et al. 1987; Kloosterman et al. 2001; Vreugdenhil et al. 2005; Andersen et al. 2007). Figure 2.5A1 illustrates the layers of hippocampal CA1 pyramidal neurons. From dorsal to ventral (top to bottom in the figure), the layers are: stratum oriens (so), pyramidal cell layer (pcl), stratum radiatum (sr), and stratum lacunosum-moleculare (sl-m). The so layer, about 300 μm thick, contains the basal dendrites of pyramidal neurons, which form a thicket-like structure with numerous branches and no clear distinction between main and secondary trunks. The pcl layer contains the somata. The sr and sl-m layers contain the apical dendrites, together spanning about 600 μm. The apical dendrites of a CA1 pyramidal neuron are long, featuring distinct main and secondary trunks that extend broadly. The neuron has a cone-shape soma with basal dendrites growing from the cone base and apical dendrites from its apex. Its axon grows from the cone base and ascends to form the alveus. In the CA1 region, the somata of pyramidal neurons average about 15 μm in diameter—smaller than those in the CA3 region, which can reach 30 μm (Andersen et al. 2007).

As shown in Fig. 2.5A2, the DG lies beneath the CA1 region, beyond the hippocampal fissure (hf). The DG comprises three layers: the molecular layer (ml), granule cell layer (gcl), and polymorphic layer (pl). A DG granule cell has a round or oval soma, with its dendrites mainly in the ml layer.

When a pulse is applied to the Schaffer collaterals—the afferent pathway in the sr layer of CA1 region (Figs. 2.4D and 2.5A1), it can activate a bundle of axons. The resulting APs travel collectively towards the synapses at the apical dendrites of CA1 pyramidal neurons. Through synaptic transmissions, a group of these downstream pyramidal neurons can be activated to fire APs synchronously. Similarly, applying a pulse at the PP—the afferent pathway to the DG—can activate PP axons and then their synapses in the ml layer of the DG (Fig. 2.5A2). Through synaptic transmissions, a group of DG granule cells can generate APs synchronously.

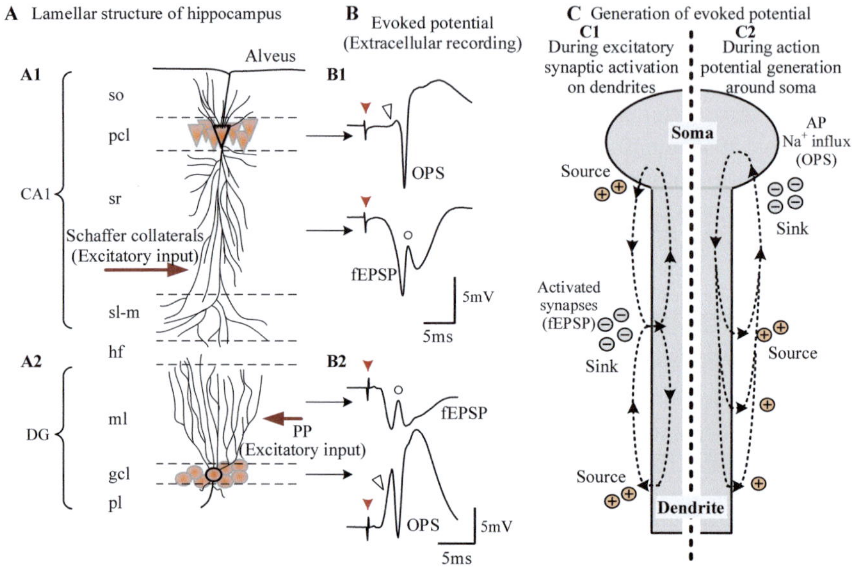

Fig. 2.5 Lamellar structures and layer-specific evoked potentials in the rat hippocampal region. **A** Lamellar structures in the hippocampal CA1 region (A1) and DG region (A2). **B** When an electrical pulse is applied to the CA1 Schaffer collaterals (B1) or to the DG PP pathway (B2), the resulting excitatory inputs can induce OPS and fEPSP in the soma and dendritic layers, respectively. **C** Schematic diagram of evoked potential generations. Small circles with "+" and "−" represent the accumulated positive and negative charges in the extracellular space, forming current "source" and "sink" respectively. The left (C1) and right (C2) diagrams, separated by a thick dashed line, respectively show the ionic flows during excitatory synaptic activation on dendrites (C1) and action potential generation at the soma (C2)

As shown in Fig. 2.5B, unique potential waveforms from activated neuron populations can be recorded in the extracellular space. In the dendritic layer receiving excitatory synaptic inputs, the recorded signal consists mainly of the field excitatory postsynaptic potential (fEPSP)—a primarily negative potential generated by synapses. In the soma layer, the recorded signal consists mainly of the negative population spike (PS), which forms from APs (Andersen et al. 1971). When excitatory synaptic inputs produce a PS, it is called an orthodromic population spike (OPS) (Kloosterman et al. 2001). The fEPSP amplitude or slope can indicate the strength of excitatory synaptic transmission, while the OPS amplitude relates closely to the number of neurons firing action potentials and their discharge synchronicity (Theoret et al. 1984).

Figure 2.5C1 illustrates ion flows during the activation period of excitatory synapses on dendrites, while Fig. 2.5C2 illustrates the period of action potential generation at soma. The diagrams explain the formation of fEPSP and OPS. When receptor ion channels open in the postsynaptic membranes, positively charged ions flow into the membranes, leaving negative charges outside. This creates the negative fEPSP outside the dendrites. Through the current loops (indicated by dashed lines with arrows in the figure), this fEPSP appears as a positive potential outside the somata (indicated by hollow triangles in Fig. 2.5B). Similarly, the negative OPS forms outside the somata when positively charged Na^+ ions flow into membranes during APs, leaving negative charges outside. The current loop causes the OPS to appear as a small positive spike superimposing on the negative fEPSP in the dendritic layer (indicated by small circles in Fig. 2.5B).

In summary, the distinct laminar structures in the hippocampal region provide unique biomarkers in evoked potentials. These potentials can serve two purposes: evaluating the excitability of neurons and their networks, and guiding proper electrode placement in electrophysiological experiments (see Sect. 3.2 for details).

2.3 Local Inhibitory Circuits in the Rat Hippocampal CA1 Region

The principal neurons in the hippocampal region—CA pyramidal neurons and DG granule cells—are responsible for information integration and transmission. These neurons connect both upstream and downstream neurons along projection pathways, while also forming local neural circuits with surrounding neurons, especially the interneurons in local inhibitory circuits. Complex regulation mechanisms within these neural networks maintain the excitation balance in principal neurons (Pelkey et al. 2017; Andersen et al. 2007; Mori et al. 2004; Sik et al. 1994; Kullmann 2011).

2.3.1 Effect of Inhibitory Circuits During Orthodromic Activations of CA1 Pyramidal Neurons

Figure 2.6A illustrates the local inhibitory circuits in the hippocampal CA1 region. Pink triangles represent pyramidal neurons, while blue ovals represent interneurons (IN). Through the Schaffer collaterals, axon terminals of upstream CA3 pyramidal neurons form excitatory synapses with both CA1 pyramidal neurons and interneurons (IN_1 and IN_2). The CA1 pyramidal neurons also form excitatory synapses with interneurons (IN_2 and IN_3), which in turn form inhibitory synapses back onto the pyramidal neurons. These connections create two types of inhibitions: feedforward (indicated by solid grey lines) and feedback (indicated by dashed grey lines).

When CA1 pyramidal neurons respond to excitatory inputs from the Schaffer collaterals, they undergo both feedforward and feedback inhibitions. Feedforward inhibition occurs when CA3 pyramidal neurons fire along the Schaffer collaterals to activate both CA1 pyramidal neurons and interneurons (such as IN_1 and IN_2). These interneurons in turn inhibit CA1 pyramidal neurons via inhibitory synapses. This process requires two synaptic transmissions. Feedback inhibition occurs when the firing of CA1 pyramidal neurons activate interneurons (such as IN_2 and IN_3),

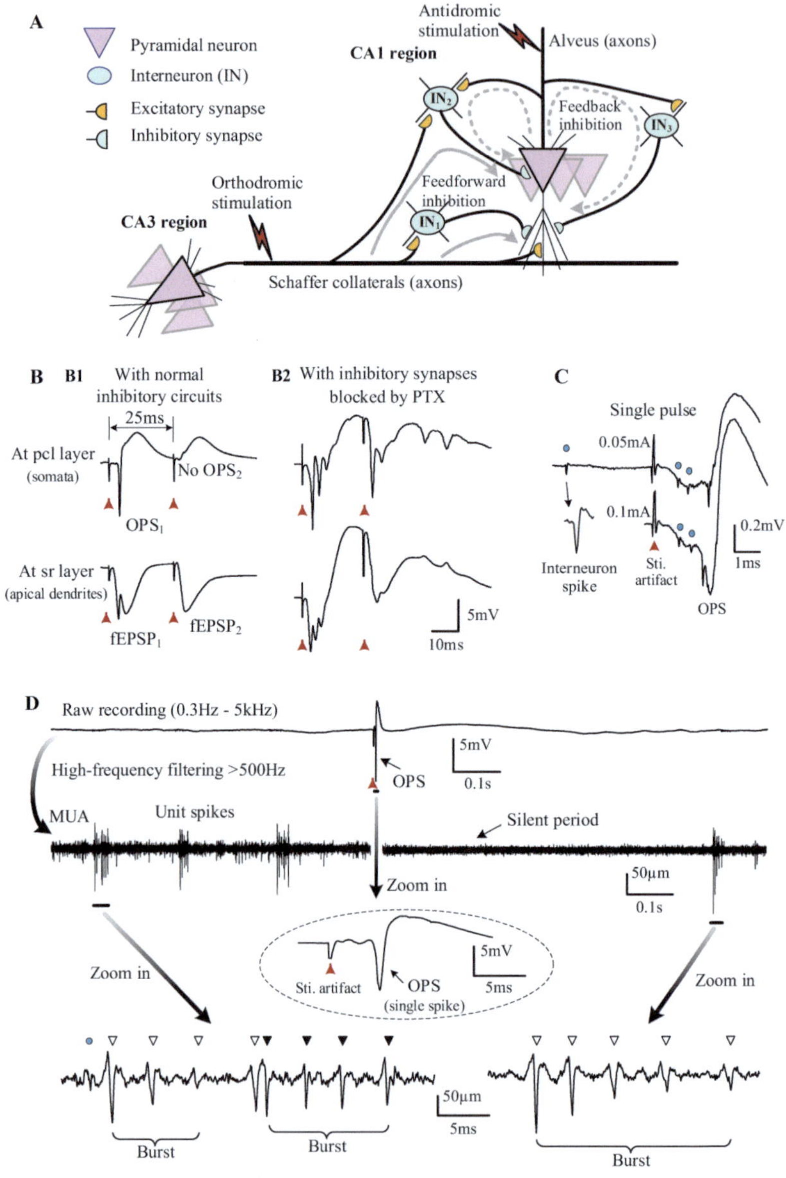

Fig. 2.6 Local inhibitory circuits in the hippocampal CA1 region. **A** Schematic diagram showing feedforward and feedback inhibitory circuits. **B** Testing the effect of inhibitory circuits by paired-pulse stimulations at the Schaffer collaterals. The four signals were recorded respectively from the pcl layer (upper) and the sr layer (lower), with normal inhibitory circuits (B1) and blocked inhibitory synapses by PTX (B2). **C** Interneuron firing induced by two single pulses at intensities of 0.05 and 0.1 mA. **D** Spontaneous burst firing of pyramidal neurons, along with induced OPS and a "silent period" following a pulse stimulation. Red small arrows in figures **B–D** indicate truncated stimulus artifacts

which then produce synaptic inhibitions back on the pyramidal neurons. This process requires three synaptic transmissions. The two types of inhibitions primarily target different sites: feedforward inhibition mainly acts on the dendrites of pyramidal neurons, while feedback inhibition acts mainly near the soma (Lipski 1981). Notably, an identical interneuron can participate in both inhibitory circuits simultaneously, like IN_2 in Fig. 2.6A.

The activation of CA1 pyramidal neurons by the Schaffer collaterals requires only one synaptic transmission, meaning these neurons are activated before inhibitions occur. This raises a question: are the inhibitions from interneurons too late to be effective? Intracellular recordings of postsynaptic potentials in a pyramidal neuron have shown that the initial EPSP rapidly reverses to an IPSP within milliseconds (Mori et al. 2004), demonstrating that inhibition indeed follows excitation. Nevertheless, this inhibition affects subsequent incoming excitations rather than the current ones. The effect of inhibitory circuits can be detected by applying paired-pulse stimulation (PPS) to the Schaffer collaterals.

When two suprathreshold pulses were applied to the Schaffer collaterals at a 25-ms interval, the first pulse induced a large OPS_1 in the CA1 pyramidal cell layer (pcl, Fig. 2.6B1). The second identical pulse hardly induced OPS_2 due to the effect from inhibitory circuits. In the apical dendritic layer (sr), however, both pulses induced large fEPSPs ($fEPSP_1$ and $fEPSP_2$). $GABA_A$ receptors mediate the fast inhibitory potentials in the CA1 region (Leung et al. 2008; Pelkey et al. 2017). When an inhibitory synapse blocker, such as the $GABA_A$ receptor antagonist picrotoxin (PTX), was applied to weaken the inhibitions, both pulses induced large OPSs with continuous discharges forming multiple peaks (Fig. 2.6B2). These negative peaks demonstrate the inherent bursting behavior of pyramidal neurons. Under normal inhibition condition (Fig. 2.6B1), the induced OPS_1 appeared as a single peak, because the effect of inhibitory circuits suppressed subsequent burst discharges.

Due to their lower activation threshold (Csicsvari et al., 1998; Buzsáki 2006), interneurons activate more quickly than pyramidal neurons when a pulse is applied to the Schaffer collaterals, as shown in Fig. 2.6C. The figure displays two single-pulse stimulations at different intensities, with the recordings aligned by their stimulus artifacts (indicated by a red arrow). Blue dots mark the firing of an interneuron. (Note: extracellular recording of an AP is called a unit spike, or simply spike; refer to Sect. 4.1 for details). The two low-intensity pulses (0.05 and 0.1 mA) activated the same interneuron, which fired at least twice in succession immediately following the pulses—its additional firing might be masked by the OPS from pyramidal neurons. If this activated interneuron participated in feedforward inhibition, its inhibition could rapidly act on pyramidal neurons.

Figure 2.6D shows spontaneous bursts from pyramidal neurons—series of APs with intervals of only a few milliseconds. The enlarged insets at the bottom use solid and hollow triangles to distinguish the bursts from two pyramidal neurons. When a single pulse was applied, it induced an OPS with one peak, followed by a "silent period" without any spontaneous neuronal firing. This silence resulted from inhibitory circuits. Particularly, feedforward inhibitions can occur immediately after the stimulation activates the pyramidal neurons (Pouille and Scanziani 2001), causing the induced OPS as a single peak.

Thus, a pulse at the Schaffer collaterals can induce two opposite effects on CA1 pyramidal neurons: excitation and inhibition (Fig. 2.6A). The inhibitory effect starts within milliseconds after pulse application and can last for hundreds of milliseconds. However, in a PPS, the evoked OPS_1 and OPS_2 can exhibit complex changes based on pulse intensities and inter-pulse intervals (IPI), rather than consistent OPS_2 decrease or disappearance. Figure 2.7 shows the effect of varying pulse intensity while keeping a constant 50-ms IPI. The strength of incoming activation determines the number of postsynaptic neuron discharges, which can be evaluated by the OPS amplitude under certain conditions (Theoret et al. 1984). At a low pulse intensity (e.g., 0.05 mA, Fig. 2.7A1), the first pulse produced either no OPS_1 or only a small one, indicating a subthreshold stimulation for most postsynaptic CA1 pyramidal neurons. However, the second pulse—at the same intensity—produced an OPS_2 with significantly increased amplitude, showing paired-pulse facilitation (PPF).

With a low-intensity PPS at an IPI shorter than 100 ms, the two evoked OPSs typically exhibit PPF. This occurs because the first pulse activates only a small number of axons in the Schaffer collaterals, causing insufficient neurotransmitter release for EPSPs to reach activation threshold. The rapid second pulse supplements neurotransmitter release, thereby enhancing EPSPs (Debanne et al. 1996; Neher and Sakaba

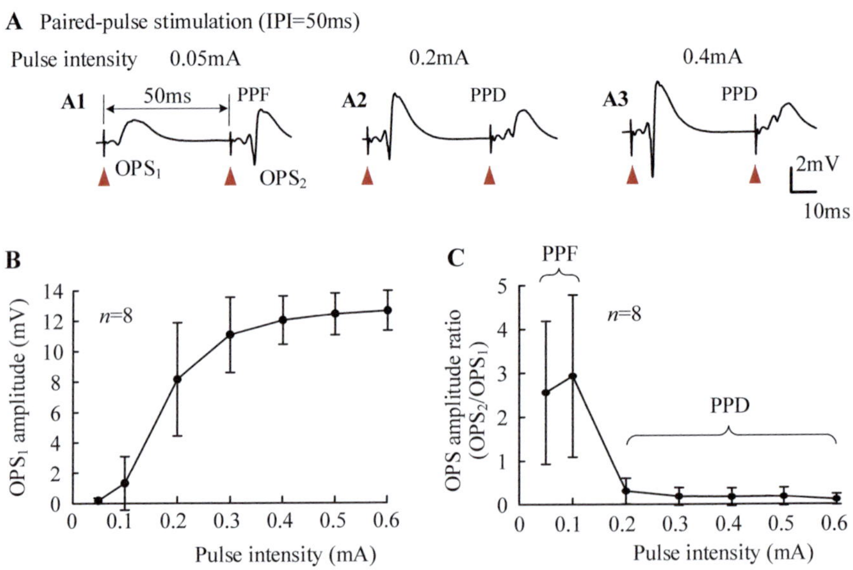

Fig. 2.7 CA1 neuronal responses to orthodromic paired-pulse stimulations (PPS) at the Schaffer collaterals using a 50-ms IPI and varying pulse intensities. **A** Examples of induced OPSs at three different intensities. Red arrows indicate truncated stimulus artifacts. **B** Changes of OPS_1 amplitudes against pulse intensity. **C** Changes of OPS amplitude ratios (OPS_2/OPS_1) against pulse intensity. n represents the number of experimental rats

2008; Regehr 2012). The integration of consecutive EPSPs can reach the threshold and trigger pyramidal neuron firing, producing a larger OPS_2. Are inhibitory circuits activated during this period? Interneurons have a lower activation threshold than pyramidal neurons, so a low-intensity pulse that fails to activate pyramidal neurons may still activate some interneurons in the inhibitory circuits, causing mild inhibition. Nevertheless, when the first low-intensity pulse fails to activate sufficient pyramidal neurons (producing no or small OPS_1), only the feedforward inhibitory circuits can be activated while most feedback inhibitory circuits remain inactive. Therefore, when the second pulse arrives, the increased EPSPs can overcome the weak inhibitions, resulting in enhanced firing rather than a decrease.

As pulse intensity increased, the OPS_1 also increased and tended towards saturation at pulse intensities above 0.4 mA (Fig. 2.7B). However, the OPS_2 decreased, causing a reduced amplitude ratio of OPS_2/OPS_1—a phenomenon known as paired-pulse depression (PPD, Fig. 2.7C). A large OPS_1 indicated that the first pulse produced synchronous APs from a large population of pyramidal neurons, which can activate interneurons in the feedback inhibitory circuits to in turn suppress pyramidal neurons (Fig. 2.6A). The feedback inhibition acted together with feedforward inhibition. Thus, when the second pulse arrived, the inhibitory effect on pyramidal neurons overwhelmed the excitatory input, preventing OPS_2 generation (Fig. 2.7A2 and A3).

Additionally, the change in OPS_1 amplitude against pulse intensity shows an S-shape form (Fig. 2.7B), reflecting both the threshold and saturation mechanisms of post-synaptic neuron activations. Postsynaptic excitatory potentials must integrate to reach the firing threshold of postsynaptic neurons to produce APs (Spruston 2008). Around this threshold, small changes in pulse intensity can cause substantial changes in neuronal discharges. Therefore, the OPS_1 amplitude changed rapidly in the 0.1–0.3 mA range. Increasing pulse intensity can expand the stimulated area and activate more axons in the Schaffer collaterals. However, once the excitatory inputs to downstream postsynaptic neurons become sufficient to produce APs, additional increase in pulse intensity has little effect. Moreover, during orthodromic activation by afferent axons, the broad distribution of axon branches and their terminals enables each axon to control multiple postsynaptic neurons, resulting in high activation efficiency (López-Aguado et al. 2002). This drives the OPS_1 quickly to a saturation level, forming an S-shape input–output relationship (Wierenga and Wadman 2003).

What happens if more pulses are applied in succession? Can inhibitory circuits keep suppressing pyramidal neurons from firing indefinitely? The answer is no. Additional pulses can trigger other mechanisms to produce more complex processes. For instance, when a GABAergic inhibitory synapse of interneuron is activated continuously at high frequencies, it can switch from inhibitory to excitatory, exciting rather than suppressing pyramidal neurons (Staley et al., 1995). As detailed in Sect. 5.3, pyramidal neurons can also show altered responses under sustained pulse stimulations.

2.3.2 Effect of Inhibitory Circuits During Antidromic Activations of CA1 Pyramidal Neurons

Unlike orthodromic activation of CA1 neurons involving synaptic transmissions, antidromic activation occurs when pulses are applied to the alveus—the efferent fiber of the CA1 region (i.e., the axons of CA1 pyramidal neurons themselves). This activation can travel back towards the somata, generating APs that form an antidromic population spike (APS). Simultaneously, the axonal activation can also orthodromically excites interneurons in the feedback circuits through synapses in axonal terminals (Fig. 2.6A). This raises a question: do PPF and PPD also occur during antidromic activation?

As shown in Fig. 2.8, the IPI of PPSs was kept at 50 ms. As pulse intensity increased, the APSs induced by both pulses gradually increased due to greater neuronal activations, but without substantial amplitude difference between the two APSs. The mean amplitude ratios of APS_2/APS_1 were slightly above 1, indicating minor PPF but no PPD. This differs from the results of orthodromic PPSs shown in Fig. 2.7. However, this doesn't mean the local inhibitory circuits are inactive. Figure 2.9 shows that after blocking $GABA_A$-mediated inhibitory synapses with PTX, the induced APSs appeared as multiple peaks rather than a single peak, similar to the same situation with orthodromic stimulation shown in Fig. 2.6B2. The multi-peak firing occurred due to the bursty feature of hippocampal pyramidal neurons. With normal inhibitory circuits, a single pulse of either orthodromic or antidromic stimulation can produce at most single-peak OPS or APS, because feedforward and feedback inhibitions can prevent subsequent burst firing.

Antidromic activation does not involve synaptic transmission and integration, making it hardly affected by inhibitory synapses (Lipski 1981). Therefore, under normal conditions, the second pulse of PPS can induce an APS as large as the first one. However, subsequent spikes in burst firing are suppressed because they depend on integration of excitatory activity in dendrites (see Sect. 2.1.3). Additionally, unlike Fig. 2.7B, the input–output curve in Fig. 2.8B shows no pronounced S-shape, indicating that no clear threshold and saturation appeared. Within this pulse intensity range of 0.05–0.6 mA, the APS_1 amplitude increased linearly with pulse intensity as more neurons became activated. With further increases in pulse intensity, the APS can eventually reach saturation (Zheng et al., 2022).

2.3.3 Analyzing Feedforward and Feedback Inhibitions Using Paired-Pulse Stimulation

Orthodromic stimulation at the Schaffer collaterals can induce APs in downstream CA1 pyramidal neurons and activate both feedforward and feedback inhibitory circuits (Fig. 2.6A), generating two inhibition components. In contrast, antidromic stimulation at the alveus can activate only feedback inhibitory circuits, producing

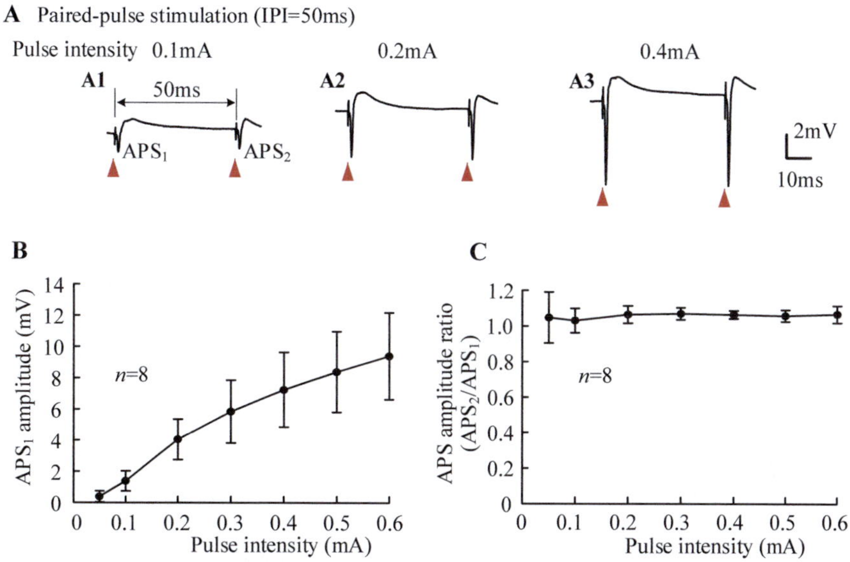

Fig. 2.8 CA1 neuronal responses to antidromic paired-pulse stimulations at the alveus using a 50-ms IPI and varying pulse intensities. **A** Examples of induced APSs at three different intensities. **B** Changes of APS_1 amplitude against pulse intensity. **C** Changes of APS amplitude ratio (APS_2/APS_1) against pulse intensity

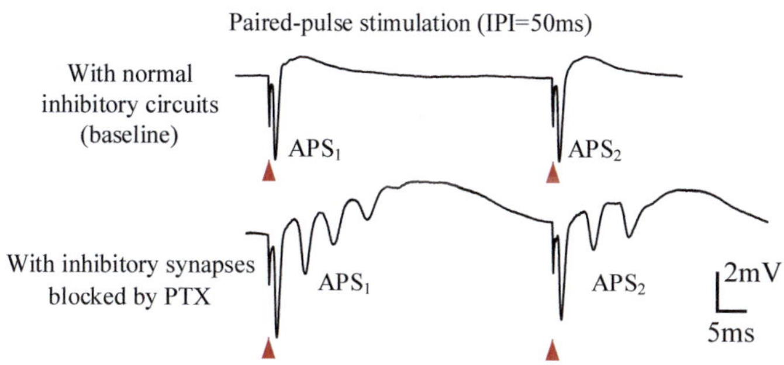

Fig. 2.9 Multiple-peak APSs induced by paired-pulses after PTX blocking inhibitory synapses

feedback inhibition. Using a combination of orthodromic and antidromic paired-pulse stimulation, we analyzed the feedforward and feedback inhibition components to address two questions (Feng et al. 2011): How do the two inhibitions change over time in regulating CA1 neuronal firing? What is the relationship between the time course of these synaptic inhibitions and the AP refractory period?

As shown in Fig. 2.10, we set the IPI between two pulses in a range of 0–400 ms, and applied both antidromic-orthodromic paired-pulse stimulation (AO) and purely orthodromic paired-pulse stimulation (OO) to examine the suppression of OPS_2

evoked by the second pulse. In an AO, the OPS_2 was affected only by feedback inhibition. In this case, the ratio $OPS_R = OPS_2/OPS_1$ reflected the strength of the feedback inhibition alone. Note: OPS_1 here was the OPS evoked by a separated single pulse, serving as a control (Fig. 2.10A). In an OO, the OPS_2 was affected by both feedforward and feedback inhibitions. The OPS_R now reflected the combined strength of both inhibitions (OPS_1 here was the evoked potential of the first pulse). A smaller OPS_R indicates a stronger inhibition. (1-OPS_R) can indicate the inhibited portion. Under certain conditions, subtracting the inhibited portion of AO stimulation from that of OO stimulation can estimate the strength of feedforward inhibition.

Figure 2.10B shows the OPS_R data for OO and AO stimulations, respectively. When the IPI was shorter than 100 ms, the mean OPS_R values were all below one, exhibiting obvious PPD. At IPI = 50 ms, the OPS_R of OO was 0.43, with an inhibited portion of 0.57 = 1–0.43, indicating that the combined feedforward and feedback inhibitions suppressed over half of the OPS amplitude. At this IPI, the OPS_R of AO was 0.71, with an inhibited portion of 0.29 = 1–0.71, indicating comparable proportions of the two inhibition types, i.e., (1–0.71) ≈ (1–0.43)/2. As IPI shortened, the proportion of feedback inhibition increased, reaching 70% at IPI = 25 ms (Fig. 2.10C). When IPI < 10 ms, the OPS_R of OO remained at 0, no evoked OPS_2. Meanwhile, the OPS_R of AO reached its lowest point at IPI = 4 ms but increased again as IPI shortened further, reaching 0.36 at IPI = 0 when the two pulses of AO were applied simultaneously. Although the two pulses were applied at the same time,

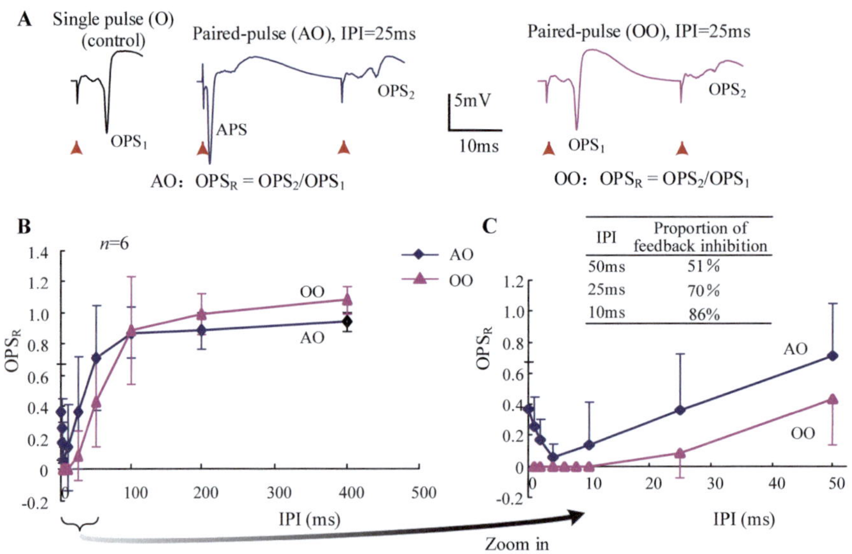

Fig. 2.10 Ratios of evoked potentials for OO and AO paired-pulse stimulations with IPIs ranging from 0 to 400 ms and a fixed 0.3mA pulse intensity. **A** Examples of evoked potentials. **B** OPS_R data of OO and AO stimulations changing against IPI. **C** Enlarged OPS_R curves in an IPI range of 0–50 ms. From Feng et al. (2011)

their evoked potentials occurred at different times due to the synaptic delay in orthodromic activation (Fig. 2.11). Additionally, the estimated 86% feedback inhibition at IPI = 10 ms may be inaccurate, as the zero OPS_R of OO made it impossible to determine whether over-inhibition existed.

The plots in Fig. 2.10B describe the OPS_R data changing against IPIs (intervals between pulses) rather than intervals between evoked potentials—inter-spike-interval (ISI). For OO stimulations, both pulses follow identical activation pathways, making their IPI equivalent to ISI. However, for AO stimulations, the antidromic activation pathway do not involve synaptic transmission. Also, length variations in activation pathways across rat experiments can result in additional differences in evoked potential latencies. Antidromic stimulation typically induced APS with a latency of 1–2 ms, while orthodromic stimulation induced OPS with a latency of 4–9 ms (Fig. 2.11A, B). With longer IPIs, these latency differences between evoked potentials become

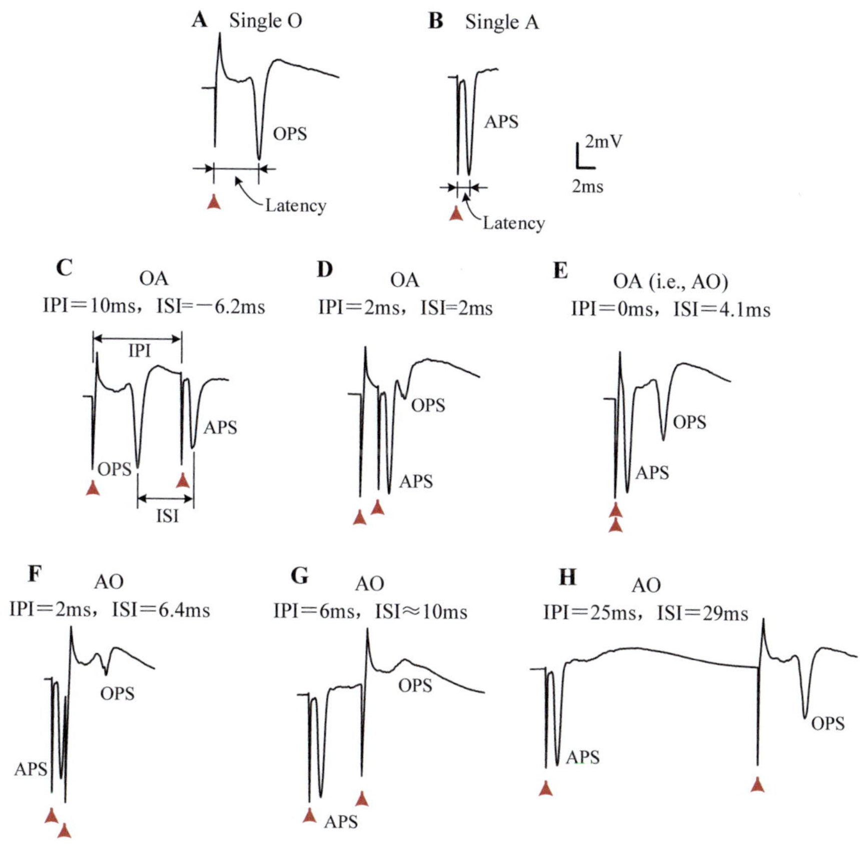

Fig. 2.11 Examples of evoked potentials in OA and AO paired-pulse stimulations at different short IPIs. From Feng et al. (2011)

negligible. However, when using shorter IPIs to study quick effects of inhibitory circuits, the latency differences can be crucial and must be considered.

As shown in Fig. 2.11E, even when the IPI of AO was 0, a substantial ISI existed between APS and OPS. The ISI was measured by the interval between the two negative peaks. We defined ISI as positive when APS preceded OPS, and negative when OPS preceded APS. Since feedback inhibition in AO stimulation is produced by the evoked APS, ISI can reflect the time course of quick effect of this inhibition more accurately than IPI. To investigate how feedback inhibition changes as ISI approaches 0, in addition to AO stimulations, OA stimulations with short IPI (where the O pulse precedes the A pulse) were also employed (Fig. 2.11C–E). Notably, in OA stimulations with very short IPIs, the evoked APS may appear before OPS, as shown in Fig. 2.11D and E.

The subplots in Fig. 2.11 show the different suppression of the second evoked potentials during OA and AO stimulations with different short IPIs. Figure 2.11A, B respectively show the evoked potentials of single O and A stimulations, serving as baseline controls. (The amplitude of control OPS was used as the denominator for calculating the OPS_R data shown in Fig. 2.12). Figure 2.11C–H show the changes of APS and OPS with ISI. When O pulse preceded A pulse by 10 ms (IPI = 10 ms), the initial OPS suppressed the subsequent APS partially (Fig. 2.11C). However, at IPI = 2 ms, although O pulse still preceded A pulse, APS occurred first and significantly suppressed the following OPS (Fig. 2.11D). Here, with ISI = 2 ms, the refractory period following APS also contributed to the OPS suppression. As IPI decreased to 0 (Fig. 2.11E), the ISI between APS and OPS increased, and the OPS rose substantially as it escaped the refractory period of APS. When ISI further increased (becoming AO stimulations), OPS decreased again (Fig. 2.11F, G). At ISI≈10 ms, OPS almost disappeared, indicating powerful feedback inhibition produced by APS (Fig. 2.11G). Only as ISI increased further did OPS gradually recover (Fig. 2.11H). These changes revealed a weak period of OPS suppression between the waning refractory period and the onset of feedback inhibition.

Fig. 2.12 Suppression of APS on OPS in AO paired-pulse stimulations with short ISIs. From Feng et al. (2011)

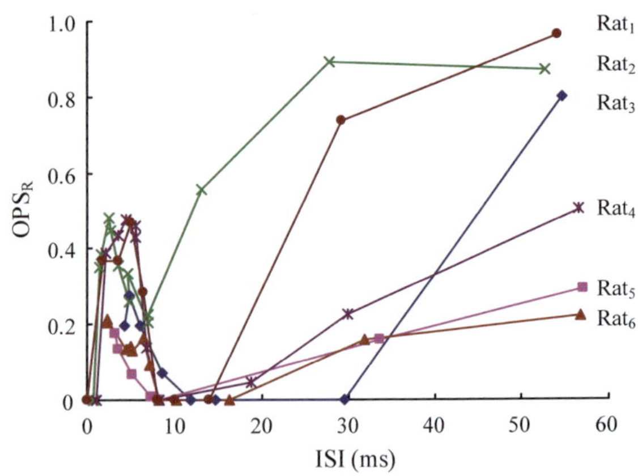

Figure 2.12 displays the OPS_R curves plotted against ISI (the interval of APS preceding OPS) in six rat experiments. Although with individual variations, the curves exhibit a refractory period inhibition at ISIs below 3 ms and a strong phase of feedback inhibition at ISIs between 7 and 17 ms.

The above experimental data indicate that the combined effect of feedforward and feedback inhibitions is strong within 50 ms and can almost completely suppress firing within 10 ms. During the 10–50 ms period, the proportion of feedback inhibition increases as time shortens. However, it decreases during the 3–7 ms period, suggesting that the activation period of feedback inhibition doesn't seamlessly connect with the AP refractory period. At this period, the supplement of feedforward inhibition becomes crucial.

When the inputs from Schaffer collaterals orthodromically activate the downstream CA1 neurons, the induction of feedforward inhibition requires two synaptic transmissions which take about 3 ms (Pouille and Scanziani 2001). Because the simultaneous induction of OPS requires only one synaptic transmission, feedforward inhibition can begin its effect about 2 ms after OPS, allowing overlap between the feedforward inhibition and the refractory period. Moreover, if the orthodromic stimulation directly activates nearby interneurons, it can produce single-synapse feedforward inhibition on CA1 pyramidal neurons more quickly. Such fast inhibition can overlap more with excitatory postsynaptic potentials induced by the same stimulus (Papatheodoropoulos and Kostopoulos 1998). The rapid and powerful effect of feedforward inhibitions plays a crucial role in enhancing the synchronicity of neuronal firing, raising the firing threshold of CA1 pyramidal neurons, and suppressing abnormal epileptiform discharges.

2.3.4 *Responses of CA3 Neurons to Paired-Pulse Stimulations*

The pyramidal neurons in the rat hippocampal CA3 region have two major axonal branches. One forms the Schaffer collaterals (thick solid lines in Fig. 2.13A), which project to the ipsilateral CA1 region. Another forms the commissural fibers (CF), which extend to the contralateral hemisphere and connect to pyramidal neurons in the contralateral CA3 and CA1 regions (thick dashed lines in Fig. 2.13A). Excitatory synaptic connections also exist between the ipsilateral CA3 pyramidal neurons. Thus, the projection range of rat CA3 pyramidal neurons is extensive. These neurons innervate both ipsilateral and contralateral hippocampal CA3 to CA1 regions. The CF, extending from bilateral CA3 regions to their contralateral sides, forms an intertwined fiber—VHC. Note: This commissural fiber projection is almost non-existent in primate brains (Andersen et al. 2007).

Based on the neuronal connections shown in Fig. 2.13B, C, when a pulse is applied to the VHC, the evoked potentials recorded in the CA3 region contain two components (Fig. 2.13D). The first is the antidromically-evoked population spike (APS)

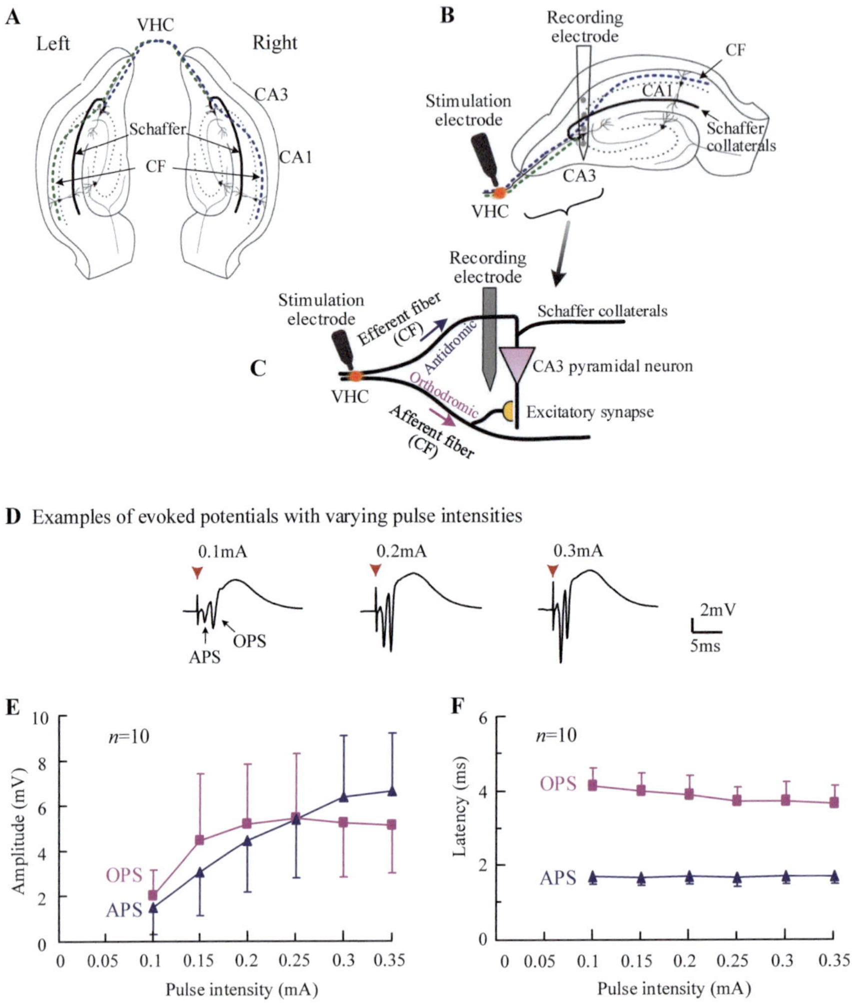

Fig. 2.13 Responses of rat CA3 neurons to VHC pulse stimulations with varying intensities. **A** Schematic diagram showing the VHC connections between left and right hippocampi and the commissural fibers (CF) within the hippocampi (green and blue dashed lines). **B** Schematic diagram showing the stimulation electrode at VHC and the recording electrode in the CA3 region. **C** Diagram showing the activations of CA3 neurons by a VHC stimulus. **D** Examples of CA3 population spikes evoked by VHC pulses with varying intensities. Red arrows indicate stimulus artifacts. **E** and **F** Changes in the population spike amplitude (**E**) and latency (**F**) against stimulus intensity ($n = 10$)

produced by the direct activation at the axons of ipsilateral CA3 neurons themselves. The second is the orthodromically-evoked population spike (OPS) produced by the activation of afferent fiber—the axons of contralateral CA3 neurons. A single pulse to the VHC can trigger both the evoked populations simultaneously. The APS appears first because its induction doesn't involve synaptic transmission, resulting in a shorter latency than the OPS. As shown in Fig. 2.13E, increasing pulse intensity from 0.1 to 0.35 mA caused growth in both APS and OPS. However, due to the refractory period—which prevents rapidly repeated firing—a large initial APS can suppress the subsequent OPS. Therefore, when pulse intensity increased beyond 0.25 mA, the OPS began to decrease rather than continue growing.

Additionally, fine-tuning the stimulation electrode position in the VHC fiber can alter the sizes of the two evoked potentials and their amplitude ratio under a same pulse intensity. Due to individual variations in rat experiments, this size change resulted in the considerable variances in statistical amplitude data as shown by the large error bars (one standard deviation) in Fig. 2.13E. For example, at the pulse intensity of 0.25 mA, the APS amplitude is 5.4 ± 2.5 mV, and the OPS amplitude is 5.4 ± 2.8 mV ($n = 10$). The variations in latencies of the two evoked spikes were much smaller (Fig. 2.13F). Within the pulse intensity range of 0.1 to 0.35 mA, the mean APS latency remained stable at around 1.7 ms, with a standard deviation of only ~ 0.2 ms. The mean OPS latency, however, decreased with increasing pulse intensity, from 4.1 ± 0.51 ms to 3.6 ± 0.50 ms ($n = 10$). This latency reduction can be caused by faster depolarization in postsynaptic neurons produced by increased excitatory inputs.

Like the hippocampal CA1 region, the CA3 region also contains GABAergic local inhibitory circuits. Applying PPS at the VHC can reveal the inhibitions through evoked potentials. Figure 2.14A shows the evoked potentials of PPSs with an intensity of 0.35 mA and IPIs ranging from 10 to 400 ms. The antidromic and orthodromic evoked potentials from the two pulses were denoted as APS_1, OPS_1 and APS_2, OPS_2, respectively. The figure shows superimposed evoked-potentials at six different IPIs, aligned to the first pulses. The first pulses consistently evoked large APS_1 and OPS_1 waveforms with constant amplitudes, while the second pulses produced varying evoked-potentials across different IPIs. At a 10-ms IPI, APS_2 decreased and OPS_2 disappeared, indicating strong inhibitions. As the IPI extended beyond 50 ms, OPS_2 became pronounced and gradually increased. Similarly, APS_2 returned to match APS_1 when the IPI exceeded 25 ms.

The amplitude ratio of OPS_2/OPS_1 showed that shorter IPIs (10-50 ms) resulted in PPD—where OPS_2 was smaller than OPS_1 (Fig. 2.14B), while longer IPIs of 100 and 200 ms produced PPF—where OPS_2 exceeded OPS_1 and even appeared with multiple peaks, as indicated by the hollow triangle on the evoked-potential at 100 ms IPI in Fig. 2.14A. At 400 ms IPI, the two OPSs approached equality. The amplitude ratio of APS_2/APS_1 showed that APS_2 was smaller than APS_1 only at very short IPI (10 ms). In this situation, the second pulse arrived within about 5 ms after the first pulse triggered APS_1 and OPS_1 consecutively. Since the neuronal membrane had not fully recovered from its refractory period, APS_2 was suppressed. At 25 ms IPI, APS_2 slightly exceeded APS_1. As IPI increased further, the two APSs approached equality.

The evoked potentials varied with both IPI and pulse intensity. At a low intensity of 0.1 mA with 25 ms IPI, OPSs exhibited PPD. However, with a longer IPI of 50 ms, OPSs showed PPF. As pulse intensity increased to 0.25 mA, OPSs exhibited PPD at both 25 and 50 ms IPIs (Fig. 2.15A). Statistical data showed that for intensities from

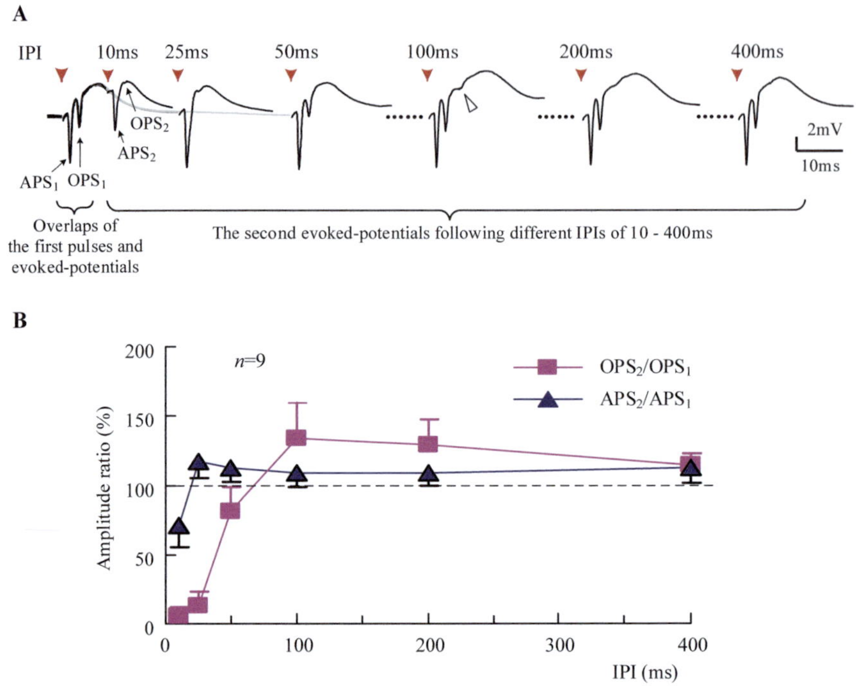

Fig. 2.14 Responses of CA3 neurons to paired-pulse stimulations (PPS) at the VHC with varying IPIs and a fixed 0.35-mA intensity. **A** Examples of evoked-potentials by PPSs with IPIs of 10, 25, 50, 100, 200 and 400 ms. The six recordings are superimposed and aligned to the first pulses (indicated by the larger red arrow). The IPI values are shown above the second pulses (indicated by smaller red arrows). **B** Changes in the amplitude ratios of OPS_2/OPS_1 and APS_2/APS_1 against IPIs ($n = 9$ rats)

0.1 to 0.35 mA, OPSs consistently exhibited PPD at the shorter 25 ms IPI (Fig. 2.15B). The PPD increased with pulse intensity; at intensities above 0.2 mA, the mean OPS_2 amplitude dropped to only about 10% of OPS_1. At 50 ms IPI (Fig. 2.15C), OPS_2 was approximately 140% of OPS_1 at 0.1 mA intensity, exhibiting PPF. As the intensity increased to 0.2 mA and above, this ratio decreased, with OPS_2 about 80% of OPS_1. Notably, APSs consistently exhibited a slight PPF with both IPIs (25 and 50 ms) at the intensity range of 0.1–0.35 mA, with APS_2 being about 120% of APS_1.

The PPD of OPS_2 in hippocampal CA3 neurons can be primarily caused by the combined effect of feedforward and feedback inhibitions similar to those in the CA1 region shown in Fig. 2.6A. The PPF of OPS_2 can result from increased neurotransmitter release and EPSP integrations from the two pulse activations. Additionally, excitatory connections among CA3 pyramidal neurons themselves may also contribute to PPF. Due to the opposing effects—PPD and PPF—the neuronal responses exhibit complex changes under different stimulation conditions. When assessing the states of neurons and local circuits, specific conditions must be carefully considered. For example, when evaluating inhibitory circuits based on PPD,

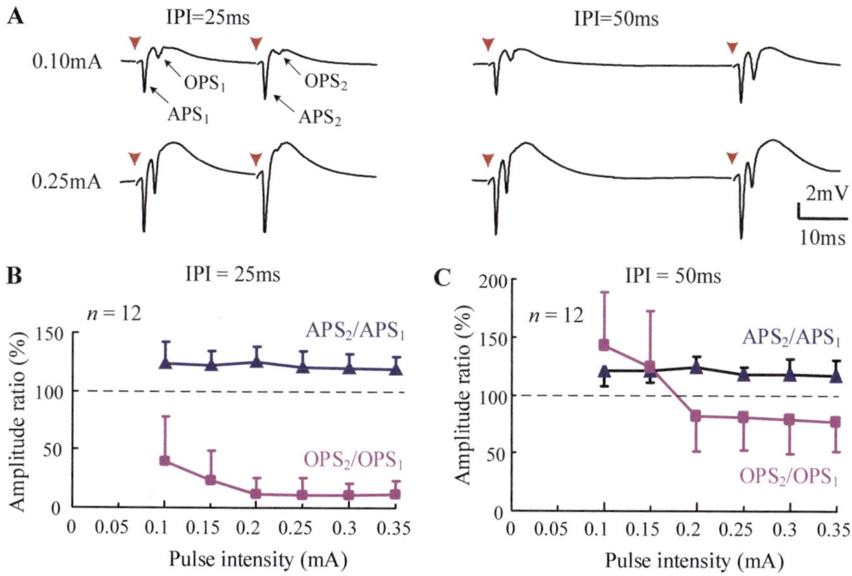

Fig. 2.15 Responses of CA3 neurons to PPSs at the VHC with varying intensities. **A** Examples of evoked-potentials at two IPIs with two intensities. **B** and **C** Amplitude ratios of OPS_2/OPS_1 and APS_2/APS_1 at 25 ms IPI (**B**) and 50 ms IPI (**C**) against intensities ($n = 12$ rats)

OPS_2 suppression should be determined using consistent IPI and pulse intensity, along with similar OPS_1 amplitudes. Otherwise, even with identical IPI and intensity, weak neuronal excitability can lead to a smaller OPS_1, resulting in weaker feedback inhibition. This may cause reduced suppression or even an increase in OPS_2 (due to PPF dominance). In such case, concluding that inhibitory circuits are impaired would be incorrect.

As shown in Fig. 2.13D, the evoked-potentials in the CA3 region produced by single-pulse stimulations at VHC contain two peaks (APS and OPS) with a latency difference of about 2 ms. This firing pattern resembles the multiple-peak discharges evoked in the CA1 region when inhibitory circuits are impaired by PTX, as shown in Fig. 2.6B2 and Fig. 2.9 (*bottom*). To verify that these two-peak potentials in the CA3 region stem from antidromic and orthodromic activations of two axon types (Fig. 2.13C), I used a calcium ion chelator (ethylene glycol tetraacetic acid, EGTA) to reduce the Ca^{2+} concentration in the hippocampus extracellular fluid ($[Ca^{2+}]_o$). This intervention can block synaptic transmissions and eliminate orthodromic activation, allowing verification of the evoked-potential sources. The experimental method is detailed in Sect. 3.3.

Figure 2.16 shows the results of this EGTA experiment. PPSs (0.35 mA intensity, 25 ms IPI) were applied to VHC at 5-min intervals. During the 30-min baseline recording with normal 2 mM $[Ca^{2+}]_o$, the first pulses produced both large APS_1 and OPS_1. After switching to a Ca^{2+}-free solution containing EGTA for 40 min, APS_1

increased while OPS_1 gradually decreased. Upon returning to the normal Ca^{2+} solution, OPS_1 gradually recovered and APS_1 returned to its baseline level (Fig. 2.16A). Throughout the 100-min recording, both APS_1 and APS_2 remained large, with stable APS_2/APS_1 amplitude ratios (Fig. 2.16B).

During the last 20 min of EGTA application—corresponding to the 50–70 min period in Fig. 2.16—the mean OPS_1 amplitude decreased significantly to 5.9% ± 4.1% ($n = 6$) of its baseline value. This indicated that most synaptic transmissions were blocked by lowering $[Ca^{2+}]_o$. In contrast, the mean APS_1 amplitude remained at 109% ± 16% ($n = 6$) of its baseline value without substantial change. Additionally,

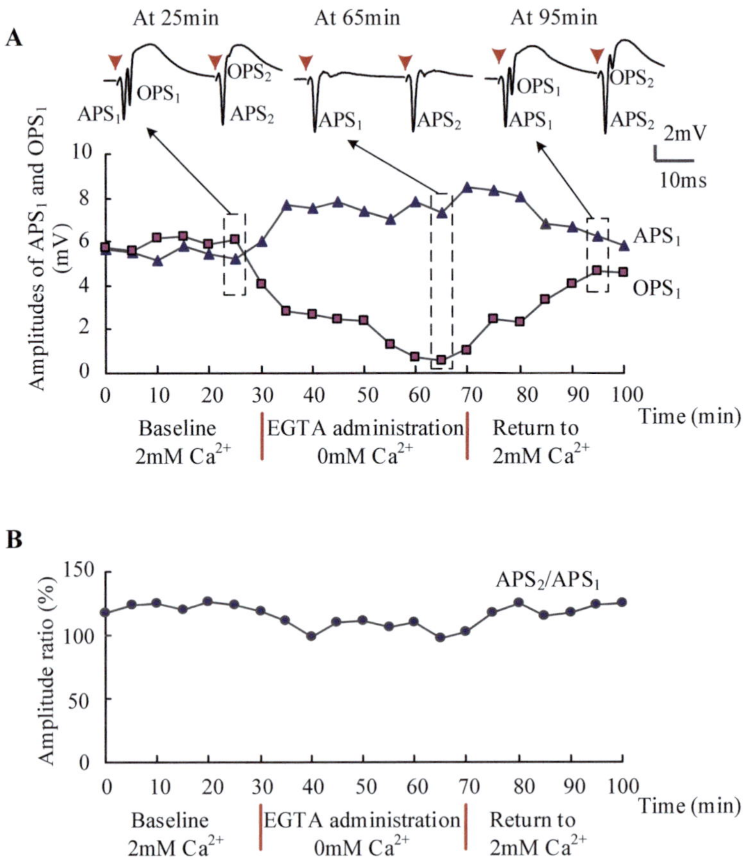

Fig. 2.16 Verification of orthodromic and antidromic evoked-potentials in the rat hippocampal CA3 region by VHC stimulation. **A** Examples of evoked-potentials (*upper*) and amplitude changes of APS_1 and OPS_1 produced by paired-pulse stimulations (0.35 mA intensity, 25-ms IPI) at VHC during the 100 min recording period—including baseline with normal solution (30 min), application of Ca^{2+}-free EGTA solution (40 min), and return to normal solution (30 min). **B** Changes in the APS_2/APS_1 amplitude ratio throughout the recording period

the mean amplitude ratio of APS_2/APS_1 (evoked by PPSs at 25 ms IPI) decreased from 129% ± 7.4% at baseline to 103% ± 5.1% ($n = 6$).

These data confirmed the distinct properties of the two components evoked by VHC stimulations. Together with other data such as the latency of evoked-potentials and the PPD and PPF produced by PPSs, we can conclude that the dual-spike potentials were generated by orthodromic and antidromic activations of afferent and efferent axon fibers. The features of these evoked-potentials, including their amplitudes, provide valuable indicators for assessing the states of neurons and their local circuits.

2.4 Summary

Neurons are the fundamental building blocks of the nervous system. Each neuron consists of a cell body (or soma), dendrites, and an axon. It receives signals through dendrite synapses and processes them using complex integration mechanisms. Once the input integration reaches activation threshold, the neuron can fire an action potential (or nerve impulse) that can travel outward along the axon. Besides forming excitatory projection connections, neurons also create local circuits. Among them, local inhibitory circuits play a vital role in maintaining the balance of neuronal excitation.

The animal experiments described in this book focus on the rat hippocampus. In this brain region, principal neurons create distinct laminar structures through their dense and orderly arrangement. The structures allow for extracellular recordings of both population spikes (PS) and field excitatory postsynaptic potentials (fEPSP) when neurons respond to electrical pulse stimulations. These evoked potentials can reach millivolts in amplitude, facilitating investigations into neurons and their circuits, as well as the modulation effects of electrical stimulations.

This chapter introduces the structure and neural circuits of the rat hippocampus, including excitatory synaptic pathways and local inhibitory circuits. It examines the waveforms, meanings, and generation mechanisms of pulse-evoked potentials—including orthodromic potentials (produced by pulses applied to presynaptic input fibers) and antidromic potentials (produced by direct axon activations). The chapter also presents our findings on the effects of feedforward and feedback inhibitory circuits in the hippocampal CA1 region by using paired-pulse stimulations. Additionally, it explores how CA3 pyramidal neurons respond to VHC stimulations.

References

Andersen P, Bliss TV, Skrede KK (1971) Unit analysis of hippocampal population spikes. Exp Brain Res 13(2):208–221

Andersen P, Morris R, Amaral D et al (2007) The hippocampus book. Oxford University Press

Bear MF, Connors BW, Paradiso MA (2016) Neuroscience: exploring the brain (4th ed.). Wolters Kluwer.
Bliss TVP, Gardner-Medwin AR (1973) Long-lasting potentiation of synaptic transmission in the dentate area of the unanaesthetized rabbit following stimulation of the perforant path. J Physiol 232(2):357–374
Bliss TVP, Lømo T (1973) Long-lasting potentiation of synaptic transmission in the dentate area of the anaesthetized rabbit following stimulation of the perforant path. J Physiol 232(2):331–356
Buzsáki G (2006) Rhythms of the brain. Oxford University Press, Oxford
Cobb SR, Buhl EH, Halasy K et al (1995) Synchronization of neuronal activity in hippocampus by individual GABAergic interneurons. Nature 378(6552):75–78
Csicsvari J, Hirase H, Czurko A et al (1998) Reliability and state dependence of pyramidal cell-interneuron synapses in the hippocampus: an ensemble approach in the behaving rat. Neuron 21(1):179–189
Debanne D, Guérineau NC, Gähwiler BH et al (1996) Paired-pulse facilitation and depression at unitary synapses in rat hippocampus: quantal fluctuation affects subsequent release. J Physiol 491(Pt 1):163–176
Feng Z, Xing H, Tian C et al (2011) Effect features of feedforward and feedback inhibitions on hippocampal CA1 region of rat. Space Medicine & Medical Engineering 24(3):167–172 (in Chinese)
Häusser M, Mel B (2003) Dendrites: Bug or feature? Curr Opin Neurobiol 13(3):372–383
Kepecs A, Fishell G (2014) Interneuron cell types are fit to function. Nature 505(7483):318–326
Klausberger T, Somogyi P (2008) Neuronal diversity and temporal dynamics: the unity of hippocampal circuit operations. Science 321(5885):53–57
Kloosterman F, Peloquin P, Leung LS (2001) Apical and basal orthodromic population spikes in hippocampal CA1 in vivo show different origins and patterns of propagation. J Neurophysiol 86(5):2435–2444
Koch C (1999) Biophysics of computation: information processing in single neurons. Oxford University Press, Oxford
Kullmann DM (2011) Interneuron networks in the hippocampus. Curr Opin Neurobiol 21(5):709–716
Leung LS, Peloquin P, Canning KJ (2008) Paired-pulse depression of excitatory postsynaptic current sinks in hippocampal CA1 in vivo. Hippocampus 18(10):1008–1020
Lipski J (1981) Antidromic activation of neurones as an analytic tool in the study of the central nervous system. J Neurosci Methods 4(1):1–32
London M, Häusser M (2005) Dendritic computation. Annu Rev Neurosci 28(1):503–532
López-Aguado L, Ibarz JM, Varona P et al (2002) Structural inhomogeneities differentially modulate action currents and population spikes initiated in the axon or dendrites. J Neurophysiol 88(5):2809–2820
Maccaferri G, Dingledine R (2002) Control of feedforward dendritic inhibition by nmda receptor-dependent spike timing in hippocampal interneurons. J Neurosci 22(13):5462–5472
Mori M, Abegg MH, Gähwiler BH et al (2004) A frequency-dependent switch from inhibition to excitation in a hippocampal unitary circuit. Nature 431(7007):453–456
Neher E, Sakaba T (2008) Multiple roles of calcium ions in the regulation of neurotransmitter release. Neuron 59(6):861–872
O'Keefe J, Dostrovsky J (1971) The hippocampus as a spatial map. Preliminary evidence from unit activity in the freely-moving rat. Brain Res 34(1):171–175
Papatheodoropoulos C, Kostopoulos G (1998) Development of a transient increase in recurrent inhibition and paired-pulse facilitation in hippocampal CA1 region. Dev Brain Res 108(1–2):273–285
Paxinos G, Watson C (2007) The rat brain in stereotaxic coordinates. Academic
Pelkey KA, Chittajallu R, Craig MT et al (2017) Hippocampal GABAergic inhibitory interneurons. Physiol Rev 97(4):1619–1747

Poirazi P, Brannon T, Mel BW (2003) Arithmetic of subthreshold synaptic summation in a model CA1 pyramidal cell. Neuron 37(6):977–987

Pouille F, Scanziani M (2001) Enforcement of temporal fidelity in pyramidal cells by somatic feed-forward inhibition. Science 293(5532):1159–1163

Regehr WG (2012) Short-term presynaptic plasticity. Cold Spring Harb Perspect Biol 4(7):a005702

Richardson TL, Turner RW, Miller JJ (1987) Action-potential discharge in hippocampal CA1 pyramidal neurons: current source-density analysis. J Neurophysiol 58(5):981–996

Sidiropoulou K, Pissadaki EK, Poirazi P (2006) Inside the brain of a neuron. EMBO Rep 7(9):886–892

Sik A, Ylinen A, Penttonen M et al (1994) Inhibitory CA1-CA3-hilar region feedback in the hippocampus. Science 265(5179):1722–1724

Spruston N (2008) Pyramidal neurons: Dendritic structure and synaptic integration. Nat Rev Neurosci 9(3):206–221

Staff NP, Spruston N (2003) Intracellular correlate of epsp-spike potentiation in CA1 pyramidal neurons is controlled by GABAergic modulation. Hippocampus 13(7):801–805

Staley KJ, Soldo BL, Proctor WR (1995) Ionic mechanisms of neuronal excitation by inhibitory GABAA receptors. Science 269(5226):977–981

Stuart G, Spruston N, Sakmann B et al (1997) Action potential initiation and backpropagation in neurons of the mammalian CNS. Trends Neurosci 20(3):125–131

Theoret Y, Brown A, Fleming SP et al (1984) Hippocampal field potential: a microcomputer aided comparison of amplitude and integral. Brain Res Bull 12(5):589–595

Vreugdenhil M, Bracci E, Jefferys JGR (2005) Layer-specific pyramidal cell oscillations evoked by tetanic stimulation in the rat hippocampal area CA1 in vitro and in vivo. J Physiol 562(1):149–164

Wierenga CJ, Wadman WJ (2003) Functional relation between interneuron input and population activity in the rat hippocampal cornu ammonis 1 area. Neuroscience 118(4):1129–1139

Yuste R, Urban R (2004) Dendritic spines and linear networks. J Physiol Paris 98(4–6):479–486

Zheng L, Feng Z, Xu Y et al (2022) An anodic phase can facilitate rather than weaken a cathodic phase to activate neurons in biphasic-pulse axonal stimulations. Front Neurosci 16:823423

Chapter 3
Experimental Methods and Setup

Abstract This chapter covers animal surgical methods, electrodes, and experimental setup for in-vivo rat neuro-electrophysiological studies. It details the use of stereotaxic instruments, methods for accurate electrode positioning, and the usages of amplifiers, recorders, and electrical stimulators. It also describes a custom-designed electrical stimulation system capable of producing diverse signals and achieving closed-loop stimulation.

Keywords Rat surgery · Electrode implantation · Hippocampus exposure · Microelectrode · Experiment setup · Amplifier · Recorder · Stimulator · Closed-loop stimulation · In vivo

This chapter presents methods for acute neuro-electrophysiological experiments in rat brains, including guidelines for using stereotaxic instruments, recording devices and electrical stimulators. Based on my practical experience, these methods—though not the exclusive approaches—can provide useful guidance for researchers, particularly beginners, in conducting similar experiments.

3.1 Surgical Procedures

Adult male Sprague–Dawley rats were used in our experiments. These rats have snow-white fur and typically weigh over 250 g at maturity. Due to their somewhat aggressive nature, handlers must wear gloves and take proper protective measures to prevent scratches or bites. For surgical preparation, a rat was anesthetized with urethane through intraperitoneal injection using a 2 mL syringe, at a dose of 1.25 g/kg. The anesthetic took effect rapidly. Once unconscious, the rat was immediately secured in a stereotaxic apparatus.

Z. Feng, *Electrical Brain Stimulation: Mechanisms and Modulation of Neuronal Activity*, https://doi.org/10.1007/978-981-95-4145-4_3

3.1.1 Rat Stereotaxic Apparatus

Figure 3.1A shows the standard dual-arm stereotaxic apparatus for rats that we used, manufactured by Stoelting Company. The apparatus includes manipulator arms, a rat adaptor, U-shape frame, and baseplate. When setting up, ensure sufficient space around the experiment table for freely operating around the stereotaxic apparatus. Before securing the rat head, remove the manipulator arms from the apparatus. Then use the left and right ear bars on the U-shape frame, along with the incisor bar and nose clamp on the rat adapter, to immobilize the rat head. Beginners typically need repeated practice to master this process. Below is the operating procedure used in our laboratory.

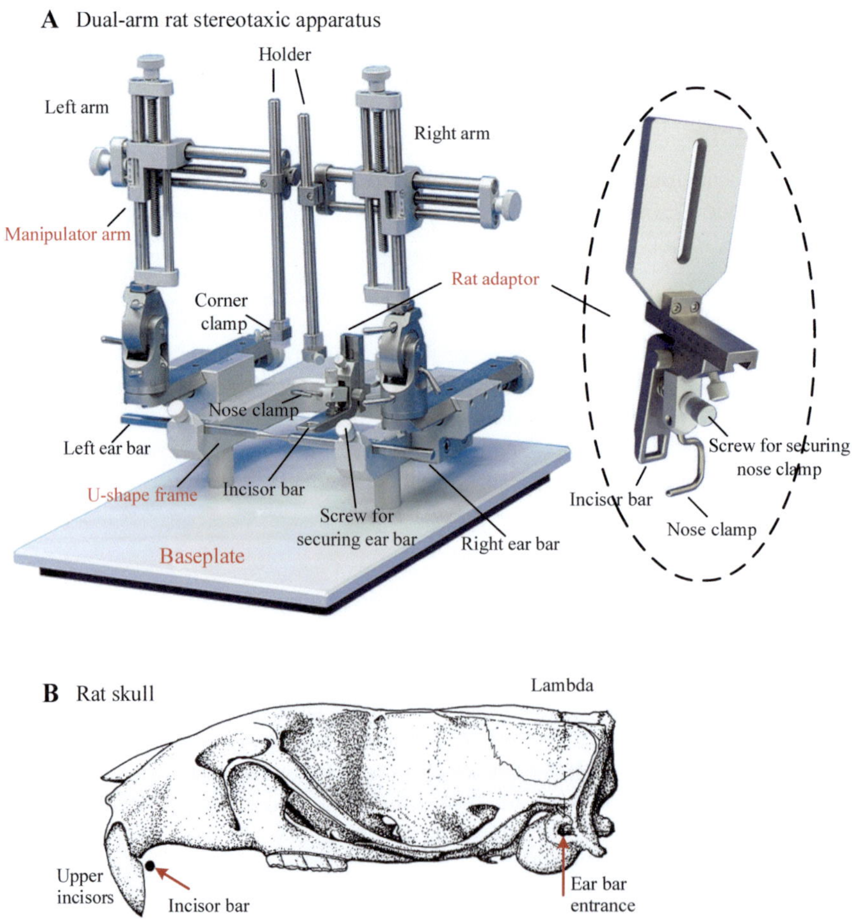

Fig. 3.1 Stoelting dual-arm stereotaxic apparatus (**A**) and rat skull (**B**)

(1) **Securing rat ears**. First, insert the two ear bars into the slots at both ends of the U-shape frame. The left ear bar remains loose, while the right one is fixed at the 5 mm mark by aligning the "5" mark on the ear bar with the "0" mark on the frame and tightening the fixing screw. Next, stand on the left side of the stereotaxic apparatus and place the rat on the baseplate. Gently hold the rat head, align its right ear canal with the tip of the fixed right ear bar, and slowly insert it. There are two types of ear bars: 18° taper and non-puncture 45° taper. If eardrum preservation is not required, use the sharper 18° ear bar—it provides easier insertion and more reliable fixation. You may hear a slight "pop" as the eardrum breaking, indicating correct positioning. After inserting the right ear bar, hold the rat head with your left hand. Using your right hand, gently push the left ear bar towards the left ear canal while adjusting the rat head slightly to align the bar with the canal. Slowly insert the bar until it enters the skull hole (Fig. 3.1B). Once in place, tighten the screw to fix the left ear bar and release the rat head from your hands. Both ear bar readings should now be between 5 and 6 mm. To center the rat head, fine-tune the ear bars to equalize the readings: Stand behind the apparatus, hold both ear bars, slightly loosen their fixing screws, and simultaneously move both ear bars slowly in the same direction to prevent dislodging. Once both readings match, retighten the screws. Finally, verify the fixation by lifting the rat tail to check whether its head can rotate smoothly around the ear bars.

(2) **Securing rat mouth**. Install the rat adapter at the front end of the U-shape frame. (For mouse experiments, use a mouse adapter instead.) Loosen the positioning screw of nose clamp to raise the nose bar, then move the incisor bar towards rat mouth. Using forceps, open rat mouth and hook its upper incisors onto the incisor bar (Fig. 3.1B). Gently pull the adapter back until the rat head is in a naturally extended position, then fix the adapter. Lower the nose clamp to press down on the rat nose. Move the rat tongue to the side against its lower jaw to ensure a clear airway. After these steps, the rat head should be immobilized. To maintain the body temperature during the experiment, wrap rat body with a diaper or use a temperature-controlled water bag.

3.1.2 Craniotomy

After securing a rat on the stereotaxic apparatus, perform the craniotomy as below. First, shave the surgical areas on its head and nose, then clean the areas with alcohol swabs. To fix reference and ground electrodes on the nose for signal recording, make a midline incision along the nasal tip using a scalpel, gently scrape away the connective tissue on the nasal bone surface, and clean it with saline solution (Fig. 3.2). Using a microdrill, make two holes on either side of the nasal bone suture, about 5 mm apart. Fix a stainless steel bone screw into each hole to serve as the reference and ground electrodes, respectively, and wire the screws to the amplifier. We used rat bone screws from BASi (Bioanalytical Systems, Inc.), measuring ~ 1 mm in diameter and

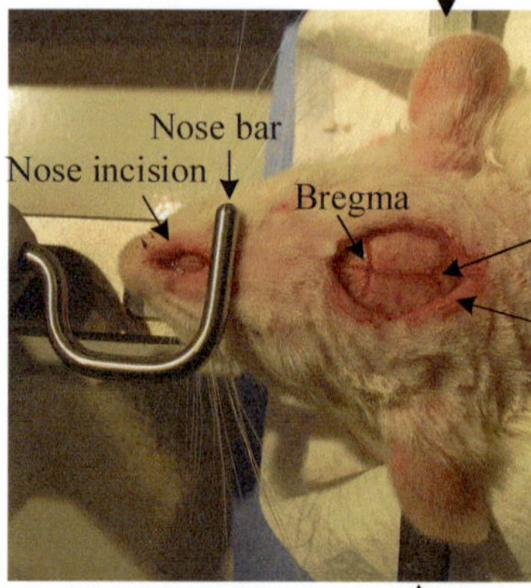

Fig. 3.2 Nasal and scalp incisions in rat surgery

2 mm in length (No. MF-5182, https://www.bioanalytical.com/products/MF-5182). Alternatively, the reference and ground electrodes can be placed on the posterior skull, provided they are positioned away from recording sites (Schjetnan and Luczak 2011).

Next, make a midline sagittal incision along the scalp, approximately 1.5–2 cm (Fig. 3.2). Remove the subcutaneous connective tissue to expose the skull, and gently scrape skull surface clean. If bleeding occurs, use a thermocautery to staunch bleeding sites. Clean the skull surface with saline solution and then let it dry. Locate the bregma and mark it with a black pen. According to electrode implantation targets, draw the outline for the cranial window. For our experiments in the unilateral hippocampal CA1 region, we typically used a window of about 8 × 5 mm on the left skull. Use an electric microdrill to grind evenly along the window outline, flushing away the drilling powder with saline solution until the thinned outline faintly reveals the red dura underneath. Switch to a small carving knife for light etching. Once the bone along the outline is almost cut through, use a bone rongeur to pry open a corner of the bone flap. Gently remove the bone flap to expose the underlying dura. Then, carefully inspect and clean bone fragments along the window edge using a bone rongeur, smoothing the edge to prevent fragments from piercing the blood vessels under the dura, which may cause bleeding during the subsequent experiment period. This smoothing process can also prevent electrode damage when inserting along the skull window edge.

3.1.3 Positioning the Manipulator Arms and Electrodes

The rat stereotaxic apparatus features two manipulator arms, one on each side (Fig. 3.1A). Additional separated micro-manipulators, such as those from World

Precision Instruments (WPI), can be placed around the stereotaxic apparatus to hold more electrodes or devices. For experiments requiring multiple electrodes, first position each manipulator arm, then attach the electrodes sequentially. To successfully insert electrodes through the small cranial window into the target sites in brain, some electrodes (holder arms) need to be adjusted to appropriate tilt angles. In our experiments, we typically inserted the recording array electrode vertically while setting other electrodes at specific angles. The entry position coordinates for electrodes on the exposed dura must be calculated accurately based on their target sites and tilt angles (as detailed in Sect. 3.2.1).

We used FHC concentric bipolar electrodes for electrical stimulations (see Sect. 3.4.2). This type of electrodes is relatively rigid and durable. Before mount it on the manipulator arm, clean the electrode by rinsing it with deionized water or gently wiping it with an alcohol disinfectant wipe. Then fix it to the manipulator using the corner clamp attached to the small cube at the holder head (Fig. 3.1A). This cube has 1 mm-spaced grooves along its three sides. Embed the electrode needle into one groove and secure it with the clamp. Finally, connect the two electrode wires to the output ports of electrical stimulator (see Sect. 3.5.4).

We used NeuroNexus microelectrode to record neural signals (see Sect. 3.4.1). Due to its extreme thinness and fragility, this electrode should be placed last. Begin by rinsing the electrode surface with deionized water. When placing the electrode, handle it carefully and prevent its tip from touching any surrounding objects to avoid breakage. NeuroNexus electrodes for acute recordings have a rod (about 2 cm long and 3 mm wide) between the electrode shank (containing contacts) and the electrode connector. Press this rod against the cube on the holder head (Fig. 3.1A), then secure it with the tightening clamp. Once secured, wire the preamplifier to the electrode connector.

Notably, when mounting an electrode, ensure that its rod is parallel with the holder rod. Otherwise, the electrode will not advance straight along the direction of holder axis (the moving direction) but instead enter the brain tissue at an angle, cutting and damaging the tissue like a knife.

Some electrodes, like the fragile NeuroNexus arrays (only 15 μm thick and made of film), are not strong sufficiently to penetrate the dura on their own. In such cases, the dura at the electrode entry site must be punctured before electrode insertion. The sharp beveled tip of a syringe needle can be used for this purpose. During puncturing, lift the electrode as high as possible to avoid accidental contact and damage. When puncturing the dura near blood vessels, be prepared to immediately control any bleeding if vessels beneath are pierced. The puncture hole is typically too small to see with the naked eyes. Slowly advance the electrode from the dura surface while carefully observing changes at the entry point. If the dura hasn't been punctured, the electrode tip will create a visible indentation when it pressing against the dura surface, indicating a need for re-puncturing. Although these operations can be performed with the naked eye, using a magnifying glass makes them easier.

After penetrating the dura and positioning all electrode tips on its surface, cover the stereotaxic apparatus with a shielding cage (or Faraday cage) and start signal recordings. While advancing the electrodes, monitor real-time changes in recorded neural signals to determine electrode positions within the brain (see Sect. 3.2.3).

The FHC electrodes we used are quite rigid and can pierce the dura on their own. However, when an electrode directly pierces the dura, its tip can visibly indent and compress the underlying brain tissue. Therefore, it is better to puncture the dura first. The perforation hole should be minimal—just large enough to allow the thin electrode entry. If the hole is too large, brain tissue may gradually protrude through the opening over extended experiment period.

After surgery, add saline or artificial cerebrospinal fluid over the cranial window to keep the exposed dura and surrounding skull surface moisture. The incised scalp forms a natural circular "dam" on the exposed skull (Fig. 3.2) that can hold fluid. Surface tension from scalp oils can prevent fluid spillage even when its level is slightly above the "dam". During experiment, replenish the fluid as needed to prevent the exposed surface from drying out.

3.1.4 Anesthesia and Euthanasia

The acute in-vivo rat experiments described in this book were conducted under urethane anesthesia. Urethane (ethyl carbamate) is an acetylcholinesterase inhibitor. It is commonly used as an anesthetic in neurophysiological experiments on rats and other small animals, providing mild and long-lasting anesthesia. At normal doses, urethane has minimal impact on neuronal activity, possibly causing slight increases in neuronal firing thresholds and reduced firing rates. It has negligible effects on both glutamate-mediated excitatory synapses and GABA-mediated inhibitory synapses. It only minimally suppresses sensory responses to stimuli such as body touch (Sceniak and Maciver 2006; Wu et al. 2007; Shirasaka and Wasterlain 1995). Additionally, this anesthetic barely affects respiratory ventilation, and artificial ventilation is not needed during experiments.

If euthanasia is required after experiments, a quick and simple method involves injecting a saturated potassium chloride solution into the rat heart (Cartner et al. 2007). The sudden increase in the extracellular concentration of potassium ions can cause immediate cardiac arrest, leading to death.

3.2 Guidelines for Electrode Implantation

3.2.1 Determining Coordinates for Electrode Positions

The bregma on rat skull typically serves as the origin of the three-dimensional coordinate system (X, Y, Z) to determine the entry point on the brain surface for each electrode. The process involves three main steps: (1) Look up the rat brain stereotaxic atlas (Paxinos and Watson 2007) for the coordinates of the target sites in brain—the desired points for the electrode tips. (2) Measure the electrode inclination angles from the experiment setup. (3) Calculate the three distances: the anterior–posterior and lateral distances between the entry point and the bregma, and the electrode advancing distance after entering the brain. These distances, typically measured in millimeters, guide the movements of the X, Y, and Z axes on manipulator arms or micromanipulators.

As shown in Fig. 3.3A, when an electrode is advanced vertically into the brain, (i.e., at a zero inclination angle), determining the coordinate data of its entry point on the brain (or skull) surface is straightforward. You simply need to look up the anterior–posterior and lateral distances between the target site and the bregma in the brain atlas. These distances are just the displacements along the X- and Y-coordinates, respectively, with a negative value indicating a position posterior to the bregma. However, as illustrated in Fig. 3.3B, when the holder (with electrode) of WPI micromanipulator forms an angle with the vertical, the displacement along its X-axis is different from the displacement of electrode tip in the anterior–posterior direction on the skull surface. The two displacements form the right and oblique sides of a right triangle, labeled in Fig. 3.3B as "X-axis displacement" and "Anterior–posterior displacement". In this case, calculating the X-axis displacement involves two steps: First, determine the "anterior–posterior displacement" from the bregma to the entry point, based on both the target site and the electrode inclination angle. Second, calculate the required "X-axis displacement" using trigonometric functions to the right triangle shown in Fig. 3.3B. It's important to note that the X-axis displacement is not equivalent to the anterior–posterior distance between the bregma and the entry point—confusing them will lead to inaccurate electrode positioning. Figure 3.3B illustrates this for an inclination angle in the X-axis direction, the same calculation method applies to an inclination angle in the Y-axis direction.

Additionally, it is important to distinguish between an intended entry point on the skull surface and one on the dura surface above the brain cortex, since skull thickness affects the calculations. Since the bregma is marked on the skull surface, for entry points on the exposed dura surface, you must subtract the skull-to-dura distance when calculating the post-entry advancing depth.

When placing multiple electrodes or other implanted devices, position them one by one at their entry points in a posterior-to-anterior order. For each electrode, first, align the electrode tip with the bregma, and read the initial coordinates in three directions on the electrode manipulator. Then, raise the electrode high enough to prevent contact with the skull and other electrodes during its movement. Finally, move the electrode tip to the entry point by adjusting the coordinates based on the calculated displacements.

Precise positioning requires meticulous attention to detail. The bregma and lambda on the rat skull should be adjusted to the same horizontal plane. Although size variations among adult rat skulls are typically negligible, body weight-related differences do exist. For accurate positioning, the distance between the bregma and interaural line needs to be corrected based on the actual skull size. The stereotaxic atlas provides a standard distance of 9.0 mm (Fig. 3.3A). When the actual distance of experimental

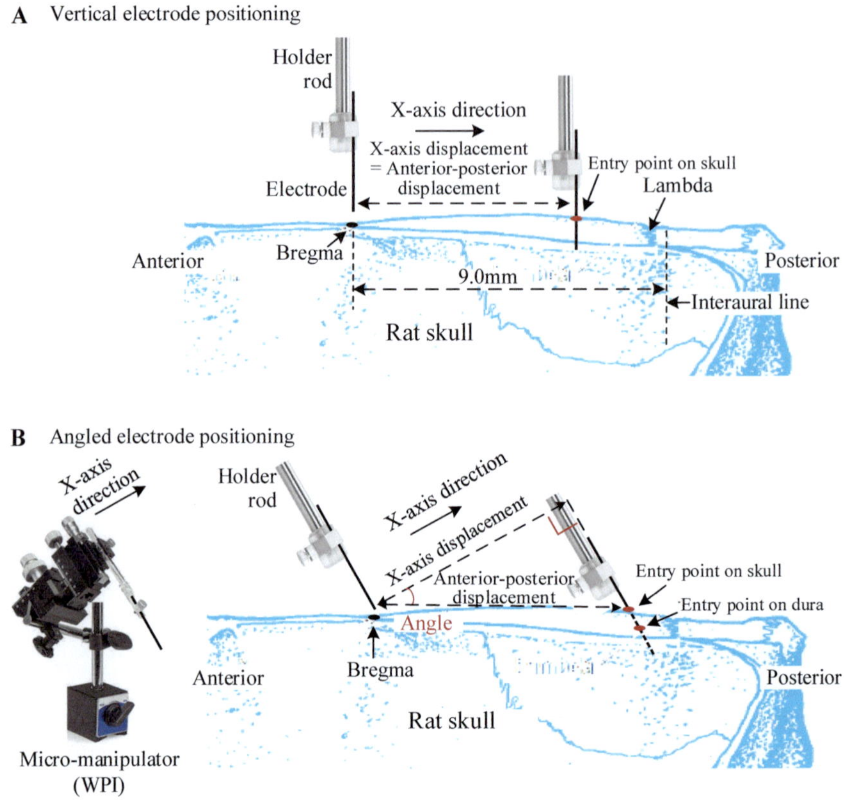

Fig. 3.3 Determining the coordinates of electrode entry points. The figures illustrate the relationship between the X-axis displacement of micromanipulator holder and the anterior–posterior displacement of electrode tip on the skull (or dura) in two cases: vertical insertion (**A**) and angled insertion (**B**), using the bregma as the reference point

rat differs significantly from this value, apply a correction factor—the ratio of the actual distance to 9.0 mm—to the anterior–posterior coordinate. For more positioning guidelines, refer to stereotaxic atlas literature. A classic reference is "The Rat Brain in Stereotaxic Coordinates" by Paxinos and Watson (2007), which is available in multiple editions.

3.2.2 Manual and Electric Micromanipulators

The rat stereotaxic apparatus shown in Fig. 3.1A is a basic setup that uses knobs to manually move the manipulator arms in three directions with a resolution of about 0.1 mm. Although this manual adjustment can meet basic requirements for recording neuronal signals such as field potentials and unit spikes, turning the knobs may cause slight vibrations in the manipulator arm and attached electrode. These vibrations can damage neurons near the advancing electrode, degrading the quality of signal recordings, particularly single-neuron firing (spikes). Consequently, recordings become limited to firing from distant neurons, resulting in smaller amplitudes and lower signal-to-noise ratios. The situation can worsen when the manipulator arm lacks sensitivity and requires greater force to move. To minimize these vibrations, manual adjustments should be as steady as possible.

RWD Life Science Co. Ltd., a manufacturer of rat stereotaxic instruments in Shenzhen, China, once developed a fine-tuning manual device modified from a spiral micrometer. This device can be easily installed on the manipulator arm. In such setup, the original Z-coordinate knob serves as a coarse adjustment to lower the electrode on the brain surface. Then the micrometer knob can provide fine adjustment for advancing the electrode in the brain. Although still manually operated, the device requires minimal force to turn the knob. The micrometer itself has both coarse and fine adjustments, greatly improving movement resolution and enhancing recording quality. Skilled operators can apply gentle force to turn the knobs, minimizing vibrations in the manipulator and electrode. One drawback is the difficulty in monitoring electrode displacement by reading the micrometer scale during implantation. Electric fine-tuning devices and those with digital displays have since provided more convenient for this manipulation.

Electric micromanipulators can minimize damage to brain tissue during electrode implantation. In addition to stereotaxic instruments with motorized manipulator arms, various standalone electric micromanipulators are available. We used an electric single-axis micromanipulator from the Finnish company Sensapex (Fig. 3.4). The device consists of two parts: a micromanipulator and a controller (Fig. 3.4A, B), achieving 10 nm step precision with a 20 mm total travel distance. The user-friendly controller can display digital readouts for displacement, speed, and other parameters. When connected to a computer via USB port, its accompanying software can control the micromanipulator movement. The micromanipulator can be mounted onto the holder of stereotaxic instrument (Fig. 3.4C). During work, the Z-coordinate knob on the stereotaxic instrument serves as coarse adjustment to position the electrode tip on the brain surface. The Sensapex micromanipulator can then drive the electrode into the brain, advancing it smoothly with consistent force and minimal vibration. Using this motorized micromanipulator, when we moved a NeuroNexus array electrode up and down in the rat brain repeatedly, we successfully recorded unit spikes from the same adjacent neurons each time the electrode returned to a same position—indicating minimal damage to neurons from the electrode movements.

An electric micromanipulator may produce high-frequency noise from their micromotor coils, which may interfere with recordings. To address this issue, these devices typically come with a grounding post. The noise usually occurs only during electrode movement when the micromotor is active, and ceases once movement stops.

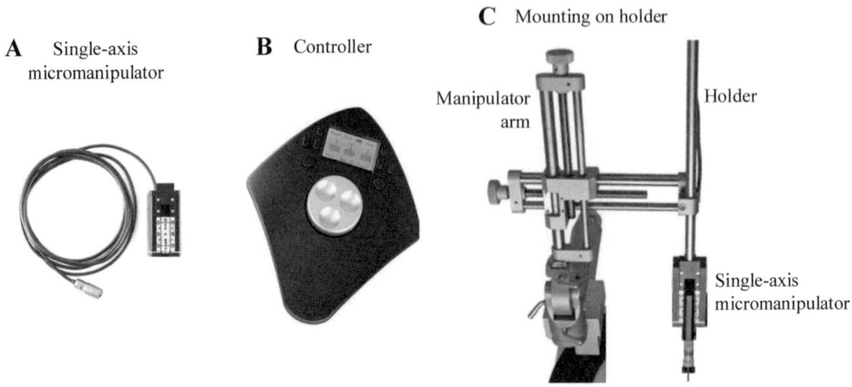

Fig. 3.4 Sensapex electric single-axis micromanipulator. **A** Single-axis micromanipulator. **B** Controller. **C** Micromanipulator mounted on the holder of a rat stereotaxic apparatus during work

3.2.3 Verification of Electrode Position During and After Experiment

The unique laminar structure of hippocampal region can facilitate accurate induction and recording of various neural signals. In in-vitro experiments using hippocampal slices, this laminar structure can be seen with the naked eyes or under a microscope, making it easy to verify correct electrode positioning. However, in in-vivo experiments with intact brains, it is impossible to directly observe recording and stimulation sites in real-time. Relying solely on post-experimental histological analysis of stained brain slice to verify electrode positions can significantly reduce experimental efficiency. Instead, the efficiency can be greatly improved by determining electrode positions through real-time signal recordings during experiments. The neuronal firing patterns and the unique waveforms of evoked potentials captured by an array electrode can enable the real-time position determination, as described below.

1. **Determining Electrode Position through Unit Spike Recordings**

As shown in Fig. 3.5, the recording electrode was a NeuroNexus 16-channel linear array (model A1 × 16-5mm-25-177). The electrode shank was 5 mm long, with 16 contacts spaced 25 μm apart, spanning 375 μm at the tip (detailed in Sect. 3.4.1). The electrode was vertically inserted into the rat brain at an entry point 3.6 mm posterior and 2.6 mm lateral to the bregma. As it was advanced, real-time recordings showed unit spikes in signals within the 500 Hz to 5 kHz frequency range. These spikes were extracellularly recorded single-neuron action potentials (see Sect. 4.1). This ongoing monitoring helped us determine the precise electrode position.

When the electrode passed vertically through the cerebral cortex into the hippocampal CA1 region (about 2 mm deep from the brain surface), it encountered the thin pyramidal cell layer (pcl) with dense somata. Figure 3.5A shows that the

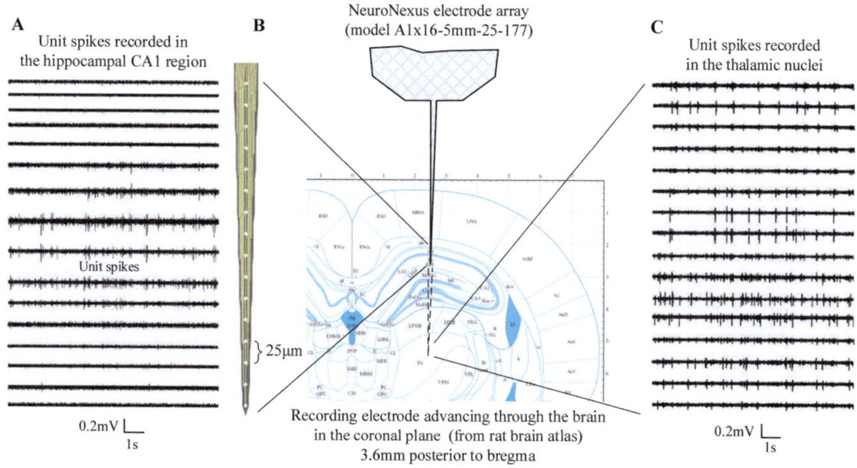

Fig. 3.5 Neural firing (unit spikes) captured by a multichannel array electrode in the rat hippocampal CA1 region and thalamic nuclei. **A** Spikes recorded in the hippocampal CA1 region. **B** Schematic diagram illustrating the electrode locations in the coronal plane 3.6 mm posterior to bregma (from the rat brain atlas). **C** Spikes recorded in the thalamic nuclei

recording contacts in this layer (channels 6–11) captured more spikes (visible as short vertical lines with varying lengths) than the contacts in the layers containing basal dendrites (above) or apical dendrites (below). It should be noticed that spikes can also be recorded beyond the pcl. Interneurons are distributed across almost all layers of the CA1 region (Klausberger and Somogyi 2008; Pelkey et al. 2017), though at a much lower density than pyramidal neurons in the pcl. Because of their low density, interneurons are not always recorded, which explains why clear spikes appeared only across the pcl region in Fig. 3.5A. However, interneurons fire at a much higher rate than pyramidal neurons, and once recorded, their spikes can appear quite dense (see Fig. 5.27A and 5.28A). In the layers of basal and apical dendrites, where pyramidal neurons' cell bodies are absent, recorded spikes typically come from the rapidly firing interneurons. Since these spikes usually originate from a single interneuron, they can show consistent amplitudes.

As the electrode advanced further, it passed sequentially through the hippocampal CA1, DG, CA3 and again DG regions (Fig. 3.5B). During this process, spikes appeared and disappeared across recording channels. When the electrode reached the thalamic nuclei (approximately 5 mm from the brain surface), spikes appeared on all 16 channels (Fig. 3.5C). This occurs because of the widespread distribution of neuronal cell bodies within the nuclei. A similar all-channel spike recording occurred when the electrode first entered the cerebral cortex beneath the dura. However, if cortical neurons had been damaged during dura piercing before electrode insertion, the recorded spikes in the cerebral cortex would be reduced or absent.

NeuroNexus array microelectrodes are extremely thin, typically 15 μm in thickness, causing minimal damage to brain tissue. These electrodes can be inserted and removed from brain tissue repeatedly while still recording spikes from the same group of neurons (Csicsvari et al. 2003). When a recording contact is close to neuronal soma, the amplitude of recorded spikes can reach hundreds of microvolts to even over one millivolt. Given that high-frequency noise in recording signals is typically less than 20 μV in amplitude, this recording quality can reach a high signal-to-noise ratio (SNR). Additionally, as the electrode advances gradually, mechanical forces can stimulate adjacent neurons, increasing their firing rates and producing dense spikes. These large, dense spikes migrate from one channel to the next as the electrode moves forward slowly. The spike migrations can indicate the survival of nearby neurons. However, damage can occur if the electrode surface is rough and exerts excessive pulling force on nearby neurons. Similarly, electrode vibration can also kill nearby neurons. In these cases, recorded spikes may gradually disappear as they migrate along channels. Importantly, spike disappearances do not always indicate neuron damage. They may result from increased distance between the recording contact and the neuron soma, or from other factors such as the neuron entering a temporary quiescent state.

NeuroNexus array electrodes are reusable, but their surfaces can become rough after multiple uses due to accumulated deposits. This roughness prevents the recording contacts from adhering closely to neurons, resulting in lower-quality signal recodings and poor signal-to-noise ratios.

2. **Determining Electrode Position Through Layer-Specific Evoked Potentials**

In addition to determining electrode position through spontaneous firing recordings, we can verify the positions of both stimulation and recording electrodes in real-time using evoked-potentials triggered by electrical pulses (Kloosterman et al. 2001; Csicsvari et al. 2003). As shown in Fig. 3.6A, in the hippocampal CA1 region, a pulse applied to the Schaffer collaterals can activate the downstream post-synaptic pyramidal neurons, while a pulse applied to the alveus can antidromically activate the pyramidal neurons somata. This setup allows recording both orthodromic and antidromic evoked-potentials from the same neuronal population using a single recording electrode.

As shown in Fig. 3.6B, the 16-channel recording electrode detected spontaneous spikes concentrated in the channels 6–11, which indicated their locations in the CA1 soma layer. A pulse (0.3 mA) to the Schaffer collaterals produced distinctive fEPSP waveforms in the apical dendritic layer (channels 14–16) and OPS waveforms in the soma layer (channels 6–8), respectively (Fig. 3.6C). The OPS latency was approximately 5.5 ms. Also, a pulse to the alveus induced APS waveforms in the soma layer (Fig. 3.6D). The APS latency was only ~ 1.5 ms, much shorter than the OPS latency. Thus, the locations of recording channels were further confirmed by these evoked-potentials.

Synchronized activation of a neuronal population can create unique evoked waveforms in the extracellular space of various CA1 layers through ion fluxes (refer to Fig. 2.5). These waveforms can provide real-time guidelines for accurately placing recording and stimulation electrodes. While the amplitude of evoked potentials (reflecting the number of activated neurons) can vary with pulse intensity (see Figs. 2.7 and 2.8), their basic waveforms remain consistent. Improper electrode positioning can distort these waveforms, indicating the need to adjust electrode positions.

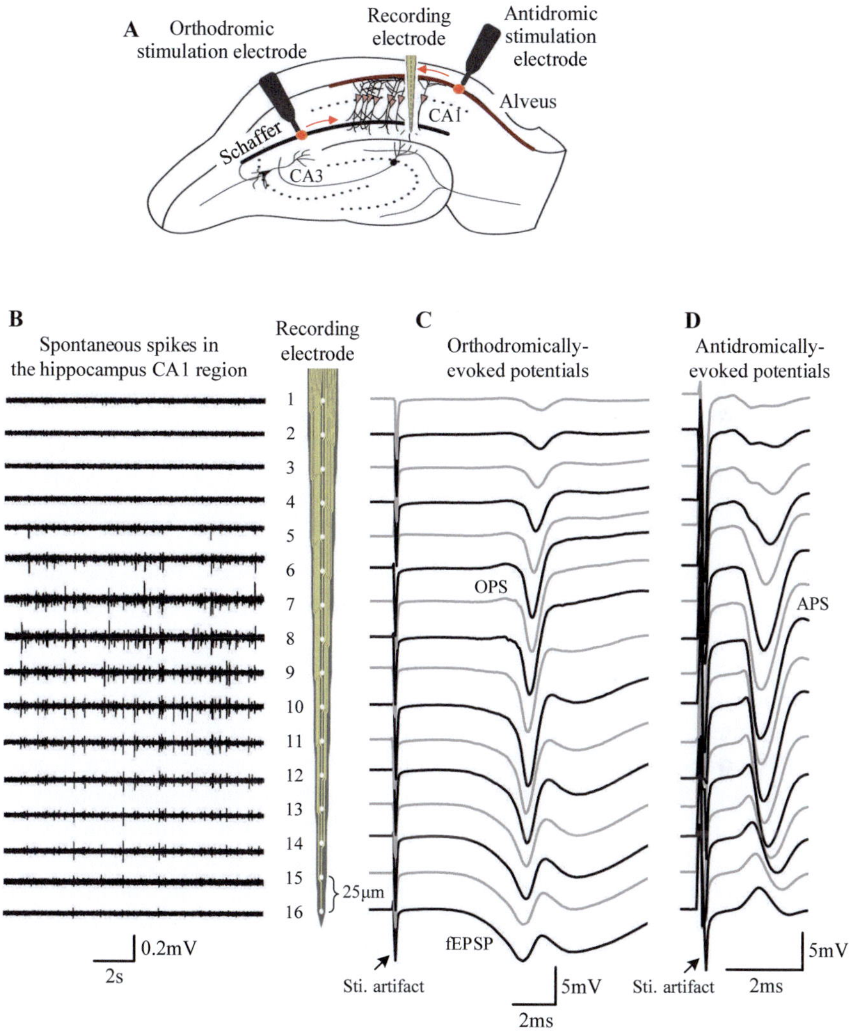

Fig. 3.6 Unit spikes and evoked potentials recorded across different layers of the rat hippocampal CA1 region using an array electrode. A schematic diagram shows the placement of stimulation and recording electrodes in the hippocampal CA1 region (**A**). Spontaneous spikes (**B**), orthodromic evoked-potentials (**C**) and antidromic evoked-potentials (**D**) were recorded in the CA1 layers. For clarity, the evoked waveforms are displayed alternately in black and grey. Note that different time scales are used in (**C**) and (**D**)

After craniotomy surgery, we usually began the experiment by inserting the recording electrode first. As the electrode advanced, we determined its entry into the hippocampal CA1 region by monitoring both the insertion depth and the sequential appearance of unit spikes across channels. Using an array electrode with appropriately spaced channels, the determination was highly reliable. We then inserted the stimulation electrodes. The bipolar stimulation electrodes we used can produce a small activation area precisely confined to the target (detailed in Sect. 3.4.2). Large-amplitude evoked waveforms can appear only when the stimulation effectively activates neurons in the recording area. Therefore, the stimulation sites needed precise alignment with the recording area that had been fixed by the position of recording electrode. During the insertion, we applied test pulses to track the position of stimulation electrode and verify its proper placement.

The alignment requirements for orthodromic and antidromic stimulations differ somewhat (López-Aguado et al. 2002). In the Schaffer collaterals projecting to the recording area, each axon terminal has multiple branches with synapses covering a broader area. Therefore, orthodromic stimulation of the Schaffer collaterals can activate neurons across a larger area in the downstream CA1 region. This means that the alignment between the orthodromic stimulation and recording area needs less precision.

When inserting the orthodromic stimulation electrode from the dorsal hippocampus, it passes through the alveus, stratum oriens (so), and pyramidal cell layer (pcl) before reaching the stratum radiatum (sr) where the Schaffer collaterals are located (Fig. 2.5A). When the stimulation electrode is in the so layer, an applied pulse can activate the axonal fibers there, specifically the commissural fibers (CF) from the contralateral hippocampal CA3 pyramidal neurons (Fig. 2.13A, B). This activation can also evoke OPS in the CA1 pcl layer. Do not mistake this for activation from Schaffer collaterals. The electrode position can be distinguished by observing the polarity of fEPSP waveforms in the so and sr layers. When stimulating the CF fibers in the so layer, the fEPSP is negative in that layer, while a positive fEPSP appears in the sr layer due to the current return loop. As the stimulation electrode advances to the sr layer to activate the Schaffer collaterals, these polarities reverse—the fEPSP becomes negative in the sr layer and positive in the so layer (Richardson et al. 1987). The polarity flip can serve as a marker for the electrode movement among layers. Additionally, the apical dendrites of pyramidal neurons in the sr layer have an extensive longitudinal distribution, enabling simultaneous fEPSP recordings across multiple channels (Fig. 3.6C).

In antidromic stimulation, when the stimulation electrode advances downward from the cerebral cortex to the hippocampus, it reaches the alveus on hippocampal dorsal surface without deep entry. Unlike orthodromic activation which spreads widely at axon terminals, antidromic stimulation can activate just one pyramidal neuron soma per axon, resulting in a smaller activation area (López-Aguado et al. 2002). Therefore, the alignment between the antidromic stimulation site and the recording site needs more precise.

In addition, stimulation electrodes can also record neural signals. Monitoring unit spikes collected at their exposed tips while advancing in the brain can help determine their positioning. Once the electrode is near its target site, switch it to stimulation mode.

After the experiment ends, the rat brain can be extracted for slice staining to verify electrode positioning. Alternatively, we developed a simple assessment method: record the coordinate data on the manipulators for each electrode, raise all electrodes from the brain, and carefully remove the rat from the stereotaxic apparatus. Then return the electrodes to their original coordinates in the brain. This can allow examination of the relative positions among electrode tips to verify placement. This method is particularly helpful for beginners analyzing the recorded evoked-potentials and spontaneous spikes.

Figure 3.7A shows a side view photograph taken during an experiment. Three electrodes protruded above the left skull, including the recording electrode (RE) and two stimulation electrodes—the orthodromic stimulation electrode (OSE) and the antidromic stimulation electrode (ASE). The RE was inserted vertically into the hippocampal CA1 region. The OSE, targeting the Schaffer collaterals, was inserted at a 30° angle to the vertical (from front to back). The ASE, targeting the alveus, was inserted at a 20° angle to the vertical (from left to right). Figure 3.7B shows a posterior view photograph taken after the experiment, with electrode positions restored after rat removal. Figure 3.7C is a side view photograph with a superimposed hippocampus schematic. These photos have some distortion due to the shooting angles. On-site examination of the relative positions between electrodes, combined with analysis of evoked-potential waveforms from orthodromic and antidromic stimulations (Fig. 3.7D), can help assess electrode positioning.

Figure 3.7D shows the recording signals from eight channels on the RE array with a 50-μm spacing. The middle channels were in the CA1 soma layer, recording OPS and APS. Upper and lower channels were in the basal and apical dendrites, respectively, with apical dendrite channels recording fEPSPs. The ASE, possibly positioned slightly deeper (see Fig. 3.7C) or aligned with the RE less precisely, produced relatively small APS waveforms. The OSE, located in the apical dendritic layer, produced large fEPSP and OPS waveforms with accurate position and alignment. Both OSE and ASE were FHC concentric bipolar electrodes with a tip length of 300 μm, an outer diameter of 250 μm, and an inner diameter of 75 μm (see Sect. 3.4.2). The tips of these electrodes in Fig. 3.7C are faintly visible. The RE tip, containing the contacts, was only about 30 μm in width—almost invisible in this photograph. At just 15 μm thick, the entire RE shank appears as a thin line in the well-focused photograph in Fig. 3.7B.

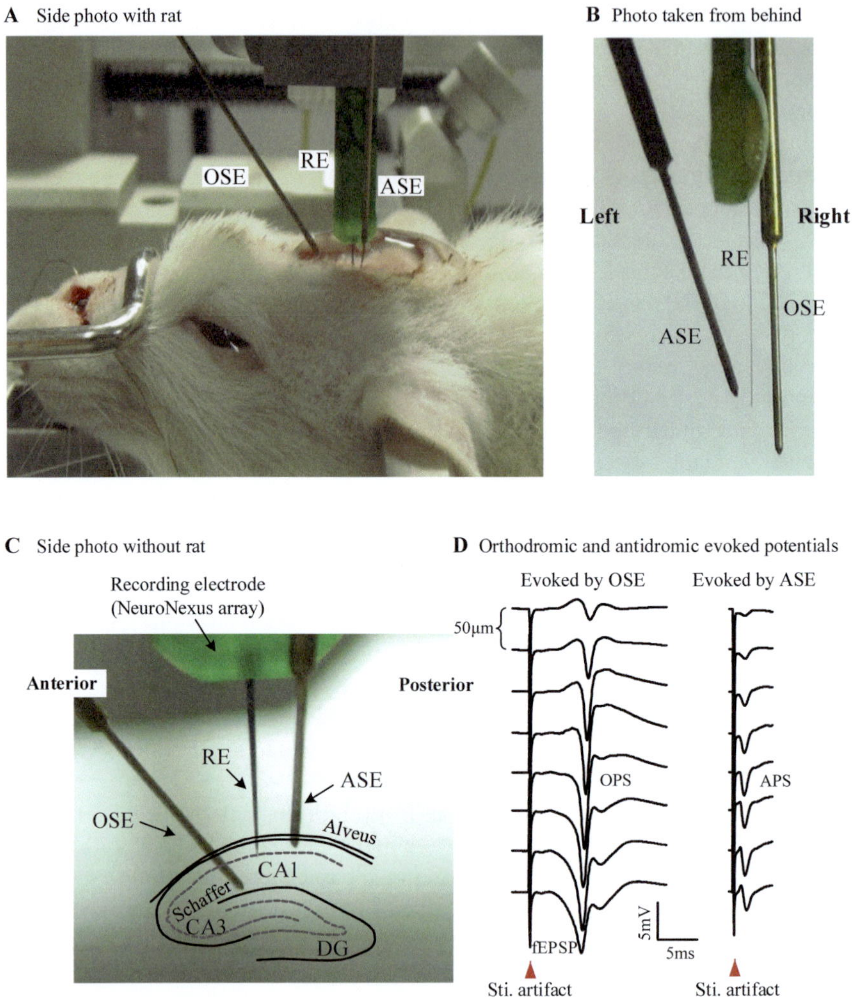

Fig. 3.7 A simple method to assess electrode placement. **A** Side-view photograph of the rat head during an experiment. **B** Posterior-view photograph of electrodes after removing rat. **C** Side-view photograph of electrode tips with a superimposed hippocampus schematic. **D** Examples of orthodromic (left) and antidromic (right) evoked potentials recorded during the experiment.

3.3 In-Vivo Experiment with Exposed Hippocampus

The unique structure of hippocampal region makes it an important subject in neuroscience research for studying neuronal activation mechanisms, synaptic transmissions, neural information processing, external stimulation effects, and pharmacological mechanisms in the nervous system. Research methods typically include in-vivo and in-vitro experiments. In-vivo hippocampal experiments offer the advantage of an intact physiological system. However, they have limitations in chemical administration. Systemic administration, such as intravenous injection, has broad effects and can cause complex reactions. Microlocal injection, where drugs act through local diffusion, offers an alternative. Yet, all these in-vivo drug administrations prevent rapid drug removal for examining recovery. Studies requiring the removal of specific chemical substances from hippocampal tissue are also difficult to perform in-vivo.

In-vitro hippocampal experiments, however, offer advantages for drug administration and chemical substance removal. A common method involves extracting the animal brain, quickly isolating its hippocampus, slicing it thinly, and incubating the slices in artificial cerebrospinal fluid with oxygen perfusion. The incubation can preserve tissue viability for several hours (van Hoeymissen et al. 2020). When cut at specific directions and proper thicknesses based on the connectivity of hippocampal neural circuits, these slices can maintain sufficient neural connections to meet research requirements (Andersen et al. 2000). Such preparations are particularly useful for studying deep mechanism in a simplified environment. However, neuronal responses in isolated slices may differ from those in intact brains. Findings from in-vitro studies often need verification through in-vivo experiments to determine clinical relevance. To overcome the limitations of in-vivo drug administration, we adopted a hybrid approach by exposing the dorsal hippocampus in the rat brain. Using this approach, we successfully verified low-calcium epileptiform activity in the in-vivo hippocampus.

In 1982, Science and Nature published findings of a remarkable phenomenon (Taylor and Dudek 1982; Jefferys and Haas 1982). When rat hippocampal slices were treated with low-calcium artificial fluid (Ca^{2+} 0.2–0.5mM) to block synaptic transmissions, the hippocampal CA1 region exhibited spontaneous, large-amplitude epileptiform discharges. They can occur periodically for several hours and are termed "low-calcium epileptiform activity". Since synaptic transmission was traditionally considered essential for synchronized neuronal activity, this discovery sparked extensive research into non-synaptic mechanisms of neuronal synchronization. However, studies were confined to in-vitro brain slices for many years (Jefferys 1995; Dudek et al. 1998; Su et al. 2001). Testing this phenomenon in intact hippocampus proved challenging because removing calcium from extracellular fluid was difficult under in-vivo conditions.

Under normal physiological conditions in the brain, the extracellular Ca^{2+} concentration ($[Ca^{2+}]_o$) is about 2mM, while the intracellular concentration ($[Ca^{2+}]_i$) is only 0.0002mM (Bear et al. 2016). Ca^{2+} ions play a crucial role in chemical synaptic transmissions. When an action potential reaches the axon terminal, it depolarizes the presynaptic membrane, opening voltage-gated Ca^{2+} channels, increasing $[Ca^{2+}]_i$ in the presynaptic axoplasm rapidly, triggering neurotransmitter release to achieve synaptic transmission (refer to Sect. 2.1.2). To maintain effective synaptic transmission, $[Ca^{2+}]_o$ must remain at a sufficient level. Interestingly, low-calcium epileptiform activity is a type of synchronized neuronal discharge that can occur without synaptic mediation.

In 2003, we adopted a special in-vivo method to observe low-calcium epileptiform activity in the intact rat hippocampus firstly (Feng and Durand 2003). The main challenge was how to reduce $[Ca^{2+}]_o$ within the hippocampus to create a low-calcium condition. We accomplished this through two steps. First, we exposed the hippocampal dorsal surface (beneath the lateral ventricle) by opening the left skull and removing a portion of the cerebral cortex above the left hippocampus by suction or excavation. The cerebral cortex removal was straightforward due to the isolation provided by the lateral ventricle, and the hippocampus remained intact during the procedure. The dorsal hippocampus, with its distinctive silvery alveus fibers, was easily identifiable. Artificial cerebrospinal fluid (ACSF) was added to the naturally formed dent above the hippocampus as a perfusion solution. Drugs can be added to the ACSF to directly act on the hippocampus. Second, we used the Ca^{2+} chelator EGTA to lower $[Ca^{2+}]_o$. Even when perfusing with Ca^{2+}-free ACSF, the $[Ca^{2+}]_o$ inside the submerged hippocampus did not decrease sufficiently to block synaptic transmissions. Therefore, we added EGTA to the Ca^{2+}-free ACSF, which entered the hippocampal tissue, bound to Ca^{2+}, and effectively reduced $[Ca^{2+}]_o$.

Figure 3.8A illustrates the placement of electrodes, including RE in the CA1 region of the exposed hippocampus, OSE at the Schaffer collaterals, and ASE at the alveus. When normal ACSF covered the hippocampal surface, the local field potential (LFP) was normal. And, a single-pulse stimulation through OSE and ASE evoked large OPS and APS respectively (Fig. 3.8B1), indicating normal synaptic transmission and neuronal excitability. After switching to Ca^{2+}-free ACSF with added EGTA, the LFP gradually increased in amplitude, and epileptiform population spikes (spontaneous PS) appeared and grew larger (Fig. 3.8B2-B4). Meanwhile, the amplitude of OSE-evoked OPS gradually decreased, became multi-peaks, and then disappeared, indicating block of synaptic transmissions. However, the ASE-evoked APS remained large and also became multi-peaks. This suggests that the decrease in $[Ca^{2+}]_o$ not only blocked synaptic transmissions but also increased neuronal excitability. After switching back to normal ACSF, both the LFP and evoked potentials gradually returned to normal levels (Fig. 3.8B5).

As shown in Fig. 3.9, a RE with 200 μm-spacing contacts was used to record LFPs and evoked potentials across CA1 layers. When Ca^{2+}-free ACSF containing EGTA was applied over the exposed hippocampus, a prolonged seizure-like discharge occurred and lasted minutes (Fig. 3.9A). This discharge was consistent with the low-calcium epileptiform activity observed in in-vitro slice experiments, characterized by slow negative waves superimposed with PS trains in the pyramidal cell layer (Fig. 3.9B). When using EGTA to lower $[Ca^{2+}]_o$ and simultaneously increasing K^+ concentration in ACSF, the discharge pattern transformed from periodic slow waves to continuous ~ 4 Hz dual-spike discharge (Fig. 3.10). This dual-spike discharge was able to persist for over an hour, providing a unique model of seizure-like activity. We also studied the propagations of the low-calcium epileptiform activity to ipsilateral and contralateral brain regions (Feng and Durand 2005a, b).

During such experiments, the solution applied over the hippocampus surface can be perfused using a method similar to in-vitro brain slice experiments (Khazipov and Holmes 2003). In our experiments, however, we simply refreshed the solution regularly (every few minutes) to maintain the necessary nutrient and drug concentrations without interfering with signal recordings. Since the blood supply system remains

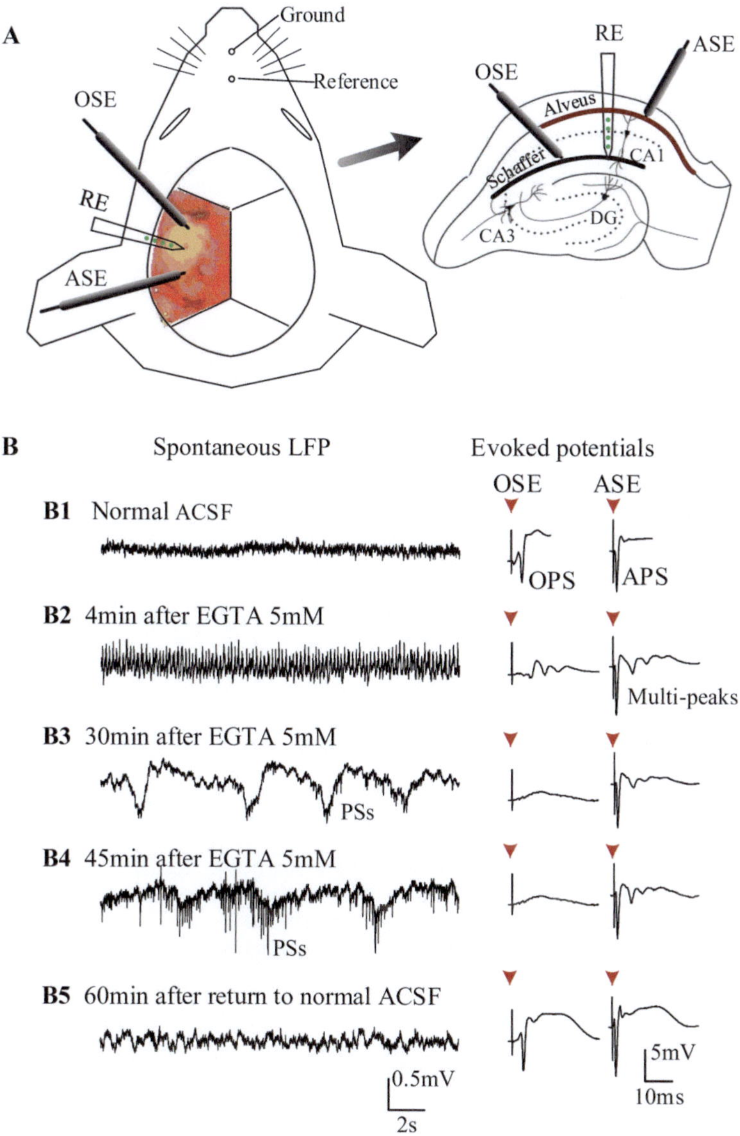

Fig. 3.8 Low-calcium epileptiform activity in exposed hippocampus produced by reducing $[Ca^{2+}]_o$ with EGTA. **A** Schematic diagram showing hippocampus exposure and electrode placement. **B** Spontaneous LFP (left column) and evoked potentials (right column) recorded in the CA1 pyramidal cell layer during different periods. Red arrows indicate stimulation artifacts. From Feng and Durand (2003)

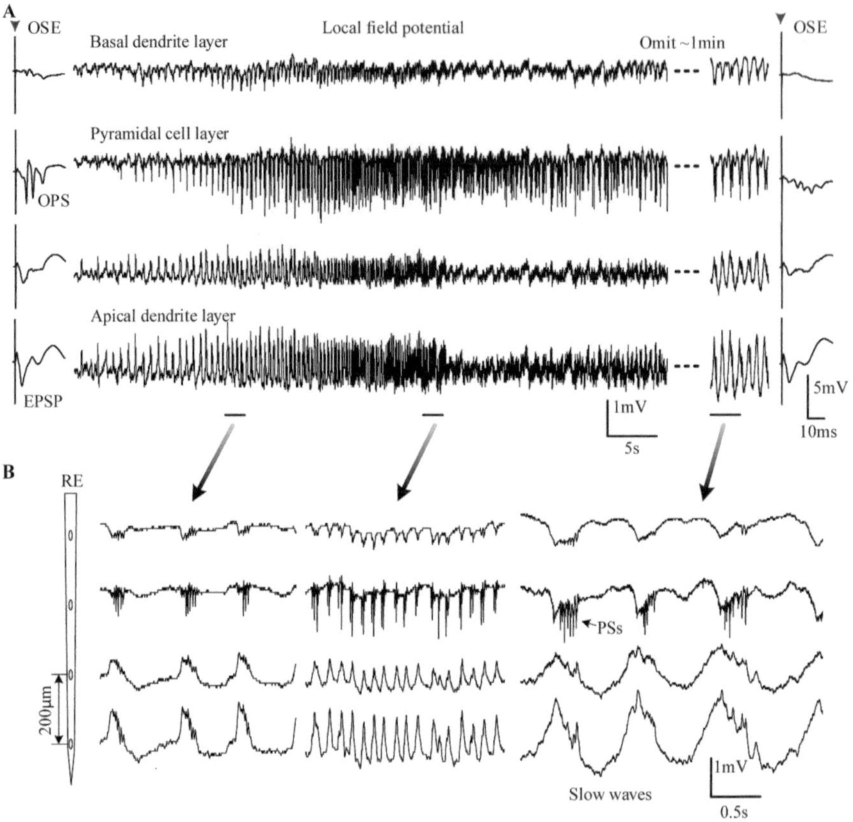

Fig. 3.9 Epileptiform activity produced in the hippocampal CA1 region in-vivo by lowering $[Ca^{2+}]_o$. **A** Epileptiform discharge recorded by four vertically aligned channels on a RE array at different CA1 layers. Insets on the left and right sides show OSE-evoked potentials immediately before and following the discharge. **B** Three enlarged segments from (**A**), with a RE schematic on the left side. From Feng and Durand (2003)

intact, these experiments require no additional oxygenation. This makes the in-vivo setup much simpler than in-vitro slice experiments, while providing better control over drug administration for quick application and clearance. Additionally, exposing the hippocampus can enable direct observation of electrode insertions, similar to experiments using an entire extracted hippocampus (Khalilov et al. 1997).

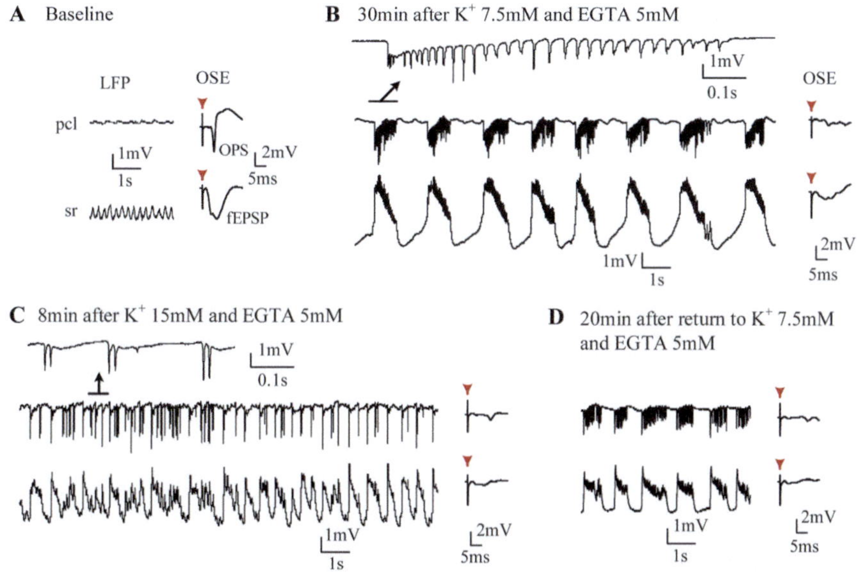

Fig. 3.10 Changes in epileptiform discharges produced by applying high concentrations of K^+ to the ACSF with EGTA. **A** Baseline recordings of spontaneous LFP and OSE-evoked potentials in the CA1 pcl and sr layers. **B** Epileptiform discharges recorded 30 min after applying K^+ 7.5mM and EGTA 5mM. **C** Dual-spike discharges recorded 8 min after K^+ increased to 15mM. **D** Epileptiform discharges recorded 20 min after K^+ returned to 7.5mM. In figures (**B**), (**C**) and (**D**), the evoked OPSs shown on the right almost disappeared, indicating block of synaptic transmissions. From Feng and Durand (2006)

3.4 Microelectrodes

3.4.1 Recording Electrodes

We performed neural signal recordings in rat brains using Michigan style microarray electrodes produced by NeuroNexus company (website: http://neuronexus.com/). These electrodes feature one or more thin-film silicon shanks with contacts coated in iridium or gold to enhance conductivity and biocompatibility. Their conductive traces are shielded by insulating layers. Standard Michigan electrodes are only 15 μm thick, minimizing tissue damage during insertion. However, these thin-film electrodes are extremely fragile and prone to breakage. Therefore, the shanks that penetrate the brain tissue typically measure only a few millimeters in length (3 or 5 mm, Fig. 3.11A). For deeper brain regions, electrodes with longer insertion shanks (e.g., 10 mm) are available, but these have an increased thickness of 50 μm.

NeuroNexus electrodes have two types of connectors (Fig. 3.11A), designed for acute and chronic experiments respectively. Acute experiments typically involve

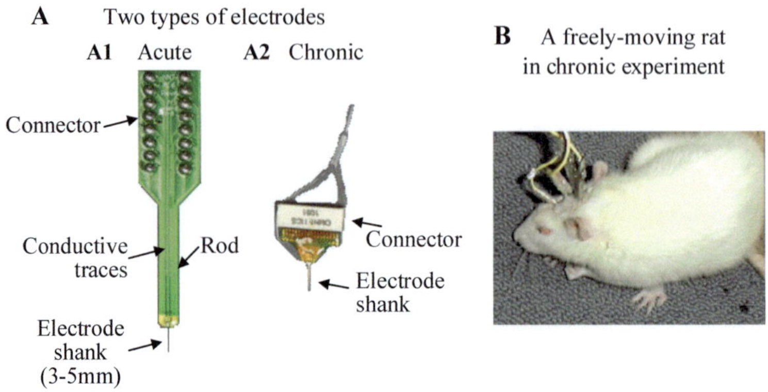

Fig. 3.11 Two types of NeuroNexus array electrodes (**A**) and a freely-moving rat with wire electrodes fixed to its skull using dental cement in a chronic experiment (**B**)

short-term recordings in anesthetized animals. With proper maintenance, acute electrodes can be reused while still yielding high-quality signals. Chronic experiments usually involve long-term recording over days or months, requiring the electrodes to be fixed to the animal skull with dental cement (Fig. 3.11B). These chronically implanted electrodes typically cannot be reused.

NeuroNexus electrodes offer a variety of contact configurations. For examples, the basic linear design has contacts evenly spaced along a shank (Fig. 3.12A). The tetrode-style configuration arranges contacts in diamond-shape groups of four (Fig. 3.12B), imitating traditional wire-made tetrodes. The "Poly2" design features two columns of contacts in a staggered pattern, creating a two-dimensional array within a single shank (Fig. 3.12C). NeuroNexus also provides customized configurations for specific needs (Fig. 3.12D). Notably, placing contacts along the shank edges can bring them closer to neurons, resulting in better signal-to-noise ratios in recordings.

These array electrodes are available in diverse specifications, varying in their shank and contact numbers, areas, spacing, and distribution patterns. Electrodes with a single contact column are one-dimensional (Fig. 3.12A), while those with multiple shanks or multiple columns on a single shank are two-dimensional (Fig. 3.12B, C). Several two-dimensional electrodes combined together can form three-dimensional recordings. The contact number (also known as channel number) includes 4, 16, 32, 64, 128, and beyond. Basic electrodes have 4 to 16 contacts on one shank (single column) and can reach up to 64 contacts (multiple columns). Contact spacing ranges from 25 to 200 μm, and shank spacing varies from 125 to 500 μm. Contact area ranges from hundreds to thousands of square micrometers. Customized specifications can be made to meet specific research requirements.

Neurons close to the electrode surface may be damaged by friction during electrode advancement. While neurons near the electrode tip or edges can remain largely

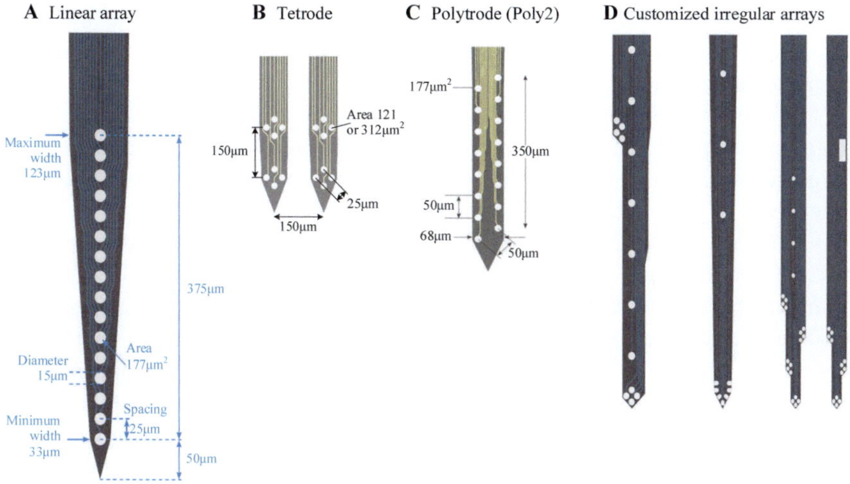

Fig. 3.12 Various contact arrangements on NeuroNexus electrodes. **A** Linear array example: No. A1 × 16-5mm-25-177. **B** Tetrode array: No. A2 × 2-tet-3mm-150-150-121/312. **C** Dual-column array: No. A1 × 16-Poly2-5mm-50s-177 (abbreviated as Poly2). **D** Four examples of customized arrays designed according to user specifications

unaffected, those along the electrode shank may be damaged by the electrode movement. Early electrodes had thicker and rougher substrates and coatings, which reduced the survival probability of nearby neurons. This made it difficult to record high-quality unit spikes from individual neurons, limiting these electrodes mainly to recording LFPs. Thanks to advances in manufacturing technology, modern electrodes can effectively record unit spikes. In our experiments, unit spikes recorded from rat hippocampal neurons occasionally reached amplitudes of up to 1 mV, indicating a close attachment of the neuronal soma to the contact.

A contact with a small area can be in close proximity to only one or two neuronal somata, resulting in just one or two types of large-amplitude spikes in a multiple unit activity (MUA) recording. This can facilitate spike detection and sorting (Buzsáki 2004; Blanche et al. 2005). In contrast, a large contact can collect unit spikes from many surrounding neurons, resulting in overlapped spikes that are difficult to identify. Therefore, electrodes with a larger contact area are suitable for recording LFPs, while those with a smaller area are better for recording unit spikes. Electrodes with a moderate contact area can accommodate recordings of both types of signals. For example, our commonly used electrode type (A1 × 16-Poly2-5mm-50s-177), abbreviated as Poly2, has a contact area of 177μm^2 (Fig. 3.12C) and can record both signals effectively. In comparison, another type of four-shank electrode (A4 × 4-4mm-200-200-1250) has a large contact area of 1250 μm^2 and can simultaneously collect overlapped spikes from many neurons, making spike identification and classification difficult. However, reducing the contact area may increase electrode impedance, leading to more noises (Bretschneider and de Weille 2006). This issue can be resolved through electrode surface modification. NeuroNexus uses iridium oxide thin film deposition for surface modification which can activate the contact surface to reduce electrode impedance.

NeuroNexus array electrodes with sufficiently large contact areas can also function as stimulation electrodes. Moreover, on a same electrode, each contact can be designated for stimulation, recording, or both—switching between the two functions as needed.

Besides the NeuroNexus electrodes originally from the University of Michigan, the Utah electrodes, developed by the University of Utah, are another prominent type of implantable array microelectrodes (Campbell et al. 1991). These Utah electrodes, also known as needle arrays, consist of multiple fine needles protruding on a silicon substrate. A typical electrode contains 100 needles arranged in a 4 × 4 mm area with 400 μm spacing. Each needle is insulated except for its platinum-coated tip, which can collect electrical signals. The needles are sturdy and resistant to breakage. In 2006, the Utah electrode made history as the first array electrode implanted in the human brain, enabling a brain-computer interface for "thought controlling machines" (Hochberg et al. 2006). However, due to the limited length of needles, this type of electrodes is primarily used for recordings from superficial brain regions and for peripheral nerve stimulations (Branner et al. 2001).

In summary, these array electrodes are implantable devices that can be used to detect extracellular signals or deliver electrical stimulations. They can simultaneously collect electrical signals from populations of neurons, including unit spikes, LFPs, and stimulus-evoked potentials. This makes them ideal tools for studying brain mechanisms, developing neural regeneration technologies, and advancing brain-computer interfaces (Buzsáki 2004; Blanche et al. 2005; Erofeev et al. 2022).

3.4.2 *Stimulation Electrodes*

We used concentric bipolar electrodes from the American company FHC (website: http://www.fh-co.com/) for precise stimulations in the rat hippocampal region, typically using the model "CBCSG75" (Fig. 3.13A). In the model name "CBCSG75", "CB" stands for concentric bipolar design, "C" represents the 250 μm diameter of outer pole—stainless steel tube, "S" represents a three-segment nested structure (inner pole, insulation, and outer pole)—each segment 100 μm long (Fig. 3.13B), and "G" represents the 75 μm diameter of inner pole—made of platinum-iridium alloy (Pt/Ir). The final "75" represents the total electrode length of 75 mm.

When applying electrical stimulation through such an electrode using a stimulator with isolator (as those introduced in Sect. 3.5.4), the stimulation current flows from the inner pole, travels about 100 μm, and then enters the outer pole. This creates an electric field confined to a small region near the electrode tip. Such localized field can enable precise stimulation of brain microstructures while minimizing interference with nearby signal recordings. It can also prevent amplifier saturation that may be caused by strong or sustained stimulation currents. Additionally, the surface area of

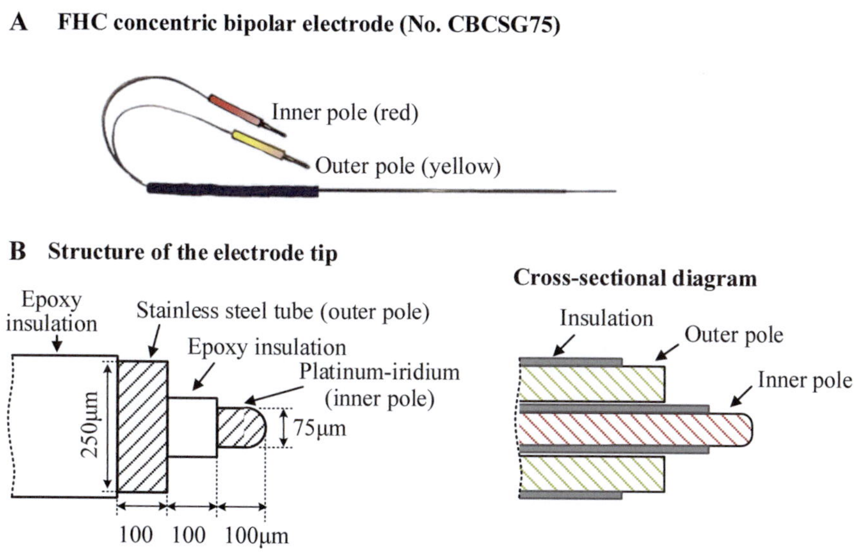

Fig. 3.13 FHC concentric bipolar electrode (CBCSG75). **A** A photograph of the electrode. **B** Electrode tip structure and its cross-sectional diagram

outer pole is about three times larger than the inner pole, resulting in a much stronger electric field around the inner pole. Therefore, the inner pole acts as the working electrode, while the outer pole serves as the return electrode for the current loop.

Stimulation safety should be assessed when applying electrical stimulations through an electrode (refer to Sect. 1.4.4). It relates to multiple factors, with Faradaic reaction posing a primary risk. This reaction can damage surrounding tissue and corrode the electrode. To ensure accurate experimental results, even in animal studies, it is better to keep the stimulation current below the Faradaic reaction threshold. For current pulse stimulation, safety can be assessed using an index k, which is calculated from the charges delivered per pulse and the electrode surface area. A stimulation is considered safe when k falls below 1.5–2.0 (Shannon 1992; Cogan et al. 2016; Gilbert et al. 2023). The k is calculated by

$$k = \lg(D) + \lg(Q) \tag{1.58}$$

Here the charge Q (μC/pulse-phase) is the product of pulse current and pulse width. The charge density D (μC/cm^2/pulse-phase) is the Q divided by the electrode surface area. The inner pole of CBCSG75 electrode has a surface area of 0.00028cm^2 (Fig. 3.13B). Appling a current pulse with a width of 100 μs and intensity of 0.3 mA results in $k = 0.47$, well below the lower safe limit of 1.5. Keeping this pulse width and increasing the pulse intensity to 1.6 mA yields $k = 2.0$ (the upper safe limit). In our rat experiments, we typically used a 100 μs pulse with an intensity less than 0.6 mA. These currents applied by such an electrode are safe even with monophasic

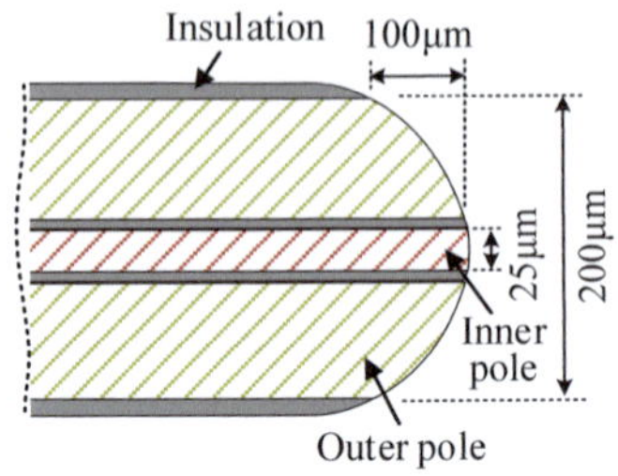

Fig. 3.14 Cross-sectional diagram of another type of FHC concentric bipolar electrodes (CBBRC75)

pulse stimulations. Let alone that we typically employed charge-balanced biphasic pulses, which further enhanced stimulation safety. Unless otherwise specified, the stimulation pulses used in this book are all biphasic current-pulses with a negative phase followed by a charge-balanced positive phase, each 100 μs wide.

Reducing electrode surface area can significantly affect safety. Figure 3.14 illustrates another type of FHC concentric bipolar electrodes (CBBRC75) with a smaller inner pole of only 0.0000049cm^2 in surface area. For a current pulse with a width of 100 μs and an intensity of 0.3 mA, $k = 2.26$, exceeding the upper safe limit of 2.0. We previously used this type of electrodes to apply continuous high-frequency pulse stimulation at 200 Hz for 1 min in the rat hippocampus to investigate whether the stimulation would cause neuronal damage (Yuan et al. 2021).

Figure 3.15A shows the experiment setup, with the stimulation electrodes (OSE and ASE) and the recording electrode (RE) positioned in the rat hippocampal CA1 region. Both stimulation electrodes were CBBRC75. The OSE delivered orthodromic high-frequency stimulation (O-HFS) and single-pulse orthodromic test stimulation (OTS) to the Schaffer collaterals. To assess the state of downstream CA1 pyramidal neurons during O-HFS, the ASE was placed at their axons (alveus) to apply antidromic test stimulation (ATS)—a single biphasic pulse every 5 s. All the current pulses had a width of 100 μs per phase and an intensity of 0.3 mA.

Figure 3.15B shows a recording from the CA1 pyramidal cell layer during 1-min 200 Hz O-HFS of biphasic pulses. The first pulse of O-HFS triggered an OPS similar to the baseline recording. Subsequently, due to local inhibitory circuits (see Sect. 2.3), the OPS disappeared. However, when an ATS was applied to the alveus at 20 ms from O-HFS onset, it still produced an APS (see the enlarged inset at bottom left of Fig. 3.15B). As the effect of inhibitory circuits diminished, PS reappeared. The ATS applied at about 5 s from O-HFS onset produced an APS along with other PSs. (Note: here "PS" refers to a population spike which lacked regular latency with a preceding pulse.) Later, HFS-induced axonal block (refer to Chap. 5) prevented O-HFS pulses from producing more PSs, but the ATS applied every 5 s was consistently able to produce a large APS. After the end of O-HFS, OTS-induced OPS gradually recovered. The results indicate that even when the k value exceeded the safety limit, sustained HFS with charge-balanced biphasic-pulses would not cause neuron damage.

However, results differed when O-HFS was switched to monophasic negative pulses while maintaining other stimulation parameters (Fig. 3.15C). The neuronal

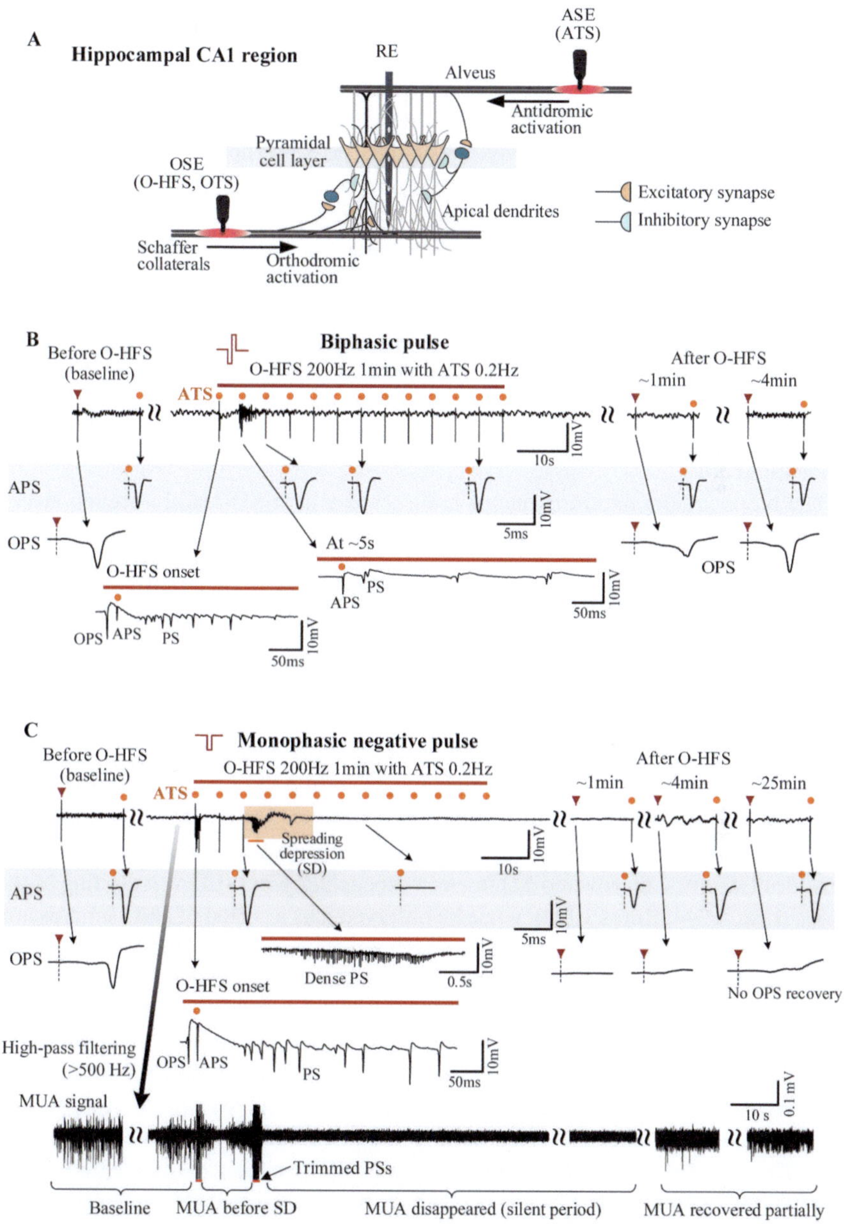

Fig. 3.15 Different neuronal responses to biphasic- and monophasic-pulse O-HFS delivered by a CBBRC75 stimulation electrode (OSE) in the rat hippocampal CA1 region. **A** Schematic diagram of electrode placements. **B** and **C** Example recordings from the CA1 pyramidal cell layer during 1-min 200 Hz O-HFS (indicated by the red horizontal bar) using biphasic-pulses (**B**) and monophasic-pulses (**C**), with pulse artifacts removed. The MUA signal was obtained by high-pass filtering of the artifact-free recording. The pink shading in (**C**) denotes an SD event. Insets show enlarged signal segments and evoked potentials (APS and OPS). Orange dots denote ATSs, while red arrows denote OTSs. Modified from Yuan et al. (2021)

response during the initial O-HFS period was similar to that of biphasic-pulse O-HFS. However, after about 10 s O-HFS, a spreading depression (SD) event occurred, characterized by dense PSs followed by a slow wave in the LFP, consistent with previous reports (Herreras and Makarova 2020; Bragin et al. 1997a, b; Herreras et al. 1994). Following the SD, during ongoing O-HFS, ATS was no longer able to produce APS. Even after O-HFS ended, it took several minutes for APS to recover gradually, while OTS-induced OPS failed to recover. This indicated that the CA1 neurons kept their ability to respond to external activation, while the directedly stimulated afferent fiber (Schaffer collaterals) was damaged and failed to activate. Additionally, during the period following SD, unit spikes disappeared (forming a "silent period") and only partially recovered after O-HFS ended (see MUA signal at the bottom of Fig. 3.15C). These results indicate that when the stimulation electrode surface area is too small, causing excessively high charge density to raise the k value beyond the safety limit, continuous HFS of monophasic-pulses can damage neural tissues such as the Schaffer collateral axons.

The hippocampal region is one of brain areas where SD readily occurs. A key feature of SD is its slow propagation. Figure 3.16A shows SD waves recorded from the CA1 soma and apical dendritic layers (pcl and sr) by two channels 150 μm apart. The SD waves propagated from the sr layer to the pcl layer at a very slow speed. Besides electrical stimulations, other inducements can also trigger SD events, such as high extracellular K^+ concentration and intense epileptiform discharges. Figure 3.16B shows an SD event induced by locally injecting a high-concentration potassium chloride solution into the hippocampus. The two recording channels were still 150 μm apart. Dense firing preceded the SD waves. The slow SD propagation can be determined through the time delay between the SD troughs recorded in the pcl and sr layers.

The SD wave in field potential is caused by massive neuronal depolarization. Aristides Leão first discovered SD in electroencephalogram (EEG) recording in the 1940s (Leao 1944; Bures 1999), when he observed it as a period of EEG suppression following epileptic seizures. SD is also associated with conditions such as migraine, ischemic brain injury, and cerebrovascular disorders (Gorji 2001; Lauritzen 1994; Parsons 2004; Kunkler and Kraig 2003). A key phenomenon accompanying SD events is the dramatic changes in extracellular ion concentrations (including K^+, Ca^{2+}, Na^+, and Cl^-). These changes disrupt the normal ion concentration gradients between intracellular and extracellular environments, causing morphological changes in neurons, such as toxic swelling (Oka et al. 2022; Martins-Ferreira et al. 2000; Kraig and Nicholson 1978; Krnjević et al. 1980; Somjen and Giacchino 1985).

Although neural damage can trigger SD, its occurrence doesn't necessarily mean damage. Neuronal activity may return to normal after SD, especially when induced by high extracellular potassium. However, a failure of neuronal activity recovery long after SD can signify neural damage, as shown in Fig. 3.15C. Therefore, the appearance of SD can serve as a warning sign for tissue damage. SD in the hippocampal region is often accompanied by dense epileptiform discharges before its slow wave (Figs. 3.15 and 3.16). Even after reversible SD, normal neuronal activity may take tens of minutes to recover. Additionally, a "silent period" without neuronal firing

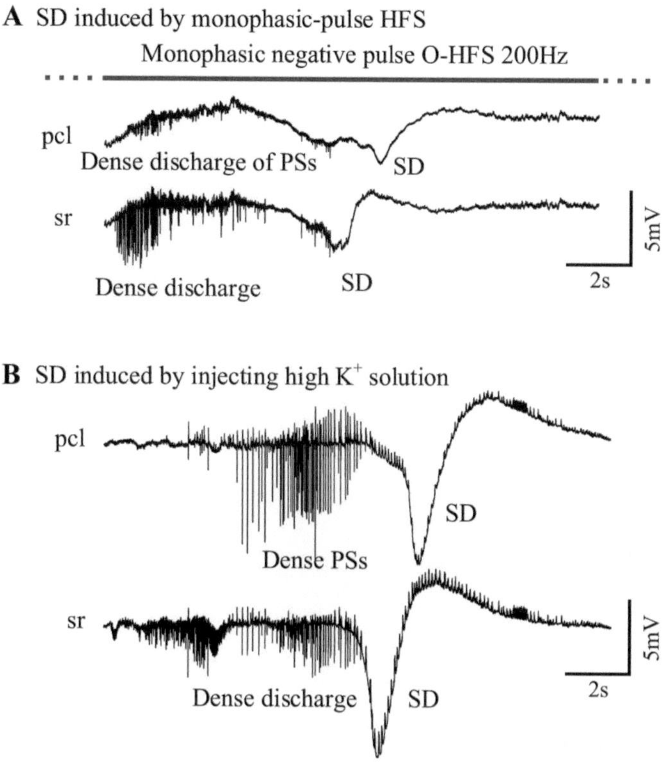

Fig. 3.16 Slow propagation of SD waves from the apical dendritic layer (sr) to the soma layer (pcl) in the rat hippocampal CA1 region. **A** SD event induced by monophasic negative pulse O-HFS. **B** SD event induced by a local injection of high-concentration potassium chloride solution in the hippocampal region

can typically occur immediately following SD waves. During this period, the field potential becomes almost flat (Fig. 3.15C). If a similar silent period suddenly appears during an experiment and the rat is still alive, you can replay the recorded signal to check for possible SD occurrence before inspecting any setup failures. If SD has occurred, the field potential and spikes may reappear over time. This situation may occur when an experiment procedure, such as electrode advancement during implantation, excessively damages neural tissue.

Note that the actual SD wave contains a negative DC shift (Herreras et al. 1994), which was filtered out by the AC coupling inputs used in our amplifier setup. As a result, the SD wave typically appeared as an alternating waveform—first a negative phase followed by a positive phase (Fig. 3.16), which was consistent with other reports (Bragin et al. 1997b).

3.5 Experiment Setup

Figure 3.17 illustrates our setup for in-vivo rat experiments. It comprises three main systems. The signal amplification and acquisition system includes a 16-channel amplifier and a PowerLab data acquisition unit. The electrical stimulation system consists of a multi-channel programmable stimulator, isolators, and a control system. The rat stereotaxic system includes a stereotaxic instrument, the animal, implanted electrodes, and a preamplifier—all enclosed within a copper mesh shielding cage.

During experiment, the neural potentials collected by a NeuroNexus array electrode are first amplified by the preamplifier put inside the shielding cage, then further amplified by the multi-channel amplifier, and finally converted to digital signals by

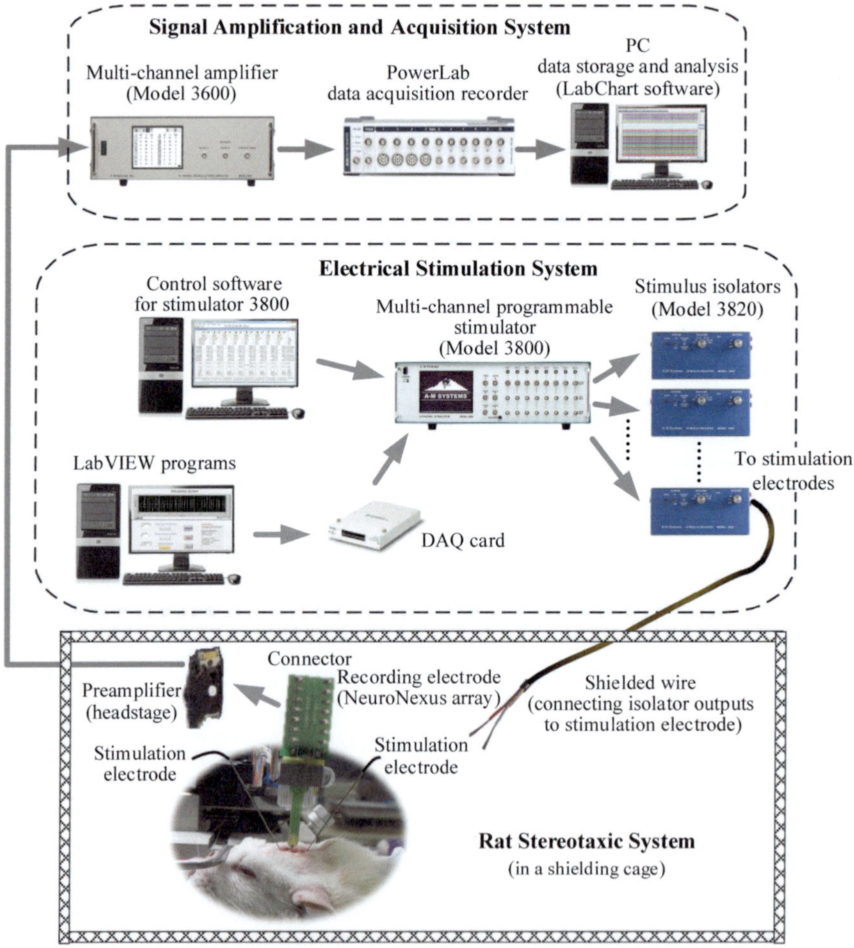

Fig. 3.17 Neuro-electrophysiological experiment setup for rats

the data acquisition unit and stored in the computer (PC). When applying electrical stimulations, the stimulator generates the required electrical pulses under software control, which then pass through independent isolators before reaching the implanted stimulation electrodes. This stimulation system can produce signals with arbitrary waveforms beyond pulses, as detailed in Sect. 3.5.5. Although Fig. 3.17 shows three PCs running different software programs, these programs can be run on a single computer. The following sections will describe these instruments and devices, along with considerations for their use.

3.5.1 Amplifier and Data Recorder

Neural electrical signals are very weak and need amplification before being converted from analog to digital. Our lab used general electrophysiological amplifier and data acquisition devices rather than specialized neural equipment. As shown in Fig. 3.17, we used a Model 3600 16-channel extracellular amplifier with headstage (preamplifier) from A-M Systems Inc. For data acquisition, we used a PL3516 PowerLab data recorder from ADInstruments Inc. (Australia). The multi-channel neural signals collected by an electrode first passed through the headstage into the amplifier, where they underwent a 100-fold amplification. They then entered the PowerLab recorder for additional amplification and analog-to-digital conversion (ADC), usually at a 20 kHz sampling rate. Because the PowerLab recorder had its own amplification, setting the preceding amplifier to a 100-fold gain was sufficient to record sub-millivolt and smaller neural signals.

The Model 3600 amplifier features low-noise and high-gain properties, designed for amplifying microvolt electrical signals (Fig. 3.18A). It can be equipped with two types of headstages (Fig. 3.18B). One is the miniature headstage used solely for recording (the small square device in the figure). The other is the larger, dual-function headstage capable of both recording and stimulation (the triangular device in the figure). The dual-function headstage offers versatility through proper wiring and settings, allowing designated channels to switch between recording and stimulation modes, enabling a single electrode contact to both collect signals and deliver stimuli. When a channel switches to stimulation mode, its corresponding contact disconnects from the recording circuit and connects to the pre-wired stimulator. This disconnection prevents stimulation signals from entering the recording circuit, protecting the amplifier from saturation and damage since stimulation signals are much larger than neural signals.

The front panel of the 3600 amplifier features a touchscreen (Fig. 3.18A), enabling settings for each channel, including high-pass and low-pass cutoff frequencies, gain, notch filter and recording mode. A-M Systems also offers the Model 1700—a four-channel amplifier with similar performance and functionality to the Model 3600, but with a traditional panel of knobs and switches instead of a touchscreen. For applications requiring fewer signal channels, the Model 1700 is a suitable alternative.

A

B

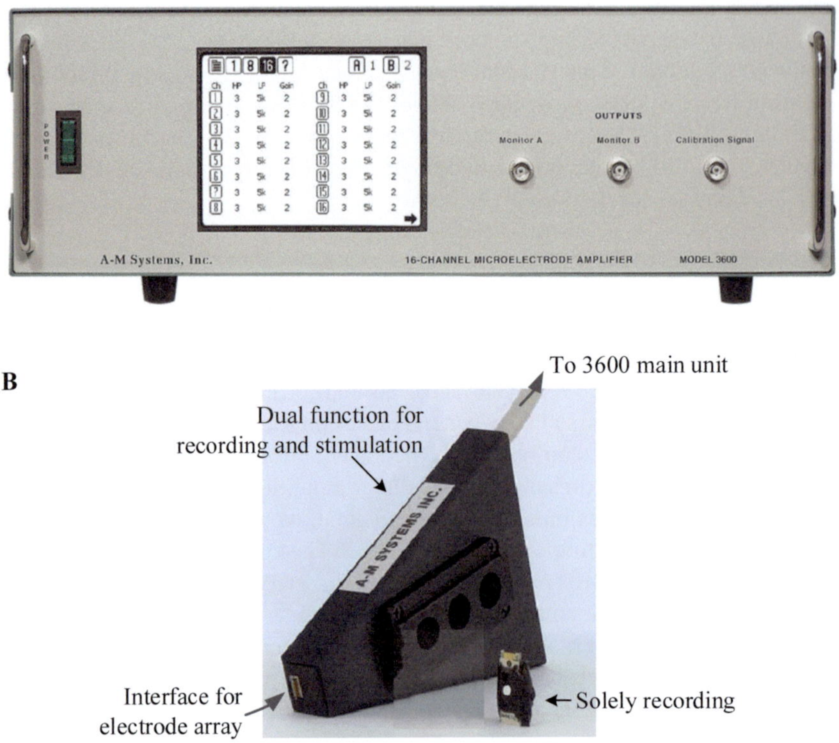

Fig. 3.18 Model 3600 amplifier (**A**) and its two preamplifier accessories—headstages (**B**)

3.5.2 Guidelines for Amplifier Usage

1. **Grounding and Shielding**

Electrical devices used in electrophysiological experiments typically require a separate ground (or "clean ground"). Do not share their ground with lighting, refrigerators, or other electrical equipment. Appropriate shielding measures are also necessary. A copper plate on the experiment table can improve shielding. Our experiment table, measuring 60 cm wide and 150 cm long, was covered with a 2 mm thick copper plate connected to the experimental ground. Ample space surrounded the table, allowing operators to work from various directions. Some equipment, like amplifiers and stimulators, occupied half the table, while the other half accommodated rat stereotaxic instruments. During experiments, we covered the stereotaxic system with a shielding cage (or Faraday cage) with an open-bottom design, made of an aluminum frame and copper mesh. The aluminum bars along the cage bottom edges contacted the copper plate on the table, providing shield grounding. Notably, only when a cage is grounded can it provide effective shielding. The copper plate can offer an additional function: operators can discharge electrostatic charges from their bodies by touching it, preventing transfers to animal and devices. This function is particularly important in cold and dry winters when static electricity easily accumulates on bodies and clothing.

Our shielding cage was custom-built using L-shape aluminum bars to form a trapezoidal cuboid frame—slightly narrower at the top than at the bottom. Copper mesh covered all sides except the open bottom. One side was designed as a liftable curtain, facilitating observations and operations during experiments. Even with this curtain raised, we still achieved excellent signal-to-noise ratios in our recordings, without significant external interference. The interior space in the cage allows for simple experimental operations. For complex procedures, such as craniotomy and electrode replacements, the cage can be completely removed for easier access.

2. **Amplifier Saturation**

It is important to be aware of the technical specifications of an amplifier before using it. For example, the Model 3600 amplifier has a maximum input voltage of only 1–2 V (determined by the headstages shown in Fig. 3.18B). Typical electrophysiological signals fall within this range. However, when input signals exceed its linear amplification range, the amplifier can saturate. Similarly, if the amplifier gain is set inappropriately, signals can be clipped due to saturation.

External noise and electrical stimulations can cause amplifier saturation. When applying electrical stimulations to a brain region being recorded, large stimulation signals may leak into the amplification circuit, causing saturation which can persist long after the stimulation ends. This amplifier malfunction can disrupt normal signal recording for minutes or even hours. More seriously, high-voltage inputs, such as a electrostatic voltage, may completely destroy an amplifier. Therefore, special care should be taken to prevent electrostatic buildup during experiment, especially in cold, dry winters when static electricity can easily accumulate and reach thousands of volts momentarily. Wearing a grounded wrist strap and keeping appropriate humidity can help reduce electrostatic risk. Before handling devices like electrode holders and manipulators, touch your free hand to a grounded copper plate to discharge any potential static electricity. Another protective approach is to turn off amplifier inputs during operations. The Model 3600 amplifier offers "ON" and "OFF" options for each channel—"OFF" disconnects the channel input. The Model 1700 amplifier has "Record" and "Stimulate" options—switching to "Stimulate" disconnects the amplification circuit. These measures can prevent amplifier damage when handling animals or adjusting electrodes during experiments. If you need to monitor recording signals during operations (when the amplification circuit must remain connected), take extra precautions against electrostatic interference. Additionally, improper connections of ground and reference wires, or their unreliable connections—such as broken wires or poor contact—can also cause amplifier saturation.

When amplifier saturation occurs, the recording signal can become a flat line or its amplitude may drop significantly. In this case, immediately disconnect the input signals or turn off the amplifier power, then check all instrument connections for troubleshooting. Be aware that severe saturation may cause unstable, low-frequency, high-amplitude drifts in recordings during the recovery period after turning the amplifier back on. Do not mistake these artifacts for actual field potentials from the brain.

3. **Testing Setup Connections Using a Calibration Signal**

A known input signal is essential for checking a newly established or updated amplification setup and its settings, or for verifying proper function after troubleshooting. Since electrophysiological amplifiers have a limited input range, using arbitrary signals could cause amplifier saturation or damage. Model 3600 amplifier features a "Calibration Signal" output on its front panel (in the right of Fig. 3.18A), which can generate a 1.0 kHz sinusoidal signal with four amplitude options: 1, 10, 100, and 1000 mV. This calibration signal provides a safe source for testing amplifier wiring. For example, after setting up the signal amplification and acquisition system as shown in Fig. 3.17, connect this calibration signal to each socket of the electrode connector sequentially. By examining the frequency and amplitude of the recorded signals in the LabChart display window—the software coming with the PowerLab recorder (refer to Sect. 3.5.3), you can verify the recording of each channel. This process validates the entire signal path from the electrode connector through amplification to sampling and recording.

4. **Noise and Interference**

Electrophysiological signals, including neural signals, are classified as weak electrical signals. Their measurements are susceptible to various types of noise and interference, including 50 Hz mains interference and pulse interference from switching electrical equipment. Using differential amplification can help eliminate interferences by recording the potential differences between the measurement and reference points, effectively removing common-mode noises.

Additionally, proper grounding and shielding are crucial for eliminating interference. The experiment setup should use an independent ground, separate from lighting and non-experiment electrical equipment. For effective shielding, place the experimental animal, along with its connecting devices and micromanipulators, inside a shielding cage. All connecting wires also need appropriate shielding. For example, use shielded cables to connect the stimulator outputs and the stimulation electrodes implanted in the rat brain. It is important to ground these shields to prevent external noise from entering the rat brain and interfering with the recorded signals. The following examples show various interferences we once encountered in our recordings in rat brains.

(1) **Mains Interference**

Mains interference in a signal is easy to identify. As shown in Fig. 3.19, a thickened signal (top left) indicates potential interference. Zooming in on the time axis can reveal the superimposed sine waves clearly (bottom). You can determine mains interference by measuring the sine frequency—typically 50 Hz, or 60 Hz in North America and some other countries.

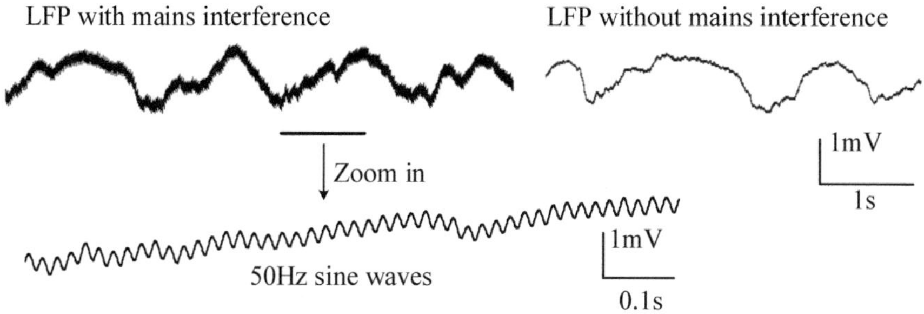

Fig. 3.19 Mains interference in an LFP signal recorded from the rat hippocampus

Many amplifiers feature a notch filter that can remove mains interference, as in the Model 3600 and 1700 multi-channel amplifiers. However, when the interference frequency overlaps with the signal frequency, filtering out the interference frequency can also cause signal loss. Furthermore, real filters are not perfect—a 50 Hz notch filter can weaken signals across a frequency range around 50 Hz, leading to more loss. For some neural signals, such as EEG and LFP, frequencies around 50 Hz are often not negligible. Therefore, careful consideration is necessary when using a notch filter to remove mains interference. Action potentials from individual neurons (unit spikes) typically have much higher frequencies—above 300 Hz—and high-pass filtering can fully eliminate mains interference in spike signals. Generally, proper grounding and shielding can suppress mains interference to an acceptable level without filtering.

(2) **Switching Interference**

Switching electrical equipment on and off or rotating electrical knobs can cause high-frequency noises. In recorded signals ranging from 500 Hz to 5 kHz, these noises may resemble unit spikes (Fig. 3.20). However, they can be readily identified in multi-channel recordings. In Fig. 3.20, the first two channels (Ch1 and Ch2) were from the right hippocampal CA3 region, with the two recording contacts only 25 μm apart—allowing simultaneous recordings of spikes from same neurons. The third channel (Ch3) recorded from the left hippocampal CA3 region, making it impossible to detect spikes from the neurons measured in the other two channels. Yet, high-frequency noises appeared simultaneously on all the three channels. Thus, spike-like signals occurring simultaneously at distant contacts can be noises, rather than real spikes.

(3) **Mobile Phone Interference**

Today, cell phones have become inseparable from our daily lives. Cell phone interference can be picked up by neural signal recording setups. The best way to prevent this is to keep cell phones away from the experimental setup. Interference can become especially severe when a phone is dialing or about to ring. When placed nearby, phones can introduce significant noise into the recorded signals.

Figure 3.21 shows the noises introduced by cell phone ringing in a recording from the rat hippocampal CA1 region. In the original signal recorded with a frequency band

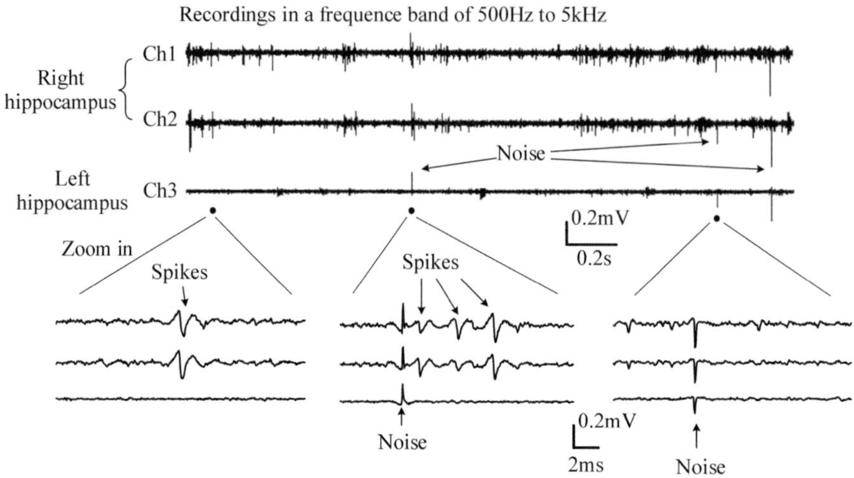

Fig. 3.20 High-frequency noises in neuronal spike recordings during electrical equipment switching

of 0.3 Hz to 5 kHz, the noises appeared as pulse trains in the enlarged inset (upper), with a pulse width of about 0.7 ms and a frequency of ~ 200 Hz. In the high-pass filtered signal (MUA signal) with a 500 Hz cutoff frequency, these noises roughly resembled neuronal spikes. However, the enlarged view showed clear differences between the noises and real spikes (bottom).

A powered-on cell phone placed near the experiment setup can generate noises even without incoming calls or conversations. To minimize their interference, keep cell phones away from the experiment region. The signal shown in Fig. 3.21 was recorded around 2010. As cell phone technology has advanced, the characteristics of the interference may have also changed.

(4) **Interference from Rat Facial Movements**

In an acute in-vivo rat experiment, the rat head was fixed in a stereotaxic apparatus (Fig. 3.7A). When recording neural signals from its brain, subtle vibrations from the head, such as mouth or face trembling, can cause interference that presents as high-frequency, cluster-like noises in recordings (Fig. 3.22A). Such noises can disrupt the recording of unit spikes in the 500 Hz to 5 kHz frequency band, since both frequency ranges overlapped. An enlarged view shows that these noises appeared as continuous oscillatory waves (Fig. 3.22B) with similar waveforms and amplitudes across adjacent recording channels. In contrast, spikes were separate waveforms with significant amplitude variations across different recording channels (Fig. 3.22C). If this type of noises is suspected during experiment, checking the rat facial movements can help confirm its presence.

Careful attention to experimental procedures can help minimize the impact of rat facial movements on neuronal spike recordings. Forceful operations with a microdrill or scalpel during craniotomy may loosen the rat fixations. Therefore, after completing

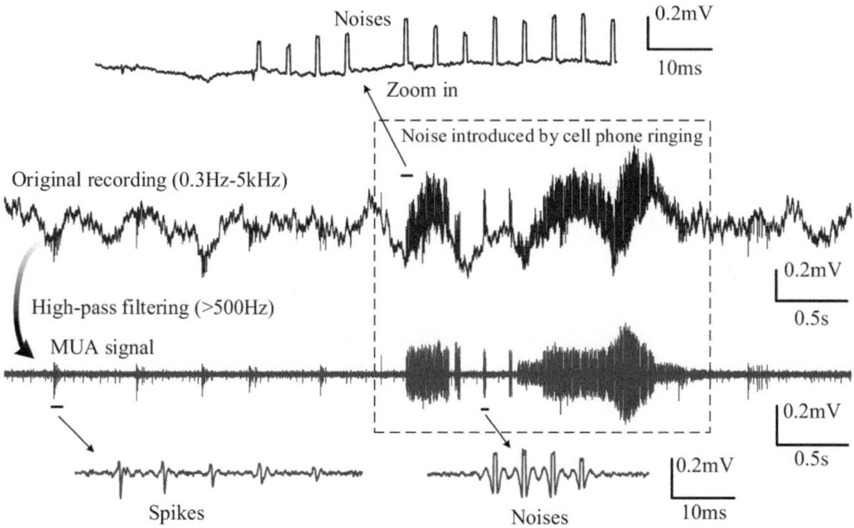

Fig. 3.21 Noises introduced by cell phone ringing in a neural signal recording from the rat hippocampal CA1 region

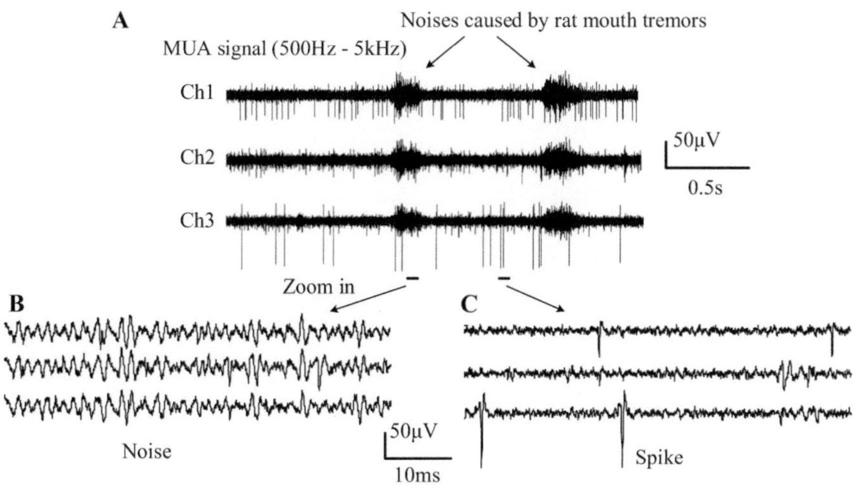

Fig. 3.22 Noises caused by rat facial movements in neuronal spike recordings

the craniotomy and before placing electrodes, carefully re-examine and reinforce all fixations, including ear bars, nose clamp, and incisor bar. This step can effectively reduce or eliminate the "cluster" noise introduced by rat facial movements during signal recordings. As the anesthetic wears off over long experiments, additional anesthetic may be needed to stop rat movements. Once the rat quiets down, the noises will disappear. Thus, the appearance of these noises can also serve as an indicator for anesthesia depth. In chronic experiments without anesthesia, electrodes are fixed to the rat skull with dental cement. When recording signals from a freely moving rat (Fig. 3.11B), its actions such as chewing food or scratching face and head may introduce noises in the recordings similar to those shown in Fig. 3.22. Do not mistake the noises as neural signals.

(5) **Crosstalk in Multi-Channel Recordings**

When recording multi-channel electrical signals, malfunctions in signal wires can lead to "crosstalk". For instance, we once encountered this issue while using an array electrode to record unit spikes in the rat hippocampal CA1 region.

As shown in Fig. 3.23, we used a NeuroNexus acute electrode (model A4 × 4-4mm-200-200-1250) with four recording shanks, each featuring four linearly arranged contacts. Figure 3.23A shows the contacts on the first shank, numbered from top to bottom as channels 1–4 (Ch1–Ch4). The contacts were spaced 200 μm apart, both between contacts on the same shank and between shanks. This substantial spacing should have prevented different contacts from recording firing (unit spikes) from the same neurons simultaneously (Buzsáki 2004). However, the recorded signals showed that the spikes in Ch1 consistently resembled those in Ch3, although with reduced amplitudes. This was puzzling given the 400 μm distance between the two channels. According to the orthodromic and antidromic evoked potentials in the hippocampal CA1 region (see Sect. 3.2.3), we determined that Ch3 was positioned in the soma layer, recording spikes from nearby neurons. Then, Ch1, located 400 μm above, was likely beyond the CA1 region. Where were the Ch1 spikes from? Why didn't Ch2—located between Ch1 and Ch3—record the same spikes (see the enlarged insets in Fig. 3.23A bottom)? How could the spikes bypass the closer contact (Ch2) to reach the more distant one (Ch1)?

We numbered the recording channels based on the contact layouts on the four shanks, which differed from the pin numbering on the connector at the electrode tail (Fig. 3.23B, bottom). After the experiment, we checked the electrode carefully. According to the site map provided by manufacturer, Ch1 and Ch3 were connected to adjacent pins 1 and 2 on the connector, respectively (the red numbers in Fig. 3.23B). We discovered a broken PCB trace connecting the Ch1 contact to its pin—the pin 1 solder pad. As a result, the pin 1 was left floating. When connected to the amplifier, this floating pin pad acted like an antenna, picking up signals from the adjacent pad of pin 2 through electromagnetic coupling, thereby producing a crosstalk signal. This explained the similarity in spike recordings between Ch1 and Ch3. The broken trace was not a manufacturing defect but resulted from repeatedly applying tape for fixation. After this incident, we no longer applied tape directly to the electrode PCB.

Crosstalk occurs through electromagnetic coupling and can only appear with high-frequency alternating signals. The faster a signal changes, the more crosstalk it can generate. Direct currents do not generate crosstalk. Since neuronal spikes contain high-frequency components and change rapidly, they can produce significant crosstalk. This is evident in the MUA signals shown in Fig. 3.23A, which fall within the 500–5000 Hz frequency band. Similarly, pulse-evoked population spikes (OPS

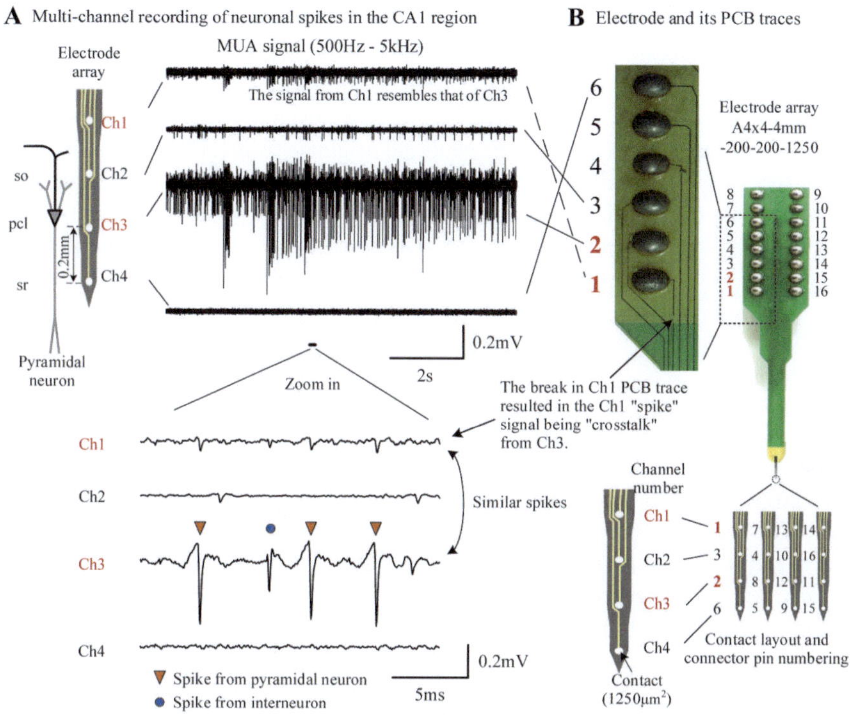

Fig. 3.23 "Crosstalk" caused by a malfunction on an electrode. **A** Spike signals (MUA) recorded by four channels with a vertical spacing of 200 μm on an array electrode in the rat hippocampal CA1 region. The enlarged inset below shows that the spikes at Ch1 were a scaled-down version of those at Ch3. **B** Part of the printed circuit board (PCB) of the electrode connector showing a broken PCB trace. The plot also shows the wiring between the contacts at the electrode tip and the solder pads on the 16-pin connector at the electrode tail

and APS) contain high-frequency components that can generate substantial crosstalk when experimental setups malfunction.

Crosstalk signals closely resemble normal signals, so be careful not to confuse these false signals with real neural signals. A crosstalk signal typically maintains a stable amplitude ratio to the original signal. To check for its presence, monitor the sequential shifting of spikes in recording channels while advancing a recording array into the brain. For example, as shown in the upper part of Fig. 3.23A, when the four-channel shank gradually enters the hippocampal CA1 region, Ch4 passes through the soma layer first and records spikes, followed by the Ch3, 2, and 1. Spike signals should appear on these channels sequentially in this order. However, when Ch1 was faulty as in this case, spikes appeared simultaneously on Ch3 and 1 when Ch3 entered the soma layer. Later, when the electrode was lowered to position Ch1 in the soma layer, no spikes were recorded because Ch3—the source of crosstalk—was already too far from the soma layer. Notably, this example shows that signal crosstalk can even occur between seemingly "unrelated" channels that are far apart, rather than between adjacent recording channels. This occurred only because the pin arrangement at the electrode connector differed from the channel (contact) arrangement at the electrode tip (Fig. 3.23B).

These above examples of interference represent just a few cases we encountered in our experiments. Noises are diverse and can originate from multiple sources, such as improper experimental procedures, amplifier equipment, and inadequate shielding measures. After establishing an experiment setup, simple tests can help identify and eliminate different types of noises, which is important for accurate neural signal recordings.

3.5.3 PowerLab Recorder and LabChart Software

After amplification, neural signals undergo ADC and subsequent analysis and processing. As shown in Fig. 3.17, we used a PowerLab recorder for signal sampling. PowerLab recorders (http://www.adinstruments.com/) are widely used in research and teaching across fields like physiology, pharmacology, pathophysiology, and neuroscience. The recorder hardware mainly comprises a CPU microprocessor, memory, analog signal amplifiers, ADC, digital-to-analog converters (DAC) and other components.

As shown in Fig. 3.24A, we used the Model PL3516 recorder, which has 16 signal channels, each equipped with its own programmable amplifier. Through the LabChart software, these amplifiers can be configured independently for settings such as gain, filter bandwidth, input coupling mode (AC or DC), sampling frequency, and processing modes. The recorder automatically grounds the input amplifiers of idle channels. Once a gain setting is changed, the amplifier is grounded to measure DC offset, enabling the software to automatically correct any DC drift and offset. The recorder contains two 16-bit DACs that can output analog voltage signals via the Output1 and Output2 ports on the panel, which can be used to trigger stimulators and other instruments. Additionally, the recorder features 8-bit digital input and output ports.

As shown in Fig. 3.24B, the recorder has LabChart software for powerful data display and analysis. We used LabChart Pro version 8, which consisted of core, module and extension components. The core components provide multi-channel recording, display, and analysis for up to 32 channels. The modules provide specialized tools such as peak analysis, spike histogram, ECG analysis, blood pressure analysis, and video capture. The extension component enables data export in various formats, including MATLAB and binary coding.

A notable advantage of LabChart is that it can be independently installed and run on computers not connected to PowerLab hardware. This is beneficial for offline analysis and processing of experimental data. Additionally, its Help menu is exceptionally comprehensive, offering both operational guidance and clear explanations of signal processing methods and principles. These features make LabChart an effective teaching tool. When using PowerLab and LabChart, consider the following points.

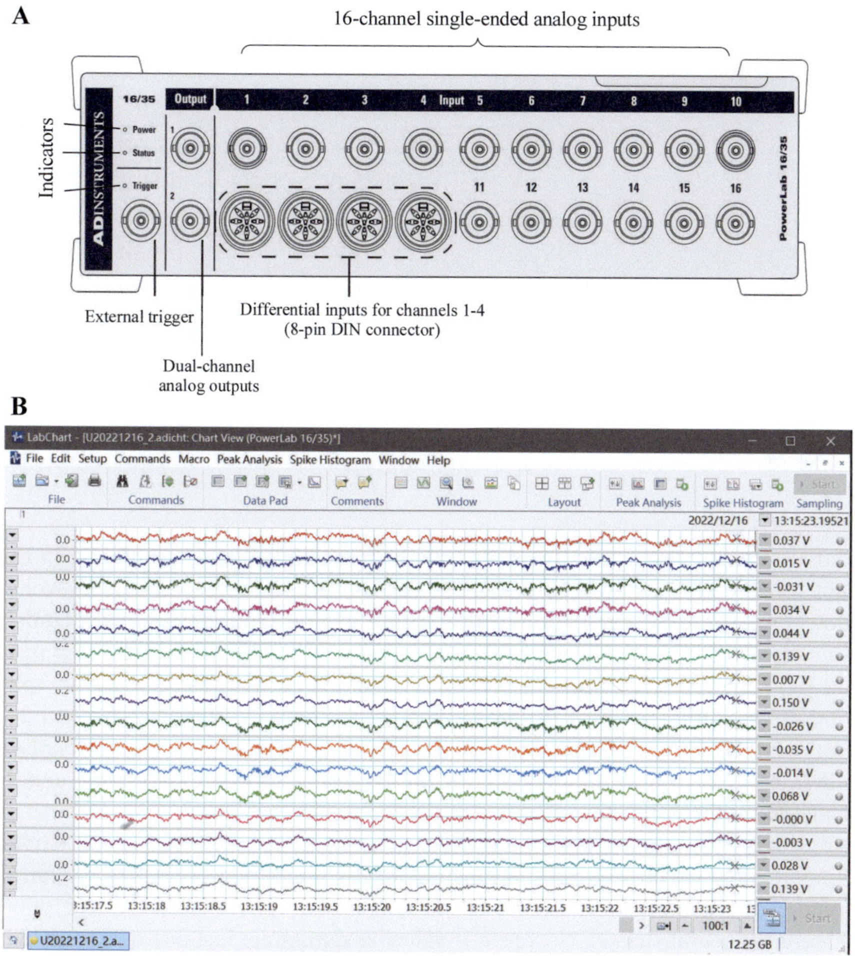

Fig. 3.24 Front panel of the PowerLab PL3516 recorder (**A**) and its LabChart software (**B**)

1. **Recording Ranges and Modes**

The amplification gain of the PowerLab recorder is software-controlled for each channel, with a maximum gain of 2000. The recorder can directly amplify and sample many types of signals without additional preamplifiers. Its gain is not presented as an amplification factor, but as an input range (or measurement range)—spanning from $\pm$ 2 mV to $\pm$ 10 V in 12 options. Users can set a desired gain by choosing an appropriate range.

The recorder features a 16-bit ADC with sampling resolution determined by the measurement range setting. For example, when the range is set to $\pm$ 10 V, the sampling resolution is 2×10 V/2^{16} = 305 μV. Table 3.1 lists the resolution and error for each range, based on the maximum value of a 16-bit binary number (approximately

Table 3.1 ADC resolutions and errors in PowerLab recorder at various measurement ranges

Range	ADC resolution	Error (RMS)
± 10 V	312.5 μV	1LSB
± 5 V	156.25 μV	1LSB
± 2 V	62.5 μV	1.5LSB
± 1 V	31.25 μV	1LSB
± 0.5 V	15.625 μV	1LSB
± 0.2 V	6.25 μV	1.5LSB
± 0.1 V	3.125 μV	1.5LSB
± 50 mV	1.56 μV	2LSB
± 20 mV	625nV	2.4 μV
± 10 mV	312.5nV	2.4 μV
± 5 mV	156.25nV	2.2 μV
± 2 mV	62.5nV	2.2 μV

64,000). Representing gains by ranges can help maximize ADC resolution. In the LabChart display window (Fig. 3.25), the areas outside the range are shaded, which allows users to easily assess amplification settings. The figure shows a signal recorded in rat hippocampal region during the application of a biphasic pulse train (0.4 mA, 200 Hz) with a measurement range of ± 2 V. The signal was displayed in two split windows with different time scales: a ~ 1.2 s segment in the right window and a ~ 6.5 ms enlarged segment in the left window which revealed clipped stimulus pulses within the ± 2 V range. Before entering the PowerLab recorder, this signal underwent 100-fold amplification, making an actual input range of ± 20 mV. This range setting was appropriate since we aimed to measure neural responses to the stimulation, treating the clipped pulse artifacts as negligible. While expanding the range would have captured the full pulse artifacts, it would have reduced the resolution of recorded neural signals. However, if valuable signals are lost due to insufficient range, immediate adjustment is necessary.

Should we always choose the maximum possible range with enough margin to prevent signal clipping? The answer is no. A larger range can result in a lower

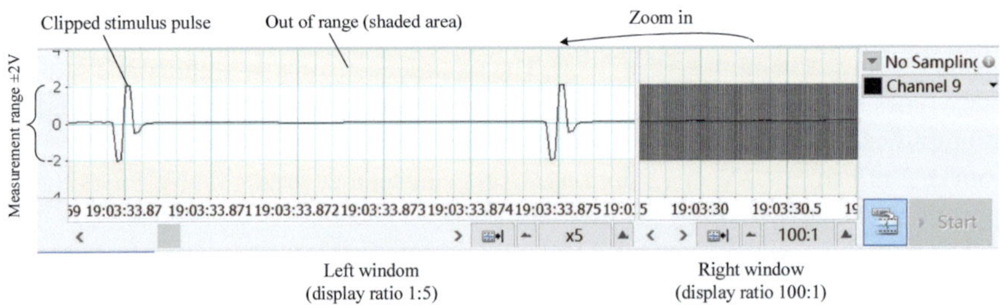

Fig. 3.25 Measurement range of a recording channel shown in the LabChart display windows

ADC resolution and poorer accuracy (Table 3.1). When PowerLab converts analog signals to digital, the amplitude can only take on a finite number of discrete steps. A 16-bit ADC divides the measurement range into 2^{16} (65,536) steps. LabChart uses the middle 64,000 steps as the nominal range, treating the extremes as out-of-range values. For a ± 10 V range, the minimum resolvable voltage is 0.3125 mV. However, if the range is set to ± 10 mV, the minimum resolvable voltage becomes 0.3125 μV (Table 3.1)—a 1000-fold improvement in ADC resolution.

For example, the amplitude of normal LFP signals ranges approximately ± 1 mV. Under anesthesia, slow waves in LFPs can be slightly larger but typically not beyond ± 2 mV. With a pre-amplifier gain of 100-fold, the input signal to PowerLab recorder ranges within ± 0.2 V. A setting of this range results in an ADC resolution of 6.25 μV (Table 3.1). When simultaneously recording LFPs and unit spikes, small spikes—such as those with an amplitude of only ~ 50 μV (5 mV after 100-fold pre-amplification)—can occupy 800 ADC discrete steps (5 mV/6.25 μV = 800). This produces a sufficient sampling precision. However, when simultaneously recording large evoked population spikes (PS) with amplitudes > 10 mV, the PowerLab range must be set to ± 2 V, resulting in ~ 50 μV spikes occupying only 80 ADC steps—a lower precision.

As shown in Fig. 3.26A, a signal can be displayed as a step curve to reveal its discrete sampling data by selecting "Use Step Interpolation" from the dropdown menu in the LabChart window. In contrast, the conventional display uses "Linear Interpolation" mode, showing a smooth curve after processing (Fig. 3.26B). Note that the dropdown menu shows the alternative mode available for selection, opposite to the current display mode.

In summary, the measurement range (i.e., amplification gain) must be set carefully. If the recorded signal approaches the shaded area and shows clipping in the LabChart display, the range must be increased. Conversely, if the signal consistently occupies only a small portion of the range, far from the shaded area, the range should be decreased to improve ADC resolution. For optimal ADC resolution, use the minimum possible range as long as signals of interest are not clipped.

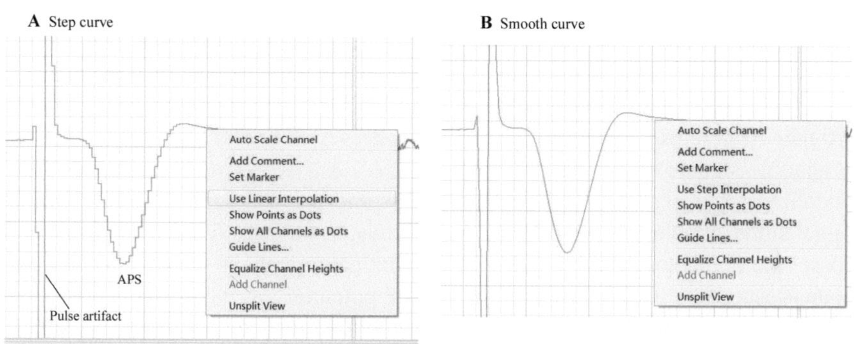

Fig. 3.26 Two display modes in LabChart—step curve (**A**) and smooth curve (**B**)

The input signals to the PowerLab recorder cannot exceed ± 15 V in amplitude. Its input impedance of about 1MΩ (10^6 Ω) and 100pF poses another limitation. When using microelectrodes to record electrophysiological signals—even metal wires with relatively low impedance for extracellular recordings—the electrode impedance is typically too high to connect directly to such a recorder. Therefore, signals from microelectrodes must first pass through a pre-amplifier with sufficiently high input impedance before connecting to the recorder. For example, the above-mentioned Model 3600 amplifier has an input impedance of about 10^{12} Ω and 50pF, making it suitable for amplifying microelectrode signals.

The amplifier parameters of the PowerLab recorder can be set in LabChart. The cutoff frequency of its low-pass filter can be set between 1 Hz and 25 kHz. Its inputs have coupling options of direct current (DC) and alternating current (AC), with AC coupling having a low-frequency cutoff of 0.15 Hz. Its frequency response (− 3 dB frequency) reaches 25 kHz for the maximum input range of ± 10 V. Essentially, each channel's input connects a 25 kHz low-pass filter. In differential input mode (Fig. 3.24A), the common-mode rejection ratio (CMRR) at 100 Hz is 100 dB. The input crosstalk is 75 dB. The root mean square (RMS) of input noise is less than 350 μV at ± 10 V input range. The maximum sampling rate varies with the number of recording channels—400 kHz for 1–2 channels, decreasing to 20 kHz per channel for 9–16 channels.

The front panel of the PowerLab recorder features two analog output channels—Output1 and Output2 (Fig. 3.24A). They can function either as two independent single-ended outputs or together as a differential output. The recorder uses a 16-bit DAC with a maximum output current of ± 50 mA and an output impedance of about 0.5Ω. Through the LabChart software, users can set the output voltage range to one of six options: ± 10, ± 5, ± 2, ± 1, ± 0.5, and ± 0.2 V. Like the ADC, the DAC resolution varies with the output range setting. Users can configure the voltage waveforms of these analog outputs in the LabChart "Stimulator Dialog" window, which offers positive, negative, differential, and independent outputs. By default, Output1 is set as positive and Output2 as negative. When operating in the differential mode, the output between the two terminals doubles the amplitude of each individual output.

The recorder offers two recording modes: continuous and trigger. The continuous mode records signals throughout the entire recording period for later analysis. The trigger mode only captures desired data, initiated by a specific signal from the "Trigger" port located on the front panel (Fig. 3.24A). This "Trigger" accepts TTL-compatible signal, with trigger thresholds of + 1.3 V for rising edge and + 1.1 V for falling edge, featuring a 0.3 V hysteresis and a 50kΩ input impedance. The trigger signal should not exceed ± 12 V, and must last at least 5 μs. When the signal crosses the trigger threshold, a yellow light illuminates on the front panel. The trigger mode is especially useful for recording evoked potentials, as it can conserve storage space by avoiding the large data produced by continuous long recordings. A simple method to use the trigger mode is to connect the recorder's analog outputs (Output1 or Output2) to the "Trigger" inputs and configure the settings in the "Stimulator

Dialog" window to produce desired trigger signal. Alternatively, a pulse signal from an external stimulator can serve as the trigger signal.

2. **Signal Acquisition with Real-time Data Analysis in LabChart**

The LabChart software offers powerful features for display, calculation, and data analysis for both online and offline situations. As shown in Fig. 3.25, its display can be split into two windows, allowing real-time recording in one window while enabling replay and analysis in the other to quickly measure and assess recorded signals. During offline analysis, these split windows can also facilitate signal comparisons across different time periods. Figure 3.27 shows another example of recording signals from the rat hippocampal CA1 region. The top 16 channels recorded wide-band (0.3–5000 Hz) signals in real-time, while the bottom 16 channels showed their high-pass filtering (> 500 Hz) results from real-time calculations. This configuration allows simultaneous online observations of both low-frequency LFPs (top 16 channels) and high-frequency unit spikes (bottom 16 channels).

The LabChart display has powerful features. Users can easily adjust zoom levels on both vertical and horizontal axes, and customize channel color, grids, and other display attributes. During recording, users can add comments to mark experimental operations, events and other relevant information in real-time. Additionally, the settings can be saved as templates for future recordings.

3. **Data Converting and Down-Sampling in LabChart**

LabChart saves recorded documents in a unique format with the file extension ".adicht". These files can be converted to other formats, such as text (.txt) or MATLAB (.mat), using the "File|Export" menu. Although decimal places in the data can be customized, adding more decimals doesn't necessarily mean increasing data accuracy. For voltage values, the appropriate number of decimal places should be determined by the ADC resolution. Similarly, for time values, the decimal places should be determined by the sampling rate.

Additionally, LabChart offers a down-sampling function to reduce the sampling rate of recorded data. When exporting a file as text using the "File|Export" menu, users can set the down-sampling factor (denoted as m) in the "Down sample by" field of the "Export as Text" dialog box. Rather than simply selecting one data for every m samples, LabChart averages m data and outputs their average as a single data. This averaging process acts as low-pass filtering (or smoothing) to reshape the data frequency-band. Following the sampling theorem—which states that the sampling rate must be at least twice the highest frequency of the original signal (known as the Nyquist frequency)—LabChart first applies a low-pass filter to limit the signal's highest frequency. This ensures accurate down-sampling that complies with the Nyquist frequency requirement.

4. **Additional LabChart Functions**

LabChart offers two viewing modes: "Chart" and "Scope," accessible through its "Window" menu. "Chart" mimics a multi-channel paper tape recorder, while "Scope"

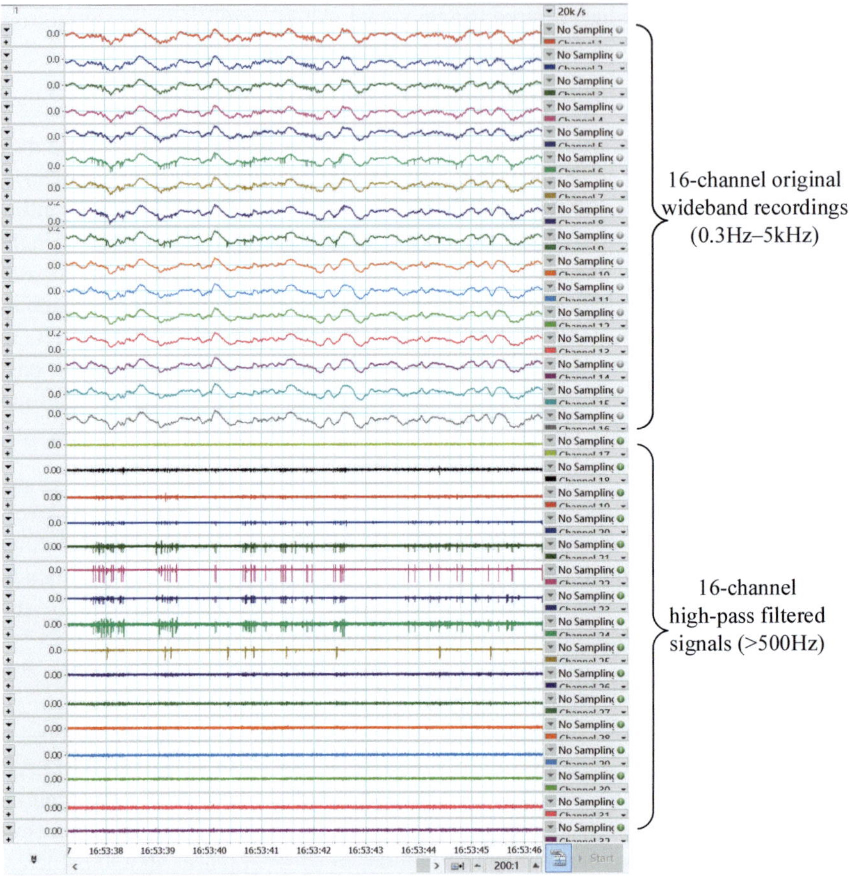

Fig. 3.27 Simultaneous displays of original signals (channels 1–16) and their real-time high-pass filtered outputs (channels 17–32) in the LabChart window

mimics an oscilloscope. These modes reflect early experimental setups, where oscilloscopes provided real-time signal observation and paper recorders and magnetic tape machines stored data. Inspired by these historical setups, LabChart developers incorporated both modes into the software and have maintained them throughout its development. It should be noticed, however, that while the older devices recorded continuous analog signals, PowerLab recorders capture digital signals. These digital signals are quantized in both amplitude and time by ADC, differing from analog signals.

Although most of our commonly used operations are in Chart mode, Scope mode offers specialized features for segmented signal display, allowing for superposition and averaging of signal segments. For example, to calculate an average waveform from 100 evoked potentials produced by pulse stimulations, simply use "Peak Analysis" to identify pulse artifacts, then let Scope to automatically align the 100 evoked potentials using these artifacts and calculate the average. Scope can display the results as superposed evoked potentials along with their average waveform, providing an intuitive comparison. Additionally, all calculation results can be exported in a text file.

"Peak Analysis" and "Spike Histogram" are two powerful LabChart modules. The Peak Analysis can automatically detect and analyze signal peaks in recordings—including unit spikes, population spikes (APS and OPS) and stimulus artifacts—by setting appropriate trigger thresholds in amplitude or slope. It can calculate multiple spike parameters, such as amplitude, width, slope and area. These measurements are useful for analyzing evoked potentials, including APS, OPS and fEPSP. The Spike Histogram can automatically identify unit spikes and create various histograms (refer to Chap. 4).

"Spectrum" menu in LabChart is a powerful tool that uses the periodogram method to estimate frequency spectra of selected signals, as detailed in Sect. 4.3. "Data Pad" is another useful tool for calculating various time and amplitude parameters. For users with programming expertise, LabChart offers macro commands to design and implement more complex processes for automated data detections and analyses. All these features are thoroughly documented in the LabChart Help guide.

3.5.4 Electrical Stimulators

The electrical stimulators used in our studies were from the A-M Systems company (http://www.a-msystems.com), including models 2100, 2200, 2300, and 3800. The Model 2100 stimulator is a single-channel pulse stimulator with electrical isolation. It can generate monophasic and biphasic pulses of voltage or current waves. The stimulator can be triggered manually or by an external electrical signal, and features a "Gate" input connector that allows an external signal to restrict its final output. Its front panel consists of three sections plus a power switch on the far right (Fig. 3.28). The left "START" section controls triggering mode; the middle "TIMING" section sets pulse parameters (delay, burst width, duration and period) using a 10 MHz internal clock; the right "OUTPUT" section provides an isolated output with controls for pulse polarity, baseline level and range. Here, users can configure output signals by selecting monophasic/biphasic and voltage/current pulses and setting pulse intensity.

As shown in Fig. 3.29, the Model 2300 stimulator is a single-channel pulse stimulus isolator, equivalent to the OUTPUT section of the Model 2100 stimulator (Fig. 3.28). The device uses built-in rechargeable batteries to produce electrically isolated output. Its front panel features a power and battery charging switch in the upper left corner. Note that stimulus signals cannot be output during charging. The panel provides options of voltage and current outputs, each with three different intensity ranges that work with the fine-adjustment knob below to set the amplitude of output pulses. The device itself cannot control pulse timing. Timing and polarity of the output pulses are controlled by external input signals through the 9-pin D-type connector in the lower left corner. The simplest control signal is a TTL-compatible

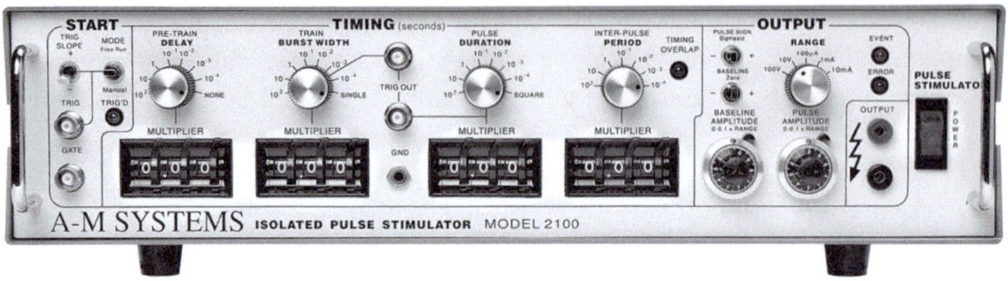

Fig. 3.28 Front panel of the Model 2100 isolated pulse stimulator

pulse to the connector's "On/Off" pin, which gates the output. In short, the 2300 stimulator provides voltage or current pulses at a set intensity but relies entirely on external signals for timing control.

As shown in Fig. 3.30, the Model 2200 stimulator resembles the Model 2300 in appearance and size. However, it is an analog stimulus isolator that can output various waveforms within its bandwidth range by following the control signal input through the "Signal in" BNC connector. In contrast, both the above-mentioned Model 2100 and 2300 are digital stimulators that can only output pulse waves. The front panel of Model 2200 allows users to select between voltage or current outputs with adjustable amplitude ranges. It also features a 9-pin D-type "Control" connector with a "Gate" pin to control the signal through the "Signal in". Its power switch and output connectors are similar to those of the Model 2300.

Figure 3.31A shows the Model 3800 eight-channel programmable stimulator. Its companion Model 3820 stimulus isolator (Fig. 3.31B) runs on rechargeable batteries. This versatile digital stimulator can produce both voltage and current pulse outputs. By connecting it to a computer via USB, users can easily configure all eight channels through software that enables flexible stimulation designs (Fig. 3.31C). It can produce monophasic and biphasic pulses with varying amplitudes and intervals. The stimulator can also create complex pulse sequences by combining up to eight programmed channels to a finally output through a single terminal.

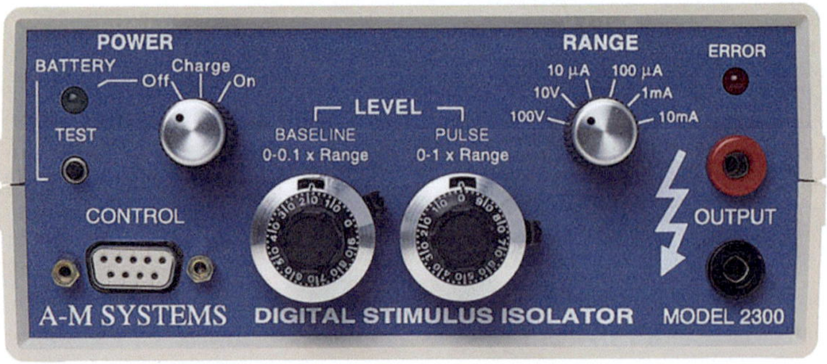

Fig. 3.29 Front panel of the Model 2300 digital stimulus isolator

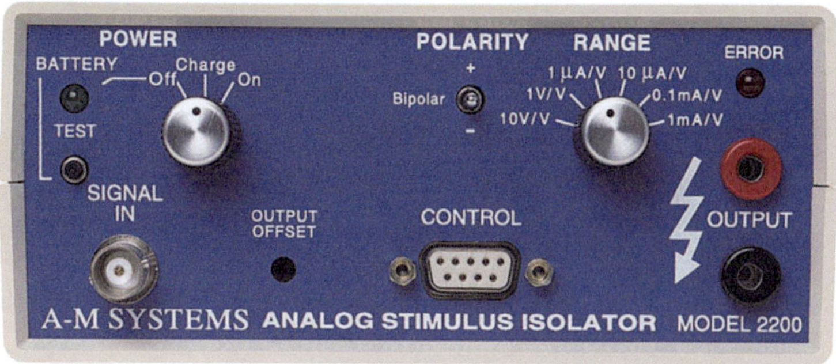

Fig. 3.30 Front panel of the Model 2200 analog stimulus isolator

Neural signals often need to be recorded simultaneously during electrical stimulations. To prevent interference, the stimulator output circuit must be electrically isolated from the recording circuit, avoiding a shared ground connection. All the above-mentioned stimulators feature such isolation. The Model 3800 stimulator is specially equipped with a 3820 isolator for each channel. These isolators run on rechargeable batteries.

Neural signals are extremely weak, typically requiring thousand-fold amplification before sampling and recording. The amplifiers have a small input range—exceeding this range can cause saturation or even damage. During extracellular electrical stimulation, the applied stimulus intensity far surpasses both neural signals and the amplifier input range. Therefore, stimulation isolation is critical to minimize interference with signal recordings. However, even with isolations, stimulus signals can still be collected by a recording electrode when the recording site is close to the stimulation site (only 1–2 mm in our experiments). This recorded stimulation waveforms, called stimulation artifacts, can have an amplitude much greater than neural signals. Before extracting and analyzing neural signals (such as unit spikes), the stimulation artifacts must be removed, as detailed in Sect. 4.5.

Off-the-shelf stimulators typically can only produce stimulus waveforms with constant or regularly varying parameters. They cannot generate arbitrary stimulation signals, such as pulse sequences with randomly varying intervals (varying frequencies), pulses with continuously varying amplitudes, and so on. These stimulators also cannot perform closed-loop stimulations. To resolve these limitations, we developed LabVIEW software for a data acquisition card to control a stimulator and create a stimulation system. It was able to output complex pulse sequences and arbitrary analog waveforms, and to perform closed-loop stimulations based on real-time recorded neural signals (Cai et al. 2018; Yang et al. 2021; Feng et al. 2012; Hu et al. 2015), as introduced briefly in the next section.

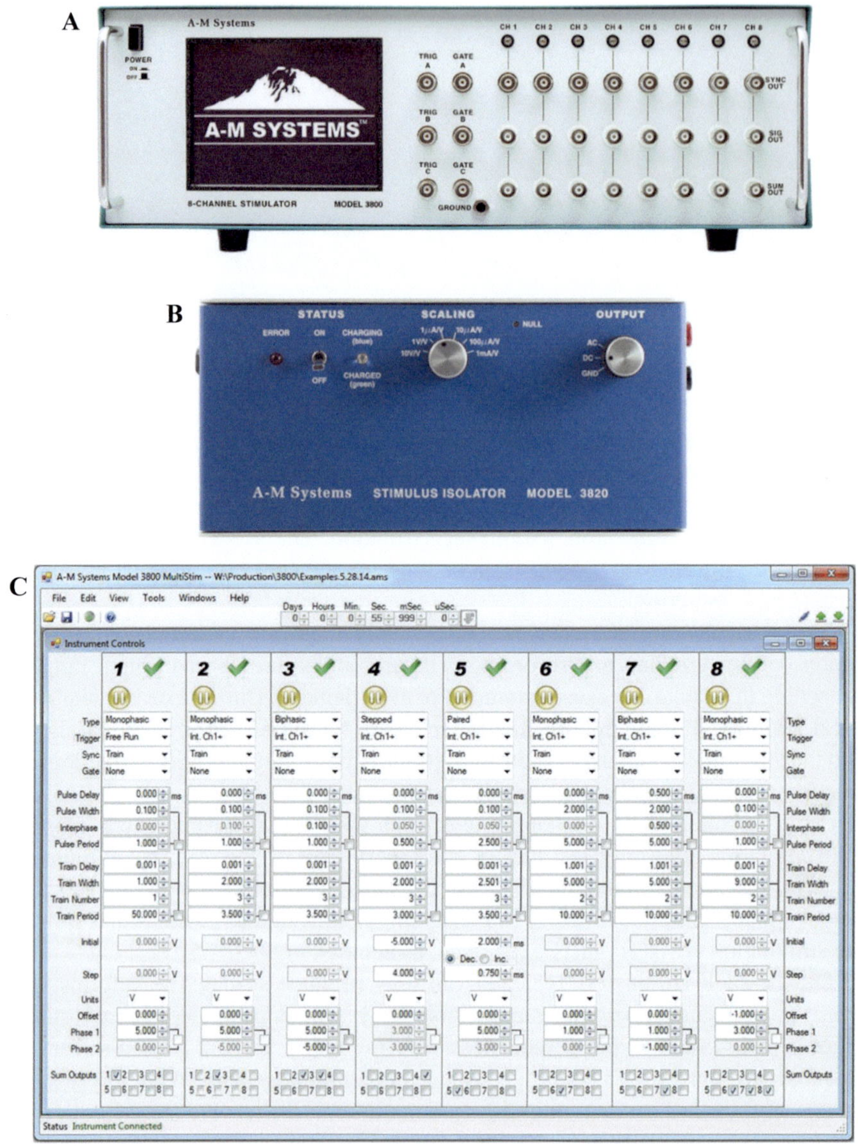

Fig. 3.31 Front panel of the Model 3800 eight-channel programmable stimulator (**A**), Model 3820 isolator (**B**), and the user interface (**C**)

3.5.5 LabVIEW-Based Electrical Stimulation System

LabVIEW, developed by National Instruments (NI), is a graphical programming tool. With an NI data acquisition card (DAQ), a customized LabVIEW program can create a virtual instrument for data acquisition and instrument control. As shown in Fig. 3.32, we designed a stimulation system comprised three devices: a personal computer (PC), a data acquisition card (USB-6251, National Instruments Inc., USA), and a Model 2200 stimulator.

The system operated as follows: a LabVIEW program controlled the DAC in the NI card to convert a digital signal into an analog one, creating the desired stimulation signal. This signal then fed into the "Signal Input" port of the 2200 stimulator, which output the final stimulation signal. Additionally, the LabVIEW program used a digital output port in the NI card to generate a TTL-compatible square wave, which connected to the stimulator "Gate" pin to control its output. When the Gate opened, the stimulator produced a stimulation signal precisely following the analog signal in the "Signal Input" port, but with an adjustable gain. To test the final stimulation current waveforms, a resistor could be connected across the stimulator output terminals to convert current into voltage (Fig. 3.32, right).

The LabVIEW program generated digital signals for desired stimulation waveforms (Yang et al. 2021). Figure 3.33 shows the method to produce a biphasic-pulse sequence with interphase gap (Fig. 3.33A). At a sampling rate of 20 kHz, the minimum pulse width of 0.05 ms met the common requirement for neuromodulation stimulation—typically above 0.06 ms (Merrill et al. 2005). The LabVIEW program created a data array for a biphasic-pulse sequence, with each element occupying a 0.05 ms duration and having a value proportional to the pulse voltage amplitude after DAC. "0"s (zero values) were inserted between adjacent phases to create the interphase gaps and pulse intervals (Fig. 3.33B). This array configuration can generate pulse sequences with varying pulse intervals (IPI), as shown below.

Figure 3.34 shows four types of IPI distributions in pulse sequences studied in brain stimulations: uniform, normal, gamma, and Poisson (Wyckhuys et al. 2010; McConnell et al. 2016; Birdno et al. 2008; Feng et al. 2019). The first three are continuous functions while the Poisson distribution is discrete, all generated using LabVIEW VIs for random numbers. These distributions are determined by specific

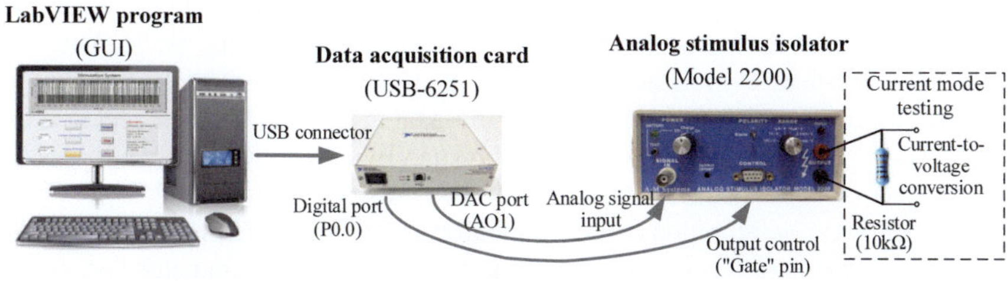

Fig. 3.32 LabVIEW-based electrical stimulation system for generating arbitrary waveforms

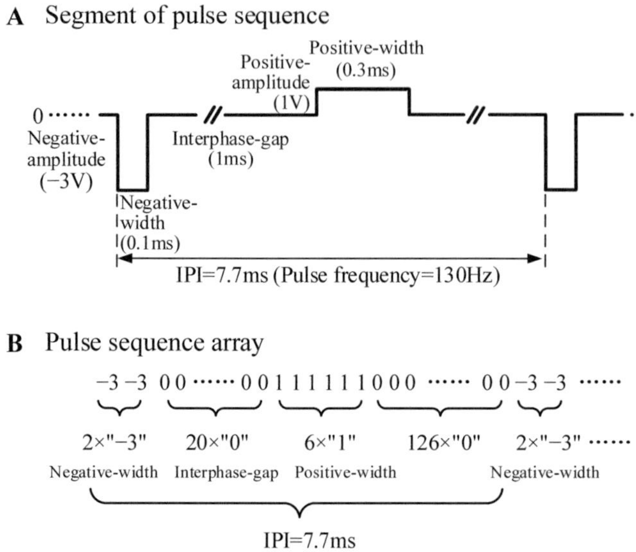

Fig. 3.33 Example of a LabVIEW data array for a biphasic-pulse sequence. A segment of pulse sequence (**A**) and its corresponding data array (**B**). From Yang et al. (2021)

parameters: uniform distribution using minimum and maximum values, normal distribution using mean and standard deviation (SD), gamma distribution using mean and coefficient of variation (CV), and Poisson distribution using mean. Note that except for the inherent discreteness of the Poisson distribution (Fig. 3.34D), the discrete points shown in the other distributions resulted from the limitation in array resolutions (Fig. 3.34B, C). Additionally, for our experimental requirements, we limited the IPI range to 1–500 ms.

The data array for a stimulation pulse sequence (Fig. 3.33B) can be generated internally by the LabVIEW program, or externally by other methods before being imported into LabVIEW. This flexibility allows users to design pulse sequences by creating text documents using common software such as Notepad, Excel or MATLAB. Beyond pulse sequences, this data array design enables the stimulation system shown in Fig. 3.32 to output any arbitrary stimulus waveforms.

Alternatively, for pure sine wave stimulation, a simple method is to use a general signal generator as the signal source for the 2200 analog stimulator, while using the "Output" ports of the PowerLab recorder (see Sect. 3.5.3) to provide a gate signal to control the stimulator output. Additionally, this analog stimulator can be replaced by a digital one (such as the 3800 stimulator shown in Fig. 3.17), if the user only needs pulse sequences with a constant intensity or a few discrete intensity steps.

The hardware configuration of the stimulation system shown in Fig. 3.32 is readily available, requiring only a general data acquisition card and a stimulator—standard equipment in most laboratories. This setup accommodates various stimulation needs, including constant and time-varying, digital pulse and analog signal stimulations. It

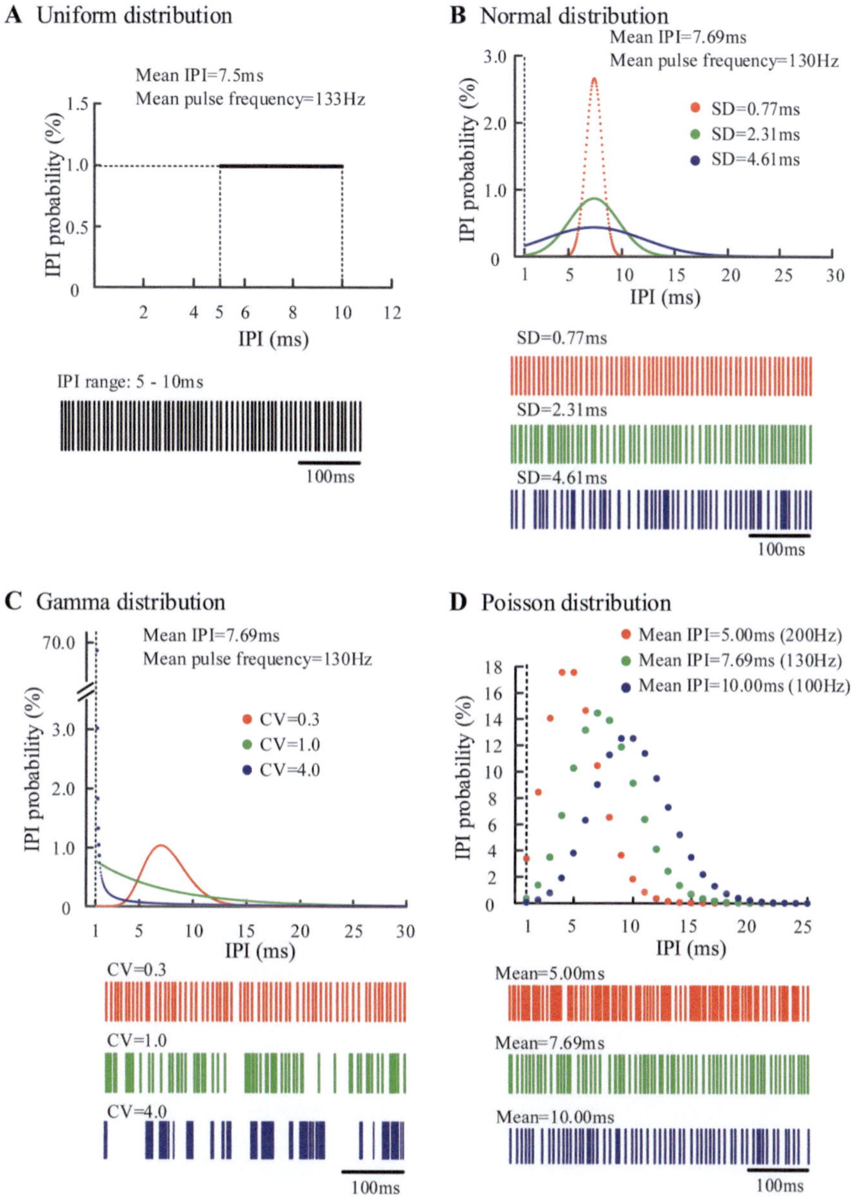

Fig. 3.34 Four distinct IPI distributions (*top*) with corresponding pulse sequence examples (*bottom*). **A** Uniform distribution. **B** Normal distribution. **C** Gamma distribution. **D** Poisson distribution. From Yang et al. (2021)

also enables the exploration and development of other new stimulation patterns, as shown in Part II of this book.

3.5.6 Closed-Loop Electrical Stimulation

Closed-loop stimulation represents a crucial advancement in neuromodulation therapy. This approach delivers electrical stimulation based on real-time brain activity, enhancing both treatment efficiency and patient safety. In epilepsy control, for instance, timely stimulation can suppress epileptiform activity more effectively when an impending seizure has been predicted through monitoring real-time brain neural signals (Sunderam et al. 2010; Smith et al. 2010). However, current off-the-shelf stimulators lack the ability to respond to neural signals for closed-loop stimulations.

We used the stimulation system shown in Fig. 3.32 to implement closed-loop stimulations (Feng et al. 2012; Hu et al. 2015). As shown in Fig. 3.35, LFP signals detected by the recording electrode (RE) in the rat hippocampal region were amplified and sampled via the ADC port in the NI card. The LabVIEW program performed real-time analysis of these signals. When an LFP signal met the preset criteria, one of the digital ports in the NI card sent a "Gate" signal to trigger the stimulator to deliver desired stimulation, which was then applied to a specific brain target via the stimulation electrode (SE). The dashed lines in Fig. 3.35 indicate the wiring for setup testing, with a PowerLab recorder simultaneously collecting neural signals (such as LFPs) and stimulation signals (either the stimulator's trigger signal or its output). Checking these recordings can verify whether the stimulation system is operating as required.

We previously used this system to apply pulse stimulations at the peak or trough of theta-rhythm waveforms in the rat hippocampus (Feng et al. 2012). The electrical activity of neuronal ensembles in brain can generate various rhythms in LFPs with amplitudes up to millivolt (Buzsáki 2006). Electrical stimulation applied at different

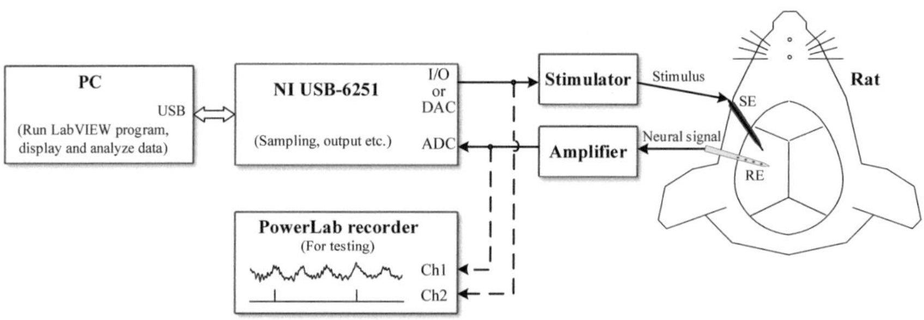

Fig. 3.35 Schematic diagram of a closed-loop stimulation system for rat experiments. From Feng et al. (2012)

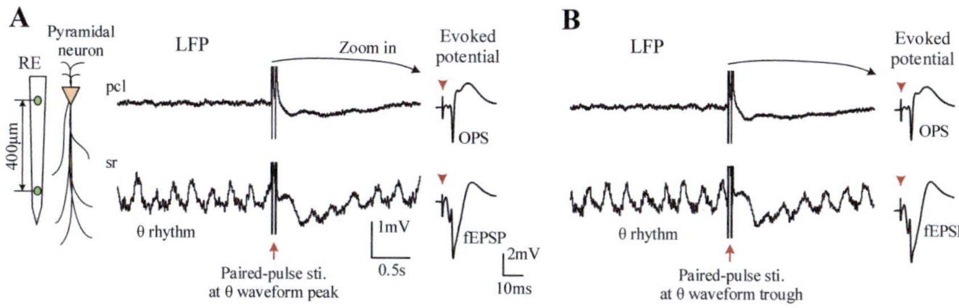

Fig. 3.36 Using closed-loop control to apply pulse stimulation respectively at the peak (**A**) and trough (**B**) of the LFP theta rhythms in the rat hippocampal CA1 region. From Feng et al. (2012)

rhythm phases may produce different effects. For example, pulses applied at the peaks versus troughs of hippocampal theta rhythms (2–7 Hz) can induce two distinct synaptic changes: long-term potentiation (LTP) and long-term depression (LTD), resulting in excitatory and inhibitory effects, respectively (Hyman et al. 2003; Huang and Kandel 2005; Huerta and Lisman 1996). To study the differences in evoked potentials at the peaks and troughs of theta rhythms, we designed a LabVIEW program to automatically identify theta rhythms and predict their upcoming peaks and troughs based on LFP amplitudes and rhythm cycles. Figure 3.36 shows typical recordings in the hippocampal soma layer (pcl) and apical dendritic layer (sr). We applied biphasic paired-pulse stimulation to the Schaffer collaterals in the rat hippocampal CA1 region, with a pulse intensity of 0.2 mA and a phase width of 100 μs. The clear theta rhythms in the sr layer before stimulation indicated that the pulses in Fig. 3.36A, B were respectively applied at a peak and a trough of theta waveforms. The enlarged evoked potentials from the first pulse show the OPS in the pcl layer and the fEPSP in the sr layer. The OPS amplitude induced at the theta peak was larger than that at the trough.

In addition, we used a threshold method to detect large epileptiform discharges, enabling closed-loop stimulation to suppress epileptiform activity (Hu et al. 2015). Figure 3.37A shows an example from our experimental results recorded in the rat hippocampal CA1 region. Once turned on, the closed-loop stimulation system automatically identified the initial PS of discharges and triggered the stimulator to deliver a 130 Hz pulse train to the Schaffer collaterals. This stimulation was able to effectively reduce the subsequent PSs in discharges.

The automatic detection in this closed-loop system can be switched on or off accordingly. When turned on, the LabVIEW program can detect the initial PS in discharges and deliver the preset pulse train with a desired delay. Once the initial PS has been detected, the detection function turns off until the pulse stimulation ends. The frequency and duration of the applied pulse train can be set as needed. Delivering stimulation before an epileptiform discharge fully develops can effectively suppress seizures. This approach can conserve stimulator power and reduce the risks of interfering with normal neurological functions and causing brain damage by continuous electrical stimulations. Using this closed-loop stimulation system, we studied the

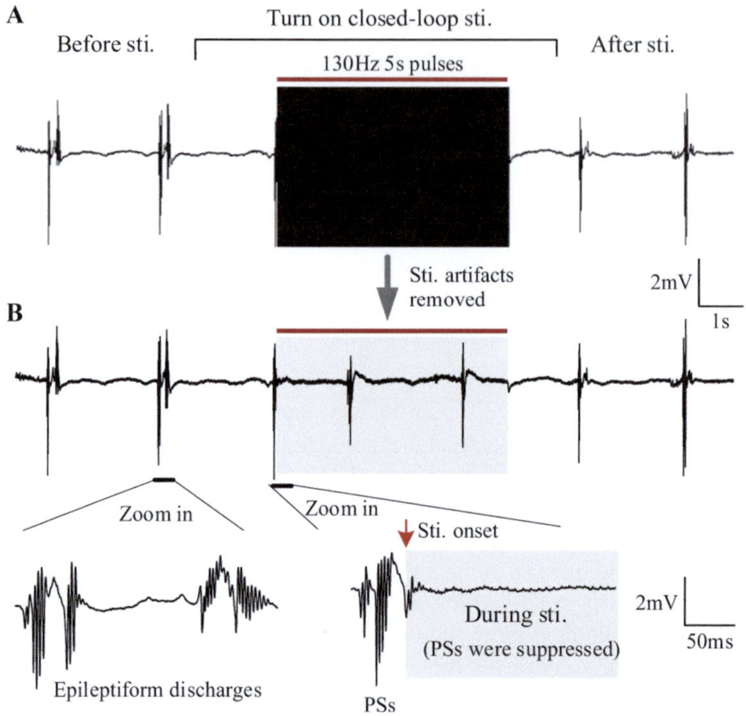

Fig. 3.37 Closed-loop high-frequency pulse stimulation applied in the rat hippocampal CA1 region to reduce epileptiform discharges. **A** A recording showing automatic PS detection and pulse train delivery controlled by a closed-loop stimulation system. The red bar indicates the stimulation period, during which stimulation artifacts obscure the neural signal in the display. **B** The recording signal after stimulation artifacts were removed, along with enlarged insets. From Hu et al. (2015)

effect of high-frequency pulse stimulation on epileptiform discharges (Cao et al. 2016), as detailed in Sect. 8.4.

3.6 Summary

This chapter covers animal surgery, electrode usage, and experimental setup for in-vivo rat neuro-electrophysiological studies. Precise electrode positioning is crucial for successful experiments. The chapter explains how to use rat stereotaxic instruments and calculate electrode implantation positions. It also details how to determine and adjust electrode placement during implantation by monitoring spontaneous and evoked neural signals, ensuring accurate positioning in target sites. In addition to common in-vivo method for hippocampal experiments, this chapter describes an alternative approach that exposes the dorsal hippocampus by removing the cerebral cortex, facilitating drug administration. We used this approach to confirm the generation of epileptiform activity when synaptic transmissions were blocked by lowering calcium concentration.

Additionally, the chapter introduces experimental devices, including NeuroNexus array electrodes (Michigan electrodes) for recording and FHC concentric bipolar electrodes for stimulating. It provides an example showing the abnormal neuronal response of spreading depression (SD), indicating potential tissue damage caused by improper use of stimulation electrodes. In addition, the chapter details the recording and stimulation setup used in our studies, including amplifiers, the recorder and its associated software, and electrical stimulators. Finally, it describes a custom LabVIEW-based stimulation system that can generate various stimulation signals and perform closed-loop stimulations.

These detailed experiment procedures, methods, and equipment guidelines provide essential guidance for researchers, especially newcomers to the field.

References

Andersen P, Soleng AF, Raastad M (2000) The hippocampal lamella hypothesis revisited. Brain Res 886(1–2):165–171

Bear MF, Connors BW, Paradiso MA (2016) Neuroscience: exploring the brain (4th ed.). Wolters Kluwer

Birdno MJ, Kuncel AM, Dorval AD et al (2008) Tremor varies as a function of the temporal regularity of deep brain stimulation. NeuroReport 19(5):599–602

Blanche TJ, Spacek MA, Hetke JF et al (2005) Polytrodes: high-density silicon electrode arrays for large-scale multiunit recording. J Neurophysiol 93(5):2987–3000

Bragin A, Csicsvári J, Penttonen M et al (1997a) Epileptic afterdischarge in the hippocampal–entorhinal system: current source density and unit studies. Neuroscience 76(4):1187–1203

Bragin A, Penttonen M, Buzsáki G (1997b) Termination of epileptic after discharge in the hippocampus. J Neurosci 17(7):2567–2579

Branner A, Stein RB, Normann RA (2001) Selective stimulation of cat sciatic nerve using an array of varying-length microelectrodes. J Neurophysiol 85(4):1585–1594

Bretschneider F, de Weille JR (2006) Introduction to Electrophysiological Methods and Instrumentation. Elsevier, Amsterdam

Bures J (1999) Leão's spreading depression of EEG activity. Brain Res Bull 50(5–6):459

Buzsáki G (2004) Large-scale recording of neuronal ensembles. Nat Neurosci 7(5):446–451

Buzsáki G (2006) Rhythms of the brain. Oxford University Press, Oxford

Cai Z, Feng Z, Hu H et al (2018) Design of a novel stimulation system with time-varying paradigms for investigating new modes of high frequency stimulation in brain. Biomed Eng Online 17(1):90

Campbell PK, Jones KE, Huber RJ et al (1991) A silicon-based, three-dimensional neural interface: manufacturing processes for an intracortical electrode array. IEEE Trans Biomed Eng 38(8):758–768

Cao J, Feng Z, Guo Z et al (2016) Mechanism of suppression of epileptiform burst by closed-loop electrical stimulation. Chin J Biomed Eng 35(1):79–87 (in Chinese)

Cartner SC, Barlow SC, Ness TJ (2007) Loss of cortical function in mice after decapitation, cervical dislocation, potassium chloride injection, and CO_2 inhalation. Comp Med 57(6):570–573

Cogan SF, Ludwig KA, Welle CG et al (2016) Tissue damage thresholds during therapeutic electrical stimulation. J Neural Eng 13(2):021001

Csicsvari J, Henze DA, Jamieson B et al (2003) Massively parallel recording of unit and local field potentials with silicon-based electrodes. J Neurophysiol 90(2):1314–1323

Dudek FE, Yasumura T, Rash JE (1998) 'Non-synaptic' mechanisms in seizures and epileptogenesis. Cell Biol Int 22(11–12):793–805

Erofeev A, Antifeev I, Bolshakova A et al (2022) In vivo penetrating microelectrodes for brain electrophysiology. Sensors (Basel) 22(23):9085

Feng Z, Chen D, Xiao Q (2012) Design of a novel closed-loop neural electrical stimulation system. Chinese Journal of Scientific Instrument 33(2):279–285 (in Chinese)

Feng Z, Durand DM (2003) Low-calcium epileptiform activity in the hippocampus in vivo. J Neurophysiol 90(4):2253–2260

Feng Z, Durand DM (2005a) Decrease in synaptic transmission can reverse the propagation direction of epileptiform activity in hippocampus in vivo. J Neurophysiol 93(3):1158–1164

Feng Z, Durand DM (2005b) Propagation of low calcium non-synaptic induced epileptiform activity to the contralateral hippocampus in vivo. Brain Res 1055(1):25–35

Feng Z, Durand DM (2006) Effects of potassium concentration on firing patterns of low-calcium epileptiform activity in anesthetized rat hippocampus: inducing of persistent spike activity. Epilepsia 47(4):727–736

Feng Z, Ma W, Wang Z et al (2019) Small changes in inter-pulse-intervals can cause synchronized neuronal firing during high-frequency stimulations in rat hippocampus. Front Neurosci 13:36

Gilbert Z, Mason X, Sebastian R et al (2023) A review of neurophysiological effects and efficiency of waveform parameters in deep brain stimulation. Clin Neurophysiol 152:93–111

Gorji A (2001) Spreading depression: a review of the clinical relevance. Brain Res Rev 38(1–2):33–60

Herreras O, Largo C, Ibarz JM et al (1994) Role of neuronal synchronizing mechanisms in the propagation of spreading depression in the in vivo hippocampus. J Neurosci 14(11):7087–7098

Herreras O, Makarova J (2020) Mechanisms of the negative potential associated with leão's spreading depolarization: a history of brain electrogenesis. J Cereb Blood Flow Metab 40(10):1934–1952

Hochberg LR, Serruya MD, Friehs GM et al (2006) Neuronal ensemble control of prosthetic devices by a human with tetraplegia. Nature 442(7099):164–171

Hu Z, Feng Z, Zheng X et al (2015) Design of a system for real-time seizure detection and closed-loop electrical stimulation. Sheng Wu Yi Xue Gong Cheng Xue Za Zhi 32(1):168–174 (in Chinese)

Huang YY, Kandel ER (2005) Theta frequency stimulation induces a local form of late phase LTP in the CA1 region of the hippocampus. Learn Mem 12(6):587–593

Huerta PT, Lisman JE (1996) Low-frequency stimulation at the troughs of theta-oscillation induces long-term depression of previously potentiated CA1 synapses. J Neurophysiol 75(2):877–884

Hyman JM, Wyble BP, Goyal V et al (2003) Stimulation in hippocampal region CA1 in behaving rats yields long-term potentiation when delivered to the peak of theta and long-term depression when delivered to the trough. J Neurosci 23(37):11725–11731

Jefferys JG (1995) Nonsynaptic modulation of neuronal activity in the brain: electric currents and extracellular ions. Physiol Rev 75(4):689–723

Jefferys JGR, Haas HL (1982) Synchronized bursting of CA1 hippocampal pyramidal cells in the absence of synaptic transmission. Nature 300(5891):448–450

Khalilov I, Esclapez M, Medina I et al (1997) A novel in vitro preparation: the intact hippocampal formation. Neuron 19(4):743–749

Khazipov R, Holmes GL (2003) Synchronization of kainate-induced epileptic activity via GABAergic inhibition in the superfused rat hippocampus in vivo. J Neurosci 23(12):5337–5341

Klausberger T, Somogyi P (2008) Neuronal diversity and temporal dynamics: the unity of hippocampal circuit operations. Science 321(5885):53–57

Kloosterman F, Peloquin P, Leung LS (2001) Apical and basal orthodromic population spikes in hippocampal CA1 in vivo show different origins and patterns of propagation. J Neurophysiol 86(5):2435–2444

Kraig RP, Nicholson C (1978) Extracellular ionic variations during spreading depression. Neuroscience 3(11):1045–1059

Krnjević K, Morris ME, Reiffenstein RJ (1980) Changes in extracellular Ca^{2+} and K^{+} activity accompanying hippocampal discharges. Can J Physiol Pharmacol 58(5):579–582

Kunkler PE, Kraig RP (2003) Hippocampal spreading depression bilaterally activates the caudal trigeminal nucleus in rodents. Hippocampus 13(7):835–844

Lauritzen M (1994) Pathophysiology of the migraine aura: the spreading depression theory. Brain 117(Pt 1):199–210

Leao AAP (1944) Spreading depression of activity in the cerebral cortex. J Neurophysiol 7(6):359–390

López-Aguado L, Ibarz JM, Varona P et al (2002) Structural inhomogeneities differentially modulate action currents and population spikes initiated in the axon or dendrites. J Neurophysiol 88(5):2809–2820

Martins-Ferreira H, Nedergaard M, Nicholson C (2000) Perspectives on spreading depression. Brain Res Rev 32(1):215–234

McConnell GC, So RQ, Grill WM (2016) Failure to suppress low-frequency neuronal oscillatory activity underlies the reduced effectiveness of random patterns of deep brain stimulation. J Neurophysiol 115(6):2791–2802

Merrill DR, Bikson M, Jefferys JGR (2005) Electrical stimulation of excitable tissue: Design of efficacious and safe protocols. J Neurosci Methods 141(2):171–198

Oka F, Sadeghian H, Yaseen MA et al (2022) Intracranial pressure spikes trigger spreading depolarizations. Brain 145(1):194–207

Parsons AA (2004) Cortical spreading depression: its role in migraine pathogenesis and possible therapeutic intervention strategies. Curr Pain Headache Rep 8(5):410–416

Paxinos G, Watson C (2007) The rat brain in stereotaxic coordinates. Academic

Pelkey KA, Chittajallu R, Craig MT et al (2017) Hippocampal GABAergic inhibitory interneurons. Physiol Rev 97(4):1619–1747

Richardson TL, Turner RW, Miller JJ (1987) Action-potential discharge in hippocampal CA1 pyramidal neurons: Current source-density analysis. J Neurophysiol, 58(5):981–996

Sceniak MP, MacIver MB (2006) Cellular actions of urethane on rat visual cortical neurons in vitro. J Neurophysiol 95(6):3865–3874

Schjetnan AGP, Luczak A (2011) Recording large-scale neuronal ensembles with silicon probes in the anesthetized rat. J vis Exp 56:e3282

Shannon RV (1992) A model of safe levels for electrical stimulation. IEEE Trans Biomed Eng 39(4):424–426

Shirasaka Y, Wasterlain CG (1995) The effect of urethane anesthesia on evoked potentials in dentate gyrus. Eur J Pharmacol 282(1–3):11–17

Smith JR, Fountas KN, Murro AM et al (2010) Closed-loop stimulation in the control of focal epilepsy of insular origin. Stereotact Funct Neurosurg 88(5):281–287

Somjen GG, Giacchino JL (1985) Potassium and calcium concentrations in interstitial fluid of hippocampal formation during paroxysmal responses. J Neurophysiol 53(4):1098–1108

Su H, Alroy G, Kirson ED et al (2001) Extracellular calcium modulates persistent sodium current-dependent burst-firing in hippocampal pyramidal neurons. J Neurosci 21(12):4173–4182

Sunderam S, Gluckman B, Reato D et al (2010) Toward rational design of electrical stimulation strategies for epilepsy control. Epilepsy Behav 17(1):6–22

Taylor CP, Dudek FE (1982) Synchronous neural afterdischarges in rat hippocampal slices without active chemical synapses. Science 218(4574):810–812

van Hoeymissen E, Philippaert K, Vennekens R et al (2020) Horizontal hippocampal slices of the mouse brain. J vis Exp 163:e61753

Wu D, Xie K, Lin L (2007) Effect of two anesthetics on activity of hippocampal neurons in mice. Laboratory Animal and Comparative Medicine 27(2):81–85 (in Chinese)

Wyckhuys T, Boon P, Raedt R et al (2010) Suppression of hippocampal epileptic seizures in the kainate rat by Poisson distributed stimulation. Epilepsia 51(11):2297–2304

Yang G, Feng Z, Zheng L et al (2021) Design and application of neural electrical stimulation system with time-varying parameters. Sheng Wu Yi Xue Gong Cheng Xue Za Zhi 38(6):1144–1153 (in Chinese)

Yuan Y, Zheng L, Feng Z et al (2021) Different effects of monophasic pulses and biphasic pulses applied by a bipolar stimulation electrode in the rat hippocampal CA1 region. Biomed Eng Online 20(1):25

Chapter 4
Neural Signal Recording and Processing

Abstract This chapter introduces methods for detecting and analyzing extracellularly recorded action potentials from individual neurons, also known as unit spikes. It describes the settings of appropriate recording frequency band, sampling rate and resolution for spike recordings. It introduces window-based detection and multiple-channel sorting algorithms for spikes, along with analysis methods for spike sequences (including inter-spike-interval histogram, autocorrelation histogram, cross-correlation histogram, and so on). The chapter also covers spectrum analysis methods for local field potentials and EEG signals. Finally, it presents the current source density analysis for tracking transmembrane current during neuronal activation and provides methods for removing stimulation artifacts from both sinusoidal and pulse electrical stimulations.

Keywords Unit spike · Spike waveform · Spike detection · Multi-channel sorting · Spike sequence analysis · Spectral analysis · Current source density · Artifact removal

The brain is the most complex structure and functional organization among living organisms. The human brain contains a hundred-billion of neurons, each forming connections with other neurons through numerous synapses—up to thousands per neuron. These synapses create extremely complex neuronal networks that can rapidly transmit and process neural information. Any brain response to external events requires information processing across numerous neurons. To reveal these complex working mechanisms, we need to collect adequate data on neuronal activity in living brains. Action potentials of neuronal firing carry vital neural information. Detecting and analyzing action potential sequences from individual neurons and neuronal ensembles form the basis for deciphering how the brain works (Hochberg et al. 2006).

This chapter introduces methods for recording and processing neural electrical signals. It covers the detection and analysis of extracellularly recorded action potentials (unit spikes), as well as power spectrum analysis and current density analysis for local field potentials. It also describes methods for removing stimulus artifacts to

Z. Feng, *Electrical Brain Stimulation: Mechanisms and Modulation of Neuronal Activity*, https://doi.org/10.1007/978-981-95-4145-4_4

extract neural signals during electrical stimulations. These methods are essential for studying neuromodulation.

4.1 Detecting and Analyzing Unit Spikes

When a neuron receives excitatory inputs or stimuli that depolarize its membrane to firing threshold, it can generate an action potential (AP). Figure 4.1A illustrates both intracellular and extracellular methods for measuring APs from individual neurons, which yield distinct potential waveforms. For distinction, we refer to the intracellular waveform as an AP and the extracellular waveform as a unit spike—or simply a spike.

Intracellular recording measures the potential difference across the neuronal membrane. The AP, recorded by a glass pipette microelectrode positioned inside the membrane, is a sharp positive-going waveform with its rising phase primarily caused by Na^+ influx (Fig. 4.1B). This Na^+ influx simultaneously reduces the positive charges outside the membrane to create a current "sink" nearby. As a result, the corresponding spike recorded extracellularly exhibits a sharp negative-going waveform (Fig. 4.1C), opposite to the intracellular recording. In theory, using a point source model, the spike waveform can be inferred as the negative first-derivative of the AP waveform (Henze et al. 2000). However, real experimental measurements have shown that this relationship only holds during the initial period of AP rising

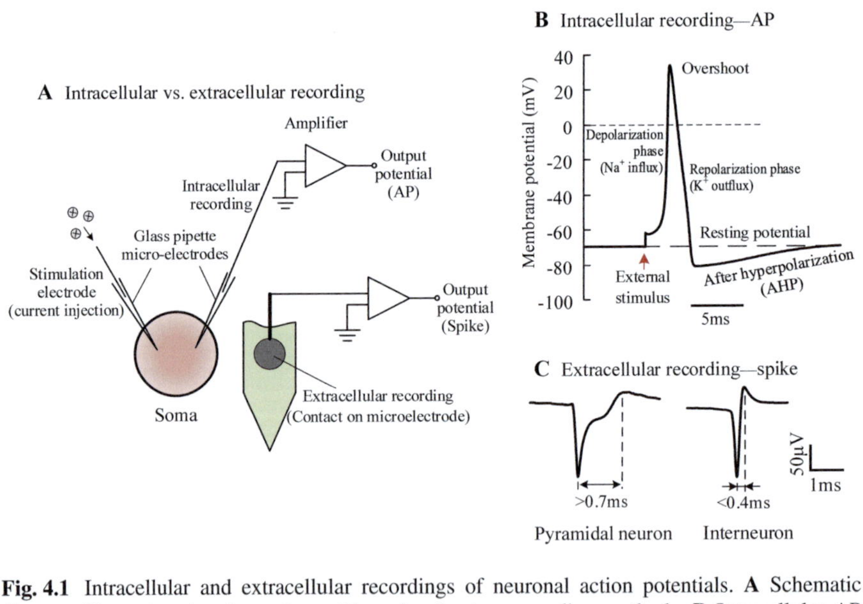

Fig. 4.1 Intracellular and extracellular recordings of neuronal action potentials. **A** Schematic diagram illustrating the electrode positions for the two recording methods. **B** Intracellular AP waveform simulated using the HH model (see Sect. 1.3). **C** Typical spike waveforms recorded extracellularly from a pyramidal neuron and an interneuron in the rat hippocampal CA1 region

phase when the Na^+ influx stems from a small area near the soma, making the point source model fairly accurate. Quickly, the ionic currents generated from other parts of the neuron—such as dendrites—can distort this first-derivative approximation (Somogyvári et al. 2005).

The diverse morphological structures of neurons result in various spike waveforms (Henze et al. 2000; Barthó et al. 2004; Huberfeld et al. 2011; Csicsvari et al. 1998). The waveform differences enable the identification of neuronal types based on their spikes. In a wideband signal recording with a frequency range from several to thousands of hertz, the rising phase width of a pyramidal neuron spike (> 0.7 ms) is significantly wider than that of an interneuron (< 0.4 ms) (Fig. 4.1C). This width index has become one of the best indices for distinguishing between the two neuron types (Barthó et al. 2004). Obviously, the extracellular recording site can significantly affect both spike amplitude and waveshape. Spikes recorded near soma have a dominant negative peak due to Na^+ influx during the initial action potential phase. In contrast, spikes recorded near dendrites exhibit a dominant positive peak due to the current outflux in its return loop (Pettersen et al. 2012; Obien et al. 2015). Furthermore, the larger soma membrane area can concentrate a greater Na^+ flow, resulting in a larger spike, while the return current around dendrites is more dispersed, producing a smaller spike. Therefore, larger spikes typically have a negative peak, while spikes with a positive peak are usually smaller.

The amplitudes of intracellular and extracellular recordings are significantly different. Intracellular AP can exceed 100 mV in amplitude, while extracellular spikes are much smaller, mostly around 100 μV or below. (Note: unless specified, spike amplitude refers to the peak-to-peak amplitude in this book.) Spike amplitude decreases rapidly and exponentially as the distance between the recording site and the neuronal membrane increases. Neurons within a ~ 100 μm radius around a recording point typically produce spikes with amplitudes ranging from tens to hundreds of microvolts (Buzsáki 2004; Henze et al. 2000; Obien et al. 2015). While an expanded radius contains more neurons, the spikes from these additional neurons have diminished amplitudes. These tiny spikes are difficult to detect and simply contribute to background activity, often considered as "noise". To record spikes with sufficiently large amplitudes, the electrode contact must be close to neuronal somata. In our experiments in the rat hippocampal region, we occasionally recorded spikes exceeding 1 mV in amplitude, primarily from pyramidal neurons. Besides distance and location, other factors can affect spike amplitude and waveform, including neuronal size and morphology, dendritic structures, ion channel distribution and density on membranes (Gold et al. 2006, 2007), as well as the size and features of electrode contacts.

Neurons generate action potentials with a well-known characteristic of "all-or-none". This means that all action potentials from the same neuron have the same amplitude, regardless of their triggers. Furthermore, the action potentials can travel outward along axons without losing strength. Neurons can encode information by varying their firing rate (frequency)—the average number of action potentials per unit time, known as frequency encoding. Other encoding modes include time, phase, and spatial encoding. All the encoding modes rely on the action potential timing rather than its amplitude or waveform. With rapid advancements in recording techniques like array microelectrodes, ADC and computer data storage, the number of neurons that can be recorded simultaneously has grown exponentially. This growth has driven continuous development of new analysis and processing methods for spike signals (Stevenson and Kording 2011; Horváth et al. 2021; Yuan et al. 2021; Buzsáki 2004; Carlson and Carin 2019; Pachitariu et al. 2024).

Due to the "all-or-none" characteristic, the amplitudes of intracellular APs from the same neuron are identical. However, the amplitudes of extracellular spikes can vary due to the following factors. (1) An electrode contact can pick up signals from many surrounding neurons locating at different distances. Nearby neurons produce large spikes with their waveforms distorted by the background noise—the numerous indistinguishable small spikes formed by the firing of distant neurons. (2) Electronic thermal noise from recording electrodes, amplifiers, and signal acquisition devices can alter recorded spikes. (3) Spikes themselves may exhibit varying waveforms during brief periods of intense firing, such as bursts. (4) Low-frequency influences from local field potentials and other potential drifts may affect spike recordings. (5) Changes in the relative positions between the electrode and recorded neuron over time, especially during long-term recordings, can affect recorded spikes.

Among these factors, the first and second are belong to noise interferences. Spikes are small in amplitude and short in duration, typically lasting no more than 2–3 ms. Their waveforms contain high-frequency components, mainly in the range of 300–5000 Hz. As a result, spike recordings are susceptible to various high-frequency noises, like electronic thermal noises and background noises of indistinguishable spikes. The amplitude of such noises usually ranges in 5 to 20 μV (Maccione et al. 2009; Nelson et al. 2008; Bretschneider and de Weille 2006). Spike recording involves a trade-off. A wider frequency band is necessary to fully preserve the spike waveform. However, noise reduction requires narrowing the frequency band, particularly by eliminating high-frequency components. The third factor is the unique variation of burst spikes, which cannot be removed and must be correctly identified. The fourth factor—low-frequency interferences—can be filtered out. The final factor—changes in electrode position—needs to be tracked and determined.

Weak neural signals collected by microelectrodes must be amplified with a gain of hundreds to thousands. The amplified signals then undergo ADC and save as digital data for subsequent analysis and processing. In this process, it is crucial to correctly set the amplifier frequency range and gain, as well as the ADC sampling frequency and resolution. In the following sections, we will first examine the real spike waveforms without noise, then investigate how frequency range and ADC settings affect spike recordings, and finally study the methods for spike detecting and sorting.

4.1.1 *Wideband Noise-Free Spike Waveforms from Signal Averaging*

To obtain real spike waveforms, the signal amplification and acquisition system shown in Fig. 3.17 was set up as following. Set the model 3600 amplifier at a frequency range of 0.3 Hz–20 kHz, much larger than spike bandwidth. Set the PowerLab recorder at a sampling frequency of 100 kHz. For spike recordings, the low-frequency cutoff doesn't need to be as low as 0.3 Hz. However, during experiments, LFP signals often need to be recorded simultaneously, requiring the retention of low-frequency signals. Figure 4.2 displays a brief neural signal recorded in the rat hippocampal CA1 soma layer with the above settings. The first row shows the original signal, dominated by slow rhythms at about 0.7 Hz that overshadows small unit spikes. The second row reveals clear spikes after LFP removal by a 300–3000 Hz bandpass digital filter in the LabChart. The enlarged inset in the upper middle clearly shows high-frequency noises with a relatively large amplitude of about 50 μV under the frequency band up to 20 kHz. These noises are reduced significantly in the filtered signal shown below. The right insets are further enlarged in time scale, highlighting amplitude and waveform differences between the filtered and original spikes. Obviously, the filtering distorted spike waveforms simultaneously with noise removal. Next, we will obtain undistorted spikes using signal averaging.

Signal averaging is widely used for denoising. When the noise is random and the wanted signal is consistent and repeatable, averaging can effectively eliminate the random noise while obtaining the deterministic signal with an improved signal-to-noise ratio (SNR). This process involves aligning repeated signals at a specific time point and then calculating their average. The increase in SNR is proportional to the square root of the number of averaged signals. For example, an auditory evoked potential (AEP) in a single trial of scalp EEG recording is typically undetectable, buried within other irrelevant EEG signals and noises. However, a clear AEP can emerge after averaging the EEG episodes from hundreds of repeated auditory stimulations (Enderle and Bronzino 2012).

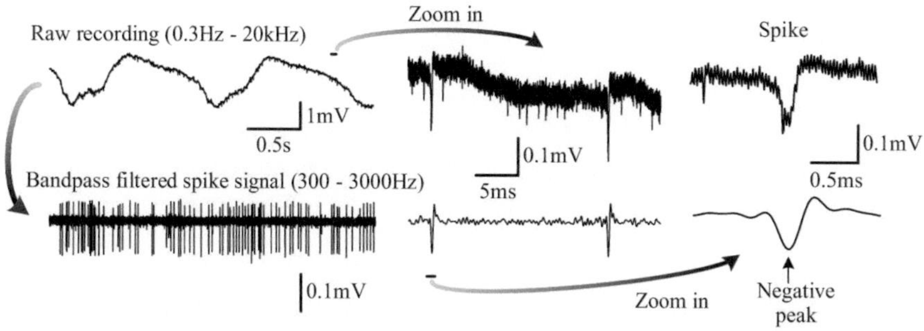

Fig. 4.2 Wideband signals (upper) recorded from the soma layer in rat hippocampal CA1 region and their bandpass filtered signals (lower). From Feng et al. (2012)

Based on the "all-or-none" characteristic of action potentials, spikes from an identical neuron can be considered as repeated deterministic signals. Note that changes in burst spikes are excluded here. Figure 4.3A displays the spikes extracted from a wideband recording of 0.3 Hz to 20 kHz, sampled at 100 kHz as the recording shown on the upper row of Fig. 4.2, and detected by using a threshold on the 300–3000 Hz bandpass filtered signal. Each spike waveform included 200 sampling points, covering 1 ms before and after the negative peak respectively. Total 1500 spikes (green waves) were superimposed in Fig. 4.3A, aligned with their negative peaks. Due to the shift of low-frequency field potentials, each spike had a potential shift. Zero-mean spike waveforms were obtained by subtracting their mean from individual spikes (Fig. 4.3B). The high-frequency noise in each zero-mean spike became evident, which was removed by signal averaging (Fig. 4.3C–F). We quantified the denoising effect using the coefficient of variation (CV) among spike waveforms.

Let x_{ij} be the jth sampling point of the ith spike waveform, M be the number of spikes, and x_j be the jth sampling point of the average waveform from$M$ spikes, then

$$x_j = \frac{1}{M}\sum_{i=1}^{M} x_{ij} \tag{4.1}$$

Let A_{PP} represent the peak-to-peak amplitude of the average waveform (x_j), and N be the length of spike waveform—the number of sampling points per spike, then

$$\mathrm{CV} = \frac{1}{A_{\mathrm{PP}}} \cdot \frac{1}{M}\sum_{i=1}^{M}\sqrt{\frac{\sum_{j=1}^{N}(x_{ij}-x_j)^2}{N-1}} \times 100\% \tag{4.2}$$

The CV is the percentage ratio of the mean spike standard deviation to the A_{PP}, representing the variability among spike waveforms. Note that the definitions of x_{ij} and x_j may vary accordingly in different CV calculations.

Due to the shifts in low-frequency field potentials (Fig. 4.3A), the CV of the 1500 original spikes reached 278% ($A_{\mathrm{PP}} = 181\ \mu\mathrm{V}$). After mean removal, the CV of zero-mean spikes dropped to 12% (Fig. 4.3B).

Using the CV index, we can examine how denoising efficiency changes with the number of averaged spikes. Averaging every 10 spikes from the total 1500 yielded 150 average spikes, reducing the CV to 3.6% (Fig. 4.3C). Additional noise reduction occurred when averaging every 100 and 500 spikes, respectively (Fig. 4.3D, E). The CV decreased rapidly as the number of averaged spikes increased (Fig. 4.3F). When more than 200 spikes were averaged, the CV fell below 1%, indicating that the obtained spikes can be treated as noise-free waveforms—real spikes. In the next section, we will use the wideband real spikes obtained through averaging to explore how frequency-band settings affect spike waveforms.

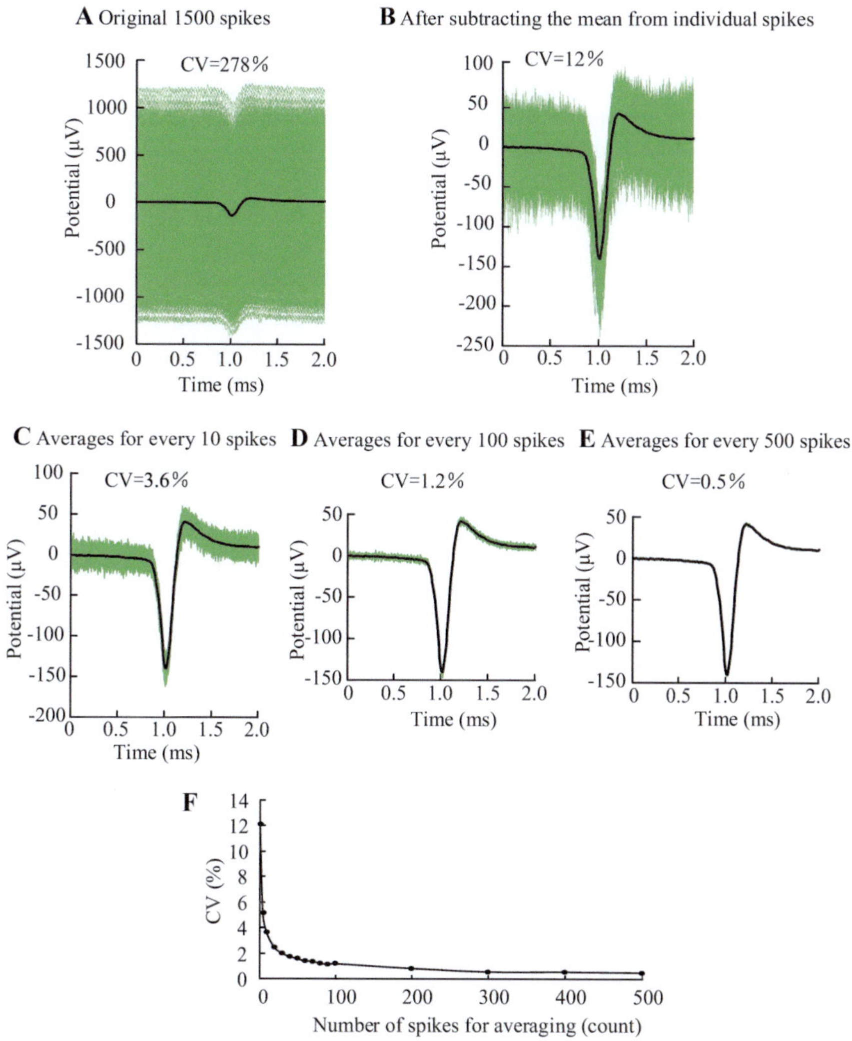

Fig. 4.3 Denoising spike waveforms through averaging. **A–E** Original wideband recordings of 1500 spikes (**A**); spikes after removing mean (**B**); and average waveforms for every 10 (**C**), 100 (**D**), and 500 (**E**) spikes. The black waveforms in (**A–E**) are the identical average spike calculated by averaging all 1500 zero-mean spikes. **F** Changes of the CV against the number of spikes for averaging

4.1.2 Effect of Frequency-Band Selection on Spike Recording

First, let's explore the selection of high-frequency cutoff (f_H). We set the amplifier's f_H respectively from high to low at 20, 10, 5, 3, and 1 kHz, while keeping the low-frequency cutoff (f_L) at 0.3 Hz and the sampling frequency at 100 kHz. Recordings were taken from the rat hippocampal CA1 pyramidal layer for about 2 min at each f_H setting. Figure 4.4A shows 10 ms recording samples and their corresponding spike SNR values under different f_H settings. The SNR was calculated as the ratio of the spike peak-to-peak amplitude (A_{PP}) to five times the noise standard deviation (Feng et al. 2012; Joshua et al. 2007):

$$\mathrm{SNR} = A_{PP}/(5 \times \text{noise standard deviation}) \tag{4.3}$$

The high-frequency noise decreased with f_H decreasing, leading to an improved SNR. However, when f_H decreased to 1 kHz, the SNR dropped again due to a substantial A_{PP} reduction.

The noise-free spike waveforms under different f_H settings were calculated from 1300 to 1800 averaged spikes. Figure 4.4B shows the noise-free spike waveforms from an interneuron (left) and a pyramidal neuron (right) recorded at five different f_H values. No significant waveform loss occurred with an f_H above 5 kHz. As f_H decreased, the waveform loss increased, especially at the negative peak of spikes. However, even when f_H dropped to 1 kHz, the repolarization phase after the "inflection point" in the pyramidal neuron's spike waveforms (Gold et al. 2007)—denoted

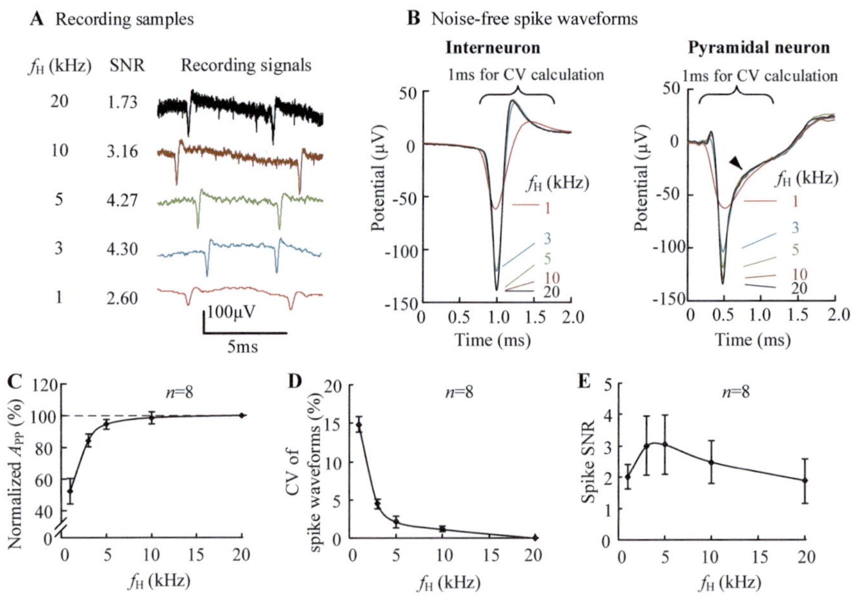

Fig. 4.4 Effects of high-frequency cutoff (f_H) on spike waveforms. **A** Representative signal episodes recorded at five different f_H settings from a same rat experiment, with their SNR values listed alongside. **B** Noise-free spike waveforms of two distinct neurons at different f_H settings. **C–E** Statistical data from eight rat experiments showing changes in the normalized A_{PP}—normalized by the A_{PP} at $f_H = 20$ kHz (**C**), CV (**D**), and SNR (**E**) against f_H. From Feng et al. (2012)

by a black arrow in the figure—retained mostly due to its relatively low frequency components.

The statistical results from eight rat experiments showed that the normalized spike amplitudes (A_{PP}) at different f_H were above 94% for $f_H \geq 5$ kHz (Fig. 4.4C). However, at $f_H = 1$ kHz, the normalized A_{PP} dropped to ~ 52%, representing approximately half loss. Using the CV index to evaluate the waveform changes at different f_H settings with the 20 kHz waveform (the real spike) as a reference (Eq. (4.2)), for $f_H \geq 5$ kHz, the CV remained below 2.5%, but increased to ~ 15% at $f_H = 1$ kHz (Fig. 4.4D). Notably, the spike SNR exhibited a non-monotonic change against f_H, peaking between 3 and 5 kHz (Fig. 4.4E). Note that the large standard deviations in SNRs (the error bars) resulted from substantial variations in spike A_{PP} values, which ranged from 90 to 205 μV across these rat experiments.

These results suggest that setting f_H at 5 kHz can maintain most high-frequency components in spike waveforms while achieving satisfactory spike SNR. Therefore, we typically set the amplifier's f_H at 5 kHz for spike recordings.

Second, let's explore the selection of the low-frequency cutoff (f_L). In wideband recordings, low-frequency field potentials can overshadow unit spikes (see Fig. 4.2, top left), making it difficult to detect spikes directly using a threshold method. Therefore, we typically filtered out low-frequency signals before spike detection. As shown in the bottom row of Fig. 4.2, a bandpass filter (300-3000 Hz) can remove low-frequency signals with the $f_L = 300$ Hz. Both the f_H and f_L cutoffs can be achieved by either analog filters in an amplifier or digital filters in software. We usually set f_H in the amplifier to eliminate high-frequency noise and avoid the need of a higher ADC sampling frequency, which could limit simultaneous multi-channel sampling. On the other hand, we typically set the low-frequency cutoff in the amplifier to 0.1 or 0.3 Hz to eliminate zero drift and record low-frequency LFPs. These settings allowed us to capture a wide range of neural electrical activity. After signal collection, we used a digital filter to remove low-frequency potentials for spike extraction. The following analysis thus focuses on how the f_L setting in digital filtering affects spike waveforms.

Again, we analyzed the spikes from both interneurons and pyramidal neurons in the rat hippocampal CA1 region. The amplifier was set to a frequency band of 0.3 Hz to 5 kHz. According to the Nyquist frequency principle, a 20 kHz sampling frequency would have been sufficient; however, we sampled the signals at 100 kHz to obtain smooth spike waveforms. Then, the 2-min recordings were high-pass filtered using the Kaiser window filter in LabChart software with a cutoff frequency (f_L) increasing from 100 to 900 Hz in 100 Hz increments to show how the f_L setting affected spike waveforms and their SNRs.

Figure 4.5A shows typical examples of noise-free spike waveforms for the two types of neurons, obtained by the averaging method described above. The wideband real waveforms are shown in red, while the other waveforms at nine different f_L settings are in blue. The waveforms at $f_L = 900$ Hz notably differed from the real waveforms. The superimposed plot of all the 10 waveforms shows how they changed as f_L increased: both their negative peak and peak-to-peak amplitudes gradually decreased, while a new positive peak emerged before the negative peak. In the pyramidal neuron spikes, the smooth posterior inflection (denoted by the black arrows in Fig. 4.5A) progressively sharpened into a positive peak. This change caused the spikes to lose their wide opening, making them increasingly similar to interneuron spikes. Both types of spikes showed a greater decrease in negative peak amplitude than in peak-to-peak amplitude, particularly at higher f_L values (Fig. 4.5B).

The CV values of the spike waveforms at each f_L were calculated by using the real waveforms as a reference. As shown in Fig. 4.5C, the CV gradually increased as f_L increased. Notably, pyramidal neurons showed significantly higher CV values than interneurons across the f_L range of 100–900 Hz. This difference resulted from the more low-frequency components in pyramidal neuron spikes. These CV values were calculated using spike waveforms spanning 3 ms for interneurons and 4 ms for pyramidal neurons, as shown in Fig. 4.5A, which were different from the 1 ms span shown in Fig. 4.4B.

Figure 4.5D shows how the SNR values of the two types of spikes changed with f_L. Spikes of pyramidal neurons had higher SNRs in the f_L range of 400-600 Hz, while

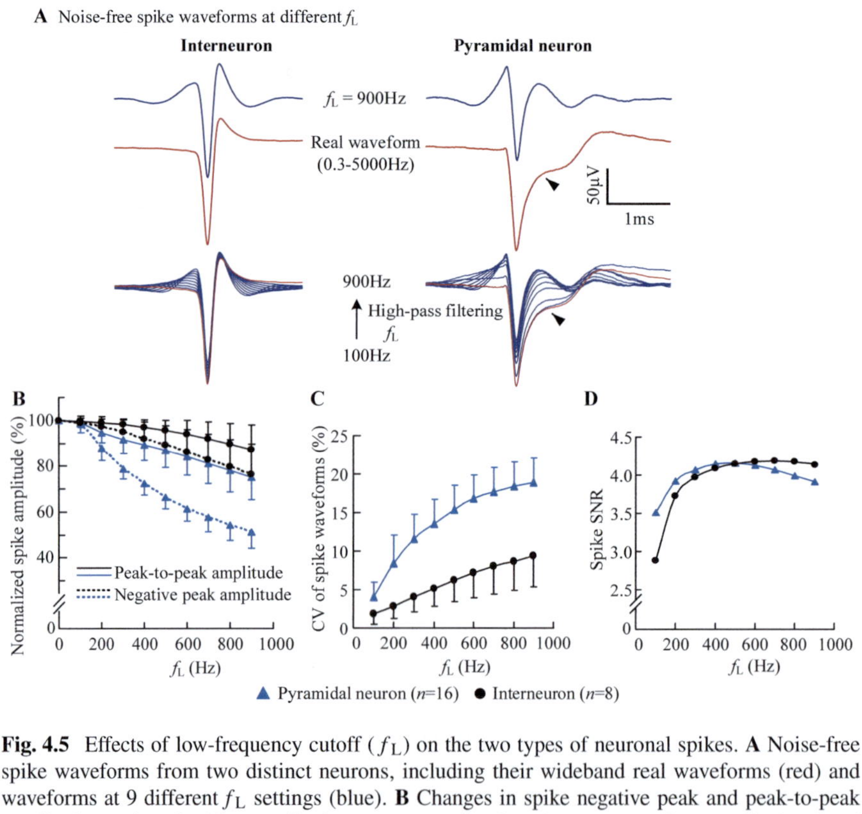

Fig. 4.5 Effects of low-frequency cutoff (f_L) on the two types of neuronal spikes. **A** Noise-free spike waveforms from two distinct neurons, including their wideband real waveforms (red) and waveforms at 9 different f_L settings (blue). **B** Changes in spike negative peak and peak-to-peak amplitudes against f_L. The amplitudes were normalized to the corresponding amplitude of real spike waveform. **C** Changes in spike CV against f_L. **D** Changes in spike SNR against f_L. From Feng et al. (2012)

those of interneurons had higher SNRs in the f_L range of 500–800 Hz. The standard deviations of these SNRs were similar to those shown in Fig. 4.4E, influenced by the various peak-to-peak amplitudes of spikes. For clearly displaying the SNR trends, the standard deviation error bars were omitted in Fig. 4.5D.

These results showed that the f_L affected the spike waveforms of pyramidal neurons more significantly than those of interneurons. The f_L range for maximum SNR was also different between the two types of spikes. Notably, waveform distortions were already evident at the f_L ranges that produced maximum SNRs. For instance, at $f_L = 600$ Hz, the pyramidal neuron spikes showed a negative peak amplitude of only ~ 60% of the real spike value due to the loss of low-frequency components, while the peak-to-peak amplitude was ~ 85%. At $f_L = 800$ Hz, the interneuron spikes maintained higher fidelity, with the negative peak amplitude at ~ 80% of the real spike value and the peak-to-peak amplitude at ~ 90% (Fig. 4.5B).

For any spike signals, a high-pass filter with a higher f_L can produce oscillations immediately before and after the spikes (Wiltschko et al. 2008). Figure 4.6 shows a burst from a pyramidal neuron. While the interval potentials between spikes were relatively flat in the original wideband signal, they became as oscillations after high-pass filtering with $f_L = 500$ Hz. It is important to distinguish these filter-produced oscillations from real neural rhythms in hippocampal LFP—such as ripples at 140–200 Hz or even 300–600 Hz (Buzsáki 2006).

A proper frequency band setting is essential for accurate spike recording and detection. In theory, a higher f_H in amplifiers is preferable for recording complete spike waveforms. However, the magnitude of electron thermal noise in electrodes and recording instruments is directly proportional to the frequency bandwidth (Bretschneider and de Weille 2006). When the high-frequency band is widened, it can at least increase by several kilohertz, causing a dramatic increase in noise. Our findings showed that at $f_H = 5$ kHz, the loss in spike peak-to-peak amplitude was only ~ 5%, the waveform CV was only ~ 2%, and the spike SNR reached its

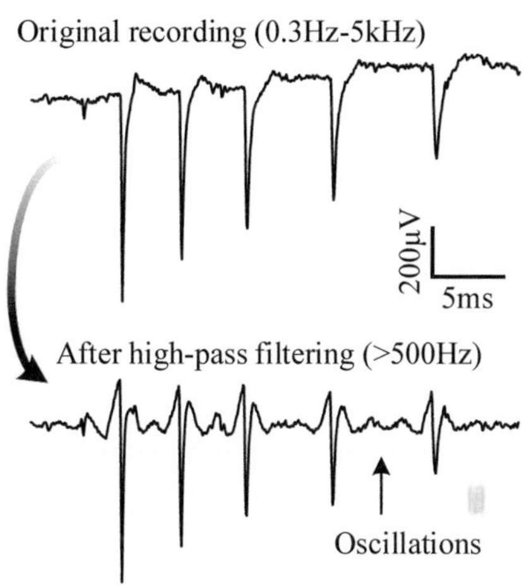

Fig. 4.6 Oscillations before and after spike waveforms caused by the removal of low-frequency components by high-pass filtering

maximum (Fig. 4.4C–E). At $f_H = 3$ kHz, although waveform distortion increased, the SNR remained comparable to that at 5 kHz. Therefore, for detecting spikes to obtain neuronal firing sequences (Lewicki 1998), setting the f_H (the low-pass cutoff) between 3 and 5 kHz is suitable. However, for analyzing subtle features of spike waveforms (Henze et al. 2000) or using them for spatial localization of cell current sources (Lee et al. 2007), a higher f_H can be required to retain higher frequency components. Actually, in such cases, both the high and low-frequency components of the spikes are required to preserve as much as possible.

A high SNR is crucial for accurate spike detection using methods like threshold detection (Musial et al. 2002), where spike distortion is a secondary concern. Due to the neuronal morphology and the formation mechanism of action potentials (Gold et al. 2007), pyramidal neuron (Pyr) spikes contain more low-frequency components than interneurons (IN) spikes. This leads to greater peak loss and SNR decrease at higher f_L values in Pyr spikes. As a result, Pyr spikes showed maximum SNR in an f_L range of 400–600 Hz, while IN spikes had a different f_L range of 500–800 Hz (Fig. 4.5D). For detecting both types of spikes, an optimal f_L can be set at 500–600 Hz.

Each recording can typically contain spikes from multiple neurons, making spike sorting necessary after obtaining spike waveforms (refer to Sect. 4.1.4). This process relies on the prerequisite that spikes from the same neuron have similar waveforms, while those from different neurons are distinct—with greater differences being preferable. Sorting accuracy depends on factors such as spike SNR and waveform fidelity. Because noise can cause variations in spike waveforms from an identical neuron, spikes with a low SNR create a more scattered distribution in the clustering space, making accurate sorting difficult. Similarly, waveform distortions—such as an increased similarity between different spikes caused by high-pass filtering (Fig. 4.5A)—can also complicate the sorting process (Feng et al. 2012).

In summary, this section explores the changes in spike waveforms from two types of neurons under different frequency cutoffs. The choice of frequency range depends on factors including the features of spike waveforms and the purpose of spike usage. In our studies on neuromodulation by electrical stimulations—covered in Part II of this book, we needed to accurately detect spikes, classify them, and distinguish their neuron types. We also needed to simultaneously record local field potentials and stimulus-evoked potentials, such as APS and OPS. For these purposes, we configured the amplifier to a wideband of 0.3 Hz to 5 kHz, used a 20 kHz sampling frequency to collect signals, and then applied digital high-pass filter in LabChart to remove low-frequency components (with an f_L typically set at 500 Hz), which produced clear spikes for subsequent detection and sorting.

Additionally, neuron types can be distinguished based on their spike waveforms and firing patterns. Our method for distinguishing pyramidal neurons from interneurons in the hippocampal region was based on the rising phase width of spike waveforms from the original wideband signals. Spikes with a rising phase wider than 0.7 ms were classified as from pyramidal neurons (Fig. 4.1C), while those narrower than 0.4 ms were considered from interneurons (Barthó et al. 2004). Figure 4.5A demonstrates that high-pass filtering, commonly used in spike detection, can distort spike rising phases—particularly narrowing Pyr spikes. Therefore, when identifying neuron types, we should use the original spike waveforms from wideband signals. Furthermore, as shown in Fig. 4.3, using averaged, noise-free waveforms can produce more accurate measurements.

4.1.3 *Effects of Analog-to-Digital Conversion on Spike Waveforms*

Modern data acquisition systems for electrophysiological signals rarely use analog devices such as magnetic tape recorders. Instead, they perform real-time ADC to sample amplified signals and store them digitally. ADC has two key specifications: sampling frequency (or sampling rate) and resolution. The sampling frequency (f_s) determines the time accuracy, which must be at least twice the highest frequency in the signal (the Nyquist frequency). In practice, an f_s of about four times or higher is typically used. Neuronal spikes require a high f_s due to their brief duration (~ 1 ms) and high frequency components.

ADC resolution (or quantization level) determines the voltage accuracy in the sampled signal. Spikes require high ADC resolution due to their small amplitudes, typically at tens to hundreds of microvolts. This resolution becomes especially critical when simultaneously recording other signals with larger amplitudes, such as population spikes (PS) above 10 mV and slow waves of spreading depression (Fig. 3.16). A 16-bit ADC converter provides 65,536 quantization levels. At a recording range of $\pm$ 20 mV, the ADC resolution is 625nV, allowing a 50 μV spike to occupy about 82 quantization levels. If the recording range is increased to $\pm$ 100 mV, the ADC resolution decreases to 3.125 μV, reducing the quantization levels for the same spike to only 16.

Let's examine the spike waveform from a pyramidal neuron shown in Fig. 4.7. The upper left plot shows the waveform with dense data points, captured at $f_s = 100$ kHz with a 0.3 Hz–5 kHz wideband. The remaining plots in the top row, enclosed in dashed boxes, display the same waveform sampled at progressively decreased f_s of 20, 10, and 5 kHz. The corresponding plots in the bottom row overlay these waveforms on the 100 kHz sampling waveform (in grey) to highlight differences. At 20 and 10 kHz, the spike waveforms are well-preserved. However, at 5 kHz, the spike negative peak degrades significantly. The steep falling phase of this spike lasts only about 0.3 ms, as shown by the shadow in the upper left plot. At 20 kHz, six data points are captured in the falling phase, while at 10 and 5 kHz, only 3 and 2 data points are captured, respectively. In the bottom left plot, even at $f_s = 100$ kHz, reducing the resolution to 10 μV results in the entire waveform to occupy only 18 quantization levels—appearing as "ladders" with a significant loss of details. Yet for spike detection, where the main goal is to determine spike timings, a lower resolution usually doesn't impact results significantly. For comparison, the resolution of the other waveforms in Fig. 4.7 is 625nV, allowing the spike to span approximately 280 quantization levels.

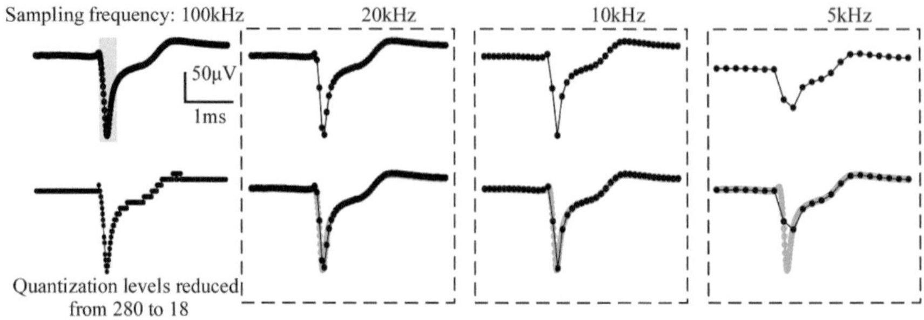

Fig. 4.7 Effects of ADC sampling frequency and resolution on the recorded waveforms of a same spike. The top row displays waveform data points of the same spike sampled at 100, 20, 10, and 5 kHz (from left to right). The bottom row, within dashed boxes, overlays each waveform from the top row onto the 100 kHz sampling waveform (in grey). The waveform in the bottom left corner has an ADC resolution of 10 μV (at 100 kHz sampling frequency), while the others have a resolution of 625 nV

4.1.4 Spike Detection and Sorting

Spike detection is to extract individual spikes from a recording signal, while spike sorting is to determine which neuron generated each spike. Each electrode contact can typically capture spikes from multiple nearby neurons. To determine the firing sequences of distinct neurons, we need to identify the origin of individual spikes based on their unique waveforms—a process known as spike sorting. Both detection and sorting are essential for obtaining the neuronal firing sequences (or neural coding information) in response to external stimuli. An electrode contact with a larger area can record more neurons simultaneously, but this creates challenges for spike analysis. When multiple neurons emit action potentials simultaneously, their spikes overlap, making extraction difficult. The more neurons recorded and the higher their firing rates, the greater the probability of producing overlapped spikes. Conversely, with fewer neurons recorded, overlapped spikes become minimal and negligible. Additionally, the contact surface acts as an equipotential conductor. If its area is too large, it can create a short-circuiting effect on the surrounding electrical fields, including those formed by spike currents. This makes it difficult to measure large spikes with such electrodes. Given these reasons, an electrode with a smaller contact is more suitable for recording spikes (refer to Sect. 3.4.1).

1. **Methods for Spike Detection**

One common method used to extract spikes from recordings is the threshold method. Other methods include the window method and template matching method.

(1) **Threshold method for spike detection**

Figure 4.8 shows an original wideband recording and its spike signal obtained by removing low-frequency potentials by digital high-pass filtering (with a cutoff frequency of 500 Hz). This filtered signal contains spikes from multiple nearby neurons, known as multiple unit activity (MUA). The spikes in MUA are clearly distinguishable. Most spikes have a negative peak greater than their positive peak. Therefore, a negative threshold can be set to capture spikes with negative peaks below it. However, setting an appropriate threshold is practically challenging. If the threshold (absolute value) is too small, noises may be mistaken as spikes. Conversely, if the threshold is too large, small spikes may be missed. It is critical to set a threshold that can minimize both errors. A common algorithm for automatic spike detection is to set the threshold at 3–5 times the standard deviation of noise or the MUA signals, sometimes with manual adjustments. Additionally, some spikes can have positive peaks greater than negative peaks. To reduce missed detections, both positive and negative thresholds can be set simultaneously.

(2) **Window method for spike detection**

A window method enables direct spike detection in original wideband recordings that contain low-frequency potentials. The method can be performed using a peak-to-peak amplitude threshold to extract spikes. To do it, a time window with a fixed width moves along the recording signal. The difference between the maximum and minimum values within the window at each position is calculated. When this difference exceeds the pre-set threshold, a putative spike is captured and its time span is then determined. The width of time window can be set to ~ 1 ms, matching typical spike widths. Using the peak-to-peak threshold, this method can evaluate both amplitude and slope changes within the signal segment in each narrow window. This effectively excludes low-frequency waves that have large amplitudes but small slopes, enabling direct spike detection from wideband signals.

When using a window method, it is critical to avoid duplicate spike detections. If the division between two adjacent window falls in the middle of a spike waveform, the peak-to-peak values in both windows may reach the threshold, causing a single spike to be mistakenly detected twice. To prevent this, a minimum time interval threshold between adjacent spike peaks can be set to eliminate any duplicate detections that occur too close together.

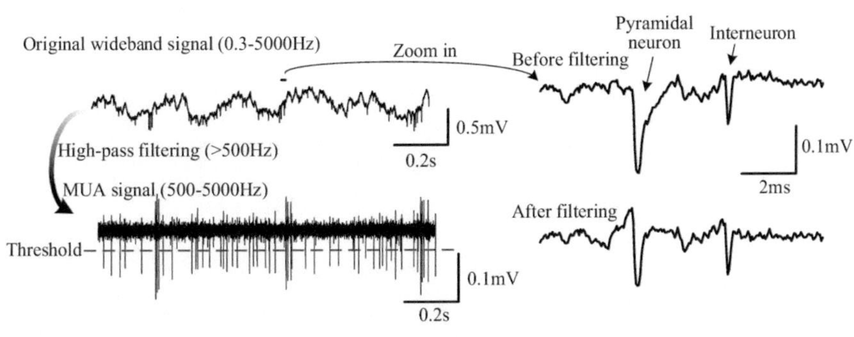

Fig. 4.8 Original wideband signal (top) and MUA signal (bottom) obtained after high-pass filtering of the original recording

Figure 4.9A shows a recording during antidromic high-frequency stimulation (A-HFS) on the alveus in the rat hippocampal CA1 region. The A-HFS was a biphasic pulse train with an intensity of only 0.05 mA and a pulse interval of 10 ms (corresponding to 100 Hz). Each pulse triggered a population spike (APS). Upon the rising phase of many APSs, an interneuron spike appeared superimposed (indicated by a blue dot). To detect these spikes using a threshold method, the stimulus artifacts were first removed from the original signal (Fig. 4.9B, see Sect. 4.5.2 for details). Then, the low-frequency potentials were removed using a high-pass filter to obtain the MUA signal shown in Fig. 4.9C. Although spikes during non-stimulation periods were readily detected by a threshold method, the high-pass filtering produced oscillations around the APS waveforms during stimulation, obscuring the following interneuron spikes and making detection difficult. To solve this problem, either remove the APS waveforms before high-pass filtering as reported (Feng et al. 2013), or detect spikes directly using a window method on the original wideband signal (Fig. 4.9A), as described in Sect. 4.1.5.

(3) **Template matching method for spike detection**

A template matching method can detect and classify spikes in the recording signal. For details, please refer to the spike sorting section below.

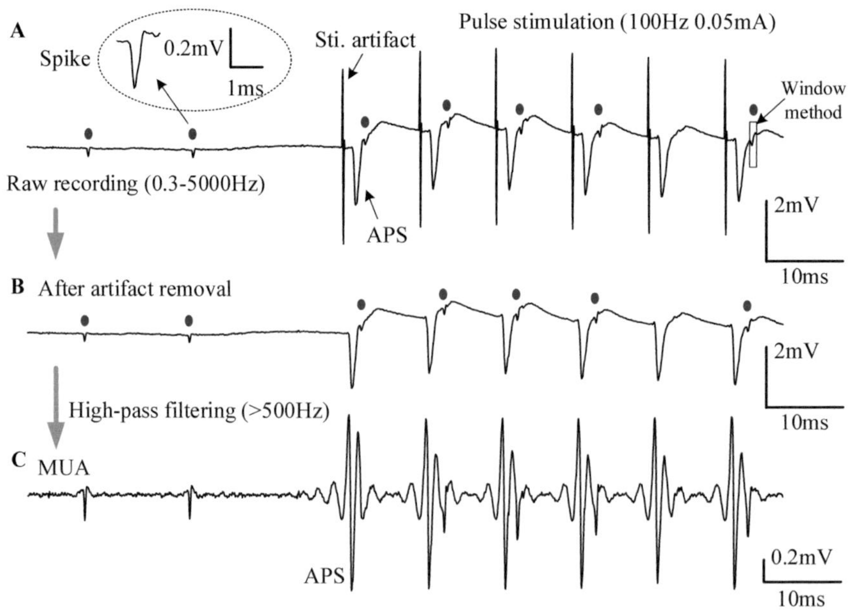

Fig. 4.9 Detection of spikes occurring immediately following APSs during A-HFS. **A** Original wideband recording with stimulus artifacts. **B** Signal with stimulus artifact removed. **C** MUA signal generated by high-pass filtering of the artifact-free signal. Blue dots in (**A**) and (**B**) denote the spikes of an interneuron

2. **Aligning and Extracting Spike Waveforms**

Threshold and window methods for spike detection only provide approximate time spans of spike waveforms. Before proceeding to spike sorting, the entire spike waveforms must be extracted by determining their start and end points. This process is known as spike alignment, which requires a reference point—typically either the negative peak or the first data point that crosses the detection threshold.

When aligning based on negative peaks, deviations may occur in the extracted spike waveforms due to differences between sampled peaks and actual peaks caused by limited sampling frequency. This issue also occurs when aligning based on threshold points. The actual intersection between the spike waveform and the threshold line may fall between two sampled points rather than exactly on one. Therefore, aligning based on the first sampled point crossing the threshold can cause deviations in extracted waveforms due to differences between the sampled points and the threshold line. To improve accuracy, the precise intersection point between each spike waveform and the threshold line can be calculated by an interpolation algorithm before alignment. Nevertheless, these alignment methods—based solely on a single sampled point—are inevitably affected by noise and limited sampling frequency. An alternative approach uses multiple data points from the major spike portion to calculate a "gravity center" point as the alignment reference, which can reduce deviations.

3. **Spike Sorting**

Spike sorting is to classify spikes to their source neurons based on the fact that spikes from the same neuron have similar waveforms. The two main spike sorting algorithms are the parameter method, which uses waveform parameters for sorting, and the template matching method, which compares each waveform with pre-set spike templates.

(1) **Parameter method**

The waveform parameters commonly used in spike sorting include amplitude measurements, such as peak-to-peak amplitude, positive peak amplitude, and negative peak amplitude (Fig. 4.10A), as well as time parameters like spike width. Figure 4.10B shows a sorting example using a single parameter—peak-to-peak amplitude (A_{pp}). TheA_{pp} data from all spikes are plotted in a histogram with its vertical axis showing the count of spikes falling in each small A_{pp} bin. The peaks in the histogram represent spikes from different neurons. When the detection threshold is set low, noise can be mistakenly detected as spikes, forming a "noise" cluster in the small amplitude range (Fig. 4.10B). These noise artifacts can be easily removed during the classification.

Parameter selection is crucial when sorting spikes based on their waveform parameters. Let's take the MUA signal shown in Fig. 4.11A as an example. It contains four distinct types of spikes. However, creating a histogram using positive peak amplitude alone yields only a single peak and fails to provide proper classification (Fig. 4.11B). This occurs because the positive peak amplitudes of the four spike types are too

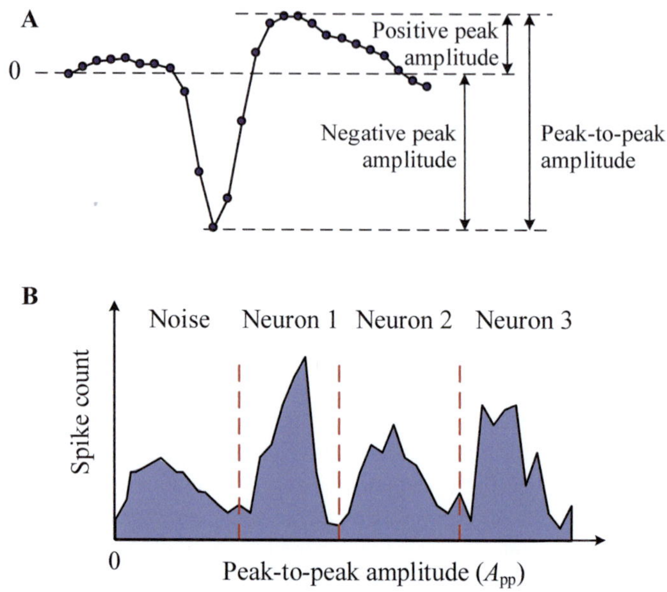

Fig. 4.10 Schematic diagram of spike sorting based on spike waveform parameters. **A** A typical spike waveform drawn by connecting sampling points (20 kHz), along with three amplitude measurements. **B** Example of a peak-to-peak amplitude histogram created from a spike recording. The four areas separated by red dashed lines represent the clusters of noise and three distinct spike types

similar to be distinguished from one another. A histogram using peak-to-peak amplitude alone can classify only three types of spikes (Fig. 4.11C), as the amplitudes of the third and fourth spike types overlap considerably. Using multiple parameters together can improve the classifications. Figure 4.11D shows a two-dimensional scatter plot using two parameters: peak-to-peak amplitude and negative peak amplitude. It clearly distinguishes all four types of spikes (encircled by red dashed ellipses). Furthermore, a three-dimensional scatter plot incorporating three parameters: peak-to-peak amplitude, negative peak amplitude, and positive peak amplitude, provides a clearer visualization of the differences among these spike types (Fig. 4.11E).

Although intuitive parameters like amplitude and width are common choices, they are not always optimal. Using all sampling points from spike waveforms as high-dimensional data for classification can yield better sorting results—though this approach requires intensive computation. Principal component analysis (PCA) offers a solution by transforming this high-dimensional data into a lower-dimensional one while preserving essential information. PCA can arrange the principal components by their variances, which correspond to component's energy and amplitude. Using the first few principal components with highest variances can achieve effective spike sorting.

(2) **Template matching method**

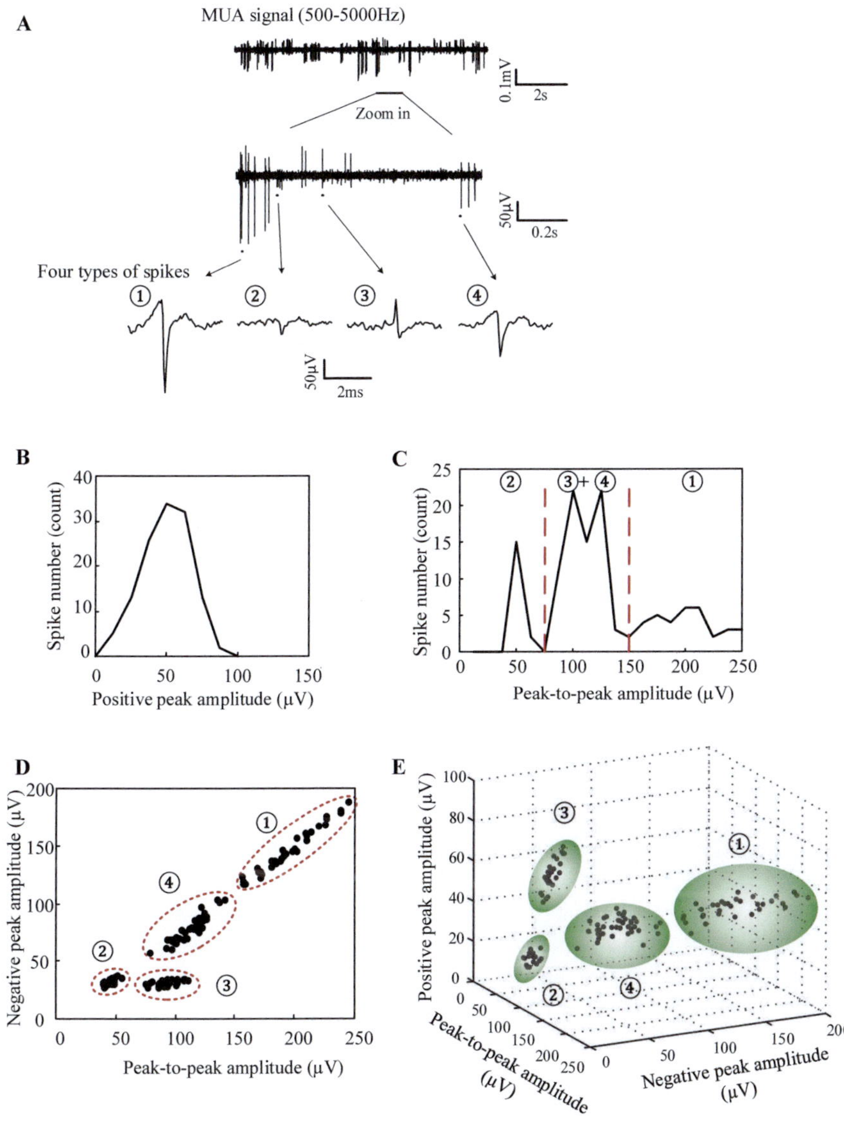

Fig. 4.11 Example of spike sorting using distinct waveform parameters. **A** A segment of MUA signal with enlarged insets showing four spike types. **B** and **C** Histograms created using single spike parameter: positive peak amplitude (**B**) and peak-to-peak amplitude (**C**). **D** Two-dimensional scatter plot created using two spike parameters of negative peak amplitude and peak-to-peak amplitude, with four spike types indicated by red dashed ellipses. **E** Three-dimensional scatter plot created using positive peak amplitude, peak-to-peak amplitude, and negative peak amplitude. Green ellipsoids denote the four spike types. Modified from Feng et al. (2007)

Spike sorting by template matching involves two main steps. First, waveform templates are created for each spike type. Second, the recorded signals are scanned and compared with each template, while the root mean square (RMS) of differences between signal sampling points in the scan window and template points is calculated at each scan movement. When the RMS value falls below a preset threshold, the signal in the scan window is identified as a spike matching the template type. This method can achieve both spike detection and sorting simultaneously on either MUA signals or original wideband signals. It can also be used on extracted spike waveforms to reduce computational cost.

A template matching method can achieve real-time spike detection and sorting during signal collection (Schaffer et al. 2021), although it requires substantial computational resources, precise spike templates, and an appropriate preset fitting threshold. In addition, the method can, to some extent, classify overlapped spikes occurring almost simultaneously from different neurons (Mokri et al. 2017). Furthermore, it can improve the detection and sorting of spikes with gradually varying amplitudes, such as in bursts (Shabestari et al. 2021).

4. **Multi-Channel Spike Detection and Sorting**

It is difficult to correctly achieve spike sorting based solely on a single channel recording signal. Each electrode contact can simultaneously capture spikes from multiple nearby neurons. For a same type of neurons at similar distances from a contact, the amplitudes and waveforms of their recorded spikes are similar. It becomes difficult to determine whether these spikes originate from one or multiple neurons based solely on the single recording. Therefore, spikes recorded simultaneously across multiple channels on an array microelectrode should be used to determine spike origins.

Figure 4.12A shows three channel recordings (Ch1-3) from an array microelectrode (A2 × 2-tet-3 mm-150–150-312, NeuroNexus) with a diamond-shape arrangement of groups of four contacts (Fig. 3.12B). The eight spikes recorded in Ch1 are too similar to distinguish when viewed alone. However, when examining all three channel recordings together, we can clearly identify that these spikes come from two distinct neurons (indicated by purple and red triangles above the signals). The spikes with red triangles show large amplitudes across all three channels, while those with purple triangles appear small on both Ch2 and Ch3, indicating that this neuron was close to contact 1 but far from the other two contacts.

Figure 4.12B shows another example of 4-channel MUA signals from a dual column array (A1 × 16-Poly2-5mm-50s-177, NeuroNexus). Colored triangles and blue dots above the signals indicate three types of spikes obtained through four-channel spike sorting for the adjacent channels (Ch4-7). When using Ch4 alone, only two spike types can be distinguished, as it cannot differentiate between the two types of triangular-marked spikes that show substantial differences on Ch6. The spikes marked by red triangles appear larger on Ch6, suggesting that the neuron was closer to the contact 6 on the array. Conversely, the neuron producing spikes with purple triangles was far from the contact 6 because its spikes hardly appear on Ch6.

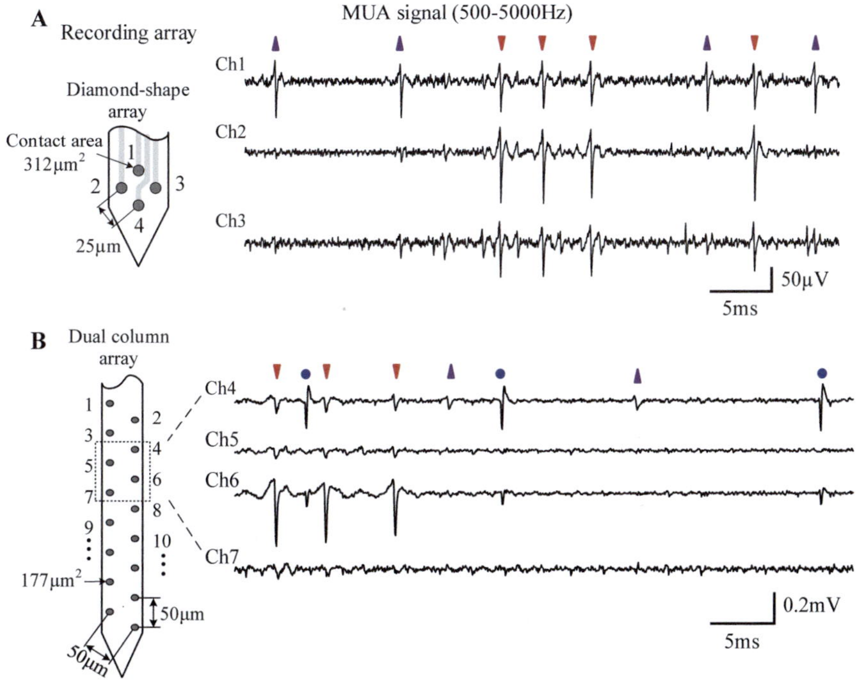

Fig. 4.12 Spike signals recorded simultaneously by densely arranged contacts on two types of array microelectrodes. **A** and **B** Examples of MUA signals recorded by contacts in diamond-shape (**A**) and dual column (**B**) arrangements

Obviously, the three spike types can be effectively sorted using just the data of a single parameter—the peak-to-peak amplitudes from the four channel recordings.

One common type of traditional recording array is the tetrode, which can be assembled by four microwires glued together (Fig. 4.13). Each wire is insulated except for the measuring contact at its tip. The amplitudes of spikes from a same neuron can vary across the four recording channels due to different distances between the neuron and the contacts. Using MUA signals from all four channels recorded simultaneously can substantially improve spike detection and sorting (Buzsáki 2004). The Neuronexus planar array electrode shown in Fig. 4.12A imitates this tetrode contact arrangement. The manufacturer, therefore, names it "tetrode" (see website: http://neuronexus.com/). The accuracy of spike sorting can be improved when multiple contacts with sufficient spatial density are used. In our laboratory, we commonly used dual column array electrodes (Fig. 4.12B). With such an electrode, MUA signals from any four adjacent channels can be used to classify spikes. Additionally, with its vertically linear arrangement spanning up to 350 μm, this type of electrodes can simultaneously record evoked potentials at different hippocampal layers, such as population spikes at the soma layer and postsynaptic potentials at the dendritic layer (Fig. 3.6).

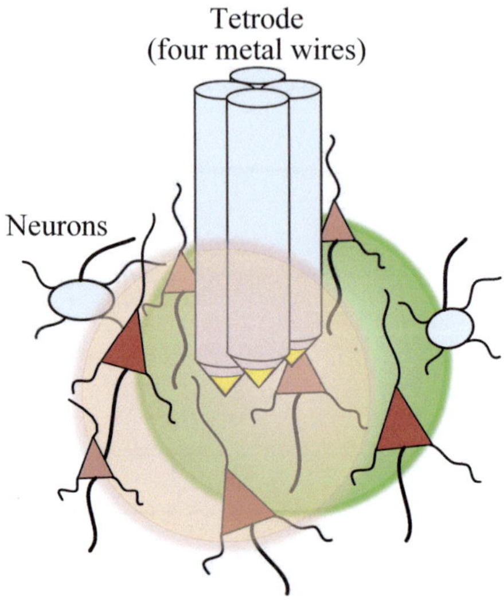

Fig. 4.13 Schematic diagram showing a tetrode made of four thin wires surrounded by neurons. Triangles represent pyramidal neurons, while circles represent interneurons. The yellow tips represent the exposed contacts. The green and pink spheres represent the spherical surfaces centered at the left and right electrode tips, respectively. Neurons of a same type—when locating along the same spherical surface with similar distances from a contact—produce similar spike waveforms, making it impossible to differentiate these neurons using single-channel spike sorting

The following are the main steps for spike detection and sorting by using multi-channel signals. First, apply a threshold method to each channel individually to detect spikes. The detection results are then merged since the number of detected spikes can vary across channels. As spikes from a same neuron can have different amplitudes across channels, spikes missed on channels with smaller amplitudes may still be detected on channels with larger amplitudes. To avoid noise, you can set a higher threshold based on the channel with larger spike amplitudes—for example, a threshold of 5 times or above the signal standard deviation. This higher threshold can reduce false positives without increasing missed detections. By combining all detected spikes from the channels and removing repeated detections based on spike timing, you can maximize spike detection. As shown in Fig. 4.12B, the spikes marked by red triangles can be detected in Ch4, Ch5, and Ch6. A merging algorithm is needed to remove repeated spikes.

After detection, proper parameters of each spike waveform are extracted for spike sorting. For instance, we employed a custom MATLAB program to extract the first principal components and the peak-to-peak amplitudes of spike waveforms from four adjacent channels to combine into eight parameters for sorting. The sorting was performed in SpikeSort3D, an open-source platform from Neuralynx Inc. (www.Neuralynx.com), which integrates the KlustaKwik clustering program (http://klustakwik.sourceforge.net/)—a widely used spike sorting tool that can also be used independently (Harris et al. 2000; Wild et al. 2012). Figure 4.14 shows an example of our

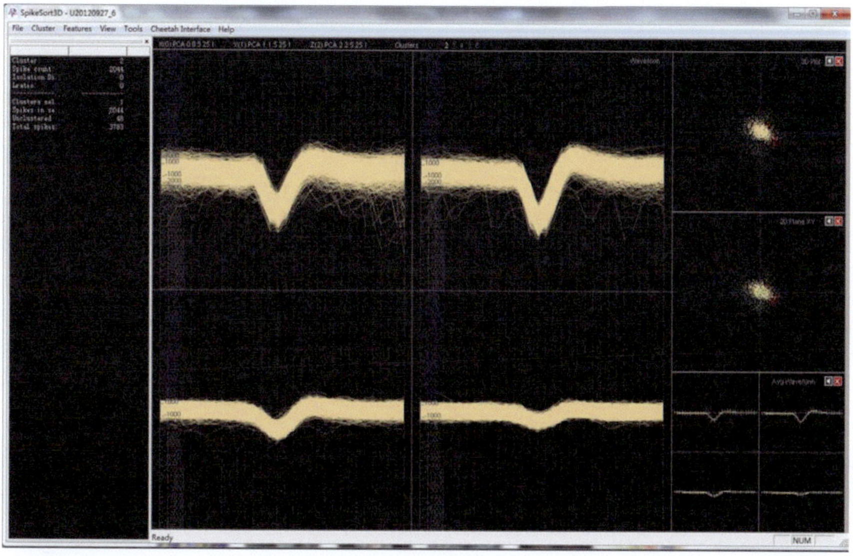

Fig. 4.14 Example of four-channel spike sorting displayed in the user interface of SpikeSort3D software

sorting results in SpikeSort3D. The four sections in the middle of the figure display the overlapped spike waveforms in four channels respectively. This SpikeSort3D user interface allows for manual adjustments to improve sorting accuracy.

In our laboratory, after collecting neural signals from the rat hippocampus using a PowerLab recorder (see Sect. 3.5.3), we selected four adjacent channels for spike detection and sorting. The main process includes: (1) *Obtain MUA signals.* High-pass filter the original wideband signals to produce MUA signals in LabChart software, with a cutoff frequency typically set to 500 Hz. Export the four-channel MUA signals into a MATLAB data file (.mat). (2) *Detect spikes and extract parameters of spike waveforms.* Run our custom MATLAB program to read the MUA data file, apply a threshold method to detect spikes, and extract the amplitudes and the first PCA components of each spike waveform across four channels. Generate a Spikesort3D data file (.ntt) consisting of these 8 parameters from all spikes. (3) *Perform spike sorting.* Open the.ntt file in Spikesort3D, adjust settings accordingly, and select the Klustakwik program for classification (Fig. 4.14). The classification can yield the spike sequences of individual neurons. Export these sequences from Spikesort3D into a text file for subsequent analyses. (4) *Verify the spike sorting results.* Combine the spike sequences with the original recording signals and MUA signals to form a LabChart file (.bin). Opening this file in LabChart as shown in Fig. 4.15 to visually check the sorting results. The left window in this figure spans ~ 4.12 s, while the enlarged right window shows ~ 0.28 s. The top four rows show the original recording signals from the four adjacent channels on an array electrode as shown in Fig. 4.12B. Spikes appear riding on ~ 4 Hz rhythmic field potentials in the left window and

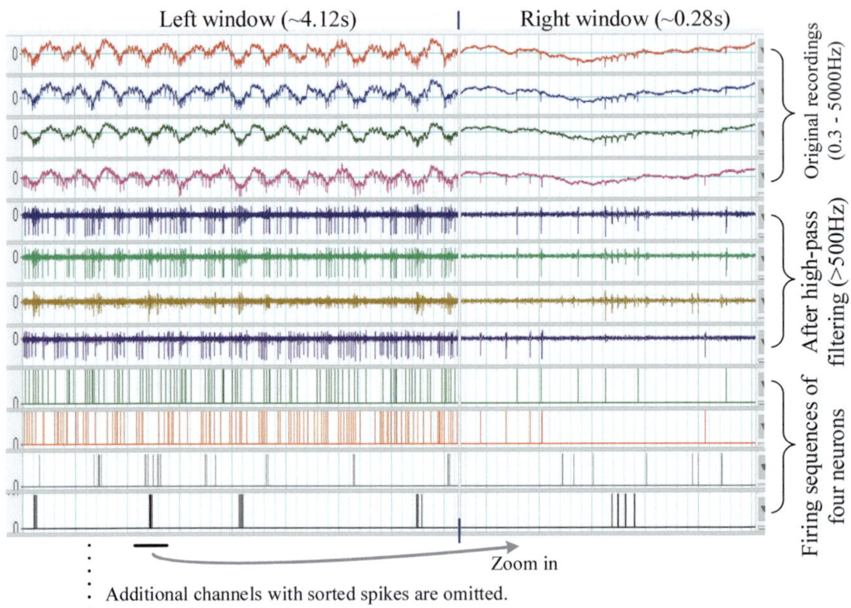

Fig. 4.15 Verification of spike sorting results in LabChart

become clearer in the enlarged signals in the right window. The middle four rows show the MUA signals with distinct spikes. The bottom four rows show parts of the sorting results—the spike sequences of four neurons, in which each vertical line (with a constant height) represents a spike whose occurrence timing aligns to the upper channels. Using the scalable display in LabChart, it is easy to examine the details of spike waveforms to verify the sorting results.

In addition to the spike detection and sorting methods described above, we have developed other approaches, including the window-based spike detection introduced in the next section.

4.1.5 Window-Based Spike Detection

Neuronal unit spikes are typically much smaller than the low-frequency field potentials. Thus, the field potentials are usually filtered out by a high-pass filter before using a threshold to detect spikes. However, a challenge can arise when the original recording contains other large potentials with a spectrum overlapping with that of spikes, such as population spikes (PS) from synchronized neuronal firing. In such case, filtering cannot remove PS waveforms without losing unit spikes, and may even introduce additional interference (see Fig. 4.9C). To solve this problem, we developed a window-based method for detecting spikes in the presence of PSs during high-frequency stimulation (HFS) (Wang et al. 2016). Its algorithm steps are as follows.

(1) Use a time window with a fixed width to segment the original signal. Set the window width (w) at 1.2 ms, which includes 25 sampling points at 20 kHz, approximately one spike width. Other window widths, such as a smaller width matching the spike falling phase, can also work.
(2) Set lower and higher amplitude thresholds for spike detection. Calculate the mean standard deviation of all window signals. Set the lower amplitude threshold (A_L) at 1.5 times the mean value. To eliminate large PS waveforms and HFS artifacts, set the higher amplitude threshold (A_H) at 0.6 mV or above. Note that both thresholds can be adjusted accordingly. This window-based algorithm can be used directly on original recording signals with stimulation artifacts, or on signals after artifact removal, as shown in Fig. 4.9A, B.
(3) Determine the data points of spike negative peaks. Scan the data in each window sequentially. Search for the minimum data in each window and measure the data range in a 0.3 ms period (i.e., 6 sampling points) before the minimum data, denoted as $A_{0.3}$. If the $A_{0.3}$ falls between the lower and higher thresholds set in step 2 (i.e., $A_L < A_{0.3} < A_H$), the 0.3 ms signal is assumed to contain the falling phase of a spike. If the minimum data within the 0.3 ms is not located at the window right boundary, then this minimum data is the negative peak point of the spike. Otherwise, continue to search into the next window until reach the negative peak. If this negative peak is the minimum data in the next window, skip the next window to avoid repeated detection of the same spike.
(4) Use the negative peak point as the spike occurrence timing to create a spike sequence.
(5) For multi-channel signals, repeat the above four steps on each channel. Then merge their spike sequences to a final sequence by removing any repeated detections (Wang and Feng 2009).

Figure 4.16A and B show recording examples during O-HFS and A-HFS in the rat hippocampal CA1 region. Removing low-frequency field potentials with a high-pass filter can cause oscillations near the evoked PS waveforms, which obscure nearby spikes (as indicated by the red dots in the figure). This makes spike detection difficult using a threshold method. Additionally, the filtering-produced oscillations may cross the threshold and be misidentified as spikes.

As shown in Fig. 4.17, the window-based algorithm enables direct spike detection on wideband signals after removing stimulus artifacts. The blue dashed lines represent window segmentations. In each window, the minimum data is initially assumed as the negative peak point of a putative spike. Then, the thresholds A_L and A_H filter out non-spike signals. The two red circles mark putative spikes with an $A_{0.3}$ between these thresholds, while the two red boxes mark a field potential with a small $A_{0.3}$ belowA_L and a PS falling phase with a large $A_{0.3}$ above A_H.

The window-based algorithm can directly detect spikes in wideband signals without the need to filter out low-frequency field potentials while effectively eliminating interference from large PSs. For spike detection during HFS periods, removing

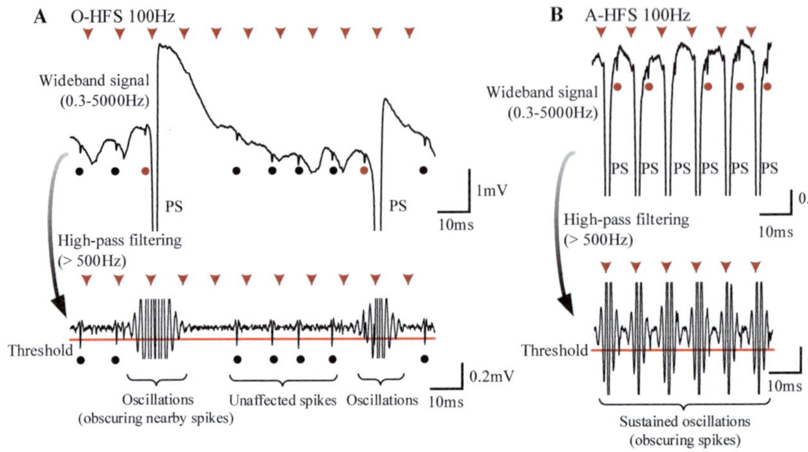

Fig. 4.16 Spike detection during O-HFS (**A**) and A-HFS (**B**) in the rat hippocampal CA1 region with evoked PSs. Red arrows above the signals denote removed stimulation artifacts. Black and red dots indicate the spikes that become highlighted and submerged, respectively, after high-pass filtering. To make small spikes visible, large-amplitude PS waveforms in both wideband and high-pass filtered signals are truncated. Modified from Wang et al. (2016)

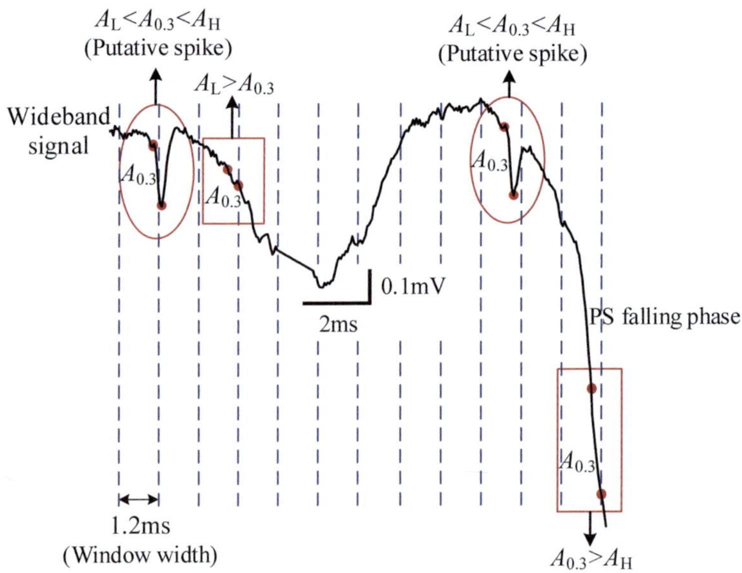

Fig. 4.17 Spike detection using a window-based algorithm directly to a wideband signal. Modified from Wang et al. (2016)

stimulation artifacts beforehand is optional. When running the algorithm directly on signals containing artifacts, the artifacts can be eliminated by properly setting the high threshold (A_H), since artifacts are typically much larger than spikes. Even if artifacts are occasionally misidentified as spikes, they can be easily excluded during spike sorting due to their narrow waveforms from ~ 100μs pulses commonly used in brain stimulations. This window-based algorithm is thus useful for detecting spikes and studying individual neuron activity, especially in recordings containing PS waveforms during HFS or epileptiform discharges. As detailed in Sect. 5.4, this method was used to analyze interneuron firing during A-HFS.

4.1.6 An Algorithm for Four-Channel Spike Detection and Sorting

Many algorithms have been developed for spike detection and sorting (Pachitariu et al. 2024; Meyer et al. 2023; Zhang et al. 2023; Carlson and Carin 2019; Lefebvre et al. 2016). We also developed an algorithm specifically for four-channel spike sorting. It combined the spikes from all four channels to create composite waveforms, which were then used to calculate PCA components for spike clustering (Wang and Feng 2009), as described below.

Since determining the true number and waveforms of various spikes in real recordings is challenging, to verify the algorithm, we created four-channel simulated data by combining spike waveforms with Gaussian white noise. The spike waveform $v_s(t)$ was produced using a combination of sinc and exponential functions:

$$v_s(t) = a\mathrm{e}^{-t}\left(\mathrm{sinc}\left(b\mathrm{e}^{-t}\right) + c\right) \tag{4.4}$$

By adjusting the constants *a*, *b* and *c*, and setting the range of time variable *t*, we created spike waveforms matching real ones. Figure 4.18 shows the two distinct types of waveforms designed to simulate spikes from two neurons. We then constructed four-channel spike signals by adding Gaussian white noise. Each channel contained spikes from both neurons, but with different amplitudes.

We created ten sets of four-channel simulated data with different SNR values based on the SNR range in real recordings. Figure 4.19 shows one set of the simulated data, including spikes from Neuron-1 (Sp1) and Neuron-2 (Sp2). Sp1 had larger amplitudes in Ch1 and Ch3, while Sp2 had smaller amplitudes across all four channels, particularly in Ch1 and Ch3. This made Sp2 difficult to detect in Ch1 and Ch3. The amplitudes of Sp1 and Sp2 were similar in Ch2 and Ch4, making them difficult to distinguish in these channels. Although the amplitude difference between the two spike types existed in Ch3, using the Ch3 signal alone for spike sorting was not feasible due to noise, as shown in the lower right of Fig. 4.19. This scatter plot was created using the first three PCA components (pc1, pc2 and pc3) of Ch3 spikes. All the 173 detected spikes were scattered and not able to cluster.

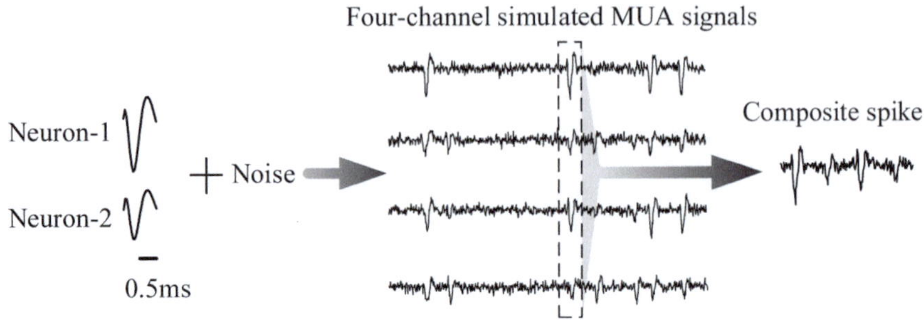

Fig. 4.18 Simulated spike waveforms, four-channel MUA signals and the creation of composite waveform. From Wang and Feng (2009)

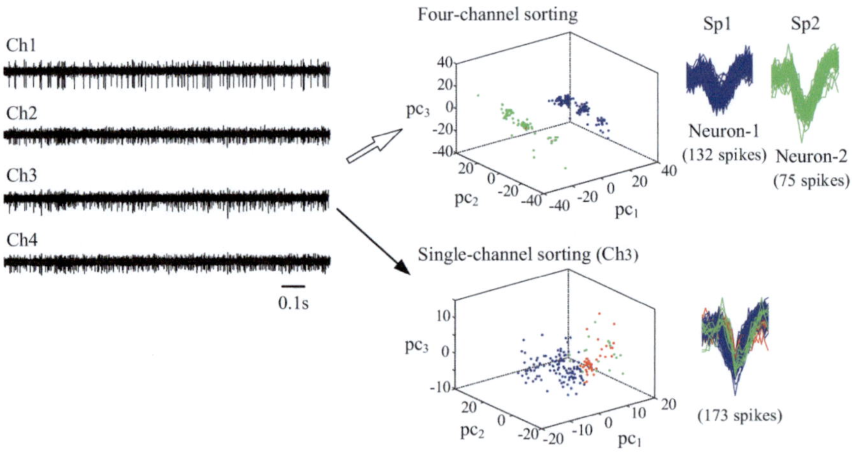

Fig. 4.19 Comparison between multi-channel and single-channel spike sorting by using four-channel simulated spike data. From Wang and Feng (2009)

For the same data set, using all four-channel signals was able to detect an additional 34 spikes, increasing the total spike count to 207. We then performed PCA analysis on their composite spikes (as shown in Fig. 4.18). The scatter plot of the first three PCA components clearly differentiated between the two spike types (Fig. 4.19, upper right). For visualization, both scatter plots in Fig. 4.19 show only the first three PCA components. However, we actually used more PCA components— achieving a cumulative contribution rate over 90%—to perform the spike sorting.

We also compared single-channel and four-channel spike sorting for experimental recordings from the rat hippocampal CA1 region. Figure 4.20 shows an example of MUA signals recorded by four contacts on a tetrode array (Fig. 4.20A). Figure 4.20B shows the superimposed spikes identified through each single-channel analysis, while Fig. 4.20C shows the results of four-channel analysis using composite spikes and their PCA components.

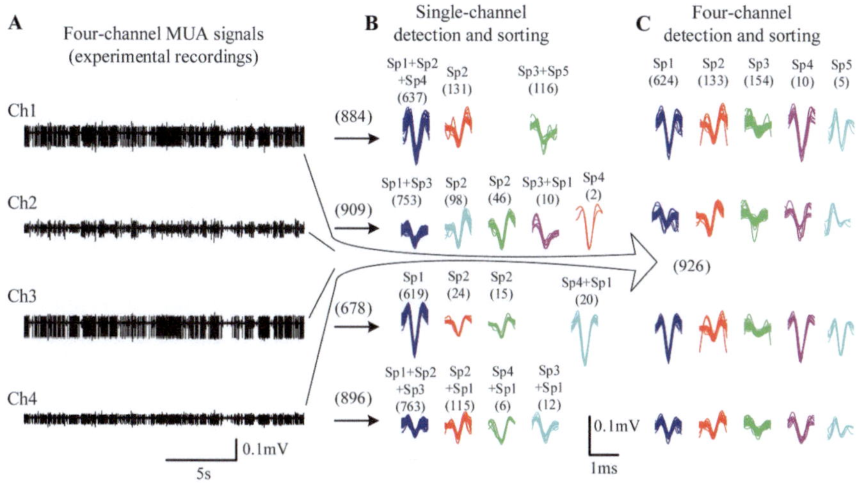

Fig. 4.20 Results of spike detection and sorting using single-channel and four-channel methods for experimental data. **A** Four-channel MUA signals recorded using a tetrode array. **B** Superimposed spikes from individual single-channel analyses. **C** Results of four-channel analysis using composite spike waveforms for spike sorting. Numbers in parentheses are spike counts. From Wang and Feng (2009)

Through careful visual inspection (see Fig. 4.15 for methodology), we identified a total of 912 spikes in the four-channel signals. These spikes originated from five distinct neurons, labeled Sp1–Sp5 with the following counts: 625 (Sp1), 133 (Sp2), 140 (Sp3), 9 (Sp4), and 5 (Sp5). Notably, Sp5 might be an overlapped waveform resulting from simultaneous firing of two neurons. Using these manual analysis results as reference, we compared single-channel and four-channel automatic analyses with a detection threshold of five times the standard deviation of the MUA signal. Table 4.1 shows the detection accuracy, classification accuracy, and number of identified neurons for both methods.

Table 4.1 Comparisons of the detection and classification from single-channel and four-channel analyses for the experimental recordings shown in Fig. 4.20A

Method	Channel No.	Detection accuracy (%)	Classification accuracy (%)	Number of identified neurons /number of actual neurons
Single-channel	1	96.9	98.3	3/5
	2	98.4	84.3	5/5
	3	74.3	94.7	4/5
	4	97.4	83.0	4/5
Four-channel	1 - 4	99.7	99.0	5/5

From Wang and Feng (2009)

In the single-channel analyses, using Ch1 alone missed spikes across all five neurons, particularly Sp3 spikes with small amplitudes. Only three spike types were identified, with the remaining two types merged into others (Fig. 4.20B). However, these merged types contained few spikes, minimally impacting classification accuracy. Using Ch2 alone resulted in a high detection accuracy, but failed to distinguish between Sp1 and Sp3 due to their similar amplitudes and waveforms. Sp2 was incorrectly divided into two types, resulting in low classification accuracy. Using Ch3 alone yielded the lowest detection accuracy, because Sp3 amplitudes were too small to detect and many Sp2 spikes were also missed. Approximately one-quarter of the total spikes were undetected. Although the classification accuracy was not low, the Ch3 analysis identified only four spike types, with one being misclassified and another partially confused. Using Ch4 alone resulted in a high detection accuracy, but similar amplitudes across the five spike types led to significant confusion, resulting in the lowest classification accuracy.

The multi-channel analysis achieved the highest detection accuracy, missing only 3 out of the total 912 spikes. These missed spikes were all Sp3 type with small amplitudes across all four channels. The analysis successfully identified all five spike types and achieved the highest classification accuracy. Note: the classification accuracy here referred to the percentage of correctly classified spikes among detected spikes, excluding missed ones.

The multi-channel method for spike detection and sorting can be used with signals from two or more channels. To record spikes simultaneously from same neurons across multiple channels, electrode contacts must be closely spaced. When a contact is within 50 μm distance from a neuronal soma, the measured spike amplitude can exceed 60 μV, facilitating detection. Contacts placed 200 μm or more apart can rarely record spikes from a same neuron (Buzsáki 2004; Blanche et al. 2005). As shown in Fig. 4.12, high-density array electrodes with closely arranged contacts can effectively record spikes from same neurons. By combining waveforms from multiple channels into composite spikes, even small or absent spikes in some channels can contribute to the distinctive features of these composite waveforms. As a result, the multi-channel sorting using PCA components can improve classification accuracy, making the method significantly superior to single-channel methods.

Note that the experimental spike waveforms in the above example were extracted from the high-frequency filtered MUA signals (Fig. 4.20A), not directly from the original wideband recordings. The filtering can distort spikes (Fig. 4.5), which may affect classification. Using original spike waveforms may improve spike sorting.

4.2 Spike Sequence Analyses

Spike timing sequences of individual neurons, also known as single unit activity (SUA), can be produced after spike detection and sorting. These sequences present firing patterns that contain both neuronal responses to external events and intrinsic neuronal characteristics. Since spike sequences act like point processes with inherent randomness, they require statistical analysis (Brown et al. 2004; Bretschneider and de Weille 2006). There are many methods to analyze these sequences to reveal distinct neuronal firing patterns, including inter-spike interval histograms and auto-correlation histograms. In addition, cross-correlation histograms can reveal relationships between different neurons, showing their interconnections and signal communications. These methods are introduced below.

4.2.1 Inter-Spike-Interval Histogram

An inter-spike-interval (ISI) histogram can help verify spike sorting results and determine neuron types, such as pyramidal neurons and interneurons. Gerstein and colleagues defined the ISI histogram mathematically as follows (Gerstein and Kiang 1960; Rodieck et al. 1962).

The spike sequence can be represented as the sum of impulse functions with a unit area:

$$f(t) = \sum_{k=1}^{N} \delta(t - t_k) \tag{4.5}$$

where $\delta(t)$ is the unit impulse function, t represents time, t_k represents the occurrence time of kth spike, and N represents the total number of spikes in the sequence. The ISI histogram, ISI(τ), is then represented as

$$\mathrm{ISI}(\tau) = \sum_{k=2}^{N} \delta(t_k - t_{k-1} - \tau) \tag{4.6}$$

Discretize the continuous time variable τ into τ_j, where $j = 1, 2, 3, \ldots$. Any τ falling within $\tau_j \leq \tau < \tau_{j+1}$ is measured as τ_j. The bin size $\tau_{j+1} - \tau_j \equiv \Delta\tau$ determines the time resolution of the histogram. In this context, ISI(τ) represents the number of intervals $(t_k - t_{k-1})$ falling within the τ_j.

Figure 4.21A shows two episodes of spike sequences respectively from a pyramidal neuron and an interneuron in the rat hippocampus. Each small bar in these raster plots represents a spike. Figure 4.21B, C are the ISI histograms revealing their firing features: (1) The action potential's refractory period appears as an absence of spikes near zero time. An absence of this gap suggests incorrect spike sorting—spikes from multiple neurons may have been grouped together. (2) These histograms show distinct firing patterns between pyramidal neurons and interneurons. The histogram of pyramidal neuron has peaks around ± 3 ms (Fig. 4.21B), signifying a burst pattern—multiple spikes firing rapidly in clusters (see the spikes indicated by red triangles in Fig. 4.12A, B). In contrast, the histogram of interneuron is relatively flat without sharp peaks (Fig. 4.21C), indicating a random and isolated firing pattern (see the spikes indicated by blue dots in Fig. 4.12B).

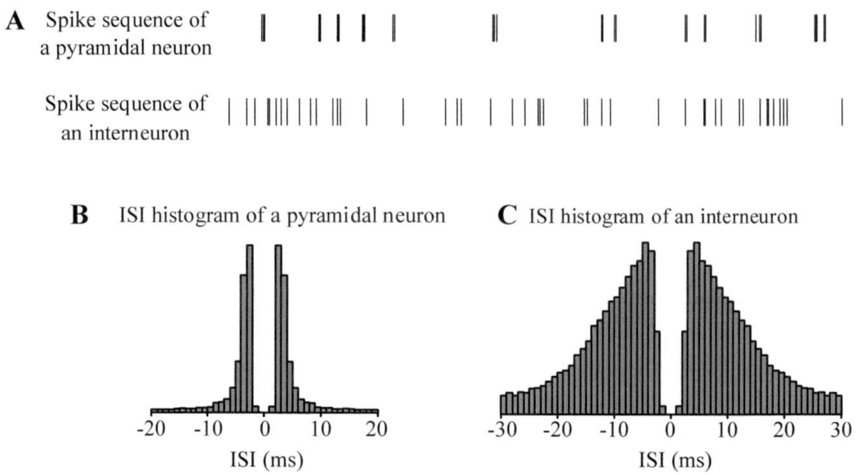

Fig. 4.21 Examples of ISI histograms from a pyramidal neuron and an interneuron in the rat hippocampus. **A** Raster plots showing distinct firing patterns of two neurons. **B** and **C** ISI histograms of a pyramidal neuron (**B**) and an interneuron (**C**)

For principal neurons involved in signal transmission, burst firing can enhance reliability by increasing neurotransmitter release. Isolated spikes can be treated as noise in some brain regions through a synaptic "filtering" effect, while burst spikes act as transmissible signals carrying information (Lisman 1997). ISI histograms can also feature other firing patterns, such as those induced by high-frequency stimulations with a phase-locked feature (see Sect. 5.3).

4.2.2 *Autocorrelation Histogram of Spike Sequences*

An ISI histogram is essentially a component in the autocorrelation function of a spike sequence. For the spike sequence described by Eq. (4.5), its autocorrelation function is (Gerstein and Kiang 1960):

$$\begin{aligned} \varphi_{ff}(\tau) &\equiv \int f(t)f(t-\tau)\mathrm{d}t = \int \sum_{k=1}^{N} \delta(t-t_k) \sum_{l=1}^{N} \delta(t-t_l-\tau)\mathrm{d}t \\ &= \sum_{k=1}^{N}\sum_{l=1}^{N} \delta(t_k - t_l - \tau) \end{aligned} \tag{4.7}$$

This double sum can be decomposed into

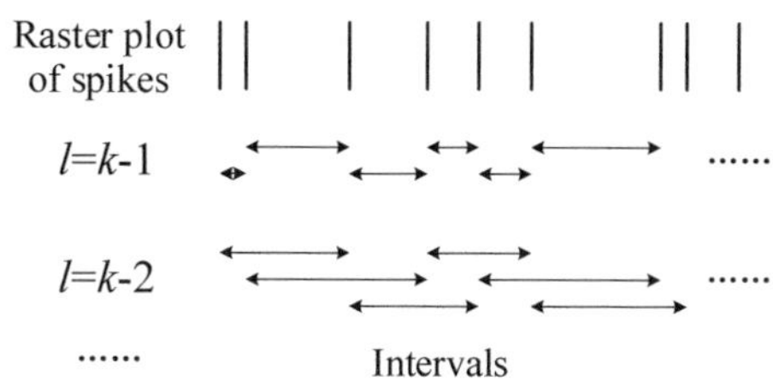

Fig. 4.22 Schematic diagram of the definitions of spike intervals at various orders

$$\varphi_{ff}(\tau) = N\delta(\tau) + \sum_{k=2}^{N} \delta(t_k - t_{k-1} - \tau) + \sum_{k=3}^{N} \delta(t_k - t_{k-2} - \tau) + \cdots \quad (4.8)$$

where each term corresponds to let $l = k$, $l = k - 1$, $l = k - 2$, ... in Eq. (4.7). Therefore, the autocorrelation function is still a sequence of δ functions. Discretize the continuous time τ into $\tau_j, j = 1, 2, 3, \ldots$. All τ values that satisfy $\tau_j \le \tau < \tau_{j+1}$ are measured as τ_j. Thus $\tau_{j+1} - \tau_j \equiv \Delta\tau$ defines a time bin that determines the resolution. Consequently, all δ functions of φ_{ff} falling between τ_j and τ_{j+1} are summed to yield the value of the autocorrelation histogram at τ_j.

The first term (where $l = k$) in Eq. (4.8) represents the constant N at $\tau = 0$, which is the total number of spikes in the sequence. When the sequence duration is known, this term reflects the average spike firing rate. The second term (where $l = k - 1$) represents the intervals between two adjacent spikes—the kth and $(k - 1)$th, which forms the ISI histogram represented by Eq. (4.6). The third term (where $l = k - 2$) is another interval histogram measuring intervals between every other spikes—the k-th and $(k - 2)$th (see Fig. 4.22). It is known as the second-order interval histogram. All subsequent terms represent interval histograms for higher-order intervals in the spike sequence.

The autocorrelation function of a spike sequence can reveal both the refractory period and rhythms in neuron firing. For example, when a hippocampal neuron fires at θ rhythms, the autocorrelation function of its spike sequence can exhibit θ rhythms.

4.2.3 Cross-Correlation Histogram of Spike Sequences

The relationship between spike sequences of two neurons, or between a spike sequence and an external event sequence, can be analyzed using a cross-correlation function. The definition of this function resembles the above-mentioned autocorrelation function. Let the spike sequence of neuron-A be

$$f(t) = \sum_{k=1}^{N_A} \delta(t - t_k) \quad (4.5)$$

and that of neuron-B be

$$g(t) = \sum_{l=1}^{N_B} \delta(t - t_l) \tag{4.9}$$

Then, the cross-correlation function between the two neurons is (Gerstein and Kiang 1960):

$$\begin{aligned} \varphi_{fg}(\tau) \equiv \int f(t)g(t-\tau)\mathrm{d}t &= \int \sum_{k=1}^{N_A} \delta(t - t_k) \sum_{l=1}^{N_B} \delta(t - t_l - \tau)\mathrm{d}t \\ &= \sum_{k=1}^{N_A} \sum_{l=1}^{N_B} \delta(t_k - t_l - \tau) \end{aligned} \tag{4.10}$$

It is still a sequence of δ functions. By discretizing the continuous time τ into small bins τ_j ($j = 1, 2, 3, \ldots$), we can create a cross-correlation histogram. As shown in Fig. 4.23, using neuron-A sequence as the reference, count the intervals between neuron-B spikes and each reference spike that fall within each $\Delta\tau$ to form the histogram. Its x-axis represents time, while its y-axis represents spike count. Dividing the spike count by the total spike number yields neuron-B firing probability against τ, producing a histogram showing how neuron-B fires in relation to neuron-A's firing.

When the spike sequences of both neurons are independent, the expected value for each $\Delta\tau$ of the cross-correlation histogram is $E = N_A N_B \Delta\tau / T$, where N_A and N_B represent the total spike numbers of the two neurons, and T represents the recording duration.

A cross-correlation histogram of spike sequences can reveal direct synaptic connections between a neuron pair (Barthó et al. 2004). When a connection exists,

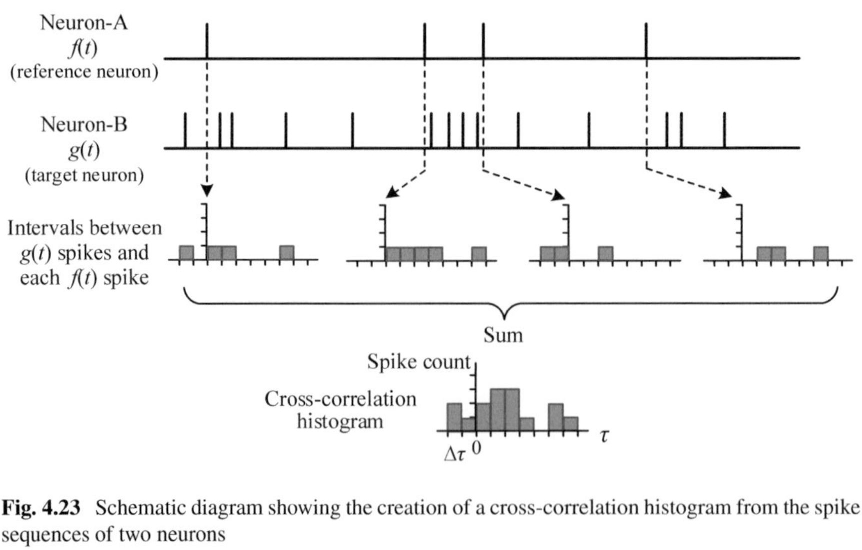

Fig. 4.23 Schematic diagram showing the creation of a cross-correlation histogram from the spike sequences of two neurons

the firing of the presynaptic neuron can either excite or inhibit the postsynaptic neuron after a delay of a few milliseconds caused by axonal conduction and synaptic transmission. For example, an excitatory connection from a presynaptic neuron (reference) to a postsynaptic neuron (target) can produce a histogram with a prominent peak around 2 ms which can represent an interneuron's firing triggered by a pyramidal neuron. In contrast, an inhibitory connection can result in a distinct gap after time zero in the histogram which indicates that the target neuron's firing is suppressed immediately following the reference neuron's firing. In addition, bidirectional synaptic connections between two neurons—where an interneuron (reference) inhibits a pyramidal neuron (target) while the latter excites the former—can show a peak to the left and a gap to the right of time zero in the histogram. This pattern reflects mutual synaptic connections, such as the local feedback inhibitory circuits in the hippocampus (see Sect. 2.3).

Nevertheless, in paired neurons with direct synaptic connections, not every spike of the presynaptic neuron can trigger or suppress the firing in the postsynaptic neuron. This is because each neuron typically has thousands of synapses receiving inputs from many other neurons. To produce an action potential, it typically requires simultaneous excitatory inputs from multiple synapses. The activation from a single presynaptic neuron can only increase or decrease the firing probability of the postsynaptic neuron. This probabilistic nature of synaptic transmission results in peaks or gaps in cross-correlation histograms.

If the reference sequence $f(t)$ is not a neuronal spike sequence but rather a series of external stimuli (e.g., a pulse train), the relationship between the spike sequence $g(t)$ and the external events $f(t)$ can also be analyzed using the cross-correlation function—Eq. (4.10). The equation can be rewritten as

$$\varphi_{fg}(\tau) = \sum_{k=1}^{N_A} \left[\sum_{l} \delta(t_l - t_k - \tau) + \sum_{l} \delta(t_l - t_k - \tau) \right]$$
$$(t_k < t_l < t_{k+1}) \ (t_l \geq t_{k+1} \text{ or } t_l \leq t_k). \tag{4.11}$$

On the right side of this equation, the first term in the square brackets represents the cumulative count of intervals between each spike in $g(t)$ and its nearest preceding stimulus in $f(t)$ that fall within each $\Delta\tau$, i.e., $t_k < t_l < t_{k+1}$. The second term represents the cumulative count of intervals between each spike and all other stimuli that fall within each $\Delta\tau$, i.e., $t_l \geq t_{k+1}$ or $t_l \leq t_k$.

When the time interval between adjacent applied stimuli is sufficiently long, the effect of one stimulus on neuronal firing can completely dissipate before the next stimulus arrives. In this case, the first term in Eq. (4.11) represents the post-stimulus time histogram (PSTH), also known as the "peri-stimulus time histogram", which is the sum of spike histograms generated by N_A repeated stimuli. A clear peak in the PSTH indicates that the neuron fires in relation to the stimulation. Conversely, a flat and uniform PSTH suggests that the neuronal firing occurs randomly, unrelated to the stimulation. The PSTH is one of the principal methods for analyzing neuronal responses to external stimuli which will be introduced in the next section, along with other methods including joint peri-stimulus scatter diagram (JPSSD) and joint peri-stimulus time histogram (JPSTH).

4.2.4 Analyses of Neuronal Responses to External Stimuli

1. **Peri-stimulus Time Histogram (PSTH)**

As defined by the first term of Eq. (4.11), the PSTH calculation involves two main steps: (1) Divide the time axis (x-axis) evenly into small bins with a resolution of $\Delta\tau$, using the stimulus timing as zero; (2) Count the spikes that fall into each $\Delta\tau$ bin across all stimulation cycles to create the histogram (Rolls et al. 1989). Figure 4.24 illustrates the responses of a neuron to 50 cycles of a 100 Hz pulse stimulation. It shows both the raster plot of neuronal firing (upper) and its PSTH (bottom). The red arrow marks the zero time point when each pulse occurred. Using a bin size of $\Delta\tau = 0.5$ ms, this PSTH reveals the temporal relationship between the neuronal spikes and the stimulation pulses, with a peak appeared at 7–7.5 ms, indicating the response latency. Notably, not every pulse triggered a spike. The PSTH can also be presented as a curve instead of a bar plot as in this figure. Additionally, the y-axis can display either spike count or firing probability—known as the post-stimulus probability distribution.

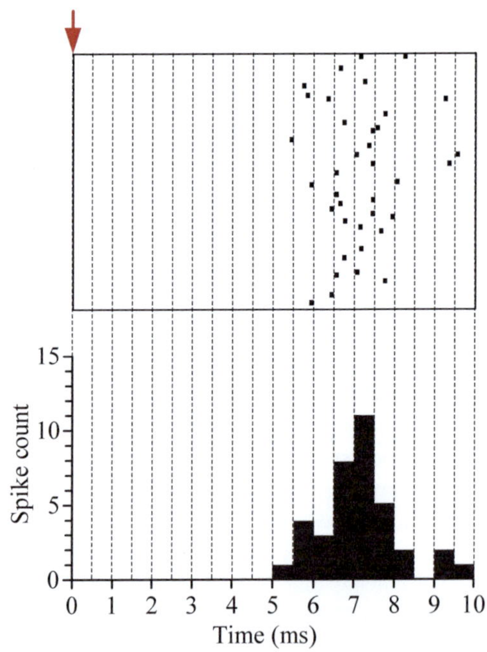

Fig. 4.24 Creation of a PSTH histogram showing neuronal responses during 50 cycles of 100 Hz pulse stimulation

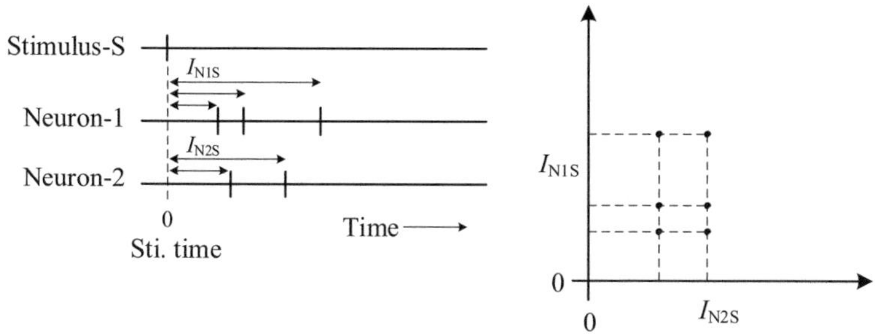

Fig. 4.25 Creation of JPSSD. *Left*: Definition of variables I_{N1S} and I_{N2S}, which respectively represent intervals between stimulus-S and the spike sequences of Neuron-1 and -2 during a stimulus cycle. *Right*: JPSSD scatter plot created using I_{N1S} and I_{N2S} variables. Modified from Gerstein and Perkel (1972)

2. Joint Peri-Stimulus Scatter Diagram (JPSSD)

When the firing of Neuron-1 and -2 are both related to a same stimulation, their relationship can be described using a joint peri-stimulus scatter diagram (JPSSD). Figure 4.25 shows the definitions of two interval variables—I_{N1S} and I_{N2S} (*left*), and the creation of JPSSD as a two-dimensional scatter plot using these variables (*right*). The I_{N1S} (y-axis) represents the intervals between spikes of Neuron-1 and stimulus-S, while the I_{N2S} (x-axis) represents the intervals between spikes of Neuron-2 and stimulus-S. With the stimulus time set as $t = 0$, when Neuron-1 has a spike at time t_{N1} and Neuron-2 has a spike at time t_{N2}, draw a dot at position (t_{N1}, t_{N2}) on the two-dimensional plot. Draw all dots sequentially within one stimulus cycle and then repeat for other cycles. The collection of all these dots forms the JPSSD.

The following JPSSD examples show various relationships between the stimulation S and the spike sequences of two neurons (N1 and N2). The JPSSDs can be analyzed along with the cross-correlation histograms of the two neurons and their respective PSTHs (Gerstein and Perkel 1972).

(1) When the two neurons fire independently and are unaffected by the stimulation, the JPSSD appears as uniformly distributed dots. In such case, the cross-correlation histogram between N1 and N2 and their PSTHs all appear flat, lacking any significant peaks.
(2) When both neurons fire independently but N2 is excited by the stimulation, a high-density band appears on the JPSSD parallel to the y-axis, indicating an increase of N2's firing with a latency. This excitatory effect also produces a peak on the N2 PSTH, while the N1 PSTH and the cross-correlation histogram between the two neurons remain flat. Similarly, when the stimulation excites both neurons (with no direct connection between them), high-density bands appear parallelly to both the y-axis and x-axis. Correspondingly, peaks appear on both neurons' PSTHs, and a peak appears near the zero time on their cross-correlation histogram, suggesting that both neurons are simultaneously controlled by the same input.

(3) When stimulation has no effect on either neuron, but N2 acts on N1 through excitatory synapses, the JPSSD displays no bands parallel to the coordinate axes but shows a high-density diagonal band parallel to the main diagonal, slightly offset to the left, indicating the synaptic connection. The band offset and width reflect the latency and duration of synaptic actions, respectively. In the cross-correlation histogram, a peak occurs near zero, which is unrelated to the stimulation input. The PSTHs of both neurons remain flat. If a third neuron simultaneously acts on the two neurons through excitatory synapses, while the stimulation remains no effect on these neurons, a high-density diagonal band will also appear, along with a peak in the cross-correlation histogram.

(4) When the stimulation excites both neurons, and N2 acts on N1 through excitatory synapses, high-density bands appear parallelly to both coordinate axes, together with a high-density diagonal band. These combined effects result in evident peaks in the cross-correlation histogram and the PSTHs of both neurons.

In summary, the JPSSD of two neurons can reveal the effects from both external stimulation and neuronal interconnections. High-density bands parallel to the coordinate axes represent excitatory stimulation effects on neurons, while a high-density diagonal band parallel to main diagonal represents synchronous firing of both neurons. A narrow diagonal band indicates a direct excitatory synapse connection between the two neurons, whereas a wide and sparse diagonal band indicates both neurons being affected by a third source.

3. **Joint Peri-Stimulus Time Histogram (JPSTH)**

Both cross-correlation histogram and JPSSD can reveal the relationships between the spike sequences of two neurons. However, a cross-correlation histogram only shows the average result of entire spike sequences. Although it can reveal whether one neuron's firing leads or lags another's, it cannot pinpoint when this relationship occurs—whether throughout the entire sequence or only during specific periods. A joint peri-stimulus time histogram (JPSTH) based on JPSSD can offer more details of dynamic changes in the neuronal relationships and in the effects of external stimulation (Gerstein et al. 1989).

4.3 Spectral Analysis and Its Applications

Frequency spectrum analysis, also known as spectral analysis, can reveal the frequency components of a signal in the frequency domain. For signals with inherent randomness, such as EEG and LFP, a spectrum can be estimated using either parametric or non-parametric methods. The non-parametric method, also called classical spectrum estimation, is widely used and employs two basic approaches (Shiavi 2007). The first one, known as the direct or periodogram method, produces a spectrum from the Fourier transform of a sampled signal based on Parseval's theorem. The second one, the indirect method, produces a power spectrum based on the principle that the power spectral density is the Fourier transform of signal's autocorrelation function. We have been using the LabChart software to calculate the spectra of hippocampal field potentials. This software provides a "Spectrum" window, which uses the periodogram method for spectral analysis. The following sections will explain the basic principle of periodogram method and then discuss the parameter settings and applications of LabChart spectrum analysis.

4.3.1 Periodogram Method for Spectrum Analysis

Let $x(n)$ be the discrete signal produced by ADC with a sampling frequency of f_s and a corresponding sampling period of $\Delta T = 1/f_s$. As a sequence of N samples, $x(n)$ can be written as $x(1)$, $x(2)$, …, $x(N)$. Its discrete Fourier transform (DFT) is:

$$X(f) = \Delta t \sum_{n=1}^{N} x(n) e^{-2\pi j f_n \Delta t} \tag{4.12}$$

The frequency range of $X(f)$ is $-f_s/2 \le f \le f_s/2$, with a resolution of $\Delta f = 1/(N\Delta t)$. The $X(f)$ is a discrete sequence of N data points: $X(1)$, $X(2)$, …, $X(N)$, evenly spaced by Δf. The plot of $X(f)$ mode ($|X(f)|$) against frequency is known as amplitude spectrum. Squaring $|X(f)|$ produces the energy spectrum. Then, dividing the energy by the signal duration $N\Delta t$ produces the power spectrum. Finally, power spectral density (PSD) is the power per unit frequency. Thus, a spectrum can be expressed in four forms: amplitude, energy, power, and PSD. For a real sequence $x(n)$ (e.g., an actual sampled signal), the negative half of its frequency spectrum mirrors the positive half, making it sufficient to calculate only the positive half.

For a random signal, estimating its spectrum from a single signal episode can yield unreliable results due to high variance. Fortunately, periodogram averaging—also known as the Bartlett approach—can effectively reduce the variance. This method divides a signal of N sampled data into K segments of length M, calculates the spectrum of each segment separately, and then averages the results. This process can reduce the variance of spectrum estimation to $1/K$ (Shiavi 2007).

The duration of a signal used to calculate the discrete Fourier transform determines the spectral frequency resolution. When the signal is segmented, it becomes shorter, resulting in a decreased spectrum resolution. Therefore, it is crucial to compromise on both spectral variance and resolution when setting the segment number. To enhance spectral resolution, segments can be overlapped and reused to yield longer segments, resulting in the modified Welch method. It divides the signal into segments with M sampled data, where the starting points of adjacent segments are apart by only D sampled data, less than M. The number of segments becomes $K = (N - M)/D + 1$, as shown in Fig. 4.26. The overlap ratio is $(M\text{-}D)/M \times 100\%$. Increasing this ratio can either raise the segment number K (with a fixed M) or extend the segment length M (with a fixed K), or increase both$K$ and M. However, an excessively high overlap

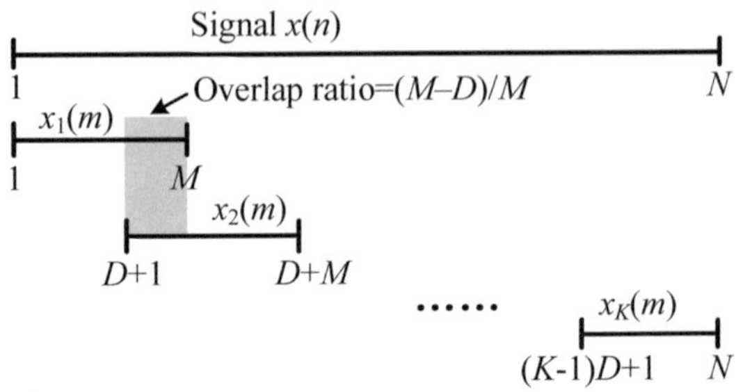

Fig. 4.26 Schematic diagram of segmentation for signal$x(n)$. The shading represents the overlapping area with an overlap ratio of $(M - D)/M$

ratio cannot further improve the spectrum variance. Typically, the overlap ratio is set at about 50%.

A sampled signal has resolution and range (or duration) in both its frequency and time domains. Understanding these parameters and their relationships is essential. In frequency domain, the frequency resolution of spectrum is determined by the duration of the signal used to calculate the Fourier transform in the time domain (i.e., the signal after segmentation). This resolution is the reciprocal of the signal duration. For example, a 10-s signal gives a frequency resolution of 0.1 Hz, while a 0.1-s signal only achieves 10 Hz—waveforms below 10 Hz cannot be accurately captured in such a short time window. Additionally, the spectrum repeats periodically with a period equal to the sampling frequency (f_s). For real signals, the basic positive frequency range is from 0 to $f_s/2$. For instance, with $f_s = 1000$ Hz, the spectrum upper limit is 500 Hz, as waveforms with a frequency beyond this limit cannot be fully captured. In the time domain, the signal resolution (sampling period) is $\Delta T = 1/f_s$—the reciprocal of the spectral period, while the signal duration is the reciprocal of the frequency resolution. These relationships between time and frequency domains exhibit symmetry.

Spectral leakage is another issue to consider when using the periodogram method for spectrum estimation. Extracting a segment from a recorded signal is equivalent to applying a rectangular window to the signal. The spectrum of this window has large sidelobes, leading to leakage and thus deviation in the estimated spectrum. To reduce the leakage, a common approach is to multiply each signal segment by a non-rectangular window with smaller sidelobes—such as Hamming or Hanning windows—before calculating the spectrum.

Popular signal processing software like MATLAB can easily calculate spectra of brain signals such as EEG and LFP. For example, MATLAB function "pwelch" can compute the PSD using the Welch periodogram method. The function's calling format is:

$$[\mathrm{Pxx}, \mathrm{f}] = \mathrm{pwelch}(\mathrm{x},\ \mathrm{window}, \mathrm{noverlap}, \mathrm{nfft},\ \mathrm{fs}) \tag{4.13}$$

In the parameter list on the right side, "x" represents the sampled signal, "window" is the chosen window with an equal length to the signal segment, "noverlap" is the overlap ratio between adjacent segments, "nfft" is the number of data points used for the Fourier transform, and "fs" is the sampling frequency of signal x. This pwelch function uses the Fast Fourier Transform (FFT) algorithm to speed up computation when "nfft" is a power of 2. Note that the window length must not exceed "nfft" to avoid errors. In the return list on the left side, "Pxx" represents the PSD of signal x, while "f" is the corresponding frequency vector. Additionally, the pwelch can provide confidence intervals for the spectrum estimation. For more details, refer to the MATLAB help documentation.

Figure 4.27 shows examples of PSD calculations for EEG signals collected from the central (C3 and C4) and occipital (O1 and O2) areas of an awake, resting adult subject with closed eyes. The EEG signals were 10 s in length with a sampling frequency of 250 Hz. Their spectra were estimated using the Welch periodogram method, with a segment length of 256 data points, a Hanning window, and a 50% overlap. When the subject was awake and quiet (Fig. 4.27A), the EEG signals showed prominent α rhythms in the 8–13 Hz range, which produced clear peaks in their power spectra (Fig. 4.27C). In contrast, when the subject performed complex mental arithmetic (Fig. 4.27B), the α rhythms were suppressed, resulting in a significant decrease in the power of α frequency range (Fig. 4.27D).

After obtaining the PSD, the mean power in each rhythmic range can be calculated based on the definitions of rhythmic frequency ranges for EEG and LFP (Feng and Zheng 2004a). For example, we can estimate the power of δ (0.5–4 Hz), θ (4–8 Hz), α (8–13 Hz), and β (14–30 Hz) rhythms. From this, the amplitude of individual rhythms can be determined. An alternative approach is to calculate the ratio of power in each rhythm range to the total signal power. This method can eliminate individual variations, facilitating comparisons among different subjects.

Long recordings of spontaneous EEG and LFP signals are typically non-stationary, changing with the behaviors and mental states of humans or animals. The above-mentioned spectral estimations are only suitable for stationary signals. Time–frequency analyses—such as short-time Fourier transform (STFT) and wavelet transform (WT)—are more suitable for non-stationary signals. STFT analyzes a signal using a shifting window with a fixed length, exhibiting time–frequency features in a two-dimensional plot. However, STFT has a trade-off: shorter signals result in poorer frequency resolution, making it impossible to achieve high resolutions in both time and frequency domains simultaneously. Unlike STFT, WT can decompose a signal using basic wavelets with varying displacement and scale factors. By using larger windows for lower frequency components and shorter ones for higher frequency components, WT can meet resolution requirements in both time and frequency domains (Clark et al. 1995; Feng 2003; Feng and Zheng 2004b). In MATLAB, the "spectrogram" function can calculate STFT, while its Wavelet Toolbox provides various functions for performing wavelet transforms.

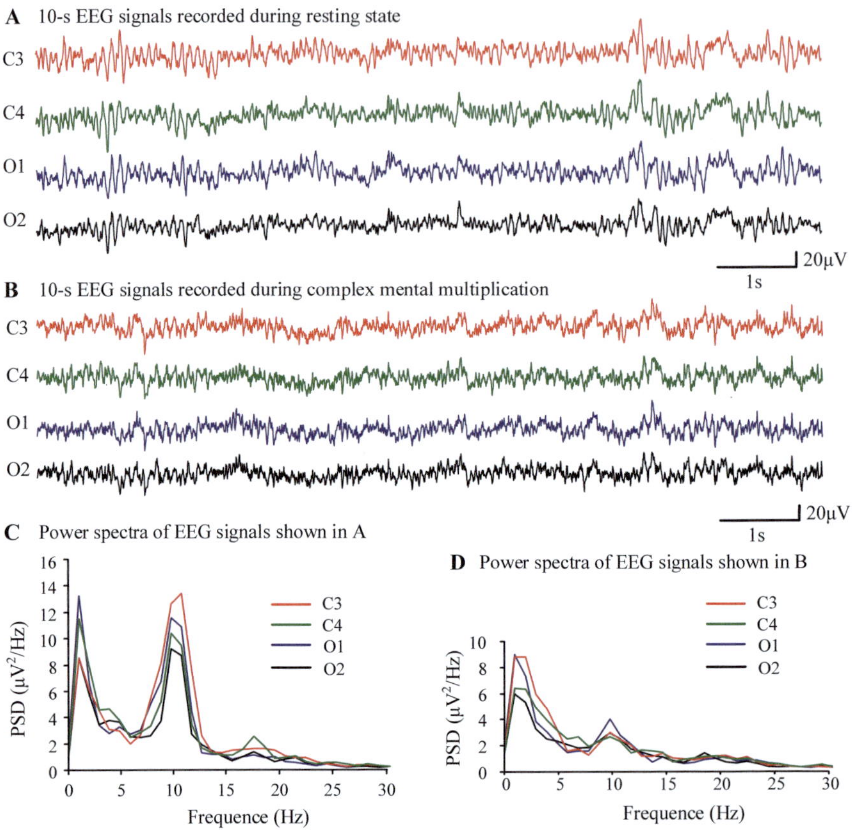

Fig. 4.27 Examples of EEG signals and their PSDs from an adult subject. **A** and **B** 10-s EEG recordings from the central (C3 and C4) and occipital (O1 and O2) areas during eyes-closed resting state under two conditions: task-free (**A**) and complex mental arithmetic (**B**). **C** and **D** PSDs of the EEG signals shown in (**A**) and (**B**), respectively

4.3.2 *Spectral Analyses in LabChart*

We used a PowerLab recorder to collect signals in the rat brain (see Sect. 3.5.3). This recorder comes with the LabChart software, which offers spectrum analysis using the periodogram method. It employs FFT to calculate the discrete Fourier transform (Eq. (4.12)). FFT requires the number of data points to be a power of 2. As shown in Fig. 4.28, the "FFT size" option in the LabChart settings dialog provides options including 32, 64… 1 K (1024), 2 K (2048)…to 128 K (131,072) in the drop-down menu, all powers of 2. This FFT size corresponds to the data length M shown in Fig. 4.26. Note that this M is the length of divided signal segments, not the total signal length. The M is crucial as it determines the spectral frequency resolution. For example, using the default FFT size of 1 K (1024 sampled data points) for a signal

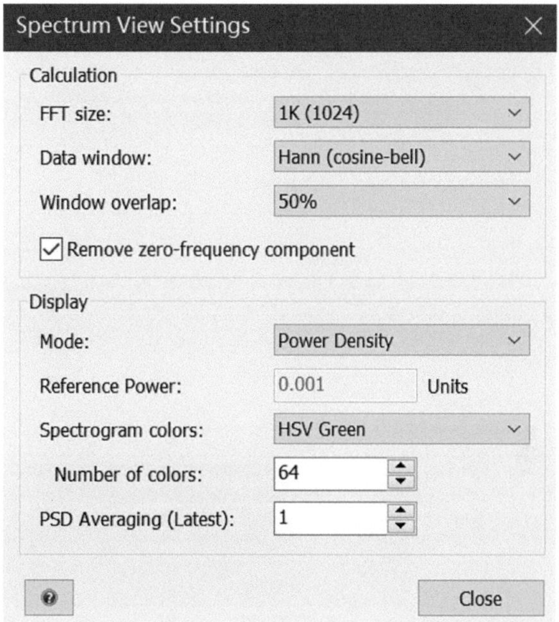

Fig. 4.28 Settings dialog for spectrum analysis in LabChart software

sampled at a 20 kHz frequency results in a spectral resolution of only 20000 Hz/1024 = 19.531 Hz. This means that the spectrum of low-frequency field potentials below ~ 20 Hz cannot be obtained. Increasing the FFT size to 32 K (32,768) or 64 K (65,536) can improve the spectral resolution to 0.610 Hz or 0.305 Hz.

The "Data window" option in the settings dialog allows user to choose a window applied to each signal segment to reduce leakage error. Available windows include Bartlett, Hamming, and Hanning windows. The "Window overlap" represents the overlap ratio between adjacent signal segments—as shown in Fig. 4.26. The "Mode" option sets the spectrum form, with options including amplitude, power, and power density.

In the cortex and hippocampus of a urethane-anesthetized rat, the recorded LFP signals typically exhibit large-amplitude slow waveforms of δ rhythms. However, when a small clip is used to slightly clamp the rat's tail, the LFPs can quickly shift to a signal dominated by small-amplitude θ rhythms (Buzsáki et al. 1983). Here, I use such a recording as an example to demonstrate LabChart spectral analysis. The upper row of Fig. 4.29A shows an LFP signal from the rat hippocampus, which spanning ~ 867 s and containing two tail-clamping events. The expanded view below clearly shows the LFP change caused by the first tail clamping.

Before entering the PowerLab recorder for sampling, the signal was magnified 100-fold by the pre-amplifier (Model 3600) with a frequency range of 0.3–5000 Hz (see Sect. 3.5 for details). Therefore, a ± 1 mV signal appears as ± 100 mV in Fig. 4.29A.

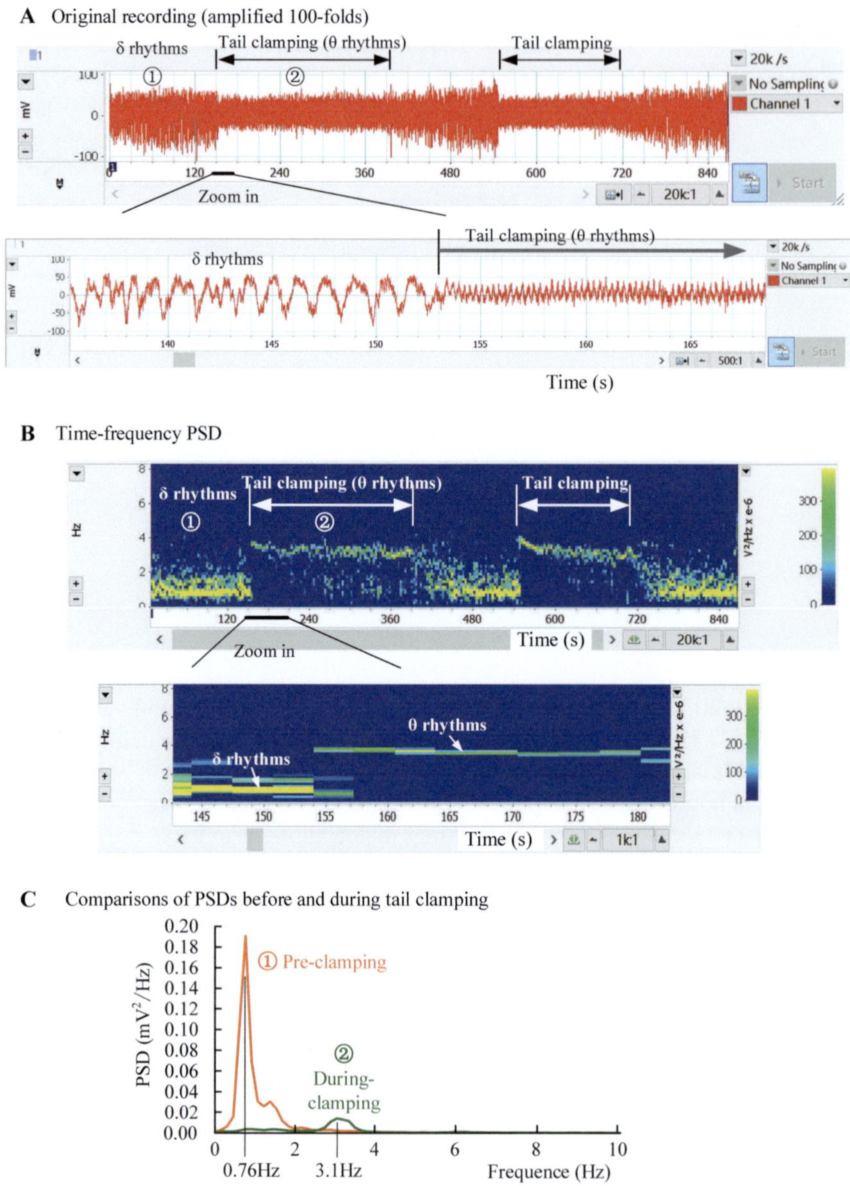

Fig. 4.29 Analysis of the frequency spectra for LFP signals recorded from rat hippocampus using LabChart. **A** Original recording of ~ 867 s (*top*) and its enlarged ~ 33 s segment (*bottom*). **B** Time–frequency PSD of the entire signal in A, with an enlarged segment showing the period from 143 to 182 s. **C** Mean PSDs for the LFP signals before (orange) and during tail clamping (green), marked by ① and ② in A and B

To perform the STFT spectral analysis in LabChart, set the FFT size to 128 K—$M = 131{,}072$. With a sampling frequency of 20 kHz, this FFT size produces a signal segment of 6.5536 s, yielding a spectral frequency resolution of 0.153 Hz. Using a Hamming window with 50% overlap can produce a time resolution of 6.5536 s/2 = 3.2768 s for the time–frequency spectrum. Figure 4.29B shows the PSD of the entire signal shown in Fig. 4.29A, demonstrating that both tail clamping events shifted the LFP from δ rhythm (~ 0.76 Hz) to θ rhythm (~ 3.5 Hz) dominance. The expanded view below focuses on the period from 143 to 182 s, clearly showing discrete spectral data with a time resolution of about 3.3 s. The large-amplitude δ rhythms (yellow) before tail clamping and the small-amplitude θ rhythms (green) during tail clamping are distinctly visible.

Figure 4.29C shows the mean PSDs of the two signals: one before tail clamping (~153 s) and the other during tail clamping (~ 262 s), marked by ① and ② in Figure 4.29A, B. The pre-clamping PSD (orange), calculated from 47 signal segments ($K = 47$), shows a δ peak at 0.76 Hz. The during-clamping PSD (green), calculated from 78 signal segments ($K = 78$), shows a θ peak at 3.1 Hz. The δ peak (0.19mV^2/Hz) is approximately 14 times that of the θ peak (0.014mV^2/Hz), indicating a 3.7:1 amplitude ratio in the original recordings. Note: the θ rhythm frequency in rat LFPs under urethane anesthesia is about 2–5 Hz, lower than the 6-9 Hz observed during awake state (Buzsáki 2002).

We have used the power spectrum analysis to study various neurophysiological phenomena in rats, including exploring changes in cortical electroencephalogram (ECoG) across different sleep stages and at various depths of anesthesia (Feng 2003; Feng and Zheng 2004a), as well as analyzing changes in hippocampal field potentials in response to high-frequency pulse stimulations (Yu et al. 2016; Ma et al. 2019).

4.4 Current Source Density Analysis

Neurons can produce electrical activity through transmembrane currents. In extracellular space, these currents create "sources" where they flow out of membrane and "sinks" where they enter membrane, as illustrated in Fig. 2.5C. Analyzing the spatiotemporal distribution of the current sources and sinks can provide insights into neuronal activity and explore the initiation and propagation of neuronal firing. However, extracellularly recorded field potentials cannot directly reflect neuronal current flow. In 1975, Nicholson and Freeman proposed a theorem of current source density (CSD) to estimate current sources and sinks using field potentials (Nicholson and Freeman 1975). The CSD calculation involves the second-order differences of a series of LFP recordings along a specific spatial axis. In early studies, when glass pipette or metal wire microelectrodes with only a single measuring contact were used for extracellular recordings, researchers had to move the electrode step-by-step to measure LFPs at different depths. This method required the LFPs to be repeatable after each electrode movement. It was therefore primarily used for analyzing repetitive evoked potentials (López-Aguado et al. 2000; Heynen and Bear 2001), not for random spontaneous potentials. This limitation has been overcome with the introduction of array microelectrode.

A linear array with multiple contacts can simultaneously record field potentials at different depths, enabling CSD analysis. This technique is widely used to analyze neural signals in both animals and humans (Townsend et al. 2002; Csercsa et al. 2010). The algorithm for estimating CSD from LFPs is essentially a "reverse problem" that requires additional constraints to solve (Pettersen et al. 2012). Let $V(t, x, y, z)$ be the LFP recorded at time t at extracellular site (x, y, z), and let σx, σy, and σz be the conductivity in the x, y, and z directions, respectively. The CSD at site (x, y, z) at time t is

$$\text{CSD}(t, x, y, z) = -\left(\sigma_x \frac{\partial^2 V}{\partial x^2} + \sigma_y \frac{\partial^2 V}{\partial y^2} + \sigma_z \frac{\partial^2 V}{\partial z^2}\right) \tag{4.14}$$

A negative CSD value indicates current flowing into the cell—extracellularly forming a sink, while a positive value indicates current flowing out of the cell—extracellularly forming a source. In brain regions with regularly arranged neurons in well-defined layers, such as the hippocampal CA1 region, one-dimensional CSD parallel to the neuronal soma-dendritic axis (z-axis) can be calculated to analyze trans-membrane currents at each layer during synchronous firing of a neuron population. The impact of extracellular conductivity changes on CSD is minimal. The conductivity values can be considered constant (Kloosterman et al. 2001; Richardson et al. 1987). Additionally, when only relative CSD values are needed for comparisons, conductivity values can be omitted. Thus, the calculation of one-dimensional CSD(t, z) in the hippocampal CA1 region can be simplified to the following second-order difference:

$$\text{CSD}(t, z) = \frac{2V(t, z) - V(t, z + N\Delta z) - V(t, z - N\Delta z)}{(N\Delta z)^2} \tag{4.15}$$

where $V(t, z)$ is the LFP at depth z at time t. Δz is the spacing between adjacent recording channels in a vertically arranged array (e.g., 50 μm), which determines the CSD spatial resolution. When N is set to 2, the LFP sampling spacing becomes 100 μm, a spacing suitable for CSD calculation (Pettersen et al. 2012).

As shown in Fig. 4.30A, we used a 16-channel array with 50 μm spacing to record the LFPs in rat hippocampal CA1 region for analyzing the initiation and propagation of the APSs evoked by alveus stimulations and the OPSs evoked by Schaffer collateral stimulations (see Sect. 2.3 for details). Signals from 9 channels (Fig. 4.30B1), spanning a ± 200 μm range around the soma layer, were used to calculate the CSD across the CA1 region—from the basal dendrite layer (so), through the soma layer (pcl), to the apical dendrite layer (sr) (Fig. 4.30B2). The APS current sink initiated above the soma layer, then propagated downward to the soma and apical dendrites while simultaneously propagated upward from the soma to the basal dendrites (Fig. 4.30B2 left). In contrast, the OPS current sink began in the apical dendrite layer and then propagated sequentially to the soma and basal dendrite layers

(Fig. 4.30B2 right). During APS generation, its negative CSD sink represents the current component from action potentials. Unlike APS, OPS is evoked by excitatory synaptic inputs on the apical dendrites of pyramidal neurons. The accompanying EPSPs produce an fEPSP current sink in the far apical dendrite layer and a passive current source near the soma layer (refer to Fig. 2.5C). Thus, the OPS current sink (indicated by the grey shadings in Fig. 4.30B2 right) rides on the fEPSP current source. These results are consistent with other reports (Kloosterman et al. 2001). Using CSD analysis, we examined how action potentials generated and propagated during high-frequency stimulation of alveus axons (see Sect. 5.2.3 for details).

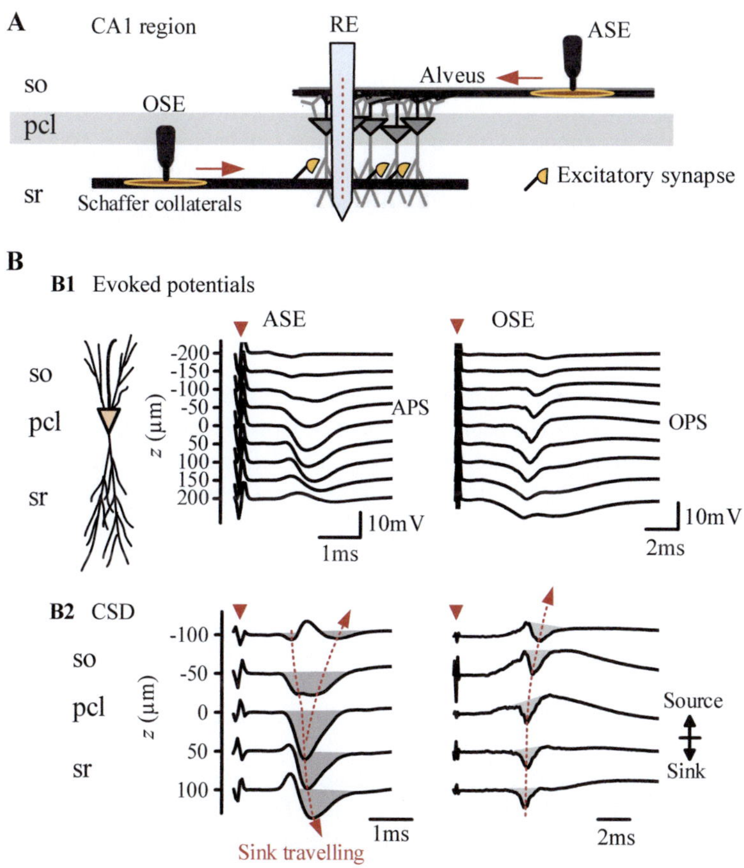

Fig. 4.30 CSD analyses of antidromic and orthodromic evoked potentials in the peri-somatic area of rat hippocampal CA1 region. **A** Schematic diagram illustrating the positions of a recording array and two stimulation electrodes. **B** Recordings of the evoked potentials (B1) and their corresponding CSDs (B2). The site of measuring contact at the soma layer—the pyramidal cell layer (pcl)—is set as the z-axis zero point. Red inverted triangles denote 0.3 mA stimulus pulses (i.e., the truncated pulse artifacts). Grey shadings in B2 represent current sinks of PSs, and red dashed lines with arrows indicate the directions of current sink propagation. From Xu et al. (2023)

4.5 Methods for Removing Stimulation Artifacts

Part II of this book will explore the effects of external electrical stimulations on neurons and their networks in the rat brain. Brain tissue acts as a volume conductor. When electrical stimulation is applied through implanted electrodes, it can generate an electric field within the brain. The strength of this field decreases rapidly over distance. When measuring neuronal responses to electrical stimulation in the brain, the recorded potentials can contain stimulation artifacts if the recording site lies within the produced electric field.

Brain tissue is not purely resistive but has passive electrical properties of both resistance and capacitance. It also has active electrical properties formed by ion channels on neuronal membranes. These electrical properties vary across anatomical structures—cell bodies (somata), axons, dendrites, and cerebrospinal fluid in the extracellular space. Therefore, brain tissue acts as a non-uniform conductor, which can be simulated by circuits with varying resistance and capacitance. When considering electrical stimulation as input and the recorded stimulation artifact as output, the artifact can typically differ from the original input waveform due to these complex tissue circuits. This difference is especially pronounced for stimulation waveforms with higher frequency components like narrow pulses. Additionally, the recorded stimulation artifacts depend on other factors, such as the structure of stimulation electrode, the stimulation site, its distance from the recording electrode, and the sampling frequency.

A stimulation artifact can serve as a useful time reference for determining the latency and timing of evoked neuronal responses. However, dense artifacts can interfere with neural signal detection and analysis. The following examples demonstrate stimulation artifacts recorded in the rat hippocampal region during sinusoidal and pulse electrical stimulations, along with methods for their removal.

4.5.1 Removing Sinusoidal Stimulation Artifacts

To investigate how the waveforms of stimulation artifacts change with the relative position and distance between stimulation and recording electrodes, we designed a simple setup to imitate stimulation and recording in rat brain. As shown in Fig. 4.31A, we filled a glass dish (75 mm diameter, 15 mm height) with physiological saline and inserted two electrodes: a bipolar concentric stimulation electrode (CBCSG75) and a 16-channel array recording electrode (A1 × 16-Poly2-5 mm-50 s-177). We connected a 50 Hz sinusoidal signal with a 5 V peak-to-peak amplitude from a signal generator (Model DG1000Z, RIGOL Technologies) to a Model 2200 analog stimulator with electrical isolation (see Sect. 3.5.4). Then, the stimulator's output—current sine waves—was connected to the stimulation electrode, with its inner pole serving as anode and outer pole as cathode.

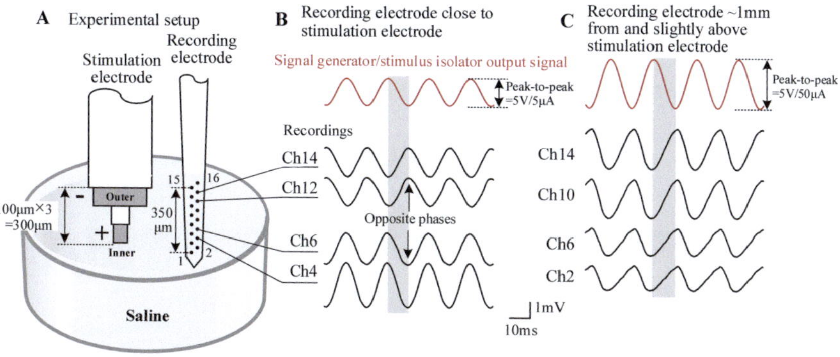

Fig. 4.31 Changes in recorded signals on an array electrode during sinusoidal current application via a stimulation electrode in physiological saline. **A** Schematic diagram of the experimental setup. **B** Signals recorded with the recording array placed close to the stimulation electrode, where the 16 contacts vertically spanned both the inner and outer poles of the stimulation electrode, and with the stimulation intensity set at 5 μA. **C** Signals recorded with the recording array positioned slightly above the stimulation electrode tip, with the stimulation intensity set at 50 μA

We set the stimulator output to a 5μA peak-to-peak amplitude and placed the recording array near the stimulation electrode, allowing the recording contacts to span both poles of the stimulation electrode (Fig. 4.31A). The recorded signals closely matched the sine waves from both the signal generator and stimulator outputs, without significant distortions (Fig. 4.31B). The phase of recorded sine waves on each channel aligned with its nearer stimulation pole. We then increased the stimulator output from 5 to 50 μA peak-to-peak sine waves and moved the recording array about 1 mm away and above the stimulation electrode. As all the recording channels were now closer to the cathode pole, the recorded sine waves aligned with the cathode pole phase, opposing to the original stimulator output signal. Additionally, the recorded waveforms became distorted.

We also used this experimental setup to explore artifact waveforms from pulse stimulations, though details are omitted here. Note that this saline model cannot fully replicate the electrical properties of brain tissue. Stimulation artifacts recorded in-vivo in rat hippocampal regions can be more complex.

Figure 4.32 shows a recording with a 2 Hz sinusoidal current stimulation (1-min, 40 μA peak-to-peak amplitude) applied to the Schaffer collaterals of rat hippocampal CA1 region. The signal was recorded in the CA1 pyramidal layer downstream of the stimulation site. The electrode placements were same as shown in Fig. 4.30A. The enlarged insets in Fig. 4.32 show that the recording during stimulation contained clear 2 Hz sinusoidal artifacts which were absent during baseline period before stimulation. These artifacts were removed by a 2 Hz notch filter, as shown in the figure. The sinusoidal stimulation produced a clear 2 Hz rhythm in neuronal firing, particularly visible in the MUA signal produced after 500 Hz high-pass filtering (Fig. 4.32, *lower right*), demonstrating the effect of stimulation modulation.

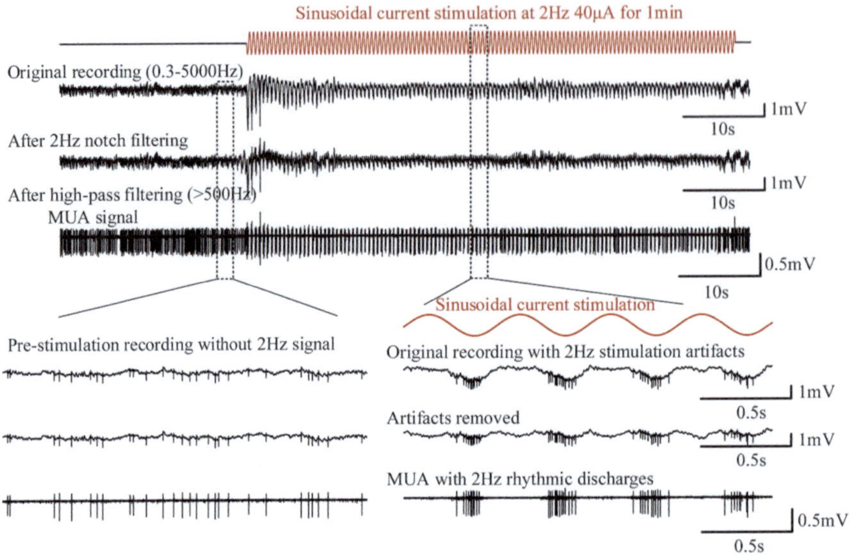

Fig. 4.32 Signal example during sinusoidal stimulation in rat hippocampal CA1 region. The red signal represents the applied sinusoidal current stimulation, i.e., the stimulator output

The sinusoidal artifact appeared as a continuous signal coexisting with neuronal responses throughout the stimulation period. In contrast, a transient narrow pulse can trigger neuronal responses with a time delay (latency), as shown by the single-pulse and paired-pulse evoked population spikes in Sect. 2.2. Even though stimulation pulses and evoked responses occur at different times, artifact removal can still be required for detecting certain neuronal signals, such as unit spikes (see Fig. 4.9). Unlike sinusoidal artifacts—which can be removed using a notch filter, pulse artifacts have a broader spectrum with high-frequency components that overlap with spike spectrum. Pulse artifacts cannot be simply filtered out (Heffer and Fallon 2008; Wichmann 2000). The next section will introduce the artifacts of narrow pulses and their removal methods.

4.5.2 Removing Artifacts of High-Frequency Pulse Stimulation

Many neuromodulation therapies, including deep brain stimulation (DBS), use pulse electrical stimulations, which often creating large-amplitude artifacts in recordings. The waveform and amplitude of these artifacts depend on multiple factors, including the structure of stimulation electrodes, the pulse waveform, intensity and frequency (rate), the relative positions of recording and stimulation electrodes, as well as the features and settings of amplifiers and stimulators. Electrical isolation between stimulation and recording circuits can greatly reduce the artifacts. With proper isolation, most stimulation current can loop through the stimulation circuit, with minimal invasion into the recording circuit. Additionally, using a bipolar stimulation electrode with two closely spaced poles (millimeter or sub-millimeter) can confine the stimulation current to a small brain area. This can also minimize stimulation artifacts in recordings, provided the stimulation site is sufficiently distant from the recording site. However, stimulation artifacts recorded by nearby electrodes can still disturb neuronal signals. This interference is particularly problematic when detecting unit spikes, which typically have only tens to hundreds of microvolts in amplitude—much smaller than stimulation artifacts. The magnitude difference between stimulation artifacts and neuronal signals can reach 4–5 orders (Wagenaar and Potter 2002).

For example, in our in-vivo experiments of electrical stimulations in rat hippocampal regions, the recording and stimulation sites were about 1–2 mm apart. Using a Model 3800 stimulator with isolators and a Model 3600 amplifier (see Sect. 3.5 for details), we applied narrow current pulses with a width of 100 μs and intensity of ~ 0.3 mA. This stimulation usually produced artifacts with several millivolts in amplitude—far greater than both unit spikes and normal field potentials. The artifacts were sometimes even larger than the population spikes (PS) generated by synchronized neuronal firing.

Large-amplitude stimulation artifacts can interfere with both neural signal display and signal extraction. Especially during high-frequency stimulation (HFS), the dense artifacts can obscure neural signals. Methods have been designed to eliminate pulse artifacts, including template subtraction, waveform fitting, and linear interpolation (Erez et al. 2010; Wagenaar and Potter 2002; Hashimoto et al. 2002). The template subtraction method involves obtaining an artifact template, which is then subtracted from the recorded signal wherever artifacts are identified. This method can recover neural signals superimposed on artifacts, provided that the recording is not truncated by amplifier saturations. In other words, the amplitudes of recorded signals with artifacts must stay within the amplifier range.

In addition, the template method requires artifact waveforms to remain constant, which is challenging in practice. First, the electrical properties of brain tissue—particularly impedance values—can change, especially during sustained HFS. This can lead to variations in artifact waveforms. Second, for narrow pulses, additional factors affect the recorded artifact waveforms, including the limitation of ADC sampling rate and the clock timing difference between the recorder ADC and stimulator.

Figure 4.33 illustrates the impact of a clock difference by an example of sampling a 10 kHz sine wave (0.1 ms cycles, the orange curve). Based on the sampling theorem, the signal is sampled at 20 kHz, yielding two samples per cycle. Assume that the clock controlling the ADC for sampling does not exactly synchronize with the clock of digital signal generator that produces the original sine wave. Taking the clock of signal generator as reference, let the actual sampling rate of sampler be 20.83 kHz rather than exact 20 kHz. This discrepancy can result in oscillation waves in the sampled data with amplitudes varying periodically at 0.83 kHz (black dots and lines in Fig. 4.33). This example exaggerates the clock difference to illustrate potential

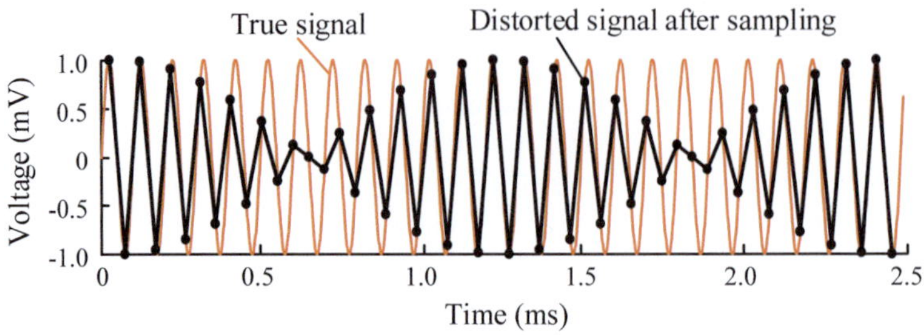

Fig. 4.33 A segment of 10 kHz sinusoidal signal (orange) and its sampled signal at 20.83 kHz (black). The black dots represent the sampled data

changes in sampled artifacts. In practice, clock differences between instruments are typically very small and negligible. Nevertheless, in long recording, even small differences can lead to significant changes in sampled stimulation artifacts, as shown in the following example.

We used a Model 3800 stimulator to generate HFS pulses applied into the rat brain. A PowerLab recorder sampled the signals collected by a recording electrode (see Fig. 3.17). Figure 4.34A shows a signal recorded in the hippocampal CA1 soma layer during a 1-min HFS of biphasic current pulses applied on the alveus. The pulses had an intensity of 0.3 mA, width of 100 μs per phase, and frequency of 200 Hz. We set the amplifier frequency band at 0.3 Hz–5 kHz and used a 20 kHz sampling rate. The dense stimulation artifacts in the original recording obscured most of the neural signal, except for the large APSs at the onset of HFS (Fig. 4.34A). The artifacts created a periodic oscillation envelope due to the clock difference between the 3800 stimulator and PowerLab sampler. This envelope had a period of about 4 s, corresponding a frequency difference of 0.25 Hz—a minuscule 0.00125% of the 20 kHz sampling rate. However, this small difference caused significant changes in pulse artifacts. With the 20 kHz sampling rate and 100 μs pulse width, the recorder captured only a few samples per pulse artifact. In the enlarged insets in Fig. 4.34B, the sampled artifact data are shown in sparse red dots. Due to the clock difference, the sampling positions on the artifacts constantly shifted, causing varying artifact waveforms and amplitudes.

By the way, the LabChart software of PowerLab recorder can display sampled signals in two modes: "Join Points with Line" and "Show Points as Dots". Figure 4.34A used the former, while the upper plot of Fig. 4.34B used the latter. The dot display mode can allow clear visualization of neural signals and on-line observation of changes in evoked potentials (APSs) during HFS. Even with HFS at a 200 Hz pulse frequency as in this example, the biphasic pulse artifacts (with a total width of ~ 0.2 ms) occupy only a small fraction of time. The artifacts contain far fewer sample data than neural signals, enabling the dot display to effectively reveal the neural signals.

These results show that sampled artifacts of narrow pulses can vary over time, even when the actual stimulation artifacts remain constant. This variation makes artifact

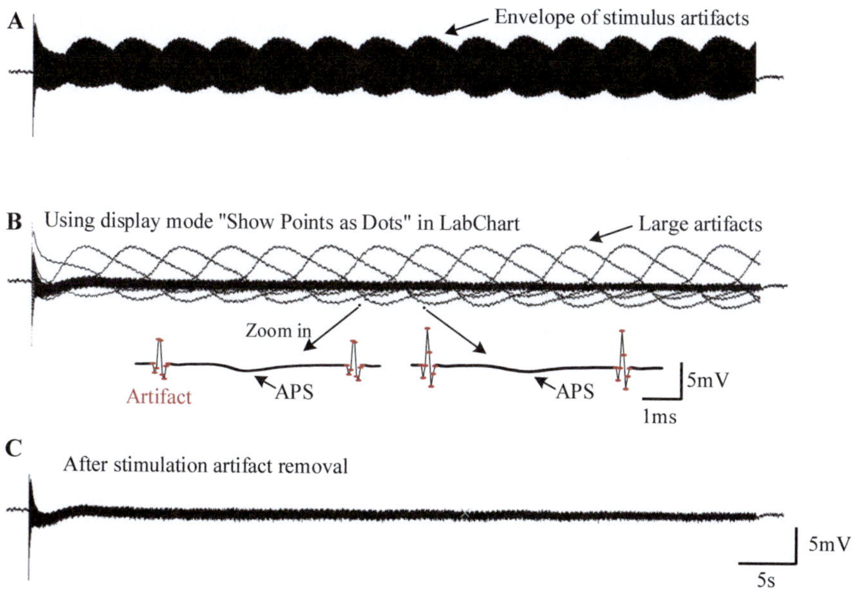

Fig. 4.34 Variations in pulse artifacts caused by sampling rate limitation and clock discrepancy. **A** Example signal recorded from the hippocampal CA1 soma layer during 1-min biphasic pulse stimulation at 200 Hz, sampled at 20 kHz. **B** Signal from (**A**) displayed in LabChart "Show Points as Dots" mode, with enlarged insets below. **C** Signal after artifact removal using linear interpolations

removal using template methods difficult. To address artifact interference, researchers have had to compromise by either reducing stimulation intensity or inferring neuronal activity during stimulation by examining signals immediately after stimulation ends (Rosa et al. 2012).

We used bipolar stimulation electrodes to confine the stimulation area in our rat experiments. However, large stimulation artifacts were still captured by nearby recording electrodes, affecting the detection of unit spikes. Figure 4.35A shows a brief segment of the original wideband recording (0.3–5000 Hz) with 100 Hz 0.3 mA HFS pulses. The pulse artifacts had a large amplitude of about 15 mV. Unit spikes with an amplitude of ~ 0.2 mV (indicated by blue dots) were only visible in the enlarged signal. These spikes were superimposed on low-frequency field potentials which needed to be removed by a high-pass filter before using a threshold method to detect the spikes. However, while filtering removed low-frequency potentials, it also altered the artifacts, making it difficult to identify spikes near the artifacts, as shown in the bottom of Fig. 4.35A. In Fig. 4.35B, each artifact segment in the original signal was replaced with a straight line connecting its endpoints before high-pass filtering was applied. This process resulted in a MUA signal with clear spikes. Figure 4.35C shows overlaid plots of the filtered signals both with and without artifacts, highlighting their differences. High-pass filtering with artifacts caused oscillations around the artifacts and interfered with spike detection. The artifact-free signal in Fig. 4.34C was also produced by this artifact removal process. The method is suitable for HFS with narrow pulses, as the artifact duration is minimal. We used it to process recordings with various high-frequency pulse stimulations reported in Part II of this book (Chaps. 5–8).

Removing an artifact by replacing it with a straight line can be accomplished by a linear interpolation algorithm, which can be implemented using a MATLAB program. Its algorithm consists of two main steps:

(1) Identify stimulation artifacts. Pulse artifacts can be detected by setting an amplitude threshold when their magnitude significantly exceeds neural signals. However, artifacts may vary in amplitude and waveform due to multiple factors, even within a same stimulation trial, as shown in Fig. 4.34A. Narrow pulse artifacts typically have a much steeper slope than neural signals. For example, Fig. 4.36 shows a recording containing an evoked APS following a biphasic-pulse artifact. Each blue dot on the recording represents a sampled datum at 20 kHz. Let an instantaneous slope represent the potential difference between adjacent data. Although an APS, formed by synchronous neuronal firing, has steep phases with notably great slopes compared to other neural signals, a pulse artifact typically exhibits even greater slopes with its maximum instantaneous slope significantly exceeding that of APS. Therefore, the instantaneous slope can be used to identify artifacts as following. First, calculate the instantaneous slope (the absolute value of the first-order difference of the sampled data). Second, identify stimulation artifacts using a preset slope threshold. Finally, determine the start and end points of the artifacts.

(2) Replace the artifact segment with linearly interpolated data—a straight line connecting the two ends of the artifact. The interpolated data are calculated as

$$x_n = \frac{x_{N-1} - x_0}{\boldsymbol{N}} n + x_0 \tag{4.16}$$

where x_0 and x_{N-1} represent the sampled data at the interpolation starting and ending points, N is the number of data on the interpolation line, and x_n represents the n-th interpolated data, $n = 1, 2, \ldots, N - 2$. The length of the interpolation segment (N) needs to be determined from the actual recorded signal. Besides the above-mentioned sampling limitations that can result in pulse artifacts different completely from the applied pulses (as shown in Figs. 4.34B and 4.36), additional factors can distort pulse artifacts and extend their duration, including transient amplifier saturation caused by excessive artifact amplitudes and the volume conductor effect of neural tissue. Consequently, the interpolation segment for artifact removal can be slightly longer than the width of applied pulse, but typically narrower than spikes (Fig. 4.35) and evoked population potentials (Fig. 4.36).

The linear interpolation method for artifact removal has a disadvantage: it can lead to neural signal loss during the artifact-occupied period. However, when the pulse amplitude exceeds the amplifier's range—causing saturation, the signal during the artifact is already truncated by the amplifier. In this case, artifact removal will not

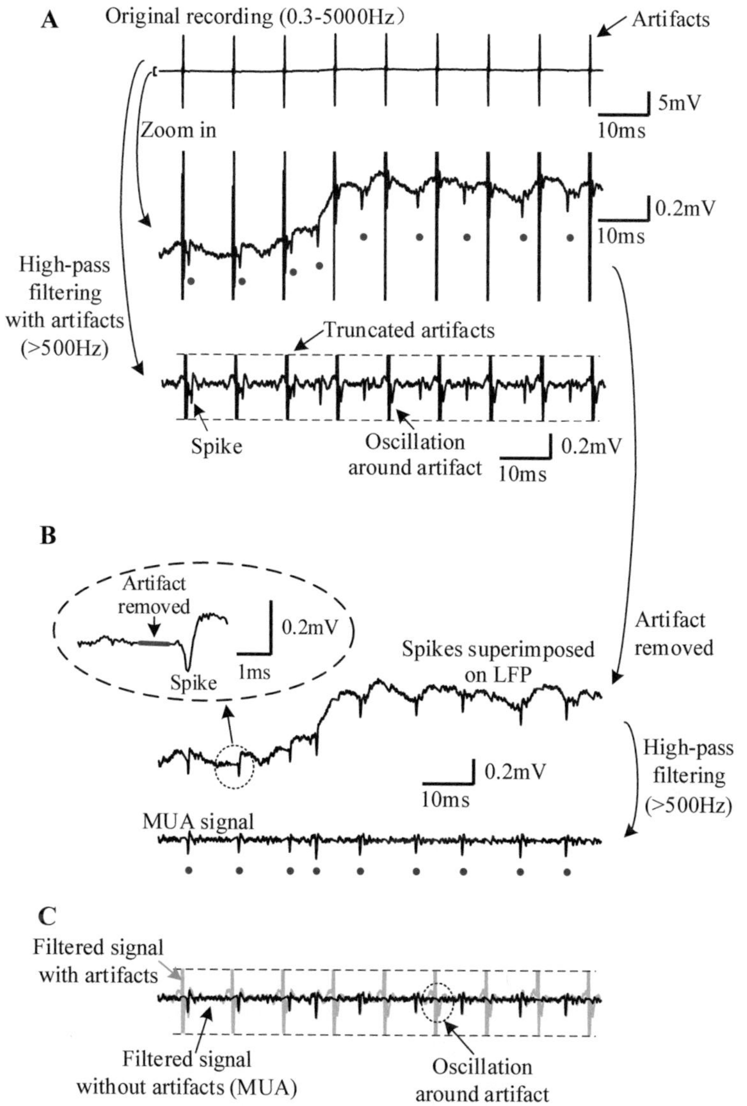

Fig. 4.35 Linear interpolation for removing stimulation artifacts of narrow pulses. **A** From top to bottom: a brief recording from the rat hippocampal CA1 region during 100 Hz 0.3 mA pulse stimulation, its longitudinally expanded signal (blue dots indicating unit spikes), and its high-pass filtered signal (with truncated artifacts). **B** The signal after artifact removal together with its MUA signal after high-pass filtering. In the dashed ellipse, the enlarged inset displays a red interpolated line replacing an artifact segment, followed immediately by a spike. **C** Overlaid plots of the filtered signals both with and without artifacts

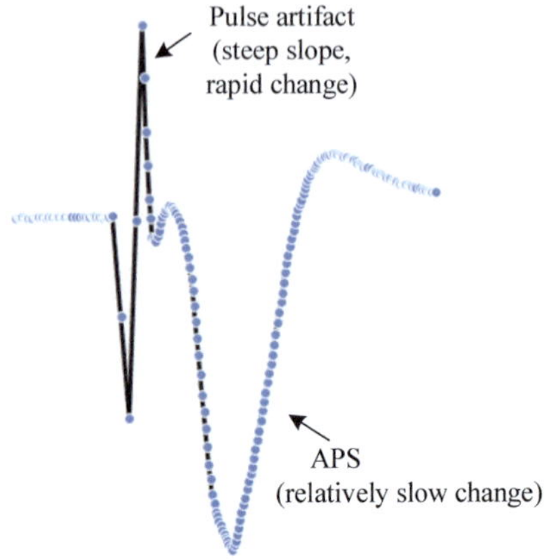

Fig. 4.36 Comparison of slopes between a pulse artifact and a neural signal (APS). Note: blue dots represent sampled data taken at 20 kHz, with 0.05 ms intervals between adjacent data

worsen the existing signal loss. Such signal truncations commonly occurred during pulse periods in our experimental recordings. High-quality neural signal recordings require a high ADC resolution (see Sect. 3.5.3). This means that the amplifier range should be set to match the maximum amplitude of neural signals, which can inevitably lead to larger stimulation artifacts being truncated (see Fig. 3.25).

4.6 Summary

This chapter first introduces methods for detecting and analyzing extracellularly recorded action potentials of individual neurons—unit spikes. Since these spike signals contain high-frequency components, their accurate recording needs an appropriate amplifier frequency band, as well as suitable sampling frequency and resolution. Different neuron types produce distinct spike waveforms, which can be used to distinguish neuron categories such as pyramidal neurons and interneurons. The amplitude and waveform of spikes also vary with the electrode position relative to the recorded neuron. These variations enable spike sorting to determine spike origin. Spike detection in recording signals can be performed using threshold, window or template methods, while spike sorting can be conducted using parameter and template methods. Accurate spike sorting requires multi-channel recordings, typically involving four-channel signals from closely spaced contacts.

The chapter also provides concrete examples of spike detection and sorting using custom-designed window method and four-channel algorithm. Additionally, it covers analysis methods for spike sequences of individual neurons, including ISI histogram, autocorrelation histogram, and cross-correlation histogram. It also explains PSTH, JPSSD and JPSTH histograms for analyzing neuronal responses to external stimulations.

This chapter then introduces spectrum analysis methods for LFP and EEG signals. Specifically, it explains the periodogram spectrum estimation method (Welch method) for random signals, discusses the relationships between resolution and range in both time and frequency domains, and demonstrates spectrum calculations using MATLAB functions and LabChart software.

Finally, the chapter introduces the current source density analysis for tracking transmembrane currents during neuron activations. It also presents methods for removing artifacts caused by electrical stimulations, including both sinusoidal and pulse artifacts. It shows how these artifacts affect spike detection during high-frequency pulse stimulations and provides their removal methods.

These approaches will be applied in later chapters to investigate the neuromodulation effects and mechanisms of electrical stimulations.

References

Barthó P, Hirase H, Monconduit L et al (2004) Characterization of neocortical principal cells and interneurons by network interactions and extracellular features. J Neurophysiol 92(1):600–608

Blanche TJ, Spacek MA, Hetke JF et al (2005) Polytrodes: high-density silicon electrode arrays for large-scale multiunit recording. J Neurophysiol 93(5):2987–3000

Bretschneider F, de Weille JR (2006) Introduction to electrophysiological methods and instrumentation. Elsevier, Amsterdam

Brown EN, Kass RE, Mitra PP (2004) Multiple neural spike train data analysis: state-of-the-art and future challenges. Nat Neurosci 7(5):456–461

Buzsáki G, Leung LW, Vanderwolf CH (1983) Cellular bases of hippocampal EEG in the behaving rat. Brain Res Rev 6(2):139–171

Buzsáki G (2002) Theta oscillations in the hippocampus. Neuron 33(3):325–340

Buzsáki G (2004) Large-scale recording of neuronal ensembles. Nat Neurosci 7(5):446–451

Buzsáki G (2006) Rhythms of the Brain. Oxford University Press, Oxford

Carlson D, Carin L (2019) Continuing progress of spike sorting in the era of big data. Curr Opin Neurobiol 55:90–96

Clark I, Biscay R, Echeverría M et al (1995) Multiresolution decomposition of non-stationary EEG signals: a preliminary study. Comput Biol Med 25(4):373–382

Csercsa R, Dombovári B, Fabó D et al (2010) Laminar analysis of slow wave activity in humans. Brain 133(9):2814–2829

Csicsvari J, Hirase H, Czurko A et al (1998) Reliability and state dependence of pyramidal cell-interneuron synapses in the hippocampus: An ensemble approach in the behaving rat. Neuron 21(1):179–189

Enderle JD, Bronzino JD. (2012) Introduction to biomedical engineering (3rd ed.). Elsevier, Amsterdam

Erez Y, Tischler H, Moran A et al (2010) Generalized framework for stimulus artifact removal. J Neurosci Methods 191(1):45–59

Feng Z (2003) Analysis of rat electroencephalogram during slow wave sleep and transition sleep using wavelet transform. Acta Biochim Biophys Sin 35(8):741–746

Feng Z, Guang L, Zheng X et al (2007) Recordings of the hippocampal field potentials and unit activity by using linear silicon electrode array. Prog Biochem Biophys 34(4):401–407 (in Chinese)

Feng Z, Wang J, Wang Y et al (2012) Selection of filtering frequencies for neuronal spike signals. J Zhejiang Univ (Eng Sci) 46(2):351–358 (in Chinese)

Feng Z, Xiao Q, Hu Z (2013) Detecting neuronal unit spikes during electrical stimulation periods. Chin J Bio Eng 32(4):403–410 (in Chinese)

Feng Z, Zheng X (2004a) The dynamic change of rat EEG's complexity and power spectrum under different anesthetized depths. Chin J Biomed Eng 23(1):87–91 (in Chinese)

Feng Z, Zheng X (2004b) Using the multi-resolution wavelet decomposition to analyze the sleep spindle in rat EEG. Chin J Biomed Eng 23(2):103–108 (in Chinese)

Gerstein GL, Bedenbaugh P, Aertsen MH (1989) Neuronal assemblies. IEEE Trans Biomed Eng 36(1):4–14

Gerstein GL, Kiang NYS (1960) An approach to the quantitative analysis of electrophysiological data from single neurons. Biophys J 1(1):15–28

Gerstein GL, Perkel DH (1972) Mutual temporal relationships among neuronal spike trains. Biophys J 12(5):453–473

Gold C, Henze DA, Koch C et al (2006) On the origin of the extracellular action potential waveform: a modeling study. J Neurophysiol 95(5):3113–3128

Gold C, Henze DA, Koch C (2007) Using extracellular action potential recordings to constrain compartmental models. J Comput Neurosci 23(1):39–58

Harris KD, Henze DA, Csicsvari J et al (2000) Accuracy of tetrode spike separation as determined by simultaneous intracellular and extracellular measurements. J Neurophysiol 84(1):401–414

Hashimoto T, Elder CM, Vitek JL (2002) A template subtraction method for stimulus artifact removal in high-frequency deep brain stimulation. J Neurosci Methods 113(2):181–186

Heffer LF, Fallon JB (2008) A novel stimulus artifact removal technique for high-rate electrical stimulation. J Neurosci Methods 170(2):277–284

Henze DA, Borhegyi Z, Csicsvari J et al (2000) Intracellular features predicted by extracellular recordings in the hippocampus in vivo. J Neurophysiol 84(1):390–400

Heynen AJ, Bear MF (2001) Long-term potentiation of thalamocortical transmission in the adult visual cortex in vivo. J Neurosci 21(24):9801–9813

Hochberg LR, Serruya MD, Friehs GM et al (2006) Neuronal ensemble control of prosthetic devices by a human with tetraplegia. Nature 442(7099):164–171

Horváth C, Tóth LF, Ulbert I et al (2021) Dataset of cortical activity recorded with high spatial resolution from anesthetized rats. Sci Data. 8(1):180

Huberfeld G, Menendez de la Prida L, Pallud J et al (2011) Glutamatergic pre-ictal discharges emerge at the transition to seizure in human epilepsy. Nat Neurosci 14(5):627–634

Joshua M, Elias S, Levine O et al (2007) Quantifying the isolation quality of extracellularly recorded action potentials. J Neurosci Methods 163(2):267–282

Kloosterman F, Peloquin P, Leung LS (2001) Apical and basal orthodromic population spikes in hippocampal CA1 in vivo show different origins and patterns of propagation. J Neurophysiol 86(5):2435–2444

Lee CW, Dang H, Nenadic Z (2007) An efficient algorithm for current source localization with tetrodes. Annu Int Conf IEEE Eng Med Biol Soc 2007:1282–1285

Lefebvre B, Yger P, Marre O (2016) Recent progress in multi-electrode spike sorting methods. Journal of Physiology-Paris 110(4):327–335

Lewicki MS (1998) A review of methods for spike sorting: The detection and classification of neural action potentials. Network 9(4):R53–78

Lisman JE (1997) Bursts as a unit of neural information: making unreliable synapses reliable. Trends Neurosci 20(1):38–43

López-Aguado L, Ibarz JM, Herreras O (2000) Modulation of dendritic action currents decreases the reliability of population spikes. J Neurophysiol 83(2):1108–1114

Ma W, Feng Z, Wang Z et al (2019) High-frequency stimulation of afferent axons alters firing rhythms of downstream neurons. J Integr Neurosci 18(1):33–41

Maccione A, Gandolfo M, Massobrio P et al (2009) A novel algorithm for precise identification of spikes in extracellularly recorded neuronal signals. J Neurosci Methods 177(1):241–249

Meyer LM, Samann F, Schanze T (2023) Dualsort: online spike sorting with a running neural network. J Neural Eng 20(5):056031

Mokri Y, Salazar RF, Goodell B et al (2017) Sorting overlapping spike waveforms from electrode and tetrode recordings. Front Neuroinform 11:53

Nelson MJ, Pouget P, Nilsen EA et al (2008) Review of signal distortion through metal microelectrode recording circuits and filters. J Neurosci Methods 169(1):141–157

Nicholson C, Freeman JA (1975) Theory of current source-density analysis and determination of conductivity tensor for anuran cerebellum. J Neurophysiol 38(2):356–368

Obien MEJ, Deligkaris K, Bullmann T et al (2015) Revealing neuronal function through microelectrode array recordings. Front Neurosci 8:423

Pachitariu M, Sridhar S, Pennington J et al (2024) Spike sorting with Kilosort4. Nat Methods 21(5):914–921

Pettersen KH, Lindén H, Dale AM, et al. (2012) Extracellular spikes and current-source density. In: Brette R, Destexhe A (eds) Handbook of neural activity measurement. Cambridge University Press.

Richardson TL, Turner RW, Miller JJ (1987) Action-potential discharge in hippocampal CA1 pyramidal neurons: current source-density analysis. J Neurophysiol 58(5):981–996

Rodieck RW, Kiang NYS, Gerstein GL (1962) Some quantitative methods for the study of spontaneous activity of single neurons. Biophys J 2(4):351–368

Rolls ET, Miyashita Y, Cahusac PMB et al (1989) Hippocampal-neurons in the monkey with activity related to the place in which a stimulus is shown. J Neurosci 9(6):1835–1845

Rosa M, Giannicola G, Marceglia S et al (2012) Neurophysiology of deep brain stimulation. Int Rev Neurobiol 107:23–55

Schaffer L, Nagy Z, Kincses Z et al (2021) Spatial information based OSort for real-time spike sorting using FPGA. IEEE Trans Biomed Eng 68(1):99–108

Shabestari PS, Buccino AP, Kumar SS et al (2021) A modulated template-matching approach to improve spike sorting of bursting neurons. IEEE Biomed Circuits Syst Conf 2021:9644995

Shiavi R (2007) Introduction to applied statistical signal analysis: guide to biomedical and electrical engineering applications (3rd ed.). Academic

Somogyvári Z, Zalányi L, Ulbert I et al (2005) Model-based source localization of extracellular action potentials. J Neurosci Methods 147(2):126–137

Stevenson IH, Kording KP (2011) How advances in neural recording affect data analysis. Nat Neurosci 14(2):139–142

Townsend G, Peloquin P, Kloosterman F et al (2002) Recording and marking with silicon multichannel electrodes. Brain Res Protoc 9(2):122–129

Wagenaar DA, Potter SM (2002) Real-time multi-channel stimulus artifact suppression by local curve fitting. J Neurosci Methods 120(2):113–120

Wang J, Feng Z (2009) A novel method for multi-channel neuronal spike detection and classification. Prog Biochem Biophys 36(5):641–647 (in Chinese)

Wang Z, Feng Z, Yu Y et al (2016) Spike potential detection method for broadband-frequency recording signals. Chinese Journal of Scientific Instrument 37(3):481–489 (in Chinese)

Wichmann T (2000) A digital averaging method for removal of stimulus artifacts in neurophysiologic experiments. J Neurosci Methods 98(1):57–62

Wild J, Prekopcsak Z, Sieger T et al (2012) Performance comparison of extracellular spike sorting algorithms for single-channel recordings. J Neurosci Methods, 203(2):369–376

Wiltschko AB, Gage GJ, Berke JD (2008) Wavelet filtering before spike detection preserves waveform shape and enhances single-unit discrimination. J Neurosci Methods 173(1):34–40

Xu Y, Feng Z, Yuan Y et al (2023) Effects on somata by high-frequency electrical stimulation at the axons of hippocampal pyramidal neurons. Prog Biochem Biophys 50(3):561–572 (in Chinese)

Yu Y, Feng Z, Cao J et al (2016) Modulation of local field potentials by high-frequency stimulation of afferent axons in the hippocampal CA1 region. J Integr Neurosci 15(1):1–17

Yuan X, Hierlemann A, Frey U (2021) Extracellular recording of entire neural networks using a dual-mode microelectrode array with 19584 electrodes and high SNR. IEEE J Solid-State Circuits 56(8):2466–2475

Zhang T, Rahimi Azghadi M, Lammie C et al (2023) Spike sorting algorithms and their efficient hardware implementation: A comprehensive survey. J Neural Eng 20(2):021001

Part II
Neuromodulation Effects of Electrical Stimulations on Axons

Chapter 5
Neuronal Responses to Axonal High-Frequency Pulse Stimulations

Abstract Among neuronal structures, the axonal membrane has the shortest chronaxie, making it most susceptible to narrow pulses commonly used in neuromodulation techniques like deep brain stimulation (DBS). This chapter details our experiments in the rat hippocampal CA1 region, which verify intermittent axonal block caused by high-frequency stimulations (HFS). The experiments include both antidromic HFS (A-HFS) applied to the axons of recorded pyramidal neurons and orthodromic HFS (O-HFS) applied to presynaptic afferent fibers. The results showed that the axonal HFS can extend the refractory period of axons and alter the excitability of somata, causing a silent period without firing immediately after HFS cessation. The experiments also showed non-uniform clustered firing produced by HFS with constant pulse intervals.

Keywords High-frequency stimulation · Axonal block · Refractory period · Clustered firing · Asynchronous firing

5.1 Introduction

The axon is a slender, specialized extension of a neuron. It acts like a cable, conducting action potentials between neurons or from neurons to effectors for information transmission. Under normal physiological conditions, an action potential typically originates at the axon initial segment (AIS) and then propagates reliably along the axon. For simplicity, this book sometimes refers to action potentials from the AIS as "originating from the cell body" to distinguish them from those that start from the middle segment of axon produced by external stimuli. Action potentials travel along the axonal membrane through local currents (Fig. 5.1A). At rest, the axonal membrane maintains a resting potential of approximately − 70 mV, with positive charge outside and negative charge inside. In the excitation area where the action potential is generating, the membrane potential momentarily reverses—becoming negative outside and positive inside. This reversal creates a potential difference between the excited and adjacent resting areas, leading to local currents. As shown by the dashed arrow

Z. Feng, *Electrical Brain Stimulation: Mechanisms and Modulation of Neuronal Activity*, https://doi.org/10.1007/978-981-95-4145-4_5

lines in Fig. 5.1A, current flows from the resting area to the excited area outside the membrane. Inside the membrane, current flows in the opposite direction to complete the current loop. In the resting area, the outward current can cause membrane depolarization. When this depolarization reaches threshold, it can trigger an action potential, turning the resting area into a new excited area. As this process continues, the action potential propagates along the axon. A myelinated axon follows the same conduction mechanism, but its action potential occurs at the nodes of Ranvier during conduction because these nodes have much lower impedance than the myelin-wrapped internodal areas. This results in saltatory conduction—the action potential jumps from node to the next.

External stimulation can produce axonal action potentials through intracellular and extracellular methods. Figure 5.1B shows intracellular induction, in which a current is injected into the axon. The induced action potential originates in the middle of the axon and then propagates bilaterally. Similarly, extracellular pulses (typically negative) can also produce bilaterally propagatable action potentials on an axon (Fig. 5.1C). However, the two methods have a key difference: intracellular stimulation causes only depolarization, while extracellular stimulation produces both

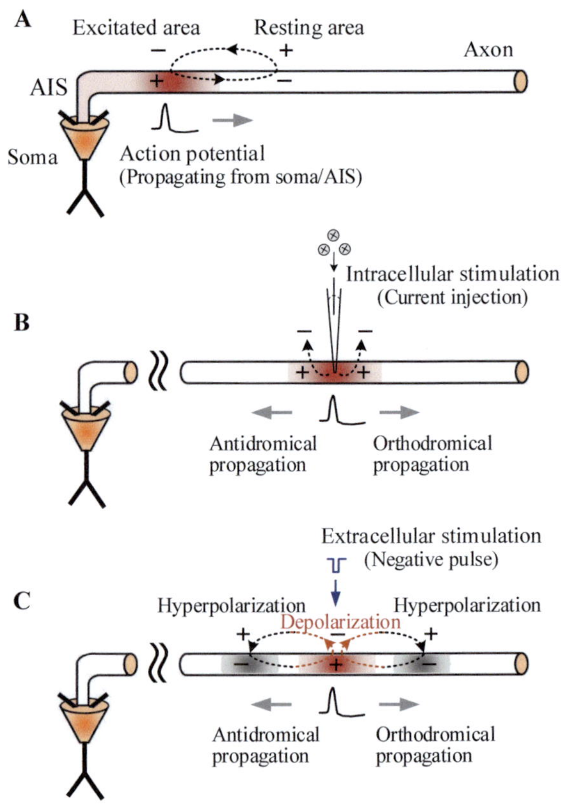

Fig. 5.1 Generation and propagation of action potentials along an axon under distinct conditions. **A** Action potential originating from the axon initial segment (AIS) under normal physiological conditions. **B** Action potential produced intracellularly in the middle of axon. **C** Action potential produced by an extracellular pulse in the middle of axon

depolarization and hyperpolarization at different sites along the axon. Immediately at the stimulation site, the transmembrane current from an extracellular negative pulse flows outward, causing depolarization. The current loop can produce inward currents and cause hyperpolarization on both flanking sides. This hyperpolarization can impede action potential conduction, resulting in a lower maximum firing rate than intracellular induction. Nevertheless, extracellular stimulation has the advantage of activating multiple axons simultaneously with a single electrode contact.

Extracellular electrical stimulation is essential for developing clinical neuromodulation techniques, as these interventions require simultaneous activations of massive neurons to achieve therapeutic effects. Many treatments specifically target axons, including spinal cord stimulation (SCS) for pain relief (Lam et al. 2023), vagus nerve stimulation (VNS) for treating refractory epilepsy, depression, and migraines (Gurbani et al. 2016; Rush et al. 2005; Silberstein et al. 2020), and functional electrical stimulation (FES) for restoring limb function. Other successful applications include cochlear stimulation for hearing restoration, sublingual nerve stimulation for sleep apnea treatment, and electrical stimulation to promote axonal regeneration in damaged peripheral nerves (Juckett et al. 2022). These approaches work by directly stimulating nerves—bundles of axons—to either block nerve signal conduction or regulate neuronal and muscular activity. Although deep brain stimulation (DBS) does not specially aim at axons, axonal activity can also play an important role in its effectiveness.

DBS typically targets specific brain regions or neural nuclei. The electrical pulses from DBS electrodes can affect various parts of surrounding neurons, including cell bodies (somata), axons, and dendrites. Axonal membranes are particularly sensitive, having the lowest activation threshold and responding most readily to narrow pulses due to their low chronaxie. Therefore, action potentials can start from the axon—even when the soma is closer to the stimulation electrode (Ranck 1975; Nowak and Bullier 1998; McIntyre and Grill 1999). Studies have shown that high-frequency pulses (around 130 Hz with 60–450 μs width) mainly activate axons rather than somata and dendrites in brain regions such as the ventral nucleus of thalamus and the medial pallidum when treating Parkinson disease or primary tremor (Holsheimer et al. 2000). Additionally, axons occupy more brain space than other neural structures. White matter, which consists of neuronal axons, accounts for about half of the human brain volume (Fields 2008). When DBS pulses are delivered, they first activate nearby axons, whether these axons originate in, terminate in, or simply pass through the stimulation area. The resulting firing travels bilaterally along the axons and can modulate the activity of neural networks. As a result, the axonal responses to high-frequency pulse stimulation play a crucial role in DBS (Lozano et al. 2019; Lee et al. 2019; Chomiak and Hu 2007; Udupa and Chen 2015).

Under normal physiological conditions, the axonal action potentials have four features: bidirectionality, no attenuation, insulation, and resistance to fatigue. Bidirectionality means that an action potential (AP) can propagate in both directions from its origin site on the axon, rather than unidirectionally from soma. The propagation toward axonal terminals is orthodromic—the normal physiological direction, while that toward the soma is antidromic. The "no attenuation" property means that

APs can maintain their strength as they travel outward. Insulation comes from the high impedance of myelin sheath which wraps around the axonal membrane and allows each axon in a bundle to conduct signals independently. Although axons typically function independently, they can form connections through specific membrane areas. For example, electrical synapses (also called gap junctions) exist between axonal membranes. These junctions allow a current to flow from highly excited to less excited axons, helping synchronize neuronal populations (Traub et al. 2002; Choi et al. 2021). Although axons are more resistant to fatigue than muscle tissue, they cannot sustain high-frequency firing indefinitely and may experience depolarized axonal block.

In theory, the maximum firing rate (frequency) of an axon is determined by the refractory period of the axonal membrane. This refractory period is typically about 1 ms, allowing axons to conduct action potentials at frequencies up to a kilohertz. Using the HH model introduced in Sect. 1.3, we can simulate continuous action potentials in an axon up to 100 Hz under ideal conditions. However, in real physiological conditions, axons cannot sustain such a high frequency for long periods. They can only maintain such rapid firing briefly. Studies have shown that the axons of brain neurons require about 100 ms for continuous and reliable action potential transmission (Chomiak and Hu 2007; Bucher and Goaillard 2011). This corresponds to a frequency of only 10 Hz.

Note that here the term "frequency" for APs refers to their occurrence rate, similarly to the "frequency" of stimulation pulses. This differs from the frequency components contained in AP and pulse waveforms, as shown by spectrum analysis in Sect. 4.3. For example, a 100 Hz pulse signal means 100 pulses per second. The spectrum of this pulse signal contains a fundamental component at 100 Hz, along with harmonic components at frequencies of 300, 500, 700 Hz and beyond. Following conventions, this book uses Hertz (Hz)—equivalent to s^{-1}—as the unit for pulse frequency. The AP firing frequency of neurons is measured in "Hz" or "count/s". The electrical stimulation pulses in this book are narrow square waveforms, typically 100 μs in width. Unless specified otherwise, current pulses are used rather than voltage pulses. Their current intensity (amplitude) is measured in milliamperes (mA).

To measure the AP propagations along axons, the most effective way is placing intracellular microelectrodes at multiple sites along an axon to detect membrane potential changes. However, measuring individual axons within the brain is challenging due to their thin, fragile nature. Current technology restricts these measurements to in-vitro experiments. For example, Radivojevic et al. (2017) measured axons of cultured rat cortical neurons using a microelectrode array (MEA) with hundreds of recording contacts to track AP propagation. Their findings showed a decreased reliability in axonal conduction when 100 Hz pulses were applied at the soma for one second. Another study on rat hippocampal slices in-vitro showed that under high potassium conditions or during epileptiform discharges, the axons in Schaffer collaterals failed to conduct APs at frequencies around 50 Hz (Meeks and Mennerick 2004; Meeks et al. 2005). These studies measured extracellular rather than intracellular APs, requiring specific constraints to verify that the measured APs came from the same axon. Therefore, such experiments are usually classified as quasi-single axon studies.

Neurons cultured separately in-vitro or in ex-vivo brain slices often have compromised neural networks that lack the integrity of intact brain networks. This integrity is essential for investigating neuromodulation therapies. Additionally, therapy evaluations require examining population-level neuronal activity, not only single-cell responses. Moreover, the stimulation effects vary across neurons within the stimulation area, as multiple factors influence the distribution of electric field around each neuron: stimulation electrode structure, stimulation model (single-pole, bipolar or multipole), and the distance between neurons and electrodes. As a result, neurons respond differently to stimulation based on their locations. Therefore, studying a single neuron or axon in isolation cannot fully reveal the complex effects of neuromodulation.

In the following chapters, I will present our findings from in-vivo experiments studying the effects of high-frequency stimulation (HFS) pulses on axons in the rat hippocampal region. This brain region has a distinct laminar structure with tightly packed neurons, making it ideal for separately stimulating at axonal fibers and recording at cell bodies (see Chap. 2 for details). Additionally, the hippocampal region is often involved in refractory temporal lobe epilepsy, Alzheimer disease and other disorders. It is also a potential target for DBS (Li and Cook 2018; Laxton et al. 2010). Therefore, studying how hippocampal neurons respond to axonal stimulations can provide insights into the modulation mechanisms of various stimulations and guide the development of new therapeutic approaches.

5.2 Axonal Block Induced by High-Frequency Sustained Stimulation

Sustained extracellular HFS can cause depolarization block at axons, preventing them from generating APs in response to each HFS pulse in in-vitro conditions (Jensen and Durand 2009; Zheng et al. 2011). However, debate has persisted about whether such axonal block can occur in the intact brain during ~ 100 Hz HFS, since axons are known to transmit APs rapidly. Additionally, if HFS-induced axonal block exists, can it simultaneously prevent both AP generation and passage? This section addresses these questions through in-vivo experiments in rat hippocampus, confirming that axonal HFS can produce intermittent block. Additional questions explored include: can axonal HFS affect the somata, and can neurons still fire during HFS-induced axonal block—and if so, how can their firing patterns change?

5.2.1 Antidromic Activation of Pyramidal Neurons by Axonal HFS in the Rat Hippocampal CA1 Region

Figure 5.2A shows the experimental configuration for investigating the effect of axonal HFS on neurons. The antidromic stimulation electrode (ASE) was placed on the alveus—the white matter covering the dorsal surface of the hippocampus. The alveus is composed of the axons of CA1 pyramidal neurons, forming the efferent pathway of hippocampus CA1 region. An array recording electrode (RE) was placed across the CA1 region upstream of the ASE. A single pulse applied to the alveus can produce an APS in the pyramidal cell layer (pcl) as detailed in Chap. 2. Using APS as an index, we can analyze how the alveus axons respond to applied stimuli. The APS can be measured by its amplitude and latency, as shown in the dashed box in Fig. 5.2A. Since the APS results from the antidromic activation of the HFS, the applied HFS is called antidromic HFS (A-HFS).

As shown in Fig. 5.2B, 1-min A-HFS at three different pulse frequencies initially produced similar APSs. As the A-HFS continued, the APS amplitudes decreased and APS latencies increased, depending on pulse frequencies. Statistical data show that by the end of A-HFS, the APS amplitude dropped to approximately 30–40% (50 Hz), 10–20% (100 Hz), and 5–10% (200 Hz) of its initial value (Feng et al. 2013, 2014). The final APS amplitude was inversely proportional to pulse frequency, with a ratio of 1:2:4 across the three frequencies. During the initial A-HFS period, both the speeds of APS amplitude decrease and latency increase were proportional to pulse frequencies, although the increments in the final APS latencies were similar across all three frequencies (Feng et al. 2014). These results indicated that A-HFS-induced APS changes correlated positively with pulse frequency.

The APS generation involves both the axon and soma (Fig. 5.2A). Changes in either of the structures could result in the APS changes. Our following experiments verified that the A-HFS-induced APS attenuation occurs due to axonal failure rather than somatic failure. The soma can remain its ability to generate APs during sustained A-HFS.

1. **Testing Soma Excitability During A-HFS Using Orthodromic Stimulation**

To assess soma excitability during A-HFS, as shown in Fig. 5.3A, orthodromic test stimuli (OTS) were applied every 5 s upstream at the Schaffer collaterals. The first OTS was applied 20 ms after the A-HFS started (Fig. 5.3B), coinciding with the third pulse of A-HFS at 100 Hz. The OPS evoked by this OTS was suppressed due to the refractory period following the large APS evoked by the A-HFS pulse, as OPS had only a slightly longer latency than APS. The fEPSP recorded at the apical dendritic layer (sr) remained similar to baseline recording (Fig. 5.3B, bottom row). As the A-HFS continued, the OTS-evoked OPS grew to baseline level while the APS decreased substantially. The statistical data in Fig. 5.3C showed the changes in APS amplitudes, OPS amplitudes and fEPSP slopes before, during and after A-HFS. After A-HFS ended, the amplitudes of APS evoked by test pulses at 1-min intervals recovered gradually, while the OPS evoked by OTS showed an initial decrease before

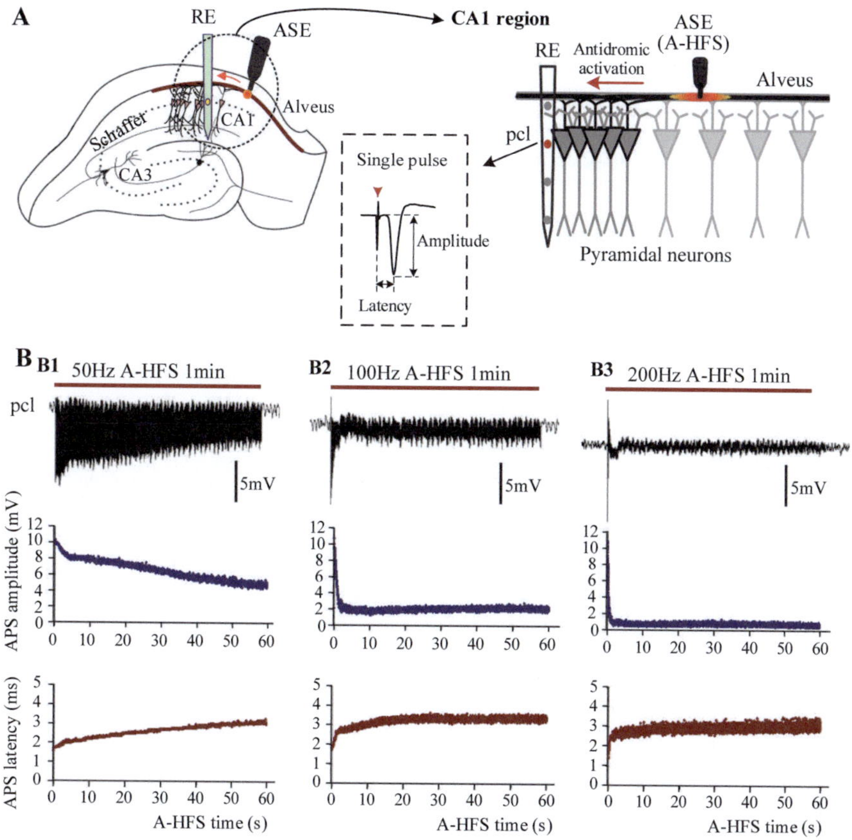

Fig. 5.2 CA1 neuronal responses to A-HFS trains with different pulse frequencies. **A** Schematic diagrams show the RE and ASE positions. A typical APS is shown in the dashed box, indicating the definitions of APS amplitude and latency. **B** Typical pcl recordings during A-HFS trains at 50, 100 and 200 Hz frequencies. The three rows (from top to bottom) show: raw recording (0.3–5000 Hz) of 1-min A-HFS with stimulation artifacts removed, and scatter plots of APS amplitudes and latencies induced by each pulse

returning to baseline level. The temporary decrease in OPS suggested a change at the somata, which is further examined in Sect. 5.2.3.

These experiment results show that the CA1 pyramidal neurons can still fire synchronously in respond to low-frequency orthodromic activations while failing to respond to high-frequency pulses at their axons. This indicates that their somata maintain firing ability during sustained A-HFS. The antidromic activation pathway of A-HFS involves only axon and soma, no synaptic transmission (Fig. 5.3A). Therefore, the APS suppression during A-HFS must stem from a failure at the axon rather than the soma. This evidence can strongly support the existence of HFS-induced axonal block.

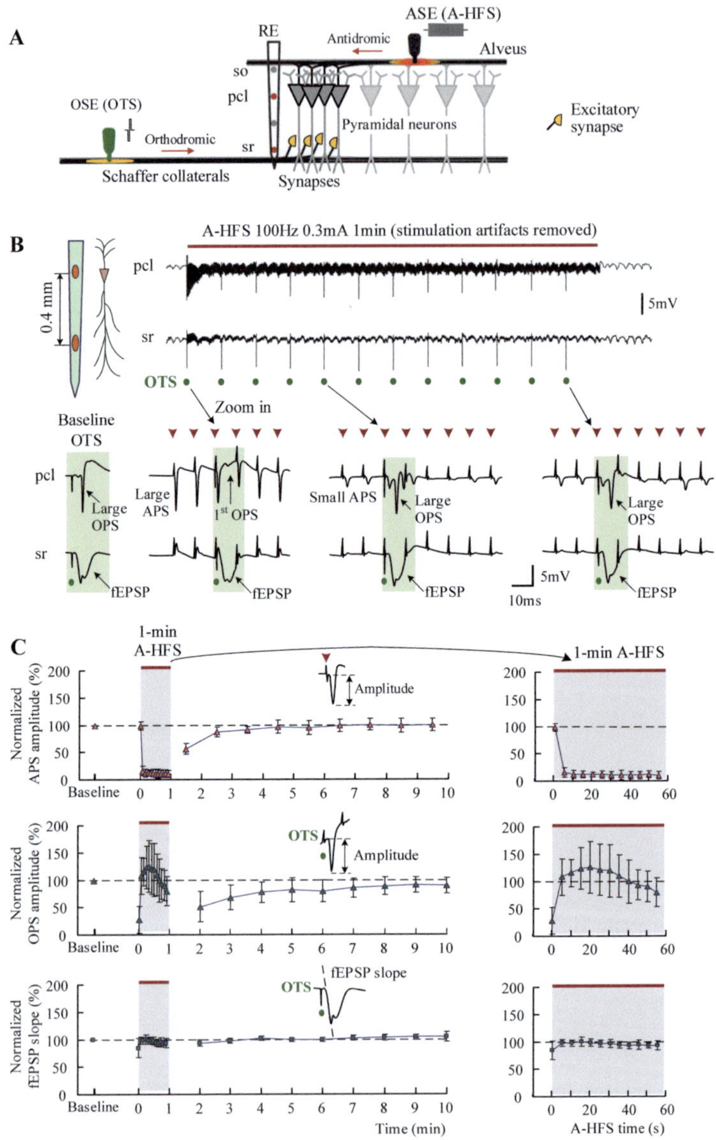

Fig. 5.3 Neuronal responses to orthodromic test stimuli (OTS) during A-HFS. **A** Schematic diagram showing the positions of the two stimulation electrodes (ASE for A-HFS and OSE for OTS) and the recording array RE. **B** Typical signals recorded simultaneously at the CA1 soma layer (pcl) and apical dendrite layer (sr) during a 1-min 100 Hz A-HFS. Pulse artifacts were removed. The enlarged insets below show the OTS-evoked OPS and fEPSP waveforms (shaded in green), with red arrows above and green dots below indicating the trimmed artifacts of A-HFS and OTS pulses, respectively. **C** Statistical data of normalized APS amplitude (upper), OTS-induced OPS amplitude (middle) and fEPSP slope (bottom) pooled from four 100 Hz and five 200 Hz 1-min A-HFS trains with 0.2 Hz OTS ($n = 9$). Grey shadings with a red bar on top indicate the A-HFS period. Note: To facilitate the comparison between APSs and OPSs during A-HFS, APSs were sampled every 5 s (20 ms before each OTS, with the first APS produced by the first A-HFS pulse). After the A-HFS ended, identical pulses were applied every minute to show APS recovery. Modified from Feng et al. (2013)

The APSs shown in Figs. 5.2 and 5.3 were all produced by the stimulation electrode ASE, which delivered A-HFS pulses at a fixed site of the axonal fiber. The APS suppression indicated a decrease in the evoked firing caused by this ASE. This decrease might only result from a trigger failure at the ASE, while the axon could still be able to transmit APs originating from other sources—meaning that the axon may not be truly blocked. To exclude this possibility, we placed a second stimulation electrode at a site sufficiently distant from the A-HFS electrode. This additional electrode delivered a test pulse to verify whether an AP generated away from the A-HFS site could pass through the site. This experiment is described below.

2. **HFS-Induced Axonal Block Prevents Action Potential Transmission**

In Fig. 5.4A, two antidromic stimulation electrodes, ASE_H and ASE_T, were placed at the alveus. The ASE_H delivered A-HFS, while the ASE_T delivered testing pulses—ATS (Fig. 5.4B). The APSs evoked by the two electrodes were termed APS_H and APS_T, respectively. Since the ASE_T was positioned farther from the recording site than the APS_H, along a same axon, an AP evoked by the ASE_T had to cross the affected site of ASE_H before reaching the soma at the recording site. The OSE was placed in the Schaffer collaterals. Except for ASE_T, the positions of the other three electrodes remained the same as in the previous experiments (Fig. 5.3A). The positioning coordinates of the four electrodes, from front to back, were: OSE at 2.2 mm posterior to bregma, 2.0 mm lateral to midline, depth ~ 2.8 mm; RE at 3.5 mm posterior to bregma, 2.7 mm lateral to midline, depth 2.1–2.5 mm; ASE_H at 4.8 mm posterior to bregma, 2.7 mm lateral to midline, depth 1.8–2.1 mm; and ASE_T at 5.8 mm posterior to bregma, 3.0–3.3 mm lateral to midline, depth 2.0–2.2 mm. The distance between ASE_T and ASE_H exceeded 1 mm. Figure 5.4A shows a photograph taken after an experiment, displaying the relative positions of the three stimulation electrodes and one array recording electrode. For details on the imaging method, refer to Sect. 3.2.3.

During baseline recording before A-HFS (Fig. 5.4C), the APS_T and the APS_H, produced respectively by single pulses from ASE_T and ASE_H, served as baseline controls. The latency (L) of APS_T was significantly longer than that of APS_H, confirming the longer distance between ASE_T and RE. When a 100 Hz A-HFS was applied by ASE_H, each pulse initially induced a large APS_H. However, the APS_H rapidly decreased to ~ 1/5 of its initial amplitude within seconds and remained stable until the end of 1-min A-HFS.

Immediately after A-HFS ended, the ASE_T delivered a test pulse 10 ms after the final A-HFS pulse. The evoked APS_T showed decreased amplitude and altered waveform compared to its baseline control. This APS_T appeared as a bimodal pattern with troughs P_1 and P_2 (denoted by two hollow arrows in Fig. 5.4C). While P_1 latency was similar to the baseline APS_T, P_2 latency was significantly longer. At 2.5 s post-A-HFS, the bimodal APS_T became more pronounced. Five seconds after A-HFS, a single pulse from ASE_H produced a partially recovered APS_H. The amplitude and latency of the APS_H then gradually returned to baseline level (Fig. 5.4D). Meanwhile, the APS_T returned to its baseline unimodal waveform as P_2 gradually shifted and merged with P_1 (Fig. 5.4E).

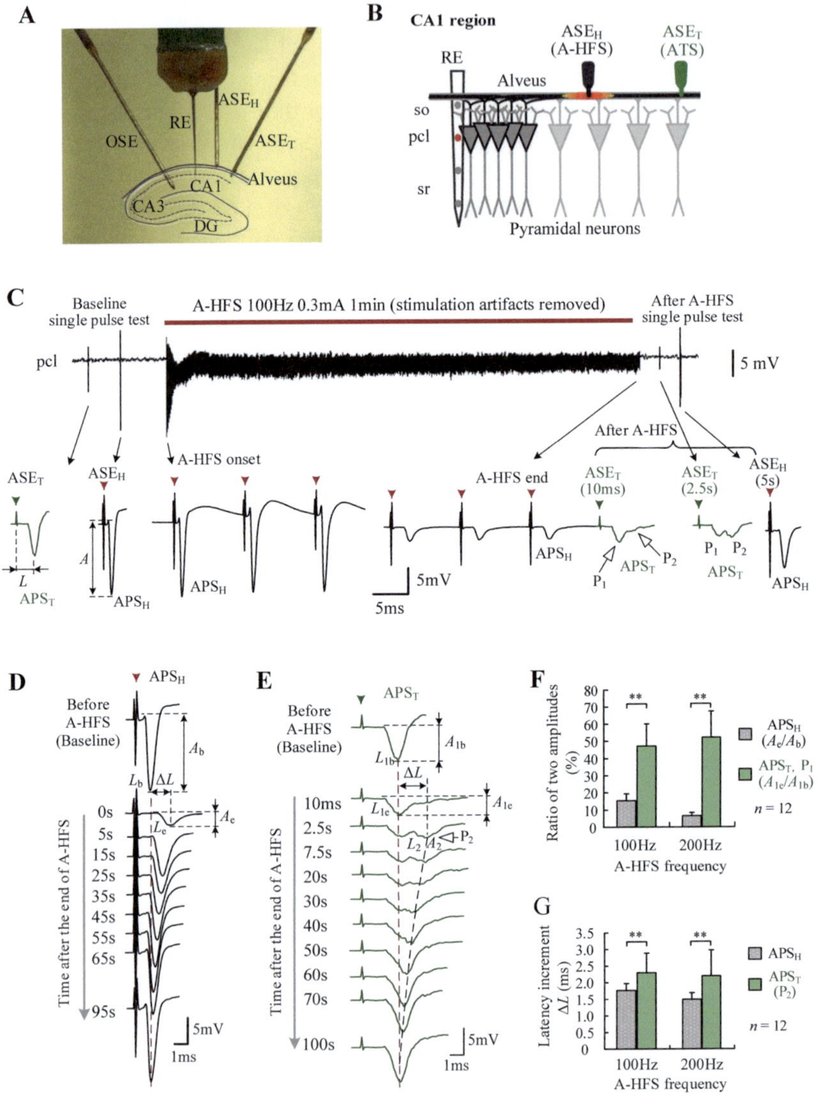

Fig. 5.4 HFS-induced axonal block preventing activation transmission across the block site. **A** A photograph showing the electrode positions. **B** Schematic diagram showing the positions of two stimulation electrodes (ASE_H and ASE_T) on the alveus and the RE in the CA1 region. **C** Typical recording in the CA1 cell layer (pcl) before, during and after 1-min 100 Hz A-HFS applied by ASE_H. The enlarged insets show APS_H and APS_T produced by the pulses applied by ASE_H and ASE_T, denoted by red and green arrows respectively. **D** and **E** Changes in the evoked APS_H and APS_T over time after the end of A-HFS, with their baseline waveforms shown at top. The evoked potentials align with the stimulus artifacts. **F** Comparisons of the amplitude ratios of APS_H and APS_T (P_1) at the end of 100 Hz and 200 Hz A-HFS to their baseline values. **G** Comparisons of the latency increments (ΔL) of APS_H and APS_T (P_2) at the end of A-HFS with two different pulse frequencies. The pulse intensity applied by both ASE_H and ASE_T was 0.3 mA, with a pulse width of 100 μs. In the figures F and G, $^{**}P < 0.01$, paired *t*-test, $n = 12$ rats. The signals of APS_H and APS_T are shown in black and green, respectively

The bimodal appearance of APS_T indicated that the activation area of ASE_T in the alveus only partially overlapped with that of ASE_H. While the ASE_T-achievable axons within the ASE_H activation area were blocked by A-HFS, those activated only by ASE_T remained unaffected. Therefore, APS_T appeared bimodally at the end of A-HFS. Its P_1 was produced by neurons with axons activated solely by ASE_T, remaining their original latency unaffected by A-HFS. The P_2 was from the axons achievable by both ASE_T and ASE_H, resulting in a prolonged latency caused by A-HFS. Once the A-HFS ended, the prolonged latency and reduced amplitude of P_2 gradually recovered, merging with P_1, as shown in Fig. 5.4E.

At the end of A-HFS, due to the unaffected part of APS_T, the ratio of APS_T amplitude to its baseline amplitude (A_{1e}/A_{1b}) significantly exceeded the value of APS_H (A_e/A_b) (Fig. 5.4F). Since ASE_T was farther from the recording site than ASE_H, the latency increment (ΔL) of P_2 in APS_T was also significantly greater than the value in APS_H (Fig. 5.4G). Note: The P_2 amplitude and latency could not be accurately measured at 10 ms after the end of A-HFS due to the extremely small P_2 in the APS_T waveform. Therefore, the P_2 measurements shown in Fig. 5.4G were taken from the APS_T produced at 2.5 s post-A-HFS (see Fig. 5.4E). Based on the "fast recovery" of APS amplitude described in Sects. 5.2.2 and 5.2.3, this latency measurement can be used to estimate the P_2 delay caused by A-HFS.

These results show that the HFS-induced axonal block not only can weaken the activation at the HFS site but also can prevent firing signals from other sites from passing through the block area. Additionally, the increased P_2 latency in APS_T indicates that the axonal block may extend beyond the site directly acted by ASE_H. Otherwise, the increment of P_2 latency should be similar to that of APS_H.

During sustained A-HFS, the increase in APS latency was always accompanied by a decrease in APS amplitude (i.e., fewer firing neurons). This raises a question: does an APS induced by a weak-intensity pulse naturally have a longer latency at baseline? In other words, when fewer neurons are activated, do axons take longer to transmit the activations antidromically to the somata? To determine whether the latency changes were caused by decreased neuron firing or axonal block, we conducted further investigations.

3. **Comparing APS Amplitudes and Latencies between Baseline and A-HFS Periods**

Figure 5.5A shows typical scatter plots of the APS amplitude (*upper*) and latency (*lower*) produced by each pulse during a 200 Hz A-HFS. Within the first second of A-HFS, the APS amplitude decreased while the APS latency increased rapidly (Fig. 5.5B). The superimposed plot of APS waveforms, extracted at five time points (0, 0.25, 0.5, 0.75 and 1 s of A-HFS), clearly shows these changes (Fig. 5.5C).

When single pulse stimulations with varying intensity were applied by the same ASE, the APS amplitude increased with the increase of pulse intensity due to the increased number of firing neurons (Fig. 5.5D, *top*). This relationship is known as the input–output (I-O) curve (see Sect. 2.3). However, the APS latency showed only slight shortening (Fig. 5.5D, *bottom*). The superimposed APS waveforms, produced

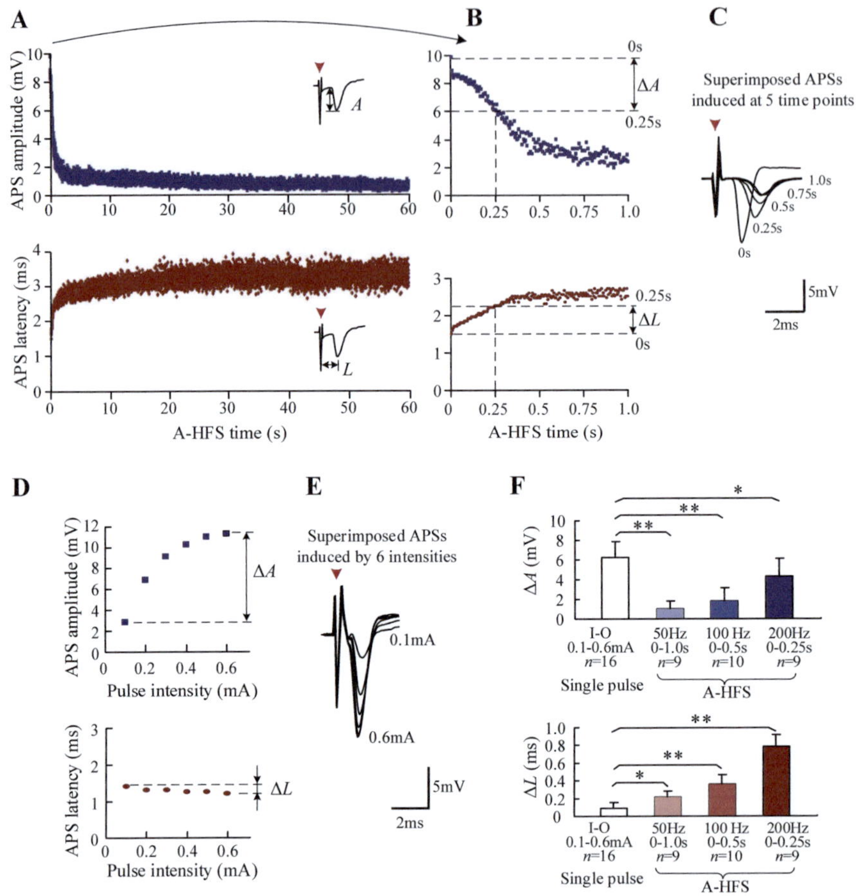

Fig. 5.5 Comparisons of APS changes caused by A-HFS and by different intensities in single pulse stimulations. **A** Typical scatter plots showing APS amplitudes (*upper*) and latencies (*lower*) produced by each pulse during 1-min 200 Hz A-HFS with 0.5 mA pulse intensity. **B** Enlarged plots of the first A-HFS second. **C** Superimposed APS waveforms extracted at five different time points. **D** Amplitudes (*upper*) and latencies (*lower*) of APSs induced by single pulses with increased intensity in an I-O test. **E** Superimposed APS waveforms induced by six pulses with intensities ranging from 0.1 to 0.6 mA shown in D. **F** Comparisons of the amplitude changes (ΔA, *upper*) and latency changes (ΔL, *lower*) between APSs induced in I-O tests and in the initial period of A-HFS. The "I-O" group represents the APS changes induced by single-pulse stimulation from 0.1 to 0.6 mA. The three A-HFS groups (50, 100 and 200 Hz) represent the differences between the 1st and 50th APS induced in initial A-HFS periods. $^{*}P < 0.02$ and $^{**}P < 0.001$, one-way ANOVA with post hoc Bonferroni tests. "*n*" represents the rat number. Modified from Feng et al. (2013)

by the pulses of six different intensities (0.1–0.6 mA), showed significant amplitude changes but minimal latency changes (Fig. 5.5E). The mean change in APS amplitudes (denoted as ΔA) in I-O curves was significantly greater than those produced during A-HFS with three different pulse frequencies (Fig. 5.5F, *top*). Conversely, the mean change in APS latencies (denoted as ΔL) in I-O curves was significantly smaller than those of A-HFS (Fig. 5.5F, *bottom*).

These results indicate that when produced by a single-pulse at baseline, the firing can take a similar time to antidromically activate the soma regardless of stimulus intensity—whether it is weak to activate only a few axons or strong to activate many. In other words, once an axon is activated, the entire activation process to the soma takes a consistent time, showing similar latency. In contrast, during the initial period of A-HFS, the APS latency increased much more than in I-O testing, even though the APS amplitude during A-HFS showed smaller changes. This substantial latency change suggests that A-HFS applied to the alveus could affect both axons and somata.

5.2.2 Extension of Refractory Period by Axonal HFS

The HFS-induced axonal block is not complete. Each A-HFS pulse can still induce a small APS. This raises a question: are these small APSs produced by a same group of unblocked axons firing persistently following each pulse, or by different groups of axons taking turns to fire? If the latter is true, this means that A-HFS could extend the refractory period of axons. Our following experiment confirms this hypothesis.

1. **Two Distinct Recovery Phases After A-HFS**

As shown in Fig. 5.6A, 1-min A-HFS trains with identical pulses but three different frequencies (50, 100 and 200Hz) were applied to the alveus. At the onset of A-HFS, each pulse produced a large APS, which then gradually decreased in size. The APS suppression occurred more rapidly and intensely at higher pulse frequencies (a and b in Fig. 5.6B). During the recovery period after the A-HFS ended, ATSs with identical pulse parameters were applied as single-pulses. Although the APS was suppressed more at the end of 200 Hz A-HFS than at 50 and 100 Hz, it unexpectedly recovered within 3 s to an amplitude comparable to those seen with 50 and 100 Hz stimulations. Furthermore, the partially recovered APSs were all similar at both 3 s and 13 s after A-HFS (c and d in Fig. 5.6B). This similarity contrasted with the frequency-dependent APS amplitudes observed at the end of A-HFS.

To determine the full recovery course of A-HFS-induced block at different pulse frequencies, we continued to apply ATS pulses for 5 min following the A-HFS (Fig. 5.7A). The ATS pulses were applied at 10-s intervals for the first 2 min (starting 3 s after A-HFS), then at 30-s intervals for the remaining 3 min, with the end of A-HFS served as time 0. The APS recovery occurred in two distinct phases—fast and slow. During the fast phase, the APS amplitude recovered quickly within seconds following A-HFS. In the subsequent slow phase, the complete recovery of APS took

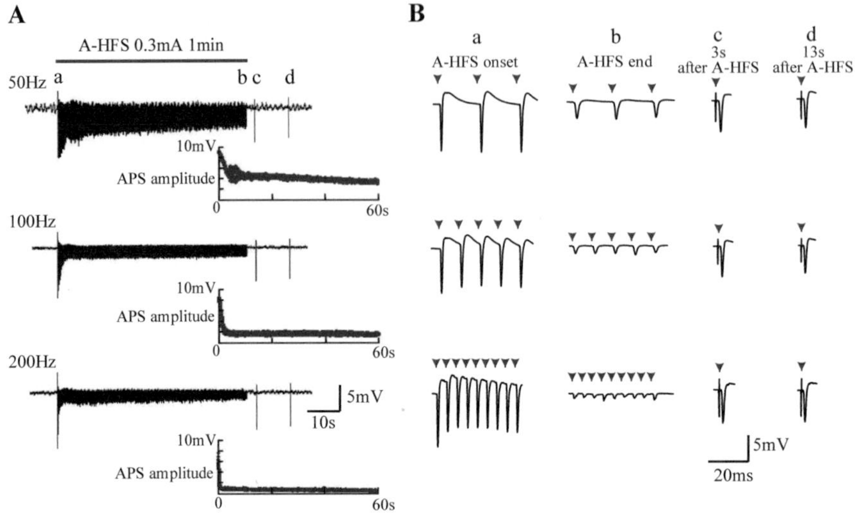

Fig. 5.6 Changes in APS amplitudes during and after A-HFS at three different pulse frequencies. **A** Typical recordings of 1-min A-HFS at 50, 100, and 200 Hz pulse frequencies, with pulse intensity of 0.3 mA. The scatter plots show APS amplitudes induced by each A-HFS pulse. **B** Enlarged APS signals induced at the onset and end of A-HFS (a and b), as well as APSs induced by single ATS pulses at 3 and 13 s after the end of A-HFS (c and d). From Feng et al. (2014)

several minutes (Fig. 5.7A). We distinguished the two phases based on the lowest point of the small trough in the APS amplitude after 3 s (Fig. 5.7A1). The mean trough time across all 70 trains of 50, 100 and 200 Hz A-HFS was (15.9 ± 5.5)s, with no significant difference among the three frequency groups. The time required for the APS to recover to 90% of baseline amplitude ($T_{90\%}$) was also not significantly different among these groups (Fig. 5.7B). The mean $T_{90\%}$ for all A-HFS was 1.20 ± 0.37 min ($n = 70$).

The fast-phase recovery of normalized APS amplitude (ΔA_1) correlated significantly with A-HFS frequency—higher frequencies resulted in greater ΔA_1 (Fig. 5.7C). However, the slow-phase recovery (ΔA_2) was independent of A-HFS frequency (Fig. 5.7D). As a result, the total recovery across both phases ($\Delta A_1 + \Delta A_2$)—i.e., the suppression of APS amplitude by A-HFS—correlated significantly with A-HFS frequency (Fig. 5.7E).

The frequency independence of slow-phase recovery indicates a commonality in neuronal responses across different A-HFS frequencies. During the late steady period of A-HFS, while the amplitude of each APS correlated negatively with pulse frequency, the number of produced APSs per second was proportional to it. This suggests that similar amounts of neuronal firing occurred during identical time periods. APS is a waveform formed by integrating the synchronous firing of a neuronal population. Both its amplitude and area can reflect the integrated AP number—and thus the number of firing neurons (Theoret et al. 1984). To compare neuronal firing during the steady period of A-HFS across different frequencies, we

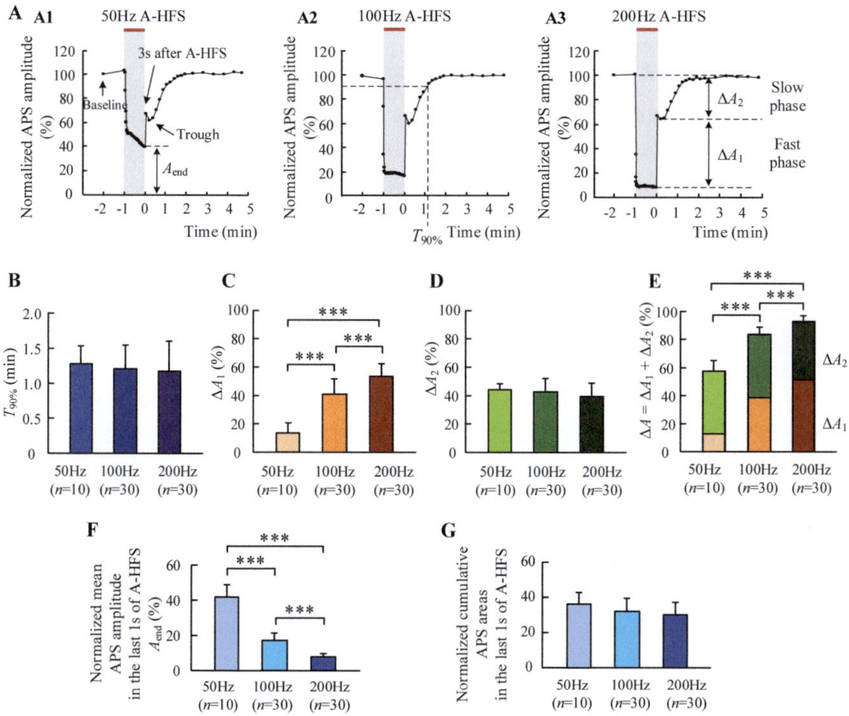

Fig. 5.7 Two distinct phases in APS recovery after the end of A-HFS. **A** Typical changes of the APS amplitudes normalized by baseline value during and following 1-min 50, 100 and 200 Hz A-HFS trains. Each data point (except the first one) during the A-HFS period represents mean APS amplitude per second. Each point before and following A-HFS represents an APS evoked by an ATS pulse. **B** No significant difference in the recovery time $T_{90\%}$ for 50, 100 and 200 Hz HFS trains. **C** The mean recovery of fast phase (ΔA_1) significantly increased with frequency. **D** No significant difference in the recovery of slow phase (ΔA_2) among the three frequency groups. **E** Total APS suppression (i.e., $\Delta A = \Delta A_1 + \Delta A_2$) significantly increased with frequency. **F** Mean APS amplitude at the last 1s of 1-min A-HFS (A_{end}) significantly decreased with frequency. **G** Cumulative APS area within the last 1s of A-HFS normalized to the area of the 1st APS of A-HFS. In **B–G**: $^{**}P <$ 0.001, one-way ANOVA with post hoc Bonferroni test. From Feng et al. (2014)

calculated the sum of APS areas in the final second of A-HFS. Although the mean APS amplitudes at 100 and 200 Hz were significantly smaller than at 50 Hz (A_{end}, Fig. 5.7F), these frequencies produced 2 and 4 times the APS counts, respectively. Therefore, the normalized cumulative APS areas showed no significant difference among the three A-HFS frequencies (Fig. 5.7G). This similarity in neuronal firing amounts may explain why the slow-phase recovery was frequency independent.

2. **Fast Recovery Following A-HFS**

In contrast to the slow-phase recovery, the fast-phase APS recovery was significantly frequency dependent (Fig. 5.7C). To investigate the fast phase with a higher time resolution, ATS was applied at 20 or 100 ms after A-HFS. To prevent interactions between these close ATS pulses, the 20 and 100 ms tests were conducted in separated A-HFS trials. For 50 Hz A-HFS, the pulse interval (IPI) was 20 ms, allowing the final pulse of A-HFS to serve as a 20 ms ATS. Similarly, the final pulses of 100 and 200 Hz A-HFS served as 10 and 5 ms ATS, respectively (Fig. 5.8A). To improve measurement accuracy, we used the mean amplitude of evoked APS in the final second of A-HFS (A_{end}) to represent the amplitude of APS evoked by these "ATS"s. We compared the A_{end} with the other APSs evoked within 3 s post-A-HFS for the three frequencies of 50, 100 and 200 Hz (Fig. 5.8B–D). The APS had already recovered to a level similar to the 3 s level across all three frequencies by 100 ms post-A-HFS. Notably, while the A_{end} of 200 Hz A-HFS was significantly smaller than those of 50 and 100 Hz A-HFS (Fig. 5.7F), at 20 ms post-A-HFS, the APS jumped to a level exceeding both 50 and 100 Hz A-HFS (Fig. 5.8E).

The results show that a substantial portion of APS suppression produced by higher frequencies (100 and 200 Hz) can recover even within 20 ms. This rapid recovery suggests that APS suppression could be related to changes in the refractory period. The A-HFS might extend the refractory period beyond the inter-pulse-interval (IPI) of 200 and 100 Hz A-HFS pulses (5 or 10 ms), causing the axons to fire action potentials intermittently in response to only a portion of A-HFS pulses. To verify this hypothesis, we examined changes in axonal refractory period during A-HFS.

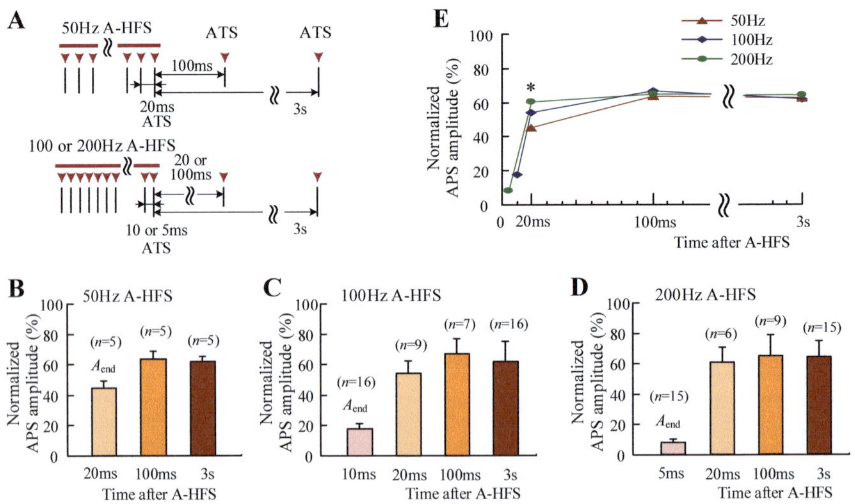

Fig. 5.8 Fast-phase of APS recovery following A-HFS at three different frequencies. **A** Schematic diagram showing the timing of ATS pulses at 20ms, 100 ms and 3 s following A-HFS. The red arrows and black vertical lines represent stimulus pulses. **B–D** Comparisons of the mean APS amplitudes during the final second of A-HFS (A_{end}) with APS amplitudes at 20ms, 100ms and 3 s following A-HFS at 50, 100 and 200 Hz, respectively. All APS amplitudes were normalized to baseline values. **E** Summary data from plots (**B–D**). Error bars are omitted for clarity since they are shown in plots (**B–D**). $*F = 5.27$, $P = 0.017$, 50 versus 200 Hz, one-way ANOVA with post hoc Bonferroni tests. From Feng et al. (2014)

3. Extension of Refractory Period

The refractory period (RP) is a phase when a neuron cannot generate a second action potential right after firing. Changes in RP can be measured using the evoked APSs from paired-pulses. To evaluate axonal RP during A-HFS, we inserted a prolonged IPI (a gap) to produce a sufficiently large APS based on the fast recovery described above.

As shown in Fig. 5.9A, at the onset of a 200 Hz A-HFS with an IPI of 5 ms, each pulse induced a large APS, indicating an RP of less than 5 ms. We denoted the amplitudes of the initial three APSs as A_1, A_2 and A_3, respectively. During the 1-min A-HFS, prolonged IPIs of 100 ms were inserted every 2 s to create 29 gaps. The amplitude of APS immediately following the gap (denoted as A_1 and indicated by a red dot in Fig. 5.9A) was significantly greater than the amplitude of the preceding APS that followed normal 5 ms IPI (denoted as A_0) (Fig. 5.9B). However, the second APS (A_2) after A_1 was significantly smaller than A_1 and nearly disappeared in the late A-HFS period. Statistical data (Fig. 5.9C) showed that during the initial period of A-HFS, the amplitude ratio of the first paired-APS (i.e., A_2/A_1 for the 1th and 2nd APS of A-HFS) was $90 \pm 11\%$ ($n = 7$), indicating thatA_2 was only slightly suppressed. However, during the final 30 s A-HFS period, the A_2/A_1 fell below 7.0%, indicating that A_2, induced after a 5 ms delay, was mostly suppressed by the gradually extended RP of A_1. For the third APS (A_3) after A_1, generated ~ 10 ms after A_1, the A_3/A_1 was around 60% in the final 30 s of A-HFS. Disregarding the ignorable A_2, this indicates that the axonal RP had exceeded 10 ms.

To further confirm the RP extension, we shortened the gaps from 100 to 20ms and inserted them in both 100 and 200 Hz A-HFS. For 100 Hz A-HFS, statistical data showed that the mean normalized amplitudes of A_1 remained above 50% ($n =$ 8, Fig. 5.10A), indicating sufficient APS recovery following the 20 ms gaps. After the initial 4 s A-HFS, the A_2 was suppressed significantly with $A_2/A_1 < 10\%$ ($n = 8$, Fig. 5.10B), indicating that the RP of A_1 exceeded 10 ms. For 200 Hz A-HFS with 20 ms gaps, the data showed that $A_2/A_1 < 6.0\%$ and$A_3/A_1 < 20\%$ for most A-HFS period with sufficiently large $A_1 > 60\%$ ($n = 6$, Fig. 5.10C, D), also indicating an extended RP beyond 10 ms.

The mechanism of RP extension produced by sustained A-HFS is not yet fully understood. Based on a mechanism of potassium ions (K^+) accumulation in peri-axonal spaces during sustained axonal membrane activation (Bellinger et al. 2008), we simulated the reactions of axonal membrane by computational modeling (Guo et al. 2018; Zheng et al. 2020). The results showed that when A-HFS pulses continuously depolarize the axonal membrane, the K^+ channels open and release K^+ into the narrow peri-axonal spaces, increasing the extracellular K^+ concentration ($[K^+]_o$). Since K^+ clearance cannot keep pace with its accumulation, $[K^+]_o$ remains high within IPIs. This elevated $[K^+]_o$ keeps the axonal membrane in a sustained depolarization state, leading to partial inactivation of membrane sodium (Na^+) channels (refer to Sect. 1.3). The inactivated Na^+ channels need multiple IPIs to recover sufficiently for the axon to fire again, resulting in an intermittent axonal block.

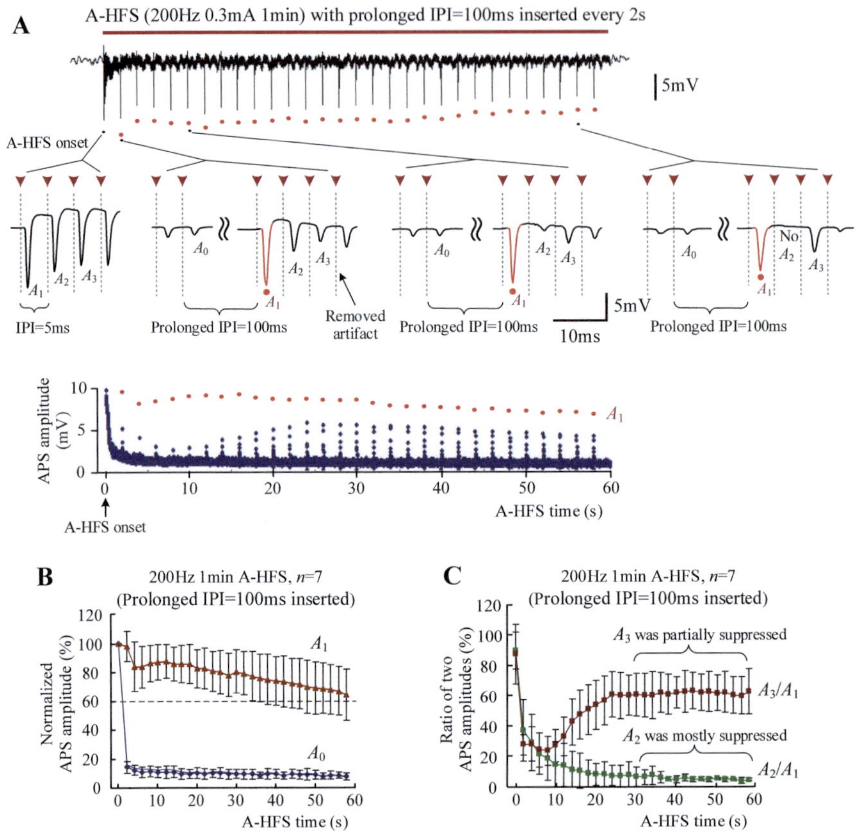

Fig. 5.9 Assessing the changes in neuronal refractory period (RP) during 200 Hz A-HFS by using the amplitude ratios of paired APS. **A** *Top*: typical neuronal responses to 1-min 200Hz A-HFS with prolonged 100 ms IPI gaps inserted every 2s. Red dots below the signal denote the evoked APSs following the gaps. *Middle*: Enlarged plots showing the evoked APSs at A-HFS onset and around three of the total 29 gaps. The APSs (A_1) following the gaps are shown in red. *Bottom*: scatter plot of APS amplitudes evoked by each A-HFS pulse, with A_1 data points highlighted in red. **B** The amplitudes of APSs before (A_0) and after (A_1) the gaps, normalized to the amplitude of the very first APS at A-HFS onset. **C** Ratios of A_2/A_1 and A_3/A_1 during A-HFS. The definitions of A_0 toA_3 are illustrated in (**A**). From Feng et al. (2014)

Our simulation results showed that small APSs induced during the steady period of 100 and 200 Hz A-HFS (with IPI of 10 and 5 ms) are generated by different neurons firing in turns, rather than the same neuronal group firing following each pulse. When A-HFS is applied to axons, it causes them to enter an intermittent block state where they can fire APs in response to only some pulses, not every pulse. Nevertheless, this intermittent firing is not necessarily evenly distributed over time. For details on the firing patterns of individual pyramidal neurons during A-HFS, please refer to Sect. 5.2.5. In addition, the above experimental results cannot rule out the possibility that sustained A-HFS can cause some neurons to become "silent" and cease firing

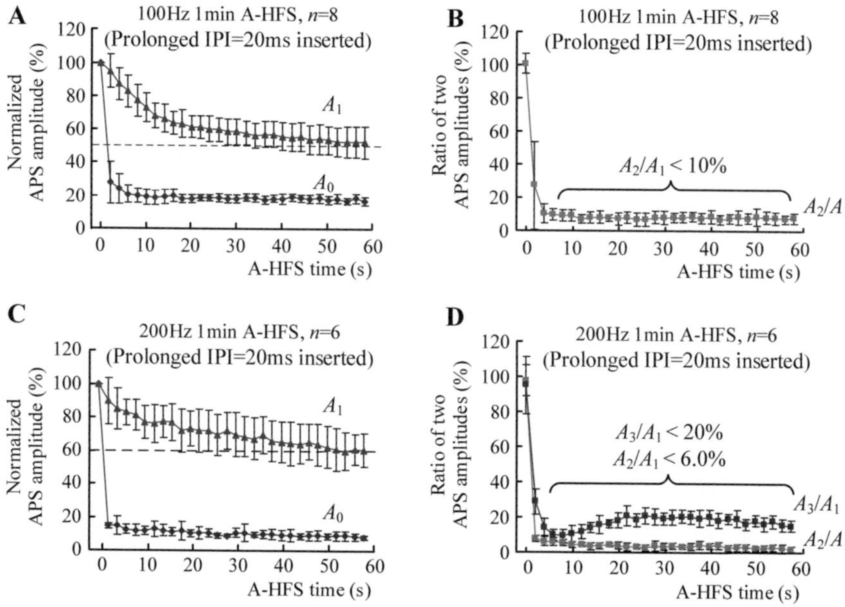

Fig. 5.10 Assessment of RP changes during 100 and 200 Hz A-HFS with 20 ms gaps. **A** Normalized amplitudes of the preceding APS (A_0) and subsequent APS (A_1) relative to the 20 ms gaps during 100 Hz A-HFS. **B** Ratios of A_2/A_1 during 100Hz A-HFS. **C** Same measurements as in (**A**), but for 200 Hz A-HFS. **D** Ratios of A_2/A_1 and A_3/A_1 during 200 Hz A-HFS. See Fig. 5.9A for the definitions of A_0 to A_3. From Feng et al. (2014)

completely. Each neuron's response to stimulation depends on its distance from the stimulation electrode. While some neurons fire intermittently due to extended refractory periods, others may become completely unresponsive, remaining silent until they are able to recover after A-HFS ends. Furthermore, A-HFS with gaps represents one of the stimulation patterns with varying frequencies (i.e., varying IPIs), which are detailed in Chap. 6.

The RP extension can explain the fast-phase recovery following A-HFS but not the slow-phase recovery that lasts several minutes, suggesting that other mechanisms could be involved in APS suppression during A-HFS. Although A-HFS was directly applied to the alveus (the axons of CA1 pyramidal neurons), the sustained stimulation could affect the somata. To investigate this issue, we used current source density analysis to examine changes in AP generation and conduction near the somata during antidromic propagation of A-HFS excitation.

5.2.3 *Effect of Axonal A-HFS on the Somata of Pyramidal Neurons*

To reveal possible somatic changes produced by sustained axonal A-HFS, we compared soma reactions between the initial and late A-HFS periods. As shown in Fig. 5.11A, to obtain large APSs during late A-HFS comparable to the initial period, we deleted one pulse every 20 pulses in the final 20 s of 1-min 100Hz A-HFS. Each pulse deletion created a 20 ms gap. Due to fast recovery, the APS (A_1) induced by the first post-gap pulse was more than twice the size of the pre-gap APS (A_0) (Figs. 5.10A and 5.11A). During the initial 2 s period of A-HFS, as the APS decreased gradually, we picked an APS with an amplitude equivalent to A_1, which we termed "A_1_equivalence". The lower left corner of Fig. 5.11A shows superimposed waveforms of five APSs to illustrate their differences: the APS induced by the first A-HFS pulse, the A_1_equivalence, the A_0, the A_1, and the recovered APS at 2 min after A-HFS ended. Statistical data in Fig. 5.11B show significant differences in the amplitudes and latencies of these APSs. While the amplitudes ofA_1 and A_1_ equivalence were similar (indicating a similar number of firing neurons), the meanA_1 latency was significantly longer than that of A_1_equivalence. However, no significant difference existed between the latencies of A_1 and A_0. We hypothesized that the extended A_1 latency might be attributed to a change in excitation conduction speed around the soma. Therefore, we utilized current source density (CSD, see Sect. 4.4) to measure the conduction speed of the APS current sink across the CA1 cell body layer (pcl).

Figure 5.11C shows evoked potentials from nine recording contacts on a 16-channel electrode array. These contacts were vertically arranged across a 400 μm span containing the pcl. This figure shows four episodes of 9-channel evoked potentials during different periods of A-HFS: the 1st APS, A_1_equivalence, A_0 (pre-gap) and A_1 (post-gap). Figure 5.11D shows the CSD curves calculated from the potential signals, with grey shadings indicating CSD current sinks. The sink travel trajectory—which can reflect action potential conduction—moved consistently across all four periods, starting above the pcl before moving sequentially downward to the soma and apical dendrites. This pattern aligns with previous findings (Kloosterman et al. 2001). The travel speed of A_1 sink was significantly slower than that of A_1_equivalence but similar to A_0 (Fig. 5.11E). This indicates that sustained A-HFS affected the somatic region by decreasing the excitation conduction speed. Furthermore, the recovery of conduction speed was slower than the restoration of activated neuron numbers.

A-HFS-induced decrease in the conduction speed of antidromic activations across somata indicated certain change. This change around somata could also affect the conduction of orthodromic activations from presynaptic neurons. To test this hypothesis, we applied orthodromic test stimulations (OTS) every 10s to the Schaffer collaterals as shown in Fig. 5.12A (refer to Fig. 5.3A for electrode positions). During A-HFS, while the amplitude of OPS induced by each OTS pulse remained its baseline level (Fig. 5.12B1), the OPS latency increased significantly (Fig. 5.12B2) and returned to baseline level about 2 min after A-HFS ended.

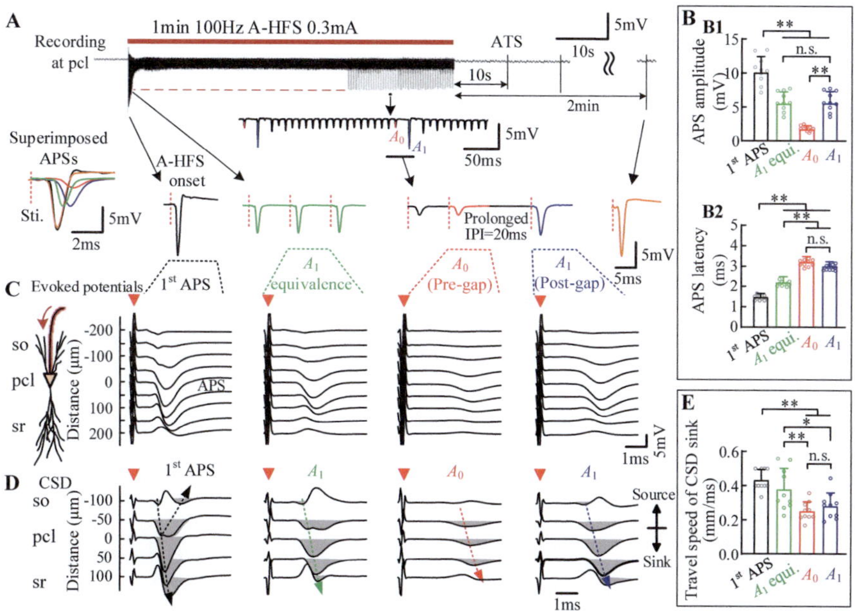

Fig. 5.11 Changes in APS current sink conduction during different periods of A-HFS. **A** Typical recording at the pcl during 1-min 100 Hz A-HFS, with 20 ms gaps inserted every 0.2 s in the final 20 s of A-HFS. The waveforms of the 1st APS, A_1_equivalence, A_0 (pre-gap) and A_1 (post-gap) during A-HFS, as well as the APS evoked 2 min after the end of A-HFS, are shown below. A superimposed plot of these APS waveforms is shown in the lower left corner. **B** Comparisons of the amplitudes (**B1**) and the latencies (**B2**) among the four APSs. **C** Examples of 9-channel evoked potentials across the pcl at the four periods. **D** CSD curves derived from the evoked potentials in (**C**), with grey shadings indicating CSD current sinks and arrows showing their travel directions. **E** Comparisons of the travel speeds of CSD sinks among the four APSs. In (**B**) and (**E**), n.s. not significant; $^{*}P < 0.05$, $^{**}P < 0.01$, one-way ANOVA with post-hoc Bonferroni tests; $n = 10$ rats. From Xu et al. (2023)

Based on the evoked OPSs recorded by the nine channels across the pcl (Fig. 5.12C), we calculated their CSD curves (Fig. 5.12D). The CSD current sinks of the OPSs originated from the apical dendrites and travelled sequentially toward the soma and then basal dendrites. During A-HFS, the travel speed of these current sinks decreased gradually and significantly, but returned to baseline level within 2 min after the A-HFS ended (Fig. 5.12E). While axonal A-HFS did not significantly affect the somatic ability to respond to OTS pulses, it did significantly slow the conduction of orthodromic activation across the somata.

The decrease of activation conduction speed suggests reduced excitability of the pyramidal neurons. However, during A-HFS, the amplitude of OTS-evoked OPS did not decrease but even slightly increased (see Figs. 5.3C and 5.12B). This could be attributed to the continuous replenishment of antidromic activations from A-HFS, which compensated for the reduced excitability of pyramidal neurons themselves. Supporting this hypothesis, we observed a temporary decrease in OPS amplitude

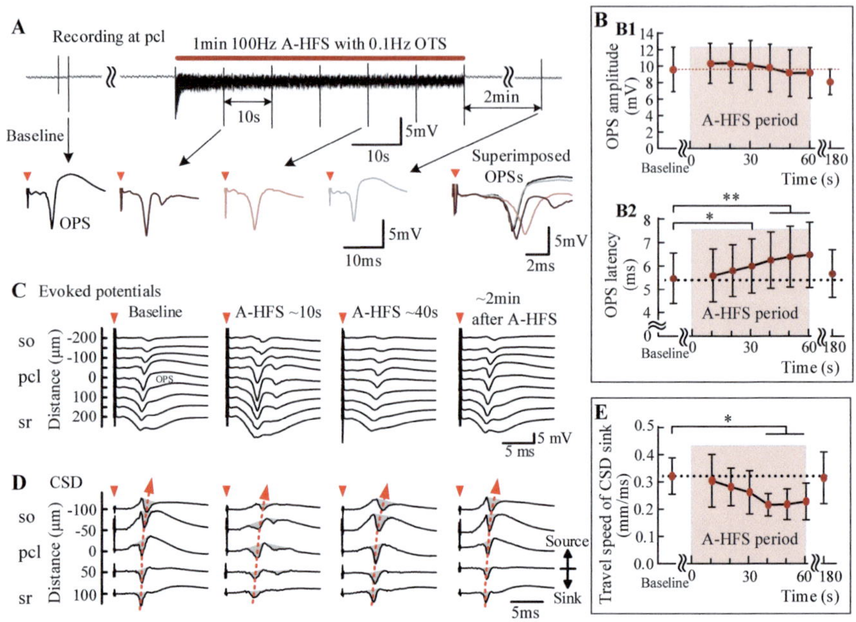

Fig. 5.12 Changes in OPS current sink conduction during different periods of A-HFS. **A** Typical recording at the pcl during 1-min 100 Hz A-HFS with OTS pulses applied every 10 s. The enlarged insets below show the OPSs evoked by OTS pulses respectively at baseline before A-HFS, at 10 s and 40 s of A-HFS and at 2 min following A-HFS. The OPS waveforms from these four periods are overlapped in the lower right corner. **B** Changes in OPS amplitudes (**B1**) and latencies (**B2**) before, during and following A-HFS. Pink shadings denote the A-HFS period. **C** Examples of 9-channel OTS-evoked potentials across the pcl at the four periods. **D** Corresponding CSD curves derived from the evoked potentials in (**C**), with grey shadings indicating current sinks and arrows showing their travel directions. **E** Changes in the travel speeds of CSD sinks before, during and following A-HFS. In (**B**) and (**E**): $^{*}P < 0.05$, $^{**}P < 0.01$, repeated-measure ANOVA with post-hoc Bonferroni tests, $n = 10$. From Xu et al. (2023)

immediately after A-HFS ended, as shown in Fig. 5.3C. To further verify this hypothesis, we investigated the firing changes of individual neurons before and after A-HFS, as shown below.

5.2.4 *Effect of Axonal HFS on Neuronal Excitability*

Figure 5.13A shows the firing of individual neurons following A-HFS. The wide-band signal from the soma layer of CA1 region (pcl) was filtered to produce MUA signal above 500 Hz. The MUA contained clear unit spikes from pyramidal neurons (Pyr) and interneurons (IN) before and after A-HFS. Although the stimulus artifacts of A-HFS were removed before filtering (detailed in 4.5.2), the evoked APSs during

A-HFS had much larger amplitudes than unit spikes, making the spikes indistinguishable during A-HFS period. Though pyramidal neurons are more numerous in the pcl than interneurons, interneurons fire at a much higher rate. Therefore, when interneuron firing is captured in the MUA signal, their spikes typically outnumber those of pyramidal neurons, as shown in Fig. 5.13A.

The signals in Fig. 5.13A revealed a notable phenomenon. Before A-HFS, spontaneous spike firing occurred in the baseline recording. During A-HFS, each pulse triggered an APS, indicating that neurons maintained continuous firing throughout the stimulation, although the APS amplitude decreased significantly in the late A-HFS period. However, immediately after A-HFS ended, a silent period without spikes appeared in the MUA signal, lasting several seconds before the firing gradually returned to baseline level (Fig. 5.13A, *bottom*).

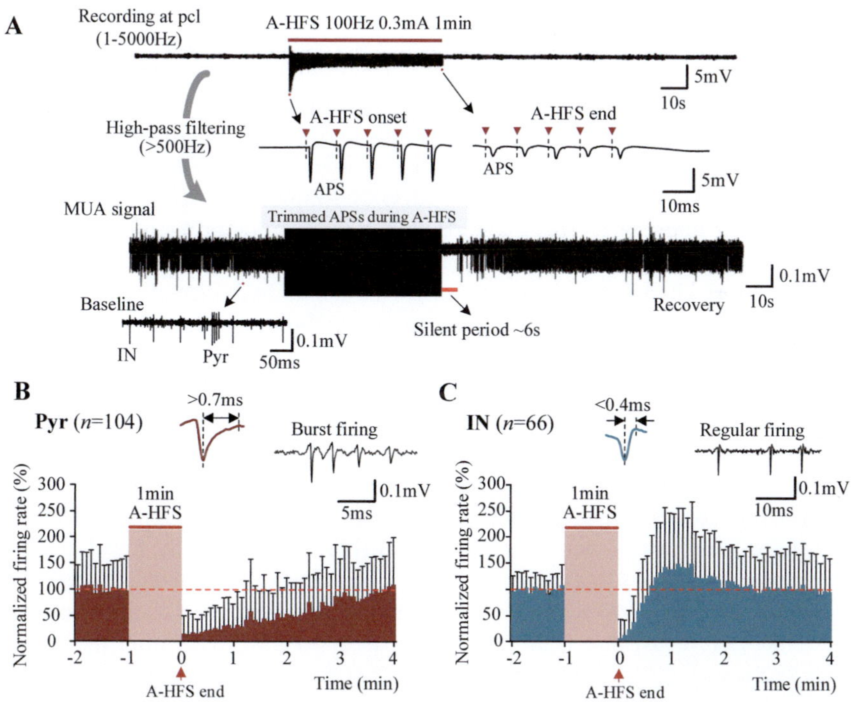

Fig. 5.13 Temporary silence of neuronal firing immediately following A-HFS. **A** *Top*: typical recording in the pcl of CA1 region before, during, and after 1-min 100 Hz A-HFS. *Middle*: Enlarged signals showing the initial large APSs and the final small APSs evoked by A-HFS pulses. Red arrows with dashed lines denote removed pulse artifacts. *Bottom*: MUA signal produced by filtering the artifact-free wide-band signal. **B** and **C** Mean normalized firing rates of pyramidal cells (Pyr) and interneurons (IN) before and after A-HFS (at 5 s time bins). Typical spike waveforms and firing patterns for both neuron types are illustrated above the plots. Red dashed horizontal lines denote the mean baseline firing rates. Due to variations in silent period durations across individual neurons (some shorter than 5 s), the plots with 5 s time bins cannot capture clear silent period. Modified from Yuan et al. (2022)

To investigate the firing of individual neurons, four-channel MUA signals recorded around the pcl were used for spike detection and sorting (refer to Sect. 4.1 for detailed method). Single unit spikes (SUA) of pyramidal neurons and interneurons were identified based on their firing patterns and spike waveforms (Barthó et al. 2004). As shown in Fig. 5.13B, C, pyramidal neurons usually fire in bursts with wider rising-phase in their spike waveforms, while interneurons fire more regularly with narrower rising-phase waveforms. We analyzed unit spikes from 104 pyramidal neurons and 66 interneurons in 40 rat experiments with 1-min 100 Hz A-HFS. During the 1-min baseline recordings before A-HFS, the pyramidal neurons showed a mean firing rate of 4.8 ± 4.6 spikes/s, while interneurons fired at 10.3 ± 9.4 spikes/s. Based on the average baseline firing rate of each neuron, we created histogram plots of normalized firing rates for both neuron types using 5 s bins (Fig. 5.13B, C). After A-HFS, pyramidal neurons showed significantly longer silent and recovery periods than interneurons ($P < 0.05$ for silent period, 21.9 ± 22.9 s vs. 11.2 ± 8.9 s; $P < 0.01$ for recovery period, 2.9 ± 1.5 min vs. 0.76 ± 0.60 min, unpaired *t*-test). Note that these mean durations were the statistical data from individual neurons, differing from the aggregated histogram data shown in Fig. 5.13B, C.

These results showed firing suppression in both pyramidal neurons and interneurons in the post-stimulation period, including a complete silent phase. It can explain the decreased amplitude of OTS-induced OPS following A-HFS, as shown in Fig. 5.3C. The temporary pause and reduction in interneuron firing indicated weakened inhibitory synaptic effects on pyramidal neurons (see Sect. 2.3 for details of local inhibitory circuits in the CA1 region). This should have increased, rather than decreased, the excitability of pyramidal neurons. Therefore, the post-stimulation firing suppression likely stemmed from decreased excitability of pyramidal neuron somata, caused by sustained A-HFS acting on their axons. However, this decreased excitability was not observed during A-HFS (Fig. 5.3C *middle line* and Fig. 5.12B1), possibly due to an offsetting effect created by the continuous A-HFS activation travelling to the somata. Once this activation ceased, decreased somatic excitability emerged immediately. In fact, during A-HFS, while OPS amplitude remained largely unchanged, OPS latency increased significantly (Fig. 5.12B2) and the travel speed of activations around the soma decreased (Fig. 5.12E). These results indicate decreased neuronal excitability. Moreover, the occurrence and recovery of this somatic change was temporally consistent with the slow phase of APS recovery after the end of A-HFS (Xu et al. 2023, and Fig. 5.7A), indicating that somatic changes may constitute another mechanism for APS decrease during A-HFS. Therefore, the effect of sustained A-HFS on axons can extend beyond the stimulation site to other neuronal structures, like somata.

5.2.5 *Clustered Firing in Pyramidal Neurons During A-HFS with Fixed-IPI*

Pyramidal neurons in the hippocampal CA1 region typically generate action potentials in a burst pattern (Fig. 5.13B). The above studies have shown that during sustained A-HFS at the axons of pyramidal neurons, each pulse can induce a small APS at the pcl. Since an APS represents integration of action potentials from a neuronal population, a regular APS sequence does not necessarily mean uniform firing of individual neurons. We next investigated the firing pattern of individual pyramidal neurons during A-HFS, to explore whether their firing maintained a non-uniform, burst-like characteristic.

Obtaining spikes from individual pyramidal neurons during their axons directly under A-HFS is challenging in intact brain, as these spikes become integrated into APS waveforms. In our in-vivo rat experiments, we typically used a bipolar stimulation electrode (CBCSG75, as detailed in Sect. 3.4.2) to deliver biphasic pulses with 100 μs width per phase and approximate 0.3 mA intensity. In normal baseline conditions, a single such pulse induced a large APS with an amplitude usually reaching ~ 10 mV. Although the APSs decreased to 1–2 mV during sustained 100 Hz A-HFS, they remained significantly larger than typical unit spikes from individual neurons, making it difficult to distinguish the spikes contained in APS waveforms. Therefore, to detect pyramidal neuron spikes during A-HFS, the APS must be minimized. While reducing pulse intensity can decrease the APS by activating fewer neurons, this approach creates a new challenge: a weaker stimulation can only produce a smaller activation scope, making it harder to reach the stimulated neurons for recording. In addition, to ensure reliable spike identification, we need to record spikes with sufficiently large amplitudes. This requires positioning the recording contact very close to a stimulated neuron's soma, demanding precise placements of both recording and stimulation electrodes.

To meet the requirement, we initially used single test pulses with a common intensity of 0.3 mA to position the stimulation electrode and the recording array to produce a satisfactory APS. We then gradually lowered the pulse intensity to around 0.05 mA and carefully adjusted the positions of both electrodes to obtain large spikes from stimulated pyramidal neurons during A-HFS. Notably, recording of large spikes in baseline spontaneous firing does not necessarily mean a successful experiment, because the neurons producing these spikes could lie outside the activation scope of the low-intensity stimulation and thus remained unmodulated by A-HFS. Such neurons would continue fire spontaneously during A-HFS. Additionally, under normal physiological conditions, not every pyramidal neuron is in a firing state—many remain silent (Andersen et al. 2007). These silent neurons could be activated and fire during A-HFS if they fall within the activation scope of A-HFS. Therefore, some firing can be unmodulated by A-HFS, while some can emerge only during A-HFS but absence in baseline recording.

Through our trials, we obtained the required spike data from individual pyramidal neurons modulated by low-intensity A-HFS. As shown in Fig. 5.14A, the spontaneous firing of pyramidal neurons in baseline recording appeared as bursts, which resulted in a clear peak at 3–8 ms in the inter-spike-intervals (ISI) histogram (Fig. 5.14B). We calculated the firing rates per second of single pyramidal neurons during 1-min baseline recordings. Figure 5.14C shows an example. The non-uniform burst firing caused substantial changes in these firing rates, characterized by a large coefficient of variation (CV). The CV—the ratio of the standard deviation to the mean—can evaluate the uniformity of neuronal firing (Di Miceli et al. 2020). A larger CV indicates more uneven firing. During 1-min baseline recordings in 18 rat experiments, we collected spontaneous firing from 27 pyramidal neurons, with a mean firing rate of 3.1 ± 1.7 spikes/s and the mean CV of the per-second firing rates of 1.2 ± 0.5.

Figure 5.14D shows the spikes of a pyramidal neuron with an amplitude of approximately 0.7 mV recorded during a 2-min train of low-intensity 100 Hz A-HFS at 0.04 mA. In the initial period of A-HFS, each evoked spike overlapped with an APS. Due to its large amplitude, these spikes were visible at the tip of APSs in the wideband raw recording signal. As the A-HFS continued and the APSs decreased, the

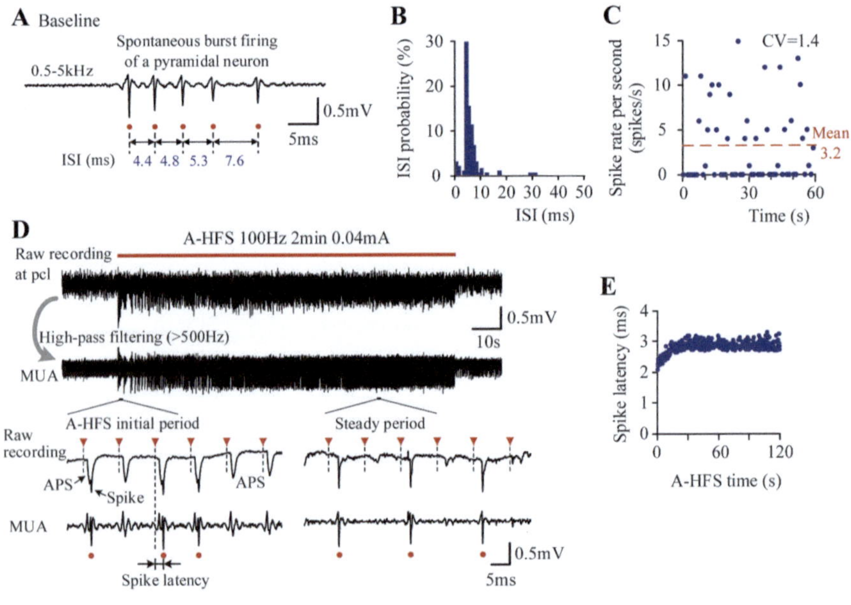

Fig. 5.14 Unit spikes of pyramidal neurons during baseline recording and during low-intensity A-HFS. **A** A typical spontaneous burst of a pyramidal neuron extracted from a baseline MUA signal with a bandwidth of 0.5–5 kHz. **B** Typical ISI histogram of baseline firing from a pyramidal neuron. **C** Scatter plot showing firing rates per second from a pyramidal neuron during 1-min baseline recording. **D** Recording example of pyramidal neuron firing during 0.04 mA low-intensity A-HFS at 100 Hz for 2 min. The first row shows the wideband raw recording at 0.3–5000 Hz. The second row shows the corresponding MUA signal (0.5–5 kHz) after high-pass filtering. The third and fourth rows show enlarged episodes of the initial and steady A-HFS periods. The red arrows with dashed lines indicate removed stimulus artifacts. The red dots indicate unit spikes. **E** Scatter plot of the spike latencies during the A-HFS shown in (**D**)

spikes became clearer. The spikes were more distinguishable in the MUA signal after high-pass filtering (denoted by red dots in the enlarged insets in Fig. 5.14D), which were strongly phase-locked with the A-HFS pulses. The spike latencies gradually increased from ~ 2 ms at the A-HFS onset to around 3 ms in a dozen seconds, and then remained stable during the late period of A-HFS (Fig. 5.14E). This pattern matched the change in APS latencies as shown in Figs. 5.2B and 5.5A, which verified that the spikes of this pyramidal neuron were produced by direct A-HFS activation on its axon.

During 28 trains of low-intensity 100 Hz A-HFS from 18 rats, we recorded 42 stimulated pyramidal neurons with spike amplitudes exceeding 300 μV to ensure accurate spike detection with minimal APS interference. We identified spikes of CA1 pyramidal neurons based on their rising phase width of > 0.7 ms (Barthó et al. 2004) and their phase-locked short latencies to the A-HFS pulses as shown in Fig. 5.14E. During A-HFS, the mean spike rate increased significantly to 15.5 ± 9.4 spikes/s (n = 42) compared to the baseline value of 3.1 ± 1.7 spikes/s (n = 27, t-test, $P < 0.01$), indicating enhanced neuronal firing by the stimulation. Meanwhile, the mean CV of the per-second firing rates during 2-min A-HFS decreased to 0.6 ± 0.3 (n = 42) compared to the baseline value of 1.2 ± 0.5 (n = 27, t-test, $P < 0.01$), indicating that the firing became uniform. Although the firing rate increased significantly during A-HFS, it was still well below the 100 Hz pulse frequency, meaning that the neurons did not fire following each pulse. Moreover, the pyramidal neurons exhibited two distinct firing patterns: non-clustered (regular) firing in 12 neurons (29%) and clustered (non-uniform) firing in the remaining 30 neurons (71%).

An example of non-clustered firing is shown in Fig. 5.15. During the 2-min A-HFS, although the spike ISIs varied as the pyramidal neuron firing at different IPIs (Fig. 5.15A), the ISI histogram was smooth (Fig. 5.15B), and the per-second firing rates were stable with a CV of only 0.6 (Fig. 5.15C). Figure 5.16 shows an example of clustered firing. During the 2-min A-HFS, the pyramidal neuron fired in alternating periods of intensive spikes and silence (Fig. 5.15A). Initially, its firing rate was higher, with clusters appearing as spikes following several successive pulses. Later, the spikes in clusters no longer followed each pulse successively, instead following every two, three, or more pulses. Two distinct peaks appeared in the ISI histogram of the clustered firing (Fig. 5.15B): one at 10–50 ms (corresponding to intra-cluster ISI) and the other at 100–200 ms (corresponding to inter-cluster ISI). The per-second firing rates changed smoothly, starting higher in the initial A-HFS period before gradually decreasing to a stable level, with a CV of only 0.7 (Fig. 5.15C). The firing rates for both non-clustered and clustered patterns followed a similar trend to the changes of APS amplitudes during A-HFS at a commonly used intensity (e.g., 0.3 mA in Fig. 5.2B). Furthermore, the ISI histogram was modulated by the10ms pulse interval and its integer multiples, indicating that the firing of pyramidal neurons was directly controlled by the stimulation pulses.

We measured the spike rates from individual pyramidal neurons during low-intensity A-HFS at frequencies of 50, 100, 133 and 200 Hz, respectively. Their mean spike rates showed no significant differences across these pulse frequencies.

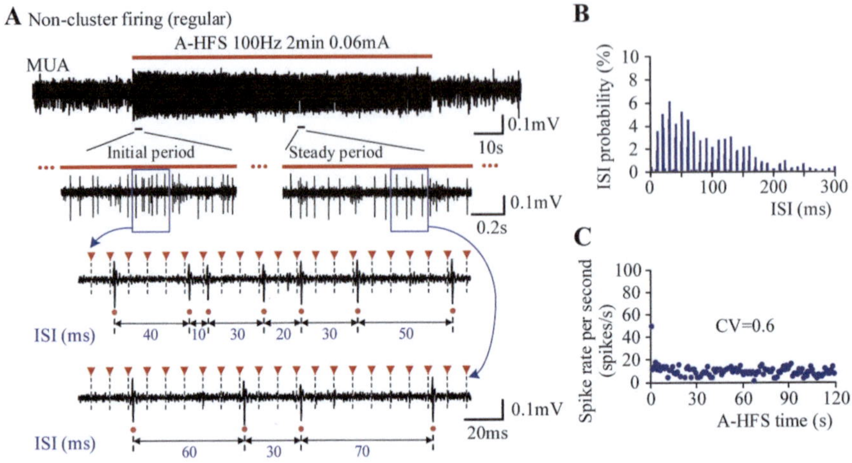

Fig. 5.15 Example of non-clustered firing of a CA1 pyramidal neuron produced by A-HFS. **A** MUA signal containing the spikes from a pyramidal neuron which was modulated by A-HFS pulses. The enlarged views at two time scales show the details of initial and steady A-HFS periods. Red dots below the signals denote the spikes, with ISI values shown in blue. **B** ISI histogram of the non-clustered spike sequence shown in (**A**). **C** Scatter plot of the per-second firing rates with a CV of 0.6. From Yuan (2023)

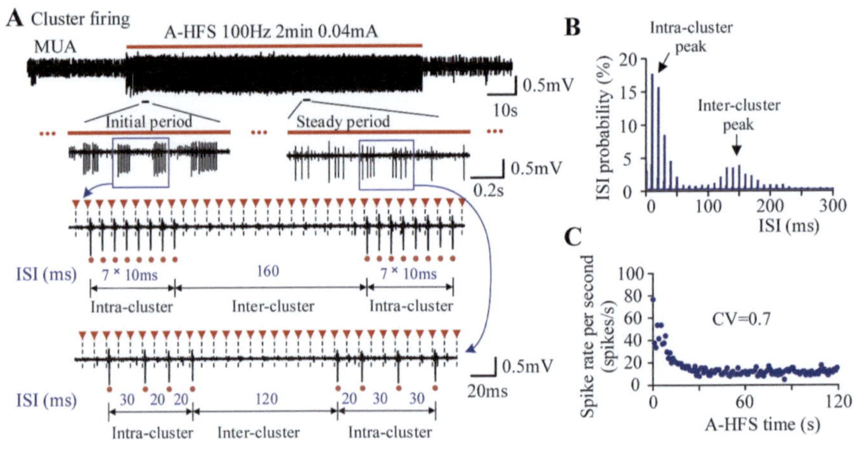

Fig. 5.16 Example of clustered firing of a CA1 pyramidal neuron produced by A-HFS. **A** MUA signal containing the spikes from a pyramidal neuron which was modulated by A-HFS pulses. **B** ISI histogram of the clustered spike sequence from (**A**), showing two distinct peaks. **C** Scatter plot of the per-second firing rates with a CV of 0.7. From Yuan (2023)

Additionally, each neuron kept its firing pattern (clustered or non-clustered) during A-HFS trains, regardless of the pulse frequency (Yuan 2023).

It is an interesting finding that A-HFS with uniform pulses can produce non-uniform clustered firing in individual pyramidal neurons, distinct from the regular APS of neuronal populations. Although this clustered firing somewhat resembled burst firing, the two patterns differed in two points. First, the amplitudes of intra-cluster spikes remained stable—the mean amplitudes of first and last spikes in clusters were 374 ± 178 μV and 336 ± 175 μV respectively ($n = 30$ pyramidal neurons). In contrast, the spike amplitudes in spontaneous bursts decreased substantially. In baseline recordings from 18 rat experiments, bursts of 3–5 spikes showed mean amplitudes of the first and last spikes of 512 ± 293 μV and 342 ± 168 μV respectively ($n = 27$ pyramidal neurons). Second, the intra-cluster ISIs were much longer than the intra-burst ISIs (Figs. 5.14A, B and 5.16A, B). These distinctions indicate different underlying mechanisms for the two firing patterns.

Intrinsic bursts of CA1 pyramidal neurons are generated in the soma through local regenerative self-excitation, triggered by a large after-depolarization (ADP). This ADP occurs through persistent Na^+ flows at or near the soma (Azouz et al. 1996; Yue et al. 2005). Consecutive spikes within a burst progressively decrease in amplitude due to increased inactivation of fast Na^+ channels (Jensen et al. 1996). Unlike these intrinsic bursts, each spike in A-HFS-induced clusters was triggered by stimulation pulse with a phase-locked latency, resulting in longer ISIs than those in bursts. The increased spike rate also indicated the excitatory modulation of A-HFS on the neuronal firing.

The results of our computational modeling revealed the following possible mechanism underlying the clustered firing (Yuan et al. 2025). During the initial period of A-HFS, intensive firing produces an intermittent axonal block caused by increased $[K^+]_o$ in the peri-axonal space and increased Na^+ channel inactivation (Guo et al. 2018; Zheng et al. 2020; Brazhe et al. 2011). This block decreases the amplitudes of evoked action potentials (AP) at the initial and adjacent axon nodes around stimulation site. The weakened AP eventually loses its ability to propagate and fails to reach the soma, resulting in pause of soma firing. This pause allows the initial AP to gradually recover and grow strong enough to trigger soma AP again. Subsequently, a cluster of APs continues due to the feedback effects of after-hyperpolarization (AHP) from the AIS and soma until the initial AP is weakened again. Additionally, our simulation showed that neurons with axons closer to the stimulation electrode tend to generate non-uniform clustered firing, while those farther away tend to generate regular firing. Nevertheless, further investigations are needed to verify the mechanism and uncover additional ones.

The absence of burst firing in the pyramidal neurons during A-HFS may result from inhibitory effects of local feedback circuits. When a pulse of A-HFS antidromically triggers somata to fire, the firing can immediately activate the interneurons in feedback inhibitory circuits to suppress subsequent burst firing (Knowles and Schwartzkroin 1981; Davies et al. 1990; Kepecs and Fishell 2014). Additionally, antidromic activation travelling along axons during A-HFS can activate interneurons through axonal branches thereby inhibiting the somata of pyramidal neurons (Ye et al. 2022; Dudok et al. 2021). As shown in Chap. 2, the experiments of single-pulse and paired-pulse stimulations have revealed that the behavior of CA1 pyramidal neurons was modulated by local inhibitory circuits, resulting in single spike in the evoked APS. When GABA antagonist (e.g., PTX) blocked the inhibitory synapses, the evoked APS appeared as burst-like firing with multiple peaks (Fig. 2.9). Therefore, the activations of interneurons in inhibitory circuits could have prevented burst firing during A-HFS.

The non-uniform clustered firing induced by a uniform incentive in the rat brain *in-vivo* provides new evidence for nonlinear dynamics of neuronal firing in intact brains. The alternating periods of clustered firing and silence resemble bistable behavior—a typical property of nonlinear bifurcation dynamics observed previously in neural computational models with high $[K^+]_o$ (Wu and Shuai 2012; Contreras et al. 2021; Hahn and Durand 2001). However, this behavior has limited support from real neurons in the brain. Additionally, the non-uniform firing originated in directly stimulated neurons can propagate to post-synaptic neurons through neuronal networks, resulting in widespread effects, as shown in our another report (Wang et al. 2024). The following sections will describe how high-frequency axonal stimulations affect post-synaptic neurons downstream of the stimulation site.

5.3 Effect of High-Frequency Axonal Stimulation on Post-synaptic Neurons

The neuronal responses to the axonal A-HFS described above were produced by antidromic propagation of axonal activation to the neuronal somata, which involving only the axon and soma of same neurons, without involving synaptic transmissions. However, the activations initiating in the middle site of axons can travel bilaterally—both antidromically toward the soma and orthodromically toward axonal terminals. The orthodromic propagation can activate downstream post-synaptic neurons through synaptic transmissions. This orthodromic pathway is more complex than the antidromic one, involving axonal conduction, synaptic transmission, integration of post-synaptic potentials, and activation of the post-synaptic neuronal soma finally. To investigate the responses of post-synaptic neurons to orthodromic high-frequency stimulation (O-HFS), we still took the hippocampal CA1 neurons as the recording target. That is, remain the same recording site as in A-HFS, but move the stimulation site to the Schaffer collaterals—the afferent axon fiber of the CA1 region thereby positioning the stimulation site upstream of the recording site. Although the axons of the Schaffer collaterals somewhat differ from those of the alveus (Andersen et al. 2007), the HFS effects on the axons remained similar. The following O-HFS experiments further demonstrate that HFS can generate axonal block.

5.3.1 Responses of Downstream Post-Synaptic Neurons to Sustained O-HFS Trains at Different Frequencies

As shown in Fig. 5.17, when 1-min O-HFS trains at 50, 100, or 200 Hz were applied to the Schaffer collaterals, the initial responses of CA1 neurons were similar, regardless of the pulse frequencies. The first O-HFS pulses induced similar orthodrminc population spikes (OPS), denoted as 1st OPS, with similar latencies. Then, in the subsequent tens of milliseconds, OPSs disappeared due to the effect of local inhibitory circuits (see Sect. 2.3 for details) before reappearing temporarily. With the lower frequency of 50 Hz, the OPSs continued for tens of seconds, sometimes persisting until the end of the 1-min O-HFS (Fig. 5.17A). The OPSs in the late O-HFS period resembled the epileptiform activity induced by epileptic agents (refer to Chap. 8), showing no stable latencies with the O-HFS pulses. At 100 and 200 Hz frequencies, no obvious OPS appeared in the late O-HFS periods (Fig. 5.17B, C). With similar pulse intensity (around 0.3 mA) and 1st OPS amplitude (around 8 mV), the mean duration with OPSs during 50 Hz O-HFS was 40.7 ± 20.8 s ($n = 8$), significantly longer than those during 100 Hz O-HFS of 9.3 ± 12.8 s ($n = 8$) and 200 Hz O-HFS of 0.7 ± 1.1 s ($n = 9$) (Feng et al. 2013). Although O-HFS at 100 and 200 Hz delivered 2 and 4 times the stimulation pulses of that at 50 Hz, they generated much less the synchronous firing—OPSs. This suggests that higher pulse frequencies can weaken synchronous responses in downstream neurons.

Similar to the A-HFS periods described in the previous sections, the axonal block caused by O-HFS at the lower frequency of 50 Hz may be weaker. This can lead to stronger activations of Schaffer collateral axons, causing increased activations of downstream neurons and resulting in a longer period of synchronous firing with OPSs. In contrast, O-HFS at higher frequencies (100 and 200 Hz) may produce deeper axonal blocks, resulting in weaker activation of downstream neurons that cannot sustain the synchronous firing.

Like the A-HFS experiment shown in Fig. 5.3, we conducted an O-HFS experiment to investigate whether post-synaptic pyramidal neurons retained their ability to fire action potentials during the late period of O-HFS when no obvious OPS occurred. In this experiment (Fig. 5.18A), while 1-min O-HFS was applied to the Schaffer collaterals, an antidromic test stimulus (ATS) was applied to the alveus every 5s to antidromically activate CA1 pyramidal neurons at the recording site. We hypothesized that if the failure of O-HFS pulses to induce OPS was due to presynaptic and/or synaptic problems, the somata of CA1 pyramidal neurons should still be able to respond to ATS activations by firing action potentials. Our results confirmed this hypothesis (Fig. 5.18B, C). When O-HFS pulses at 200 Hz no longer induced OPSs, ATS pulses persisted to induce large APSs. This indicated that the absence of OPS during late O-HFS was not caused by failures in the post-synaptic neurons themselves. The O-HFS activation pathway involves afferent axons (Schaffer collaterals), pre- and post-synaptic structures, and the integration of post-synaptic potentials.

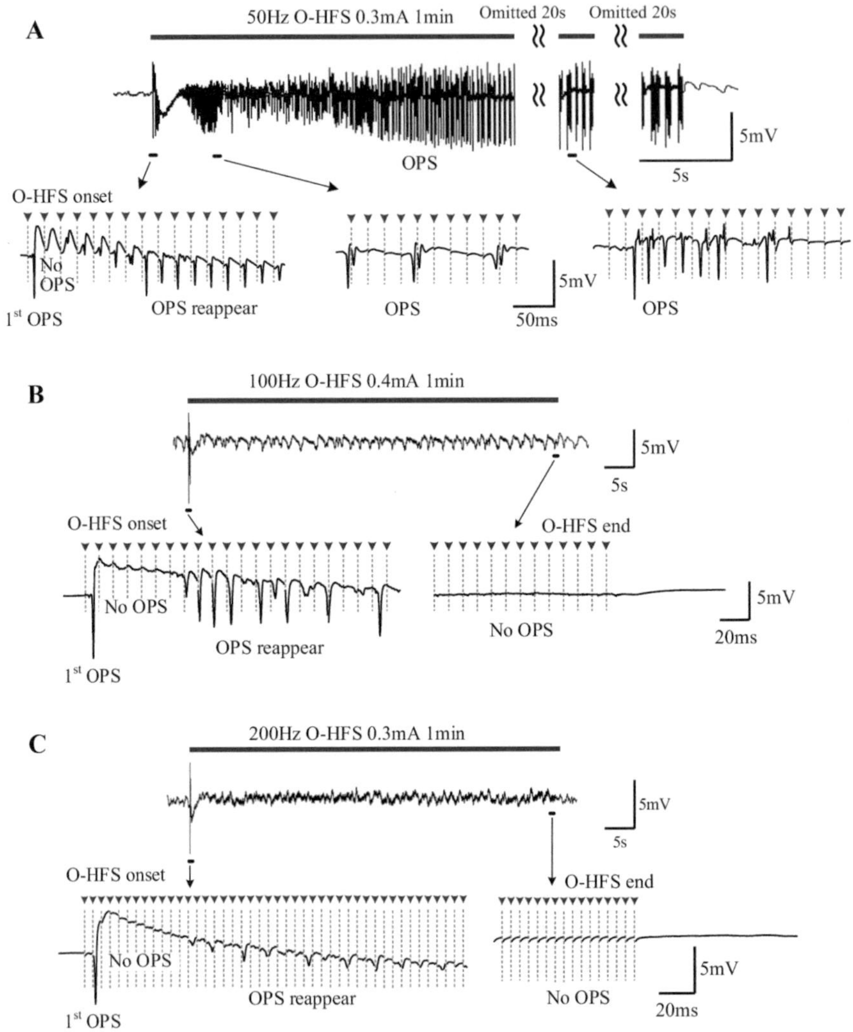

Fig. 5.17 Neuronal responses recorded in the pcl of rat hippocampal CA1 region during O-HFS at three different pulse frequencies. **A–C** Examples of 1-min O-HFS trains at 50 (**A**), 100 (**B**) and 200 Hz (**C**), with enlarged insets showing evoked OPSs

While this experiment cannot pinpoint the exact failure locations, multiple mechanisms may be involved and act jointly: axons under O-HFS can experience depolarization block, and presynaptic structure at axonal terminals may fail to transmit activation reliably due to neurotransmitter depletion from intensive synaptic activations by O-HFS (Anderson et al. 2006; Iremonger et al. 2006).

Since A-HFS on the alveus axons of CA1 pyramidal neurons can induce axonal block (Sect. 5.2), the O-HFS on the Schaffer collaterals (the axons of CA3 pyramidal

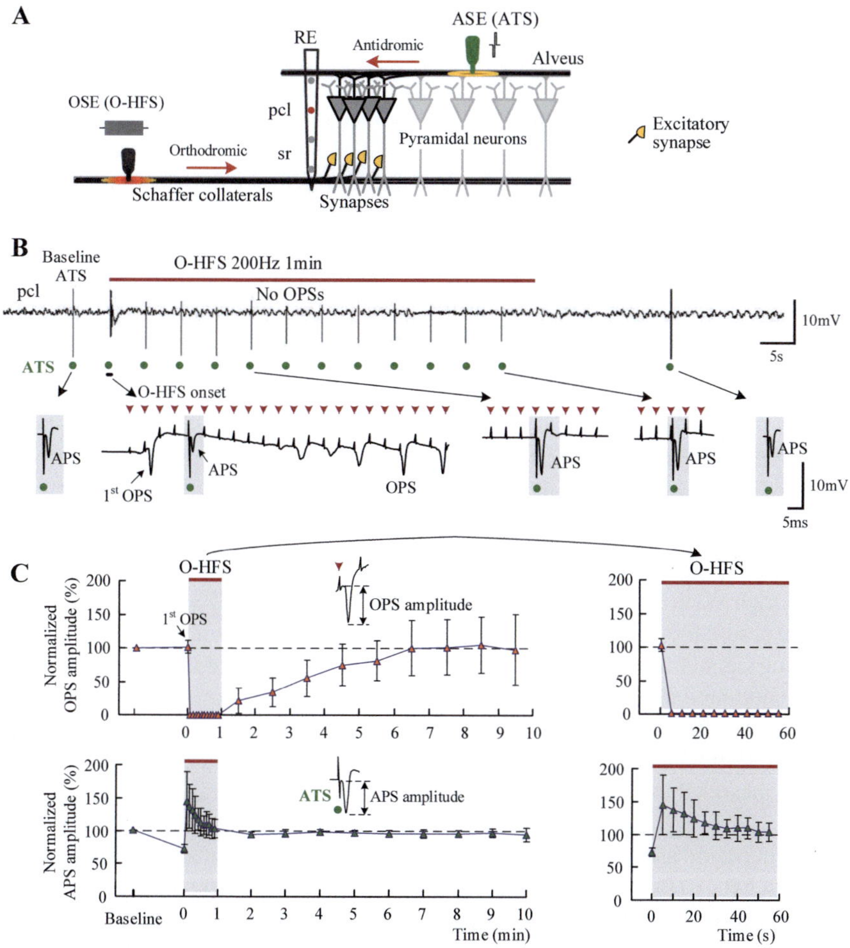

Fig. 5.18 Neuronal responses to antidromic test stimuli (ATS) during O-HFS. **A** Schematic diagram showing the electrode positions. **B** Typical recording at pcl with 1-min 200 Hz O-HFS (stimulus artifacts removed). The enlarged insets below show the ATS-evoked APSs (shaded in grey). Red arrows above the waveforms denote the trimmed artifacts of 200 Hz O-HFS pulses, while the green dots below denote the ATS artifacts. **C** Changes in mean normalized amplitudes of OPS (upper) and APS (bottom) pooled from four 100 Hz and six 200 Hz 1-min O-HFS trains with 0.2 Hz ATS (n = 10). APSs were induced every 5 s during O-HFS and every minute after O-HFS. For the comparison between OPSs and APSs, during O-HFS, OPS samples were taken every 5 s (20 ms before each ATS, with the 1st OPS induced by the first O-HFS pulse). After O-HFS ended, identical orthodromic stimuli were applied per minute to show OPS recovery. Modified from Feng et al. (2013)

neurons) may also induce axonal block. The two axon types have some differences: the alveus axons have thin myelin sheaths, while the Schaffer collateral axons are unmyelinated (Debanne et al. 2011; Andersen et al. 2007). Nevertheless, studies have shown that unmyelinated axons can also generate and even be more prone to axonal block (Chomiak and Hu 2007; Kim et al. 2012; Burke et al. 2004). Therefore, we tentatively assume that O-HFS on the Schaffer collaterals can induce more severe axonal block, resulting in no synchronous firing in the downstream neurons during the late steady period of O-HFS at pulse frequencies over 100 Hz. However, this does not necessarily mean that O-HFS has no activation effect on these neurons, it may drive the neurons to fire asynchronously. To verify the hypothesis, we investigated the behavior of individual neurons during O-HFS, as shown below.

5.3.2 Asynchronous Firing in Downstream Post-Synaptic Neurons Responding to O-HFS

As shown in Fig. 5.19A *top*, when we applied 1 min 100 Hz O-HFS, no obvious OPS events occurred in the late 40 s period except for the initial O-HFS period. We termed this late period the "steady period". After stimulation artifact removal, the wideband recording signal was filtered to produce MUA signal. During the steady period, MUA spikes continued until the end of O-HFS with a firing rate exceeding baseline level (Fig. 5.19A *bottom*). However, upon the end of O-HFS, a silent period without spikes occurred, after which spikes gradually recovered (Fig. 5.19B). Statistical data showed that the mean MUA rate during the late 40 s O-HFS period was 77 $\pm$ 44 spikes/s, significantly higher than the baseline rate 26 $\pm$ 21 spikes/s (n = 29, Fig. 5.19C). The silent period after O-HFS lasted 15.6 $\pm$ 7.7 s. Within 2 min after O-HFS ended, the mean MUA rate returned to baseline level. Note: The silent period here likely stemmed from CA3 pyramidal neurons' silence following A-HFS of their axons (as the Schaffer collateral stimulation was antidromic when viewed from the somata of these CA3 neurons), similar to the firing silence of CA1 pyramidal neurons following alveus A-HFS shown in Sect. 5.2.4. Without the inputs from upstream CA3 pyramidal neurons, the downstream CA1 neurons thus ceased firing.

The silent period after O-HFS indicated that the MUA appearance during the O-HFS was produced by the stimulation pulses. Moreover, during the late O-HFS period, when OPS events were absent due to axonal block, the MUA increased rather than decreased compared to baseline (Fig. 5.19C), because the axonal block was intermittent, not complete. Since the activated axonal terminals of the Schaffer collaterals can excite both pyramidal neurons and interneurons in the post-synaptic region (refer to Fig. 2.6A), the MUA signal should have contained the firing from both neuron types. We next classified the mixed unit spikes to investigate their firing alterations respectively.

1. **Increased Firing of Both Pyramidal Neurons and Interneurons by O-HFS**

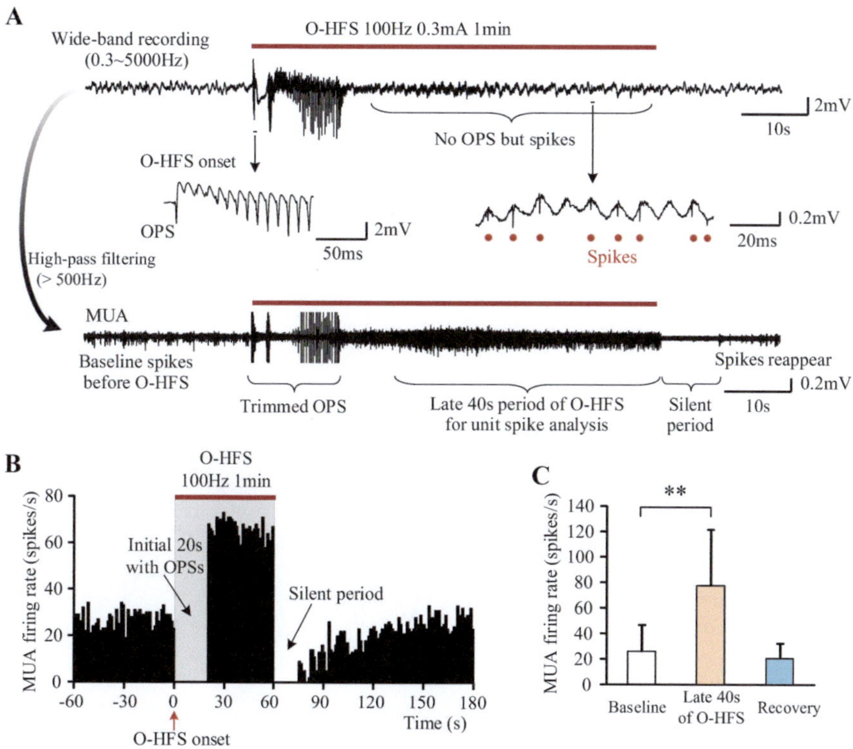

Fig. 5.19 Increased firing in the downstream CA1 neurons by 100 Hz O-HFS at the Schaffer collaterals. **A** Example recording of 1-min 100 Hz O-HFS with stimulus artifacts removed (top), its enlarged episodes (middle), and its MUA signal (bottom). Red dots denote the unit spikes. **B** Peri-stimulus histogram of MUA firing rates of the signal shown in (**A**) (bin size = 1 s). The initial 20 s O-HFS period was skipped due to the interference of OPS events. **C** Comparison of mean MUA firing rates among the three periods: baseline before O-HFS, during O-HFS, and after O-HFS (measured from the second minute after the end of O-HFS to skip the silent period). $^{**}P < 0.001$, paired t-test, $n = 29$. Modified from Feng et al. (2017)

By analyzing MUA signals during 1 min 100 Hz O-HFS from 29 rat experiments, we identified SUA spikes from 214 neurons, including 154 (72%) pyramidal neurons and 60 (28%) interneurons. During the steady period (the late 40 s O-HFS period), the mean firing rate of pyramidal neurons was 5.7 ± 6.5 spikes/s, significantly higher than their baseline rate of 1.5 ± 2.0 spikes/s (Fig. 5.20A1). The mean firing rate of interneurons during the steady period was 24.5 ± 28.3 spikes/s, also significantly higher than their baseline rate of 8.8 ± 11.1 spikes/s (Fig. 5.20B1).

Based on firing rates during O-HFS, we defined three neuronal response types: “excited” neurons (with a rate exceeding 120% of baseline), “inhibited” neurons (with a rate below 80% of baseline), and “no response” neurons (with a rate between 80 and 120% of baseline). Their distributions among the 154 pyramidal neurons and 60 interneurons are shown respectively in Fig. 5.20A2, B2. Even for the excited

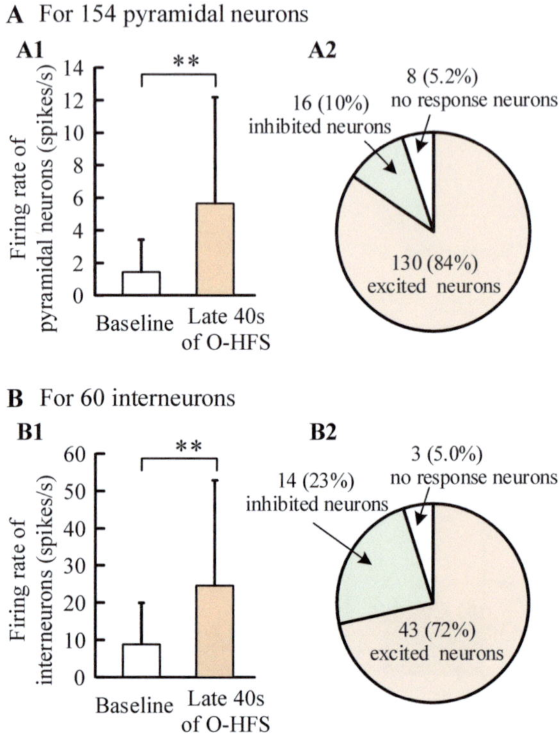

Fig. 5.20 Changes in the firing rates of pyramidal neurons and interneurons induced by 1 min 100 Hz O-HFS. **A** Firing changes in pyramidal neurons. **A1**: Comparison of mean firing rates between baseline (pre-O-HFS) and during steady O-HFS period (late 40 s). **A2**: Three response types in the 154 pyramidal neurons (excited, inhibited, and no response) defined by thresholds of ± 20% differences from baseline rate. **B** Firing changes in interneurons. **B1** and **B2**: Corresponding analysis for 60 interneurons, matching (**A1**) and (**A2**) respectively. In (**A1**) and (**B1**), $^{**}P < 0.01$, Wilcoxon signed rank test. Modified from Feng et al. (2017)

neurons, the mean firing rates of pyramidal neurons (6.1 ± 5.6 spikes/s) and interneurons (41.7 ± 28.2 spikes/s) were far below the O-HFS pulse frequency of 100 Hz. Similar results were observed during the steady period of 1 min O-HFS at a higher frequency of 200 Hz (refer to Feng et al. 2017).

2. **Firing Patterns during Sustained O-HFS**

Under normal physiological conditions, pyramidal neurons commonly fire action potentials in a burst pattern, while interneurons fire in a regular pattern (Barthó et al. 2004; Csicsvari et al. 1998). As shown in Fig. 5.21A1, in the baseline recording before O-HFS, the 130 "excited" pyramidal neurons exhibited burst firing, resulting in a peak around 5 ms in their average ISI histogram, with a cumulative ISI probability of 45 ± 25% in the range of 0–8 ms. However, as shown in Fig. 5.21A2, during the steady period of 100 Hz O-HFS, the pyramidal neurons no longer fired in bursts. The ISI probability in the range of 0–8 ms dropped to only 4.6 ± 3.4% without obvious peak. Instead, small peaks appeared in the ISI histogram at integer multiples of the 10 ms pulse interval of O-HFS, indicating that stimulation pulses modulated the firing of pyramidal neurons.

Unlike the pyramidal neurons, in baseline recording, the average ISI histogram of 43 "excited" interneurons was relatively flat, with a cumulative ISI probability

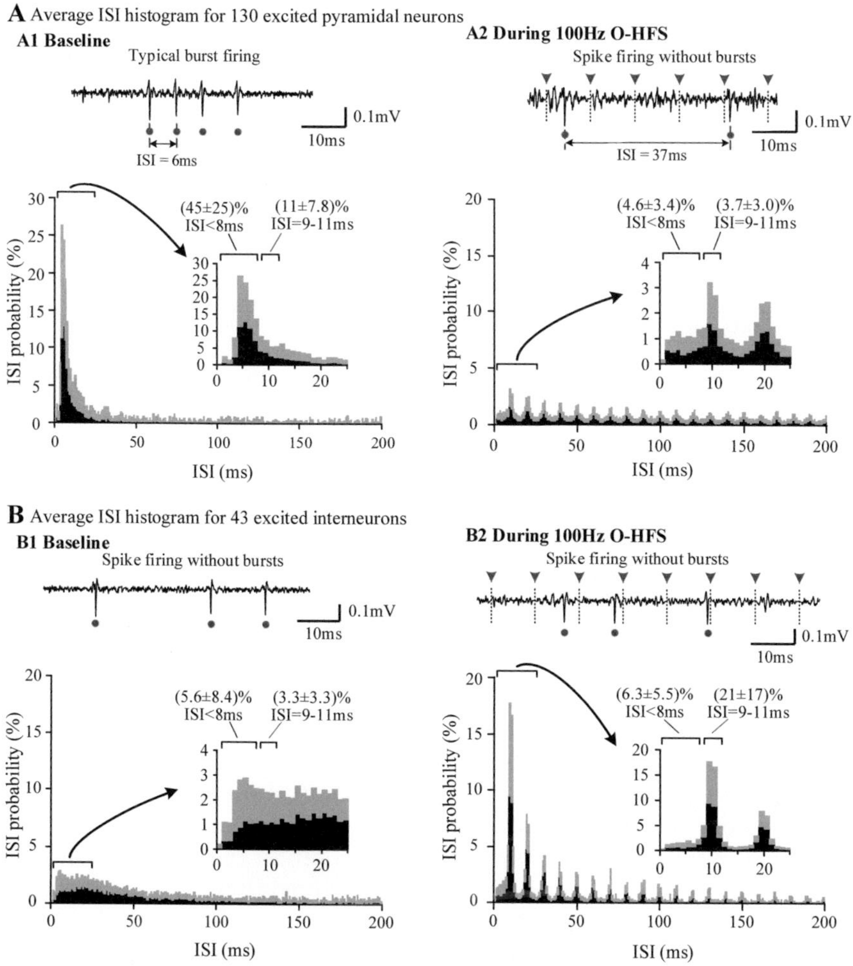

Fig. 5.21 Changes in the firing patterns of pyramidal neurons and interneurons by O-HFS. **A** and **B** Comparisons of the average ISI histograms of the 130 "excited" pyramidal neurons (**A**) and 43 "excited" interneurons (**B**), showing baseline vs. steady period of 100 Hz O-HFS. Each plot includes a typical episode of unit spikes, an average ISI histogram (with grey shading indicating the range of one standard deviation), and an enlarged view of the 0–25 ms ISI range. The cumulative probabilities in the two ISI ranges (0–8 ms and 9–11 ms) are denoted. From Feng et al. (2017)

of only 5.6 ± 8.4% in the 0–8 ms range (Fig. 5.21B1). However, during the steady period of 100 Hz O-HFS, peaks appeared around integer multiples of the 10 ms pulse interval in the ISI histogram (Fig. 5.21B2). In the 9–11 ms ISI range (containing the first peak), the cumulative ISI probability of 21 ± 17% was significantly greater than the 3.3 ± 3.3% at baseline ($P < 0.01$, Wilcoxon signed rank test, $n = 43$), indicating the strong modulation effect of O-HFS pulses on the interneuron firing. Since the firing rate of interneurons (~ 40 spikes/s) was much higher than that of pyramidal neurons (~ 6 spikes/s), the interneuron spikes represented by the 21% probability in the 9–11 ms ISI range was much more than the corresponding 3.7% of pyramidal neuron spikes. This firing difference between the two neuron types may be caused by the lower action potential threshold of interneurons (Csicsvari et al. 1998).

These results indicate that O-HFS can modulate and increase neuronal firing. Unlike single pulse stimulation, the absence of evoked-population spikes (PS) during steady O-HFS period suggests that sustained axonal HFS can weaken the synchronization in pulse-evoked neuronal firing. We next examined the changes in neuronal firing synchrony during O-HFS.

3. **Asynchronous Firing during Sustained O-HFS**

As described in Sect. 4.1.4, we used recordings from four adjacent channels on the electrode array to classify and distinguish spikes from individual neurons. To study firing synchrony between neurons, we analyzed spikes from pairs of interneurons simultaneously recorded on these four channels. The two interneurons in each pair were denoted as "Neuron-1" and "Neuron-2" (Fig. 5.22A). During baseline recording, the two neurons fired asynchronously (Fig. 5.22B1). We denoted the interval between adjacent spikes of the paired neurons as Δt. The Δt histogram was flat without obvious peak (Fig. 5.22B2). The synchronous firing ratio of the paired neurons was only 8%. Note: we defined synchronous firing as spike pairs with $\Delta t <$ 2 ms, and the synchronous firing ratio between Neuron-1 and Neuron-2 as $F/(F_1 \cdot F_2)^{1/2}$—where F was the firing rate of spike pairs with $\Delta t < 2$ ms, F_1 and F_2 were the firing rates of Neuron-1 and Neuron-2 respectively (Quian Quiroga et al. 2002).

As shown in Fig. 5.22C1, an orthodromic test stimulus (OTS) of 50μA (weak intensity) induced paired neurons to fire synchronously with a latency difference of $\Delta t \approx 1$ms. This synchronous firing indicated that the stimulation at the Schaffer collaterals activated both neurons simultaneously (Fig. 5.22C2). The latencies of induced spikes were less than 5 ms, indicating only monosynaptic transmission involved in the activation pathway (Buzsáki 1984). In 40 repetitions of the weak OTS (single-pulse stimulation), the synchronous firing ratio was ~ 85%, resulting in a peak near 0 ms in the Δt histogram (Fig. 5.22C3). Note: to prevent from generating population spikes, we applied the OTS at a very low intensity. At such intensity level, both neurons did not always fire simultaneously—in the remaining ~ 15% of trials, at least one neuron failed to respond to the OTS.

Unlike the weak OTS, during the steady period of 1 min 100 Hz O-HFS, even with a strong intensity of 0.3 mA, the pulses rarely induced paired neurons to fire synchronously (Fig. 5.22D1). Small peaks with intervals of about 10 ms appeared in the Δt histogram, indicating the modulation effect of 100 Hz O-HFS on neuronal firing (Fig. 5.22D2), but with a synchronous firing ratio of only 9%. Among the 5 pairs of interneurons obtained from 5 rat experiments, the mean synchronous firing ratio during O-HFS was 10.4% ± 12.4%—significantly lower than the 86.0% ± 3.3% induced by weak OTS, but close to the baseline ratio of 6.8 ± 7.1% (Fig. 5.22E). This result indicates that sustained O-HFS can generate asynchronous firing in postsynaptic interneurons.

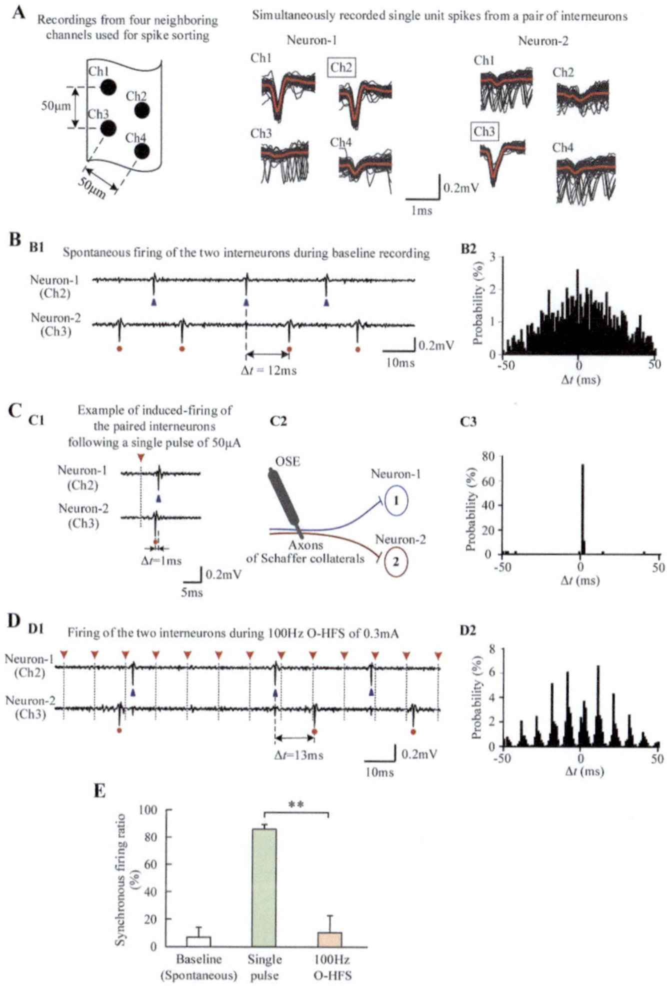

Fig. 5.22 O-HFS-induced asynchronous firing of paired interneurons. **A** Unit spikes from two interneurons (Neuron-1 and Neuron-2) recorded simultaneously by four adjacent channels in the electrode array. The spike waveforms from both neurons in four channels (Ch1-4) were superimposed, together with their averaged waveforms shown in red. The neurons showed largest spike amplitudes in Ch2 and Ch3 respectively, so these channels were used to analyze the firing relationship between the paired neurons. **B** Baseline firing. **B1**: Asynchronous firing of the paired neurons. **B2**: Histogram of firing probability versus interval (Δt) between adjacent spikes from the paired neurons. **C** Synchronous firing of the paired neurons induced by a single pulse at a weak 50 μA intensity. **C1**: Example of evoked spikes from both neurons. **C2**: Schematic diagram showing simultaneous excitation of the neurons by the orthodromic stimulation of Schaffer collaterals. **C3**: Histogram of Δt from 40 single-pulse trials. **D** Example of asynchronous firing of paired neurons during steady period of 100 Hz O-HFS (**D1**) and corresponding Δt histogram (**D2**). **E** Comparison of synchronous firing ratios of paired neurons among baseline firing, single-pulse evoked firing, and the firing during steady O-HFS period. $^{**}P < 0.001$, paired t-test, $n = 5$ pairs. From Feng et al. (2017)

Stimulating the Schaffer collaterals can activate both pyramidal neurons and interneurons through monosynaptic transmissions. Interneurons have a lower action potential threshold than pyramidal neurons. Additionally, interneurons are fewer in number, comprising only about 10% of CA1 neurons, and are sparsely distributed across CA1 layers (Andersen et al. 2007; Csicsvari et al. 1998). Due to these characteristics, pairs of innervated interneurons can be reliably activated by single OTS pulses with weak intensity in baseline trials without inducing PS events. The spikes of paired interneurons can be clearly identified following the weak pulses, without contamination from PS or from spikes of pyramidal neurons (Fig. 5.22C). However, it is difficult to activate paired pyramidal neurons simultaneously without inducing PS events, because of their densely packed population and higher action potential threshold. Therefore, we could not repeat the baseline OTS experiment for paired pyramidal neurons by using the extracellular recording approach.

Nevertheless, we calculated a synchronous firing ratio for each pair of simultaneously recorded neurons (Feng et al. 2017). For example, four pyramidal neurons obtained in the same experiment formed 6 pairs, with each neuron participating in three pairs. From the baseline recordings of 29 rat experiments containing 3–11 pyramidal neurons, we obtained a total of 684 pairs. Similarly, for the "excited" neurons (including both pyramidal neurons and interneurons) collected during O-HFS at 100 and 200 Hz, the same pairing analysis was performed. The mean synchronous firing ratios during the steady O-HFS period were small for both neuron types and both pulse frequencies (Fig. 5.23)—less than 3% for pyramidal neurons and less than 7% for interneurons. They all showed no significant differences from the baseline values.

These results show that single-pulses applied to afferent axons can induce synchronous firing in downstream post-synaptic neurons, whereas sustained O-HFS can produce asynchronous firing. The asynchronous firing of post-synaptic neurons can reflect asynchronous excitatory inputs from presynaptic neurons due to HFS-induced intermittent axonal block in the afferent pathway.

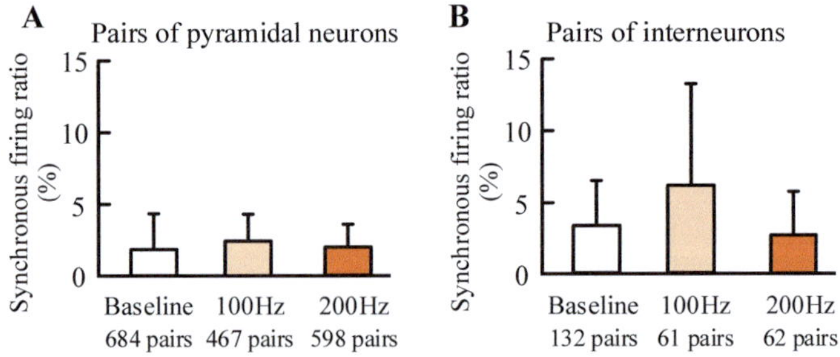

Fig. 5.23 Comparisons of synchronous firing ratios for paired pyramidal neurons (**A**) and paired interneurons (**B**) among baseline firing and during the steady period of 100 and 200 Hz 1 min O-HFS trains. From Feng et al. (2017)

5.3.3 Higher-Frequency Axonal HFS Can Increase Firing Randomness Rather Than Firing Rate

Clinical application of deep brain stimulation (DBS) for disorders, such as Parkinson's disease, has shown that HFS with pulse frequencies above 90 Hz usually has better therapeutic effects, while frequencies below 60 Hz are usually ineffective or even detrimental (McConnell et al. 2012; Birdno and Grill 2008; Kuncel et al. 2006). Previous studies have also shown that HFS at around 130Hz is more effective in controlling epilepsy than lower frequencies (Vonck et al. 2013; Jobst 2010). Although many studies on DBS mechanisms have focused mainly on whether target neurons are excited or inhibited based on changes in firing rates (Vitek 2002; Florence et al. 2016), the desynchronization effect of HFS has also gained significant attentions (Medeiros and Moraes 2014; Popovych and Tass 2014). Many neurological disorders, such as motor disorders and epilepsy, are associated with increased synchronous and rhythmic oscillations in the firing of neuronal populations (Hammond et al. 2007; Gatev et al. 2006; Rampp and Stefan 2006; Jiruska et al. 2013). Studies have shown that HFS can reduce the pathological oscillations and synchronous firing of target neurons (Eusebio et al. 2011; Wingeier et al. 2006; Deniau et al. 2010; Medeiros and Moraes 2014). Therefore, the therapeutic effect of DBS may stem from a desynchronization effect of HFS, rather than just changing neuronal firing rates (Hashimoto et al. 2003; McCairn and Turner 2009). In our O-HFS experiments in rat hippocampal CA1 region, we investigated the changes in asynchronous effects of HFS at different pulse frequencies by analyzing the distributions of evoked spikes.

As shown in Fig. 5.24, during O-HFS with pulse frequencies at 50, 100 and 200 Hz respectively, we recorded both the evoked OPS in the pyramidal cell layer (pcl) and the fEPSP in the apical dendritic layer (sr). The first O-HFS pulses always induced large OPS and fEPSP, indicating a highly synchronized response of downstream neurons (enlarged view ① in Fig. 5.24A–C). The O-HFS period with OPS events was longer at 50 Hz than at higher frequencies of 100 and 200Hz. After OPS disappeared in the late O-HFS period, unit spikes continued in the pcl layer. Meanwhile, small oscillations exhibited in the sr layer, pacing with the O-HFS pulses and particularly evident at 50 and 100 Hz frequencies (enlarged view ②). These oscillations reflected fluctuations in post-synaptic potentials produced by the O-HFS activations from the Schaffer collaterals. As pulse frequency increased, the oscillation amplitudes decreased substantially. Let A_1 represent the amplitude of fEPSP induced by the first O-HFS pulse and A_2 represent the mean amplitude of the oscillations in the final second of O-HFS. At 50, 100 and 200 Hz, A_2 was only about 15%, 2%, and 0.2% of A_1, respectively (Fig. 5.24D). This indicated that higher pulse frequencies resulted in smaller fluctuations in post-synaptic potentials by each pulse. Note: A_2 was not the magnitude of post-synaptic potential (fEPSP). The fEPSPs induced by O-HFS pulses fused together as shown in the recordings at the CA1 sr layer in the enlarged view denoted by ①. Our AC coupling recording could not capture the DC offset component in the fused waveforms.

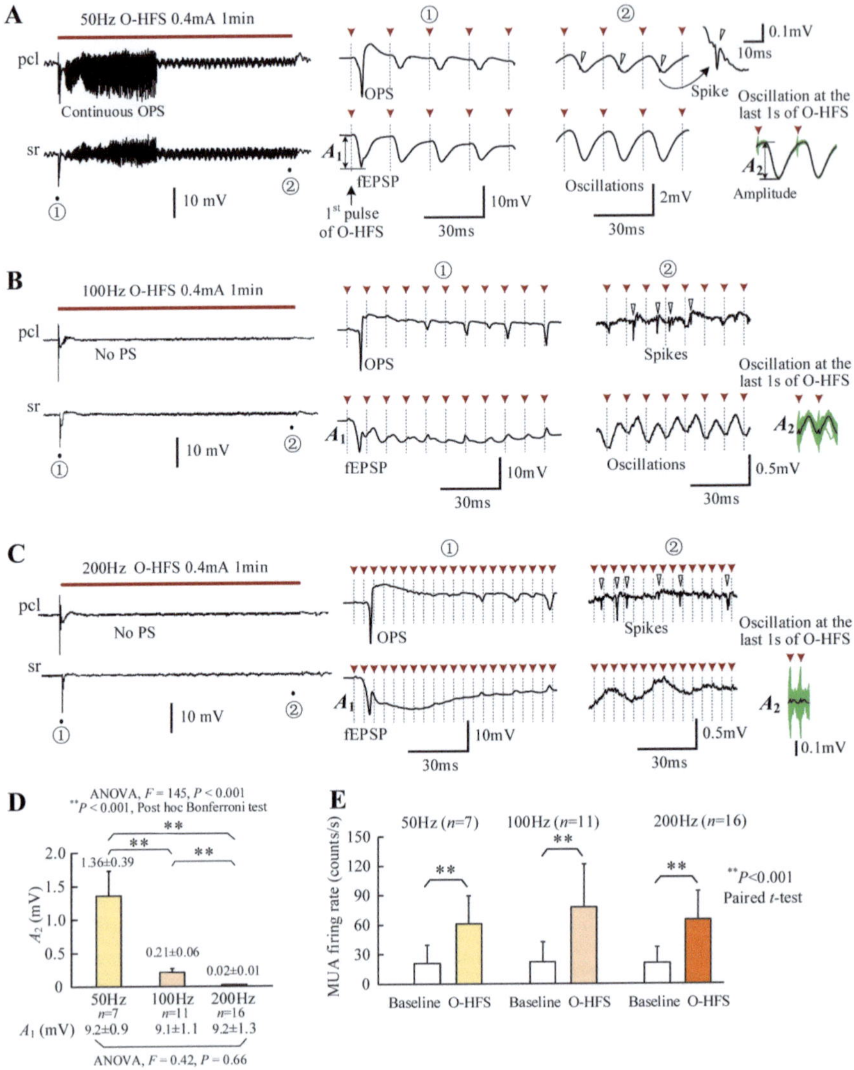

Fig. 5.24 Changes of evoked potentials during O-HFS with three different frequencies. **A–C** Typical recordings in the CA1 pyramidal cell layer (pcl) and stratum radiatum (sr) collected by two contacts separated by 0.2 mm during 1 min O-HFS at pulse frequencies of 50 (**A**), 100 (**B**) and 200 Hz (**C**), respectively. The neuronal responses at the O-HFS initial and late periods (denoted by "①" and "②") are enlarged. **D** Comparison of the amplitudes of sr oscillations (A_2) in the final O-HFS second among the three pulse frequencies. With the similar amplitudes at the O-HFS onset (A_1, listed below), the mean A_2 was suppressed more at higher frequencies. The A_2 values were calculated by superposing and averaging the inter-pulse signals in the final second of O-HFS, as shown in the right insets of figures **A-C**. Green waveforms are superposed signals, while black ones are average waveforms. Two repeated inter-pulse signals are shown in each inset to highlight oscillations. Red arrows denote removed stimulus artifacts. **E** Comparisons of MUA rates between baseline and late 30 s periods of 1 min O-HFS at the three pulse frequencies respectively. Modified from Wang et al. (2018)

Comparing the firing rates of MUA recorded at the pcl layer during the late 30 s of 1 min O-HFS showed that the neuronal firing at the three stimulation frequencies significantly exceeded baseline levels. However, the mean MUA rates showed no significant differences across the three frequencies (Fig. 5.24E).

These results showed that varying pulse frequencies from 50 to 200 Hz had no significant effect on the MUA firing rates in downstream neurons, despite the four-fold difference in the electrical energy delivered by the O-HFS trains. Nevertheless, the fluctuations in post-synaptic potentials were frequency dependent (Fig. 5.24D), suggesting possible differences in firing patterns. This led us to hypothesize that the extra electrical energy from higher frequency pulses could be used to modulate firing timing rather than firing rates. To investigate this, we analyzed the post-stimulation time histogram (PSTH) of MUA during O-HFS at three different pulse frequencies.

As a control, we created a mimic PSTH for baseline recording by adding virtual 10 ms intervals (100 Hz) to the baseline MUA signal. The PSTH showed an even distribution, indicating random spontaneous firing (Fig. 5.25A). During the steady period of 50 Hz O-HFS, the PSTH of MUA was highly non-uniform with a sharp peak around 10 ms in the 20 ms inter-pulse-intervals (IPI). Most spikes occurred between 8 and 13 ms (Fig. 5.25B1). The PSTH distributions became progressively more uniform as pulse frequency increased (Fig. 5.25B2, B3). In these PSTHs, spike counts were measured across different time spans—20, 10 and 5 ms (4:2:1) for 50, 100 and 200 Hz respectively. To compare them with a same time span, we divided the PSTHs of 100 and 200 Hz into two and four equal segments respectively, then connected them to form PSTHs with a 20 ms time span, matching the 50 Hz O-HFS. We also created a mimic PSTH of baseline MUA with 20 ms intervals to serve as a control (Fig. 5.25C). These PSTHs with identical time spans clearly showed that increasing pulse frequency flattened the MUA distribution in the intervals, indicating weakened phase-locking between neuronal firing and O-HFS pulses.

We used the index of peak coefficient ($\Delta C/C_{\text{ave}}$) to measure the non-uniformity of PSTH distribution (Fig. 5.25B). Here, C_{ave} represented the mean of PSTH and ΔC was the difference between the peak and mean PSTH values. A higher PSTH peak resulted in a greater peak coefficient, while a flat PSTH yielded a peak coefficient close to 0. In addition, we used the duty ratio (DR) of excitatory phase—the percentage of excitatory bins in PSTH—to measure the concentration of spikes in the inter-pulse-intervals. An excitatory bin was defined as a PSTH bin (0.5 ms) with a spike count above 120% of the mean value of baseline mimic PSTH (Wang et al. 2018). Our statistical analysis showed that as pulse frequency increased, the mean peak coefficient decreased significantly (Fig. 5.25D), while the DR of excitatory phases increased significantly (Fig. 5.25E). These results indicated that with similar MUA rates produced by O-HFS (Fig. 5.24E), a lower frequency of 50 Hz generated phase-locked firing, whereas higher frequencies of 100 or 200 Hz produced more random firing.

An MUA signal contains action potentials from multiple neurons. Its distribution may not represent the firing patterns of individual neurons. A uniform MUA pattern could result from either random firing of individual neurons or varying firing timings among neurons that maintain different phase-locked firing with O-HFS pulses. To

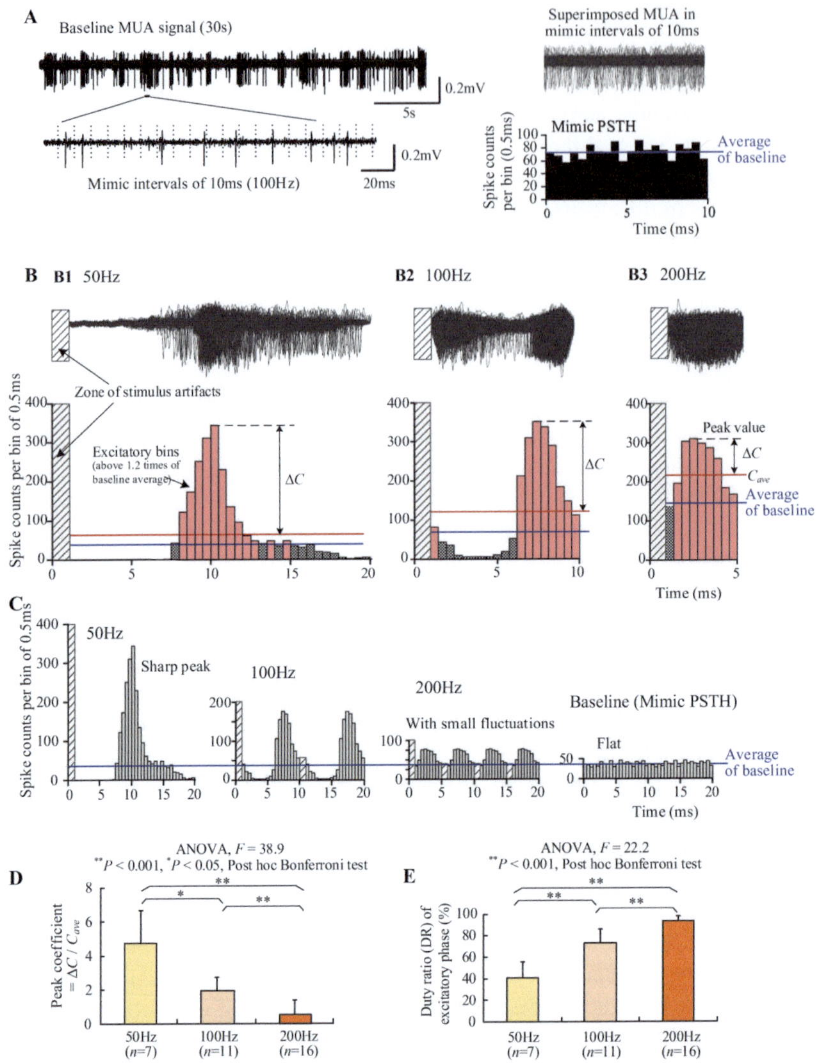

Fig. 5.25 Changes in PSTHs of MUA signals induced by O-HFS at various pulse frequencies. **A** Creation of mimic PSTH for 30 s baseline MUA signal to serve as a control. *Left*: typical baseline MUA signal (30 s) divided into virtual 10 ms intervals. *Right*: superimposed signals of all 10 ms intervals from the 30 s MUA (top) and the corresponding mimic PSTH (bottom). The blue horizontal line denotes the average PSTH value. **B** Typical plots of MUA PSTH for the final 30 s of 1 min O-HFS at 50, 100, and 200 Hz pulse frequencies. *Top*: superimposed signals of all inter-pulse intervals. *Bottom*: corresponding PSTH plots. In these PSTHs, red and blue horizontal lines denote their own averages (C_{ave}) and the baseline averages, respectively. ΔC represents the difference between peak and average values. Pink bins indicate PSTH values exceeding 1.2 times the baseline average (termed as excitatory bins). **C** Comparisons of PSTH distributions under an identical 20 ms interval by dividing the PSTHs of 100 and 200 Hz into two and four equal segments, respectively. **D** and **E** Comparisons of peak coefficient ($\Delta C/C_{ave}$) and duty ratio (DR) of excitatory phases among O-HFS with three pulse frequencies. From Wang et al. (2018)

clarify this, we examined the PSTHs of single unit activity (SUA) from individual neurons after spike sorting. As O-HFS frequency increased, the PSTHs of SUA signals from both pyramidal neurons and interneurons were similar to those of MUA (Wang et al. 2018), indicating that the randomization observed in MUA distributions at higher O-HFS frequencies stemmed from the random firing of individual neurons. The results suggest that axonal HFS with a sufficiently high frequency can generate random rather than phase-locked activation in downstream neurons, leading to asynchronous firing.

The above investigations show that O-HFS can induce a frequency-dependent axonal block in the Schaffer collaterals at pulse frequencies ranging in 50–200 Hz. During this type of axonal block, O-HFS can still activate downstream post-synaptic neurons and increase their firing rates, indicating partial or intermittent axonal blocks. As pulse frequencies increase, neuronal firing becomes more random. Although the mean firing rates during O-HFS showed no significant differences across this frequency range, the rate peaked at 100 Hz and declined slightly at 200 Hz (Fig. 5.24E). Previous studies have shown that HFSs in peripheral nerves with pulse frequencies up to thousands of hertz can completely block axonal membranes from firing (McGee et al. 2015). This suggests that pulse frequencies above 200 Hz may lead to further decreases in neuronal firing rates.

5.4 Activation of Interneurons in Local Inhibitory Circuits by HFS at Efferent Fibers

Section 5.3 describes how O-HFS modulates downstream post-synaptic neurons, including pyramidal neurons and interneurons. The O-HFS was applied at the Schaffer collaterals—one of the major afferent pathways in the hippocampal CA1 region. Besides the excitatory projection pathways, neurons also form connections through local circuits, such as the inhibitory circuits in the hippocampal CA1 region described in Sect. 2.3. As shown in Fig. 5.26A, we hypothesized that antidromic activation from the alveus (the CA1 efferent fibers) could activate interneurons in local inhibitory circuits in addition to activating the pyramidal neurons' somata. To test this hypothesis, we analyzed the SUA from interneurons (IN) during A-HFS at the alveus.

The excitability of pyramidal neurons (CA1 principal neurons) is regulated by both local feedforward and feedback inhibitory circuits composed of interneurons (Pelkey et al. 2017). As shown in Fig. 5.26A, interneurons in the feedforward circuits receive activation from the Schaffer collaterals alongside pyramidal neurons. The activated interneurons then inhibit the pyramidal neurons through inhibitory synapses. In the feedback circuits, interneurons are activated by adjacent pyramidal neurons and subsequently inhibit pyramidal neurons in return (see Sect. 2.3 for details). In the A-HFS experiment, the stimulating electrode (ASE) was placed about 1.3 mm away

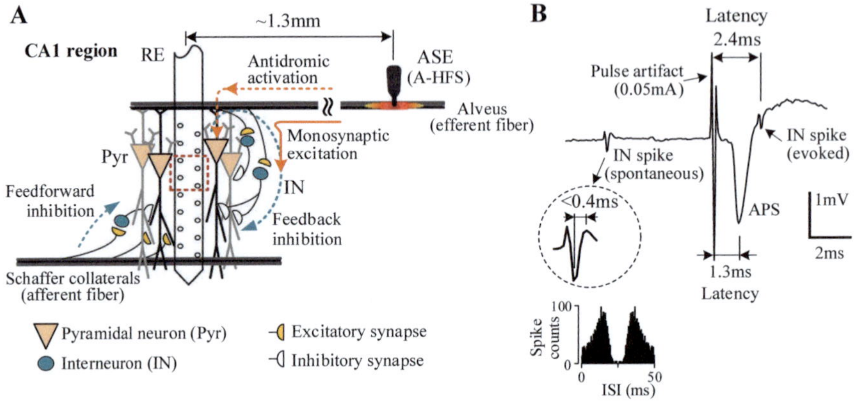

Fig. 5.26 Responses of interneurons (IN) to antidromic stimulation at the CA1 alveus. **A** Schematic diagram of the local inhibitory circuits and electrode positions. Pyramidal neurons (Pyr) receive inhibitory inputs from both feedforward and feedback inhibitory circuits. INs in these inhibitory circuits are activated by excitatory synapses and then act on Pyr somata or dendrites through inhibitory synapses. **B** IN spikes and an APS evoked by a single-pulse applied at the alveus. The enlarged spike waveform and the ISI histogram of the IN firing from baseline recording (shown in the lower left) were used to identify the IN firing. From Ye et al. (2022)

from the recording electrode (refer to Sect. 5.2.1 and Fig. 5.4A for electrode positions). Therefore, the activation produced by A-HFS on the alveus may antidromically propagate along the axons to the vicinity of the recording electrode (RE) to then activate nearby interneurons recorded by the RE.

As shown in Fig. 5.26B, a single 0.05 mA pulse applied to the alveus antidromically activated a population of pyramidal neurons near the RE, generating an APS with a latency of ~ 1.3 ms. Immediately following this APS, an IN spike appeared with a latency of ~ 2.4 ms. This latency of IN spike was longer than the APS latency and aligned with previous reports of monosynaptic transmission delays (Csicsvari et al. 1998), indicating that the IN was in a feedback inhibition circuit. Although the IN spike had a much smaller amplitude than the APS, their latency difference made the IN spike clearly distinguishable after the APS (Fig. 5.26B).

However, when an IN spike occurred too close to an APS, it could be covered by the APS. To ensure accurate IN spike detection, only INs with sufficiently large spikes (> 100 μV) were used. Due to the similar spectral ranges of unit spikes and population spikes, we could not extract IN spikes by using a filter to remove LFP and APS (refer to Sect. 4.1.4). Instead, we used a window detection method to extract spikes. As described in Sect. 4.1.5, spikes were directly detected in the raw wideband (0.3–5000 Hz) recordings from four adjacent channels and then used for spike sorting to isolate unit spikes from individual neurons. Finally, we identified IN spikes based on two features: a rising phase less than 0.4 ms and a relatively flat ISI histogram in baseline firing (Fig. 5.26B, lower left).

We obtained unit spikes from one IN in each of the 19 rat experiments. Their firing triggered by the alveus stimulation had a mean latency of 2.7 ± 0.45 ms, indicating monosynaptic activation. The mean spike amplitude of these INs reached 287 ± 110 μV in the maximum amplitude recording channel. However, even these "super large" spikes would be submerged in the large APSs generated during the initial A-HFS period. Therefore, to investigate IN firing through the entire A-HFS period, we used weak-intensity A-HFS of 0.06 ± 0.02 mA in 9 rats. In the remaining 10 rats, we applied strong-intensity A-HFS of 0.33 ± 0.08 mA to examine IN firing during the steady A-HFS period when APS declined.

Figure 5.27 shows an example of low-intensity A-HFS. In the 2 min baseline spontaneous MUA signal, IN spikes were clearly distinguishable (Fig. 5.27A) with a mean firing rate of 9.4 Hz. During the subsequent 2min 100 Hz A-HFS with a weak-intensity of 0.05 mA, each pulse induced an APS with an amplitude of approximately 1.9 mV at the initial period, followed most by an IN spike (Fig. 5.27B). The coupling ratio between the IN firing and the A-HFS pulses was ~ 80%. As the A-HFS continued, the induced APS declined (Fig. 5.27C), accompanied by decreases in both the IN firing rate and its coupling ratio (Fig. 5.27B, *bottom right*).

The two-dimensional raster plot showed that the IN initially fired in phase-lock with the pulses during the first second of A-HFS (Fig. 5.27D, left). Subsequently, the firing gradually dispersed and decreased in rate (Fig. 5.27D, middle and right). The raster plot showed the timing of each IN spike within the 10ms A-HFS inter-pulse-intervals, known as the post-simulation time (PST). The PST data were used to create the PSTH (see Sect. 4.2.4 for details). The PSTHs (bin = 0.5 ms) for the IN spikes during the first second and late 60 s of A-HFS periods are shown in the left and right plots of Fig. 5.27D. To quantitatively assess the phase-locking between IN spikes and A-HFS pulses, we calculated the interquartile range (IQR) of the PSTH data, denoted as IQR_{PST}, representing the range containing the middle 50% of spikes in the PSTH. A smaller IQR_{PST} value indicated stronger phase-locking between spikes and pulses.

As shown in the left of Fig. 5.27D, during the first second of A-HFS, the example IN had a firing rate (FR) of 47 Hz, with spikes concentrating within 3.0–5.5 ms of the 10 ms IPI. Its IQR_{PST} was only 0.83 ms, indicating strong phase-locking between the IN firing and A-HFS pulses. However, during the final 60 s of 2 min A-HFS, the mean firing rate decreased to 13.8 Hz—though still above the baseline rate of 9.4 Hz. The IQR_{PST} increased to 1.63 ms, with some spikes no longer phase-locked to the pulses. Additionally, as the APS latency gradually increased (Fig. 5.27C), the latency of IN spikes also increased from 3.00 ms at the onset of A-HFS to a mean value of 6.75 ms in the final 60 s of A-HFS, as shown by the peak time of the PSTH plotted in the right of Fig. 5.27D.

Statistical data in Fig. 5.27E, F show that the mean IN firing rate during the first A-HFS second was significantly higher than both baseline and the late 60 s A-HFS period. The mean rate in the late 60 s was also significantly higher than baseline. Additionally, the mean IN spike latency in the late 60 s A-HFS was significantly longer than the value in the first second (Fig. 5.27G), while the IQR_{PST} also showed a significant increase (Fig. 5.27H). These results indicate that axonal block produced by sustained A-HFS reduced the stimulation activation on the INs.

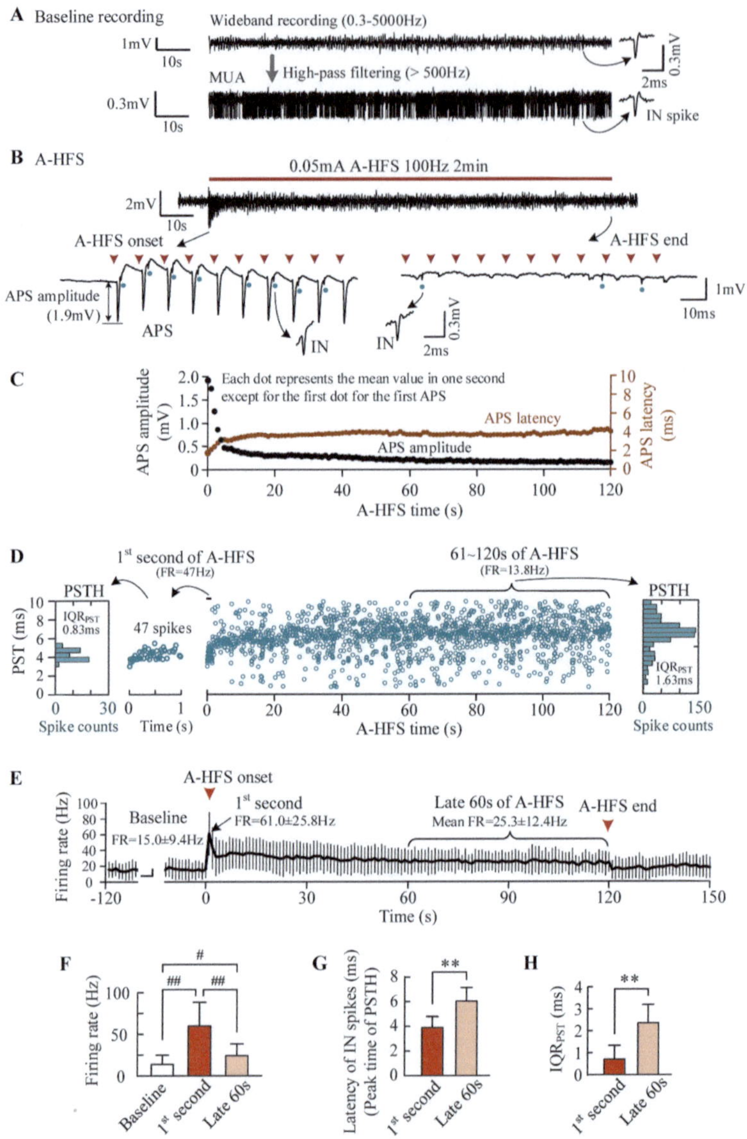

Fig. 5.27 Modulation of weak A-HFS on IN firing. **A** Typical wideband baseline recording and its high-pass filtered MUA signal with clear IN spikes. **B** Evoked APSs and IN spikes during 2 min A-HFS with a weak-intensity of 0.05 mA. Blue dots denote IN spikes. **C** Scatter plots showing APS amplitudes (black) and latencies (brown) during A-HFS. **D** Two-dimensional raster plot of IN spikes during the A-HFS shown in (**B**), together with the PSTH plots of the first second and late 60 s period (61–120 s) of A-HFS shown on the left and right, respectively. **E** Mean IN firing rates per second before, during and after weak A-HFS ($n = 9$ INs). The error bars represent one standard deviation. **F–H** Comparisons of the mean IN firing rates (**F**), the mean IN spike latencies (**G**) and the mean interquartile range (**H**, IQR_{PST}) of IN PSTHs. $^{\#}P < 0.05$, $^{\#\#}$ $P < 0.01$, repeated measurements one-way ANOVA with post hoc Bonferroni tests, 9 rats. $^{**}P < 0.01$, paired t-test, 9 rats. From Ye et al. (2022)

To enhance the A-HFS effect on INs, we increased the pulse intensity to around 0.3 mA, which was strong enough to induce an APS with about 3/4 maximum amplitude in single-pulse stimulation. While IN spikes were distinct in baseline recording (Fig. 5.28A), the large evoked APSs obscured the IN spikes during the initial period of 100 Hz A-HFS (Fig. 5.28B). After about ten seconds of A-HFS, as the APS declined (Fig. 5.28C) and spike latency increased, the evoked IN spikes became visible following APSs (denoted by blue dots in the enlarged insets of Fig. 5.28B). The IN firing rate was 92 Hz at the 11th second of A-HFS with an IQR_{PST} of 0.60 ms in the corresponding PSTH (Fig. 5.28D). This high rate largely persisted until the end of A-HFS, resulting in a mean rate of 78.0 Hz and a still small IQR_{PST} of 0.75 ms in the late 60 s A-HFS. From the 11th second to the late 60 s of A-HFS, the mean APS latency increased slightly from 3.10 to 3.43 ms (Fig. 5.28C), while the mean latency of IN spikes increased from 5.25 to 6.25 ms.

Figure 5.29 shows comparisons between the weak- and strong-intensity A-HFS groups. The mean APS amplitude induced by the first pulse was significantly greater in the strong-intensity group than the weak-intensity group (Fig. 5.29A). During the steady A-HFS period (late 60 s), despite significant decreases, the APS amplitude difference between the two groups persisted (Fig. 5.29B), indicating the substantial difference in the numbers of activated pyramidal neurons. Although A-HFS doubled the APS latency, both weak- and strong-intensity groups showed similar APS latencies since they shared the same activation pathway (Fig. 5.29C, D).

Neither IN spike amplitudes nor baseline IN firing rates showed significant differences between the two groups (Fig. 5.29E, F). However, during the late 60 s of A-HFS, the strong-intensity group showed a significantly higher mean IN firing rate (Fig. 5.29G) and stronger firing phase-locking, evidenced by its significantly smaller IQR_{PST} value (Fig. 5.29H). Since both groups shared the same activation pathway, the latencies of IN spikes showed no significant differences between the groups (Fig. 5.29I). However, IN and APS had different activation pathways, leading to a significant difference in their latency increments. Throughout the weak-intensity A-HFS period with distinguishable IN spikes, the mean latency increment of IN spikes (3.41 ± 0.61 ms) significantly exceeded the APS latency increment (1.71 ± 0.45 ms) ($n = 9$, Fig. 5.29J). This shows that the IN activation pathway—involving synaptic transmission—produced longer delays than the antidromic activation pathway of pyramidal neurons.

Unlike the downstream IN activation by O-HFS at the Schaffer collaterals described in Sect. 5.3, A-HFS at the alveus (the CA1 efferent fiber) activates the INs located upstream of the stimulation site. Based on the CA1 local circuits shown in Fig. 5.26A, this A-HFS activation on IN may involve the following process: First, the stimulation activates the alveus axons to generate excitation that propagates antidromically along the axon. This excitation then turns to axonal branches and propagates in an orthodromic direction to the synapses at the axonal terminals. Through synaptic transmissions, the IN in the feedback inhibition circuit is activated (as denoted by the solid orange curve in the figure). Due to the intermittent depolarization block on the axons caused by sustained A-HFS, the excitation propagation to the IN reduces, resulting in a significant decrease in the IN firing rate during the

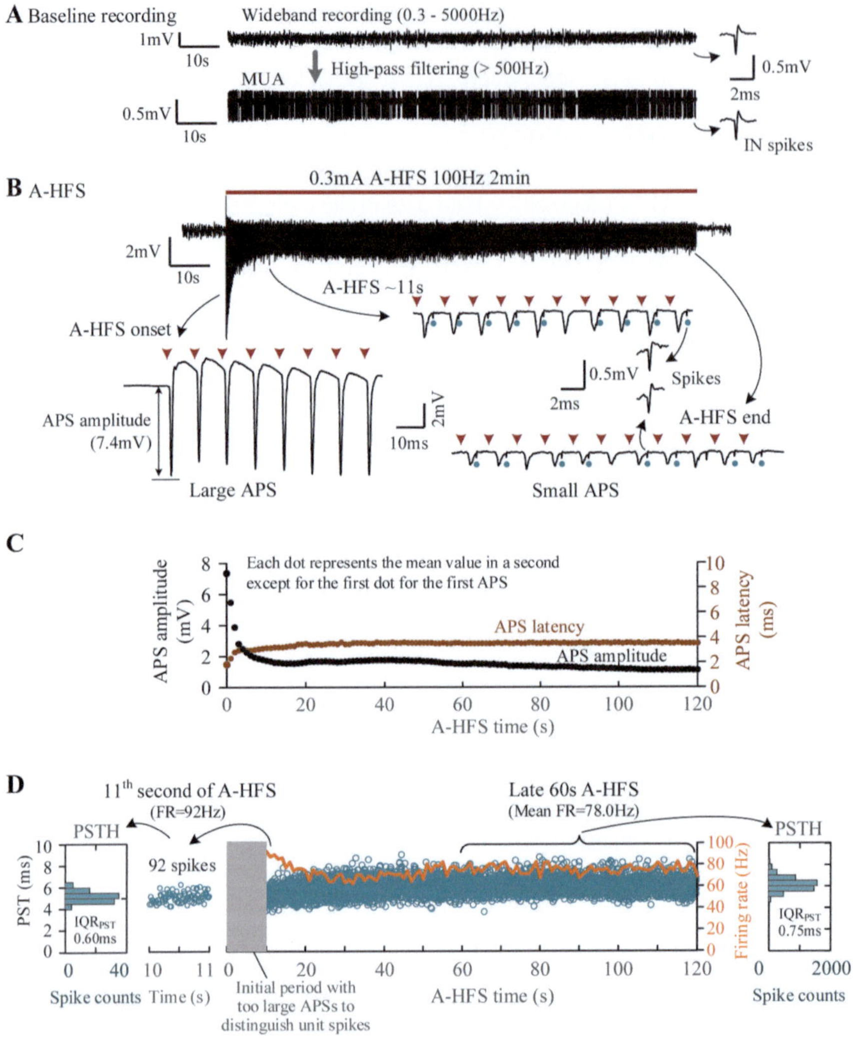

Fig. 5.28 Modulation of strong A-HFS on IN firing. **A** Typical wideband baseline recording and its high-pass filtered MUA signal with distinct spikes from an IN. **B** Evoked APSs and IN spikes during 2 min A-HFS with a strong-intensity of 0.3 mA. **C** Scatter plots showing the APS amplitudes (black) and the latencies (brown) during A-HFS. **D** Two-dimensional raster plot of the IN spikes during the A-HFS shown in (**B**), together with per-second firing rates (orange curve) and the PSTH plots for the 11th second (left) and late 60 s period (61–120 s) (right). Modified from Ye et al. (2022)

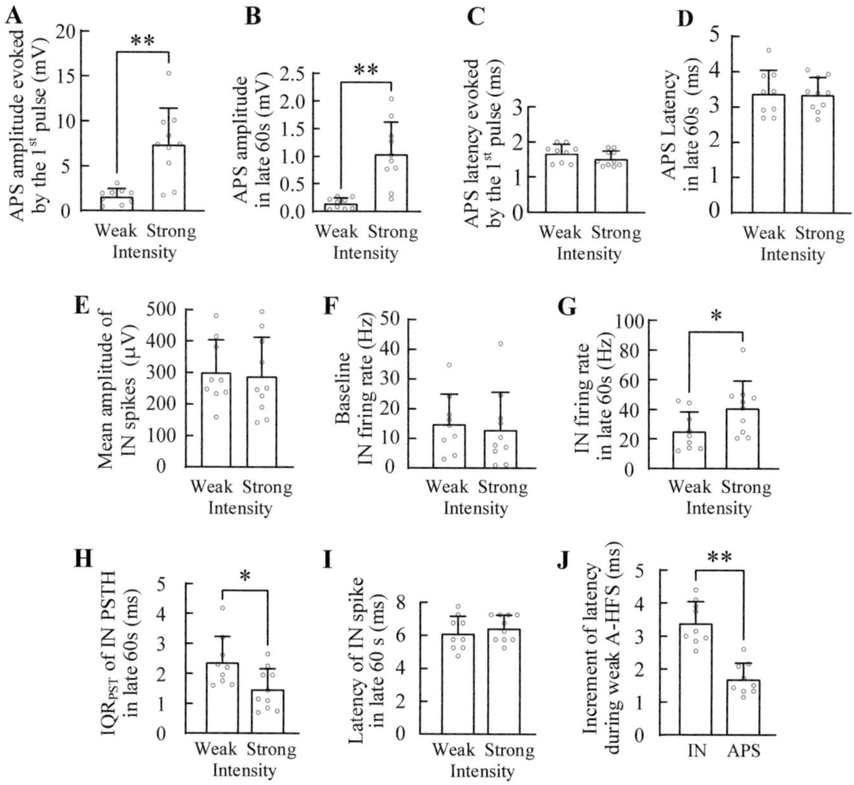

Fig. 5.29 Comparisons between the two A-HFS groups with weak- and strong-intensity pulses. **A–D** Comparisons of mean initial APS amplitudes (**A**), mean APS amplitudes in late 60 s of A-HFS (**B**), mean initial APS latencies (**C**), and mean APS latencies in late 60 s of A-HFS (**D**). **E–G** Comparisons of IN spike amplitudes (**E**), baseline IN firing rates (**F**), and IN firing rates in late 60 s of A-HFS (**G**). **H** Comparison of IQR_{PST} of IN firing in late 60 s of A-HFS. **I** Comparison of the IN spike latencies in late 60 s of A-HFS. **J** Comparison of latency increments between IN spikes and APSs evoked during weak A-HFS. For IN spikes, the latency increment was the peak time of PSTH in late 60 s minus the latency following the 1st pulse; for APS, it was the mean latency in late 60 s minus the latency following the 1st pulse. $^{*}P < 0.05$, $^{**}P < 0.01$, in (**A-I**) unpaired *t*-test, in (**J**) paired *t*-test, $n = 9$ rats for weak-intensity group, $n = 10$ rats for strong-intensity group. From Ye et al. (2022)

steady A-HFS period compared to the initial period (Fig. 5.27). Nevertheless, IN can still maintain a firing rate of ~ 40 Hz during the steady period of strong-intensity A-HFS (Fig. 5.29G), achieving a coupling ratio of ~ 40% between the IN firing and the 100 Hz pulses. However, the APS amplitude in the steady A-HFS period, which reflects the number of firing pyramidal neurons, drops to only ~ 15% of the initial value (Fig. 5.29A, B).

During the steady period, despite its weakened effect, A-HFS can still activate INs sufficiently to fire at a high rate. This may stem from IN characteristics. While INs comprise only about 10% of total neurons in the hippocampal CA1 region (Andersen et al. 2007), each IN receives excitatory inputs from many pyramidal neurons, and it possesses a low activation threshold (Csicsvari et al. 1998). As a result, an IN requires only a few simultaneous excitatory inputs to fire. Even weak-intensity A-HFS can produce IN firing rates up to ~ 60 spikes/s in the first second of stimulation (Fig. 5.27F), while strong-intensity A-HFS can sustain high IN firing rates throughout its steady period.

The HFS-induced partial axonal block can extend the latency of neuronal responses (both APS and IN spike). While this latency extension remains consistent along the same activation pathway, IN responses show longer extensions because its activation pathway involves additional parts such as axonal branches and synaptic transmissions.

5.5 Summary

Among neuronal structures, the axonal membrane has the shortest chronaxie, making it most sensitive to the narrow pulses commonly used in neuromodulation techniques like DBS. Axons can conduct action potentials quickly and reliably. However, under sustained HFS, axons can experience depolarization block, preventing them from generating action potentials in response to every stimulation pulse. This chapter details our experiments in the rat hippocampal CA1 region to verify intermittent axonal block produced by HFS at frequencies ranging from tens to hundreds of hertz. We found that when A-HFS on the alveus—the axons of CA1 pyramidal neurons—fail to induce synchronized action potentials in their somata through antidromic activation, single-pulse stimulations on their afferent fibers can still orthodromically activate these somata to fire through dendrite excitatory inputs. This finding can clearly demonstrate axonal block produced by sustained HFS. Furthermore, we revealed that this axonal block can not only weaken the excitation from the HFS electrode but also block excitation from other sites passing through the blocked region.

Our experiments using A-HFS with inserted gaps showed that axonal HFS can extend the refractory period of axons, resulting in intermittent axonal block. Additionally, sustained HFS at axons can alter the somata excitability, slowing activation conduction around the soma and causing a silent period of firing immediately after A-HFS cessation. Moreover, with the intermittent axonal block, the stimulated pyramidal neurons can exhibit non-uniform clustered firing even under the stimulations of periodic pulses at constant intervals. This non-uniform response may stem from the nonlinear dynamic properties of neurons. Nevertheless, the HFS-induced clustered firing differed from the spontaneous burst firing commonly observed in pyramidal neurons.

Experiments of O-HFS at the Schaffer collaterals showed that HFS-induced axonal block can attenuate and smooth the excitatory stimulation effect on the downstream postsynaptic neurons, resulting in asynchronous firing after initial synchronous firing of OPS events. Although weakened, sustained O-HFS can still elevate the firing rates of postsynaptic neurons (both pyramidal neurons and interneurons) in the O-HFS frequency range from 50 to 200 Hz, while neuronal firing can become more random and less synchronized at higher frequencies. The silent period of neuronal firing following O-HFS can verify that the increased firing during O-HFS is stimulation-induced.

During steady HFS periods with different pulse frequencies (50, 100 and 200 Hz) and their corresponding electrical energies (1×, 2×, and 4×), neither A-HFS nor O-HFS showed significant differences in the amounts of evoked neuronal firing. However, higher frequencies did produce greater firing dispersity. This finding aligns with observations that neuronal desynchronization needs a sufficiently high frequency when treating Parkinson's disease by DBS (Brown et al. 2004). Thus, in HFS with a higher-frequency, some of the electrical energy may serve to randomize firing rather than increase it. The mechanism by which sustained HFS achieves this function can be the intermittent depolarization block of neuronal membranes, with a large amount of electrical energy from the stimulation being used to maintain this block state.

Many brain diseases involve synchronized, rhythmic neuronal firing. For example, in Parkinson's disease, movement disorders are associated with increased synchronous bursts and low-frequency oscillatory discharges in basal and thalamic neurons (Birdno and Grill 2008; Gale et al. 2008). Epilepsy is typically characterized by excessive synchronous discharges in neuronal populations (Le Van Quyen et al. 2003; Lopes da Silva et al. 2003). Studies have suggested that desynchronizing neuronal discharges may be a key mechanism for DBS therapy (Medeiros and Moraes 2014; McConnell et al. 2012; Wilson et al. 2011). Effective DBS treatment for movement disorders has been considered using HFS to create new firing patterns that replace the pathological oscillations and synchronous activity (Llinás et al. 1999; Birdno and Grill 2008). Similarly, certain stimulation therapies have controlled seizures by desynchronizing epileptic activity in neuronal networks (Cota et al. 2009, Medeiros and Moraes 2014). Our experimental results in this chapter confirm the asynchronous effect of HFS and support desynchronization as a mechanism of DBS action.

Furthermore, our experiments showed that both O-HFS at afferent fibers and A-HFS at efferent fibers can increase the firing of interneurons. These interneurons can regulate pyramidal neurons in a wide area through feedforward and feedback inhibitory circuits (refer to Sect. 2.3). Particularly, the A-HFS effect on interneurons suggests a potential DBS target along the hippocampal efferent pathway to control disorders like epilepsy. The hippocampus is frequently involved in epilepsy, with its CA3 region being particularly prone to epileptiform activity due to abundant excitatory connections among CA3 pyramidal neurons (Andersen et al. 2007). The abnormal activity can spread to the CA1 region and subsequently to other brain regions through the alveus efferent fibers. By activating interneurons, electrical stimulation at the efferent fibers may inhibit CA1 pyramidal neurons, potentially interrupting and eliminating the spread of epileptiform activity. Further research is needed to validate this hypothesis.

References

Andersen P, Morris R, Amaral D et al (2007) The hippocampus book. Oxford University Press

Anderson TR, Hu B, Iremonger K et al (2006) Selective attenuation of afferent synaptic transmission as a mechanism of thalamic deep brain stimulation-induced tremor arrest. J Neurosci 26(3):841–850

Azouz R, Jensen MS, Yaari Y (1996) Ionic basis of spike after-depolarization and burst generation in adult rat hippocampal CA1 pyramidal cells. J Physiol 492(Pt 1):211–223

Barthó P, Hirase H, Monconduit L et al (2004) Characterization of neocortical principal cells and interneurons by network interactions and extracellular features. J Neurophysiol 92(1):600–608

Bellinger SC, Miyazawa G, Steinmetz PN (2008) Submyelin potassium accumulation may functionally block subsets of local axons during deep brain stimulation: a modeling study. J Neural Eng 5(3):263–274

Birdno MJ, Grill WM (2008) Mechanisms of deep brain stimulation in movement disorders as revealed by changes in stimulus frequency. Neurotherapeutics 5(1):14–25

Brazhe AR, Maksimov GV, Mosekilde E et al (2011) Excitation block in a nerve fibre model owing to potassium-dependent changes in myelin resistance. Interface Focus 1(1):86–100

Brown P, Mazzone P, Oliviero A et al (2004) Effects of stimulation of the subthalamic area on oscillatory pallidal activity in Parkinson's disease. Exp Neurol 188(2):480–490

Bucher D, Goaillard JM (2011) Beyond faithful conduction: Short-term dynamics, neuromodulation, and long-term regulation of spike propagation in the axon. Prog Neurobiol 94(4):307–346

Burke D, Cappelen-Smith C, Kuwabara S (2004) Conduction block in demyelinated axons precipitated by normally innocuous physiological processes. Suppl Clin Neurophysiol 57:191–194

Buzsáki G (1984) Feedforward inhibition in the hippocampal-formation. Prog Neurobiol 22(2):131–153

Choi U, Wang H, Hu M et al (2021) Presynaptic coupling by electrical synapses coordinates a rhythmic behavior by synchronizing the activities of a neuron pair. Proc Natl Acad Sci USA 118(20):e2022599118

Chomiak T, Hu B (2007) Axonal and somatic filtering of antidromically evoked cortical excitation by simulated deep brain stimulation in rat brain. J Physiol 579(Pt 2):403–412

Contreras SA, Schleimer JH, Gulledge AT et al (2021) Activity-mediated accumulation of potassium induces a switch in firing pattern and neuronal excitability type. PLoS Comput Biol 17(5):e1008510

Cota VR, Medeiros Dde C, Vilela MR et al (2009) Distinct patterns of electrical stimulation of the basolateral amygdala influence pentylenetetrazole seizure outcome. Epilepsy Behav 14(Suppl 1):26–31

Csicsvari J, Hirase H, Czurko A et al (1998) Reliability and state dependence of pyramidal cell-interneuron synapses in the hippocampus: An ensemble approach in the behaving rat. Neuron 21(1):179–189

Davies CH, Davies SN, Collingridge GL (1990) Paired-pulse depression of monosynaptic GABA-mediated inhibitory postsynaptic responses in rat hippocampus. J Physiol 424:513–531

Debanne D, Campanac E, Bialowas A et al (2011) Axon physiology. Physiol Rev 91(2):555–602

Deniau JM, Degos B, Bosch C et al (2010) Deep brain stimulation mechanisms: beyond the concept of local functional inhibition. Eur J Neurosci 32(7):1080–1091

Di Miceli M, Husson Z, Ruel P et al (2020) In silico hierarchical clustering of neuronal populations in the rat ventral tegmental area based on extracellular electrophysiological properties. Front Neural Circuits 14:51

Dudok B, Szoboszlay M, Paul A et al (2021) Recruitment and inhibitory action of hippocampal axo-axonic cells during behavior. Neuron 109(23):3838-3850.e8

Eusebio A, Thevathasan W, Gaynor LD et al (2011) Deep brain stimulation can suppress pathological synchronisation in Parkinsonian patients. J Neurol Neurosurg Psychiatry 82(5):569–573

Feng Z, Wang Z, Guo Z et al (2017) High frequency stimulation of afferent fibers generates asynchronous firing in the downstream neurons in hippocampus through partial block of axonal conduction. Brain Res 1661:67–78

Feng Z, Yu Y, Guo Z et al (2014) High frequency stimulation extends the refractory period and generates axonal block in the rat hippocampus. Brain Stimul 7(5):680–689

Feng Z, Zheng X, Yu Y et al (2013) Functional disconnection of axonal fibers generated by high frequency stimulation in the hippocampal CA1 region in-vivo. Brain Res 1509:32–42

Fields RD (2008) White matter matters. Sci Am 298(3):42–49

Florence G, Sameshima K, Fonoff ET et al (2016) Deep brain stimulation: more complex than the inhibition of cells and excitation of fibers. Neuroscientist 22(4):332–345

Gale JT, Amirnovin R, Williains ZM et al (2008) From symphony to cacophony: pathophysiology of the human basal ganglia in Parkinson disease. Neurosci Biobehav Rev 32(3):378–387

Gatev P, Darbin O, Wichmann T (2006) Oscillations in the basal ganglia under normal conditions and in movement disorders. Mov Disord 21(10):1566–1577

Guo Z, Feng Z, Wang Y et al (2018) Simulation study of intermittent axonal block and desynchronization effect induced by high-frequency stimulation of electrical pulses. Front Neurosci 12:858

Gurbani S, Chayasirisobhon S, Cahan L et al (2016) Neuromodulation therapy with vagus nerve stimulation for intractable epilepsy: a 2-year efficacy analysis study in patients under 12 years of age. Epilepsy Res Treat 2016:9709056

Hahn PJ, Durand DM (2001) Bistability dynamics in simulations of neural activity in high-extracellular-potassium conditions. J Comput Neurosci 11(1):5–18

Hammond C, Bergman H, Brown P (2007) Pathological synchronization in Parkinson's disease: networks, models and treatments. Trends Neurosci 30(7):357–364

Hashimoto T, Elder CM, Okun MS et al (2003) Stimulation of the subthalamic nucleus changes the firing pattern of pallidal neurons. J Neurosci 23(5):1916–1923

Holsheimer J, Demeulemeester H, Nuttin B et al (2000) Identification of the target neuronal elements in electrical deep brain stimulation. Eur J Neurosci 12(12):4573–4577

Iremonger KJ, Anderson TR, Hu B et al (2006) Cellular mechanisms preventing sustained activation of cortex during subcortical high-frequency stimulation. J Neurophysiol 96(2):613–621

Jensen AL, Durand DM (2009) High frequency stimulation can block axonal conduction. Exp Neurol 220(1):57–70

Jensen MS, Azouz R, Yaari Y (1996) Spike after-depolarization and burst generation in adult rat hippocampal CA1 pyramidal cells. J Physiol 492(Pt 1):199–210

Jiruska P, de Curtis M, Jefferys JGR et al (2013) Synchronization and desynchronization in epilepsy: controversies and hypotheses. J Physiol 591(4):787–797

Jobst B (2010) Brain stimulation for surgical epilepsy. Epilepsy Res 89(1):154–161

Juckett L, Saffari TM, Ormseth B et al (2022) The effect of electrical stimulation on nerve regeneration following peripheral nerve injury. Biomolecules 12(12):1856

Kepecs A, Fishell G (2014) Interneuron cell types are fit to function. Nature 505(7483):318–326

Kim E, Owen B, Holmes WR et al (2012) Decreased afferent excitability contributes to synaptic depression during high-frequency stimulation in hippocampal area CA1. J Neurophysiol 108(7):1965–1976

Kloosterman F, Peloquin P, Leung LS (2001) Apical and basal orthodromic population spikes in hippocampal CA1 in vivo show different origins and patterns of propagation. J Neurophysiol 86(5):2435–2444

Knowles WD, Schwartzkroin PA (1981) Local circuit synaptic interactions in hippocampal brain slices. J Neurosci 1(3):318–322

Kuncel AM, Cooper SE, Wolgamuth BR et al (2006) Clinical response to varying the stimulus parameters in deep brain stimulation for essential tremor. Mov Disord 21(11):1920–1928

Lam CM, Latif U, Sack A et al (2023) Advances in spinal cord stimulation. Bioengineering 10(2):185

Laxton AW, Tang-Wai DF, McAndrews MP et al (2010) A phase I trial of deep brain stimulation of memory circuits in Alzheimer's disease. Ann Neurol 68(4):521–534

Le Van QM, Navarro V, Martinerie J et al (2003) Toward a neurodynamical understanding of ictogenesis. Epilepsia 44(Suppl 12):30–43
Lee DJ, Lozano CS, Dallapiazza RF et al (2019) Current and future directions of deep brain stimulation for neurological and psychiatric disorders. J Neurosurg 131(2):333–342
Li MCH, Cook MJ (2018) Deep brain stimulation for drug-resistant epilepsy. Epilepsia 59(2):273–290
Llinás RR, Ribary U, Jeanmonod D et al (1999) Thalamocortical dysrhythmia: a neurological and neuropsychiatric syndrome characterized by magnetoencephalography. Proc Natl Acad Sci USA 96(26):15222–15227
Lopes da Silva F, Blanes W, Kalitzin SN et al (2003) Epilepsies as dynamical diseases of brain systems: basic models of the transition between normal and epileptic activity. Epilepsia 44(Suppl 12):72–83
Lozano AM, Lipsman N, Bergman H et al (2019) Deep brain stimulation: Current challenges and future directions. Nat Rev Neurol 15(3):148–160
McCairn KW, Turner RS (2009) Deep brain stimulation of the globus pallidus internus in the Parkinsonian primate: local entrainment and suppression of low-frequency oscillations. J Neurophysiol 101(4):1941–1960
McConnell GC, So RQ, Hilliard JD et al (2012) Effective deep brain stimulation suppresses low-frequency network oscillations in the basal ganglia by regularizing neural firing patterns. J Neurosci 32(45):15657–15668
McGee MJ, Amundsen CL, Grill WM (2015) Electrical stimulation for the treatment of lower urinary tract dysfunction after spinal cord injury. J Spinal Cord Med 38(2):135–146
McIntyre CC, Grill WM (1999) Excitation of central nervous system neurons by nonuniform electric fields. Biophys J 76(2):878–888
Medeiros Dde C, Moraes MF (2014) Focus on desynchronization rather than excitability: a new strategy for intraencephalic electrical stimulation. Epilepsy Behav 38:32–36
Meeks JP, Jiang X, Mennerick S (2005) Action potential fidelity during normal and epileptiform activity in paired soma–axon recordings from rat hippocampus. J Physiol 566(2):425–441
Meeks JP, Mennerick S (2004) Selective effects of potassium elevations on glutamate signaling and action potential conduction in hippocampus. J Neurosci 24(1):197–206
Nowak LG, Bullier J (1998) Axons, but not cell bodies, are activated by electrical stimulation in cortical gray matter. Exp Brain Res 118(4):489–500
Pelkey KA, Chittajallu R, Craig MT et al (2017) Hippocampal GABAergic inhibitory interneurons. Physiol Rev 97(4):1619–1747
Popovych O, Tass PA (2014) Control of abnormal synchronization in neurological disorders. Front Neurol 5:268
Quian Quiroga R, Kreuz T, Grassberger P (2002) Event synchronization: a simple and fast method to measure synchronicity and time delay patterns. Phys Rev E Stat Nonlin Soft Matter Phys 66(4 Pt 1):041904
Radivojevic M, Franke F, Altermatt M et al (2017) Tracking individual action potentials throughout mammalian axonal arbors. eLife 6:e30198
Rampp S, Stefan H (2006) Fast activity as a surrogate marker of epileptic network function? Clin Neurophysiol 117(10):2111–2117
Ranck JB (1975) Which elements are excited in electrical stimulation of mammalian central nervous system: a review. Brain Res 98(3):417–440
Rush AJ, Marangell LB, Sackeim HA et al (2005) Vagus nerve stimulation for treatment-resistant depression: a randomized, controlled acute phase trial. Biol Psychiatry 58(5):347–354
Silberstein SD, Yuan H, Najib U et al (2020) Non-invasive vagus nerve stimulation for primary headache: a clinical update. Cephalalgia 40(12):1370–1384
Theoret Y, Brown A, Fleming SP et al (1984) Hippocampal field potential: a microcomputer aided comparison of amplitude and integral. Brain Res Bull 12(5):589–595

Traub RD, Draguhn A, Whittington MA et al (2002) Axonal gap junctions between principal neurons: a novel source of network oscillations, and perhaps epileptogenesis. Rev Neurosci 13(1):1–30

Udupa K, Chen R (2015) The mechanisms of action of deep brain stimulation and ideas for the future development. Prog Neurobiol 133:27–49

Vitek JL (2002) Mechanisms of deep brain stimulation: Excitation or inhibition. Mov Disord 17(S3):S69–S72

Vonck K, Sprengers M, Carrette E et al (2013) A decade of experience with deep brain stimulation for patients with refractory medial temporal lobe epilepsy. Int J Neural Syst 23(1):1250034

Wang Z, Feng Z, Wei X (2018) Axonal stimulations with a higher frequency generate more randomness in neuronal firing rather than increase firing rates in rat hippocampus. Front Neurosci 12:783

Wang Z, Feng Z, Yuan Y et al (2024) Dynamics of neuronal firing modulated by high-frequency electrical pulse stimulations at axons in rat hippocampus. J Neural Eng 21(2):026025

Wilson CJ, Beverlin B, Netoff T (2011) Chaotic desynchronization as the therapeutic mechanism of deep brain stimulation. Front Syst Neurosci 5:50

Wingeier B, Tcheng T, Koop MM et al (2006) Intra-operative STN DBS attenuates the prominent beta rhythm in the STN in Parkinson's disease. Exp Neurol 197(1):244–251

Wu X, Shuai J (2012) Multistability in a neuron model with extracellular potassium dynamics. Phys Rev E Stat Nonlin Soft Matter Phys 85(6 Pt 1):061911

Xu Y, Feng Z, Yuan Y et al (2023) Effects on somata by high-frequency electrical stimulation at the axons of hippocampal pyramidal neurons. Prog Biochem Biophys 50(3):561–572 (in Chinese)

Ye X, Feng Z, Wang Z et al (2022) Activating interneurons in local inhibitory circuits by high-frequency stimulations at the efferent fibers of pyramidal neurons in rat hippocampal CA1 region. Brain Sci 12(10):1350

Yuan Y (2023) Investigation of firing patterns and post-stimulation effects of pyramidal neurons under axonal high-frequency stimulation in rat hippocampus. PhD dissertation, Zhejiang University, Hangzhou (in Chinese)

Yuan Y, Feng Z, Wang Z (2025) Cluster neuronal firing induced by uniform pulses of high-frequency stimulation on axons in rat hippocampus. IEEE Trans Biomed Eng 72(3):1108–1120

Yuan Y, Feng Z, Yang G et al (2022) Suppression of neuronal firing following antidromic high-frequency stimulations on the neuronal axons in rat hippocampal CA1 region. Front Neurosci 16:881426

Yue C, Remy S, Su H et al (2005) Proximal persistent Na^+ channels drive spike afterdepolarizations and associated bursting in adult CA1 pyramidal cells. J Neurosci 25(42):9704–9720

Zheng F, Lammert K, Nixdorf-Bergweiler BE et al (2011) Axonal failure during high frequency stimulation of rat subthalamic nucleus. J Physiol 589(Pt 11):2781–2793

Zheng L, Feng Z, Hu H et al (2020) The appearance order of varying intervals introduces extra modulation effects on neuronal firing through non-linear dynamics of sodium channels during high-frequency stimulations. Front Neurosci 14:397

Chapter 6
Neuronal Responses to Axonal HFS with Time-Varying Parameters

Abstract This chapter presents our findings from different HFS paradigms using time-varying parameters, including gradient changes in pulse frequency and intensity, alternating pulses with dual parameters, inserting and deleting pulses in regular stimulation, and randomly varying inter-pulse-intervals within a small range. Under HFS-induced axonal block, stimulations with time-varying parameters can produce diverse neuronal responses. This creates opportunities for developing novel stimulation paradigms that can address a wide range of neuromodulation needs.

Keywords Time-varying stimulation · Ramping intensity · Bifurcation · Pulse interval · Randomly varying intervals · Constant stimulation

Developing new stimulation waveforms and paradigms is crucial for improving neuromodulation efficacy and expanding its applications (Grill 2018; Gilbert et al. 2023). Theoretically, electrical stimulation can modulate various neural behaviors in the brain, potentially alleviating symptoms of neurological diseases. While conventional DBS has effectively treated Parkinson's disease and motor disorders since 1980s (Benabid et al. 1987), its application to other brain diseases remains limited. This limitation stems from challenges including the diverse pathological mechanisms of different brain diseases, our incomplete understanding of complex DBS mechanisms, and the lack of diverse stimulation paradigm options for various brain disorders.

Conventional DBS has been using stimulation paradigms with pulse sequences at frequencies about 100 Hz or higher. Although the pulse parameters—including width, intensity, and frequency—can be programmed, they typically remain constant once set. With the diverse pathogenesis of brain diseases, exploring varied stimulation paradigms with different "effects" and "dosages" is crucial to extend treatment options for DBS applications. A promising approach is to design new stimulation paradigms with time-varying parameters. Since neurons respond nonlinearly to electrical pulses, time-varying stimulations can produce diverse therapeutic effects, creating numerous possibilities for new treatment approaches. However, the

Z. Feng, *Electrical Brain Stimulation: Mechanisms and Modulation of Neuronal Activity*, https://doi.org/10.1007/978-981-95-4145-4_6

nonlinear properties of the nervous system also present challenges and uncertainties in exploring time-varying stimulations. We have conducted a series of studies on time-varying stimulations to understand neuronal responses and their underlying mechanisms, as described below.

6.1 Initial Neuronal Responses Shaped by Gradually Varying Stimulations

As described in Sect. 5.2, during the initial period of A-HFS at 100 and 200 Hz, there is a process where evoked large APSs decrease until neuronal responses reach steady state after several seconds due to HFS-induced axonal block. Similarly, during the O-HFS described in Sect. 5.3, initial large OPSs occur before neuronal responses reach steady state with no more PSs but only increased MUA. This brief initial transition is known as the "onset response" in peripheral nerve stimulation. It occurs because the applied HFS starts as a step excitation input from zero, producing a transient step response. According to the principles of control theory, a system responds to an external step input with two phases: a transient phase followed by a steady phase. When peripheral motor nerve fibers receive stimulation pulses at kilohertz frequencies, their responses appear as synchronized axon firing at onset before transition into a steady phase of complete axonal block. This synchronous onset firing can result in tonic spasms in innervated muscles and cause pain (Bhadra and Kilgore 2005).

Although the onset response in DBS has not attracted much attentions, transient side effects like sensory abnormalities have been observed in clinical DBS treatments (Kuncel et al. 2006), which may be caused by onset responses corresponding to the synchronous firing appearing as large population spikes in the initial A-HFS and O-HFS periods (Chap. 5). For common DBS durations that last for hours or days, this transient phase can be too brief to generate a notable impact. However, for stimulation modes that switch frequently between "ON" and "OFF" states—such as in adaptive or closed-loop DBS (Ansó et al. 2022; Hosain et al. 2014)—the transient phase could have potential impact due to its occurrence at each onset of "ON" state.

Time-varying stimulations with real-time adjustments to pulse intensity and frequency may prevent excessive onset responses. For instance, for high-frequency sinusoidal stimulations at peripheral nerves, using higher frequency and greater intensity in the initial period can speed up the generation of axonal conduction block, shortening the duration of the onset response (Gerges et al. 2010; Bhadra et al. 2009). However, other animal studies and computational simulations have shown that sinusoidal stimulations with linearly varying intensity cannot prevent onset responses (Miles et al. 2007). To investigate whether time-varying parameters could eliminate the onset responses in axonal HFS pulse stimulations, we used a self-custom stimulation system (refer to Sect. 3.5.5) to test different time-varying A-HFS stimulations in the rat hippocampal CA1 region (Cai et al. 2017, 2018).

6.1.1 A-HFS with Ramping Pulse Intensity

Figures 5.2 and 5.7 in Chap. 5 show the changes in the APSs induced by antidromic activation of CA1 pyramidal neurons during constant A-HFS applied at alveus. The initial APS amplitude at the onset of 100 Hz A-HFS was approximately 5 times its steady amplitude in the late A-HFS period. To prevent the large initial APSs, we used a modified A-HFS train including an initial 10 s ramp-up pulse intensity linearly from 0.02 to 0.2 mA before remaining constant at 0.2 mA for 50 s until the end of stimulation (Fig. 6.1A). The small 0.02 mA initial intensity hardly induced any APSs. As the intensity gradually increased, the evoked APS gradually appeared with increasing amplitude. However, no large APS appeared throughout the entire 1 min A-HFS. The peak APS (only ~ 3 mV) appeared at ~ 6 s when the intensity increased to ~ 0.13 mA (Fig. 6.1B). During the late 40 s period with the fixed 0.2 mA intensity, the APS amplitudes matched those evoked by the control A-HFS with a fixed 0.2 mA intensity from beginning to end, in the same rat experiment. But, the initial APS evoked in the control A-HFS exceeded 10 mV (Fig. 6.1B).

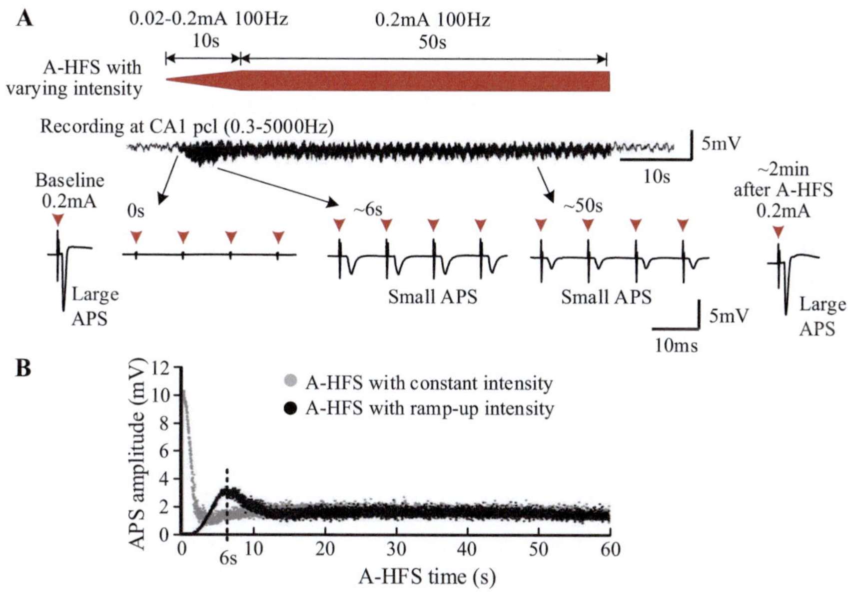

Fig. 6.1 Comparison of neuronal responses to A-HFS trains with ramp-up versus constant intensity. **A** Top: Diagram of 100 Hz 1 min A-HFS with ramp-up intensity during the initial 10 s followed by constant intensity for the rest 50 s. Middle: APS signal recorded in the CA1 pyramidal cell layer (pcl) with artifacts removed. Bottom: Enlarged APS waveforms at baseline, and at onset (0 s), ~ 6 s and ~ 50 s of A-HFS, along with recovered APS at ~ 2 min after the end of A-HFS. Red arrows indicate pulse artifacts. **B** Scatter plots of the APS amplitudes during both A-HFS paradigms. Modified from Cai et al. (2017)

Statistical data showed that the maximum APS during ramp-up A-HFS appeared within the initial 10 s period, reaching only 41 ± 7.7% of the maximum APS amplitude induced by constant A-HFS ($n = 8$ rats). Between the A-HFS period from 10 to 60 s, the average APS amplitudes showed no significant difference between the two groups. Therefore, gradually increasing pulse intensity from a lower to required level can prevent excessively synchronized firing of neuronal populations at A-HFS onset.

While the duration of ramp-up transition period in Fig. 6.1 was set as 10 s, shortening this duration could speed up the extension of stimulated area to a desired range. We compared the changes in APS amplitudes with three different transition periods of 1, 5 and 10 s (Fig. 6.2A). With a 1 s transition, the peak APS only decreased to ~ 90% of the control, significantly greater than with 5 s (~ 60%) and 10 s (~ 40%) transitions (Fig. 6.2B). While merely shortening the transition caused large APS to reappear, simultaneously increasing pulse frequency could suppress the APSs. With a 1 s transition, increasing the pulse frequency from 100 to 400 Hz resulted in significantly smaller peak APS than that of 100 and 200 Hz (Fig. 6.2C and D). Therefore, we next tested a paradigm using higher pulse frequency and ramp-up intensity in a shortened transition period at the beginning of A-HFS.

6.1.2 *Effects of A-HFS with Both Ramp-Up Intensity and Varying Frequency*

As shown in Fig. 6.3, we designed a 100 s A-HFS train where the first 10 s used a 400 Hz frequency with a ramp-up intensity from 0.03 to 0.3 mA, followed by 90 s at constant 100 Hz and 0.3 mA. No large APS appeared during the initial 10 s period. However, when the frequency dropped suddenly from 400 to 100 Hz at 0.3 mA intensity, a large APS occurred due to the abrupt IPI change from 2.5 to 10 ms. This APS reached approximately half the baseline amplitude. Then the evoked APS gradually decreased to a steady level.

To prevent the large APSs triggered by the abrupt frequency change, we designed an A-HFS train with three distinct periods: an initial 1 s period at 400 Hz with intensity ramping up from 0.03 to 0.3 mA, followed by a 10 s period at 0.3 mA with frequency ramping down from 400 to 100 Hz (using a reciprocal frequency decay curve from linearly increasing IPIs, as detailed in Cai et al. 2018), and a final 49 s period of constant stimulation at 0.3 mA and 100 Hz—completing a total 60 s A-HFS train.

Figure 6.4A shows an example of neuronal responses to this A-HFS paradigm. During the initial 1 s period, the maximum APS amplitude was only 3.2 mV. In the subsequent 10 s period, as frequency gradually decreased, the APS amplitude increased slightly. During the final 49 s period with constant stimulation parameters, the APS amplitude remained small (the black scatter plot in Fig. 6.4B). In contrast, the control A-HFS with constant parameters of 0.3 mA and 100 Hz produced a

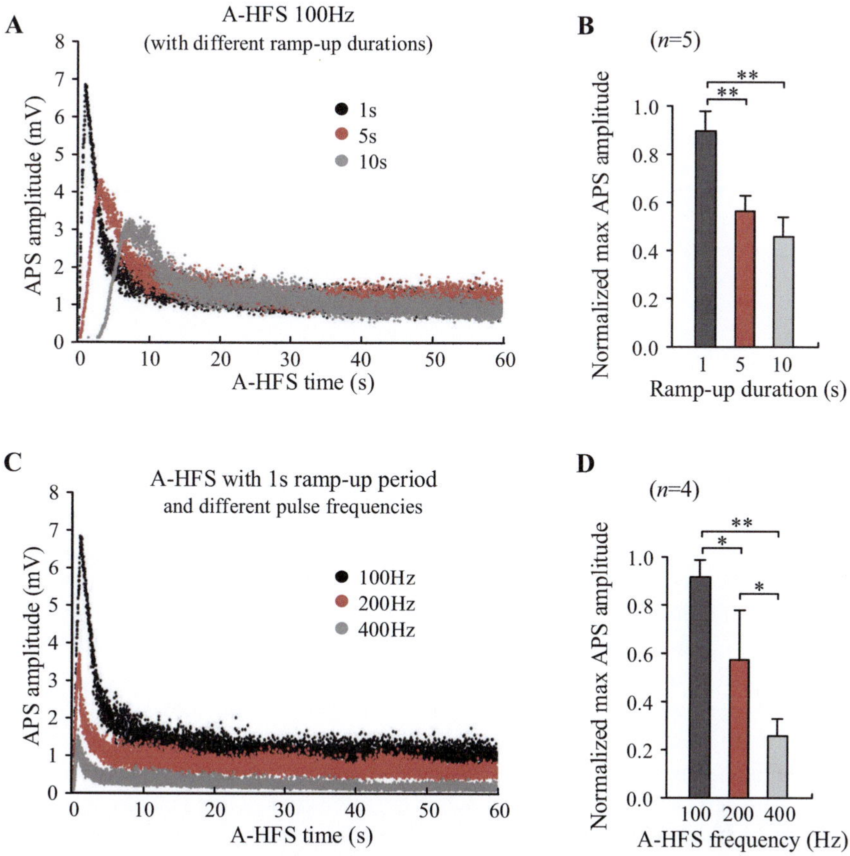

Fig. 6.2 Comparisons of APS amplitudes during A-HFS trains with different ramp-up durations and different pulse frequencies. **A** Scatter plots showing APS amplitudes during three A-HFS trains (100 Hz, 60 s) with different initial periods of 1, 5, and 10 s using ramp-up intensity (0.03–0.3 mA), before maintaining constant 0.3 mA intensity for the rest stimulation. **B** Comparisons of peak APS amplitudes (normalized by max APS in constant-intensity A-HFS) among the three A-HFS paradigms shown in (**A**). $^{**}P < 0.01$, ANOVA with post-hoc Bonferroni test, $n = 5$ rats. **C** Scatter plots showing APS amplitudes during three A-HFS paradigms at frequencies of 100, 200, and 400 Hz, each with a 1 s initial ramp-up period (0.03–0.3 mA). **D** Comparisons of normalized peak APS amplitudes among the three A-HFS paradigms shown in (**C**). $^{*}P < 0.05$, $^{**}P < 0.01$, ANOVA with post-hoc Bonferroni test, $n = 4$ rats. From Cai et al. (2017)

maximum APS amplitude of 10.3 mV at onset (the grey scatter plot in Fig. 6.4B). The late 49 s period showed no significant difference between control and varying A-HFS. Statistical results showed that during the initial 11 s transition period, the varying A-HFS induced a maximum APS amplitude of only 33 ± 5.0% ($n = 7$ rats) of the value of constant A-HFS (0.3 mA, 100 Hz). The mean APS amplitudes during the late 49 s period (from 11 to 60 s) were similar between both paradigms (Fig. 6.4C). The APS recovered quickly within 2 min after A-HFS ended, indicating

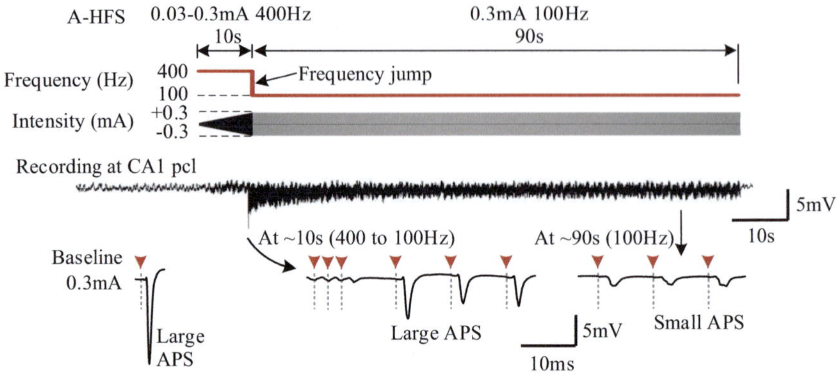

Fig. 6.3 A-HFS with an initial period of higher frequency and ramp-up intensity. Top: Schematic diagram of the A-HFS paradigm. Middle: APS signal recorded in the CA1 pcl. Bottom: enlarged APS waveforms at baseline, around the frequency shift and during steady A-HFS period. Red arrows with dashed lines indicate removed pulse artifacts

that the 400 Hz transition period stimulation is safe, despite requiring higher electrical power consumption.

The above results indicate that using ramped intensity and frequency during the initial A-HFS period can reshape the neuronal responses, preventing synchronous firing of large neuronal populations at stimulation onset. During the transition period with ramp-up intensity, the initial weak intensity activated too few axons to generate large APSs. As intensity increased, the stimulated area gradually expanded and more axons were activated. However, even when the intensity reached that of constant A-HFS, the evoked APS remained much smaller than that at the onset of constant A-HFS. Additionally, the peak APS amplitude was smaller with either a slower intensity increase (Fig. 6.2A and B) or a higher pulse frequency (Fig. 6.2C and D). These responses can be explained by HFS-induced axonal block. By the time pulse intensity reaches and maintains its upper level, axonal block has already developed in the early stimulated axons. Each pulse can then only activate a subset of stimulated axons, keeping the APS amplitude well below baseline level. While a slower ramp-up intensity can provide sufficient time for more axons to gradually enter block states, it also slows the stimulation to reach the desired action area. Higher pulse frequencies can enable axons to enter their block states more quickly (refer to Sect. 5.2). Therefore, using a higher frequency can allow to shorten the initial transition period of ramped intensity. Adding a subsequent period with ramp-down frequency can reduce the frequency to the desired level. This approach—combining two transition periods with varying parameters—can quickly achieve a sufficiently large stimulated area while preventing synchronous firing of large neuronal populations (Fig. 6.4).

Similar stimulation paradigms have been used in peripheral nerve stimulations to minimize onset responses (Bhadra et al. 2009; Bhadra and Kilgore 2005; Miles et al. 2007; Gerges et al. 2010). However, this approach has not been applied to clinical brain stimulations like DBS therapy. The highly synchronized neuronal firing

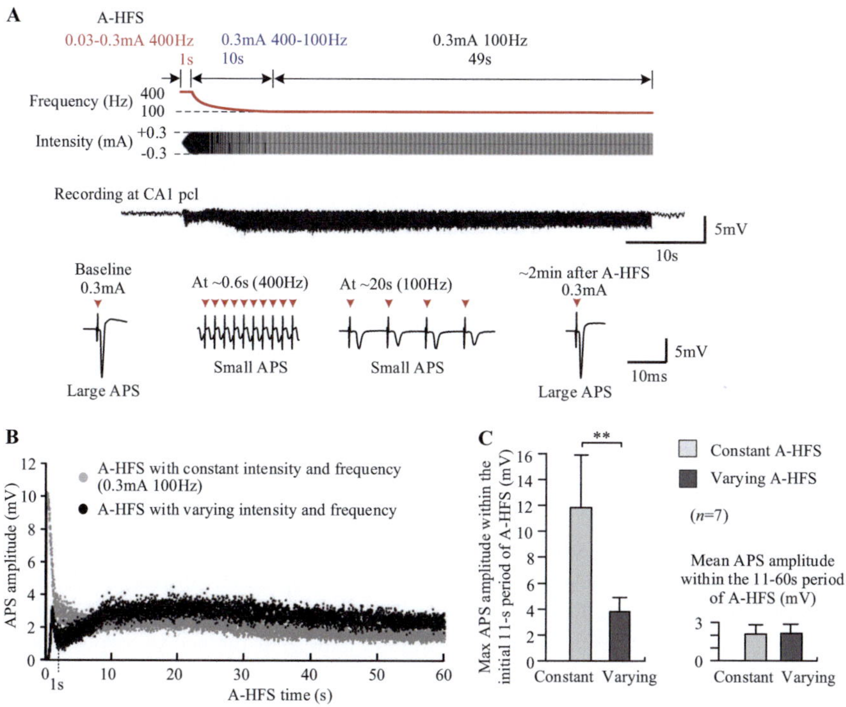

Fig. 6.4 Neuronal responses to A-HFS trains with varying intensity and frequency. **A** Top two rows: schematic diagrams of 60 s A-HFS comprising three periods—a 1 s ramp-up intensity (0.03–0.3 mA) at 400 Hz, a 10 s ramp-down frequency (400–100 Hz) at 0.3 mA, and a final 49 s period at constant 0.3 mA and 100 Hz. Bottom two rows: recording of an entire A-HFS with enlarged views of APS waveforms at baseline, 0.6 and 20 s from A-HFS onset, and the recovered APS at 2 min following A-HFS. **B** Scatter plots of the APS amplitudes evoked by each pulse during A-HFS with constant versus varying parameters. **C** Comparisons between the two A-HFS paradigms for peak APS amplitudes within the initial 11 s period (*left*) and mean APS amplitudes during the 10–60 s A-HFS period (*right*). $^{**}P < 0.01$, paired *t*-test, $n = 7$. From Cai et al. (2017)

produced by HFS resembles epileptiform activity that could be harmful to the brain. Clinical observations have shown that frequent ON/OFF stimulations at intervals shorter than one minute can reduce DBS efficacy in controlling the symptoms of Parkinson's disease (Khandhar et al. 2005). Similarly, periodic DBS is less effective at relieving patient tremors compared to conventional DBS (Kuncel et al. 2012). While the mechanisms behind these phenomena remain unclear, the onset response may be a contributing factor. If this is true, addressing the onset responses can be crucial for improving adaptive stimulation treatments that require frequent ON/OFF switching.

6.2 Nonlinear Neuronal Responses to Dual-Parameter Stimulation

As described in Sect. 5.2, the magnitude of APS induced by each pulse during the steady period of constant A-HFS was inversely proportional to pulse frequency and directly proportional to pulse interval (IPI). The ratio of IPIs at 50, 100 and 200 Hz is 4:2:1. With similar baseline APS amplitudes, the ratio of APS amplitudes during the steady A-HFS periods at these three frequencies was also approximately 4:2:1 (Fig. 5.7F). Within a certain range of pulse intensities, the APS amplitude produced by a single pulse increased nearly linearly with pulse intensity. (Fig. 2.8). However, during non-steady periods—such as the transition periods with ramp-up intensity shown in Sect. 6.1—the APS amplitude did not change monotonically with pulse intensity (Figs. 6.1 and 6.2). Nevertheless, during these transition periods with gradually varying pulse parameters, the changes in APS amplitudes were smooth rather than abrupt.

Due to nonlinear properties, neuronal firing can present nonlinear behaviors such as bifurcation, chaos and stochastic resonance (Fan and Holden 1993; Deco et al. 2008; De Maesschalck and Wechselberger 2015). Among these, bifurcation is characterized by abrupt changes. It occurs when a small change in inputs or in circumstances causes a dramatic shift in firing rate or temporal pattern as the neuron approaches a specific state so called bifurcation point (Yang et al. 2006; Rabinovich et al. 2006; Atherton et al. 2016). For example, a slight increase in stimulation intensity from 0.3 to 0.4 mA can trigger a sudden jump in neuronal firing rate from $\sim$30 to $\sim$ 80 spikes/s, transforming the firing pattern from regular to bursting (Yang et al. 2006). Such changes can occur when the neuronal membrane is sufficiently depolarized to a bifurcation point through increased extracellular potassium concentration ($[K^+]_o$) or decreased calcium concentration ($[Ca^{2+}]_o$) (Yang et al. 2006; Gu and Chen 2014; Jia et al. 2017). In contrast, when the neuronal state is far from the bifurcation point, the same increase in stimulation intensity can raise the firing rate by only $\sim$ 5 spikes/s. Sustained pulse stimulation in the brain can also induce bifurcation in neuronal firing. For instance, fluctuations in axonal membrane potential or in synaptic inputs can cause a neuron to switch between high-rate firing and complete silence, with no intermediate firing levels (Zhang et al. 2020, 2019; Gu et al. 2015). Computational models have successfully reproduced these nonlinear bifurcation changes in neuronal firing (Atherton et al. 2016; Hahn and Durand 2001; Paydarfar et al. 2006; Ma and Tang 2017).

These previous studies on neuronal firing bifurcations have focused on single neurons rather than neuronal populations. We hypothesized that when A-HFS drives the axonal membrane into intermittent depolarization block, the neuron state might approach certain bifurcation point. Such conditions could mean that small adjustments in stimulation parameters would lead to bifurcation changes in the evoked APSs. We tested the hypothesis as below.

6.2.1 *Bifurcation Responses of Neuronal Population to A-HFS with Bi-IPI*

To investigate how changes in pulse intervals affect neuronal population responses, we still examined the neuronal responses to antidromic stimulations (Fig. 6.5A). As controls, we first evaluated the APS amplitude ratios (APS_2/APS_1) produced by paired-pulses with various IPIs (Fig. 6.5B). With IPIs ranging in 5–50 ms and an identical 0.3 mA pulse intensity, both pulses in the pairs produced large APSs of similar amplitude. Even with an IPI as short as 5 ms, the amplitude of the second APS (APS_2) was above 85% of the first APS (APS_1), indicating minimal impact of APS_1 on APS_2. When the IPI exceeded 12 ms, the APS_2/APS_1 ratio approached 100%, showing no effect of APS_1 on APS_2. This occurred because the refractory period of CA1 pyramidal neurons is shorter than 5 ms under baseline conditions. Additionally, the activation of neuronal soma through antidromic stimulation on its own axon does not involve synaptic integration. As a result, antidromic activation is hardly affected by local inhibitory circuits. The antidromic response thus differs from the orthodromic response of paired-pulse stimulation applied at the Schaffer collaterals, which is strongly affected by inhibitory circuits (refer to Sect. 2.3).

As shown in Fig. 6.6A, during 1 min A-HFS with a fixed frequency of 66.7 Hz, large APSs appeared only at onset, despite the IPI being as long as 15 ms. After ~ 1 s A-HFS, the APS amplitudes decreased markedly, and after~10 s A-HFS, they declined to ~ 25% of the initial amplitude (Fig. 6.6B). Interestingly, during the 1-10 s A-HFS period when APS amplitudes decreased substantially, a bifurcation emerged between adjacent APSs. Larger (~ 4 mV) and smaller (~ 1 mV) APSs appeared alternatively, despite the constant IPI. In the study of 16 rats applied with mono-IPI A-HFS at 66.7 Hz, such APS bifurcation appeared in 6 (38%) rats and lasted only several seconds, after which the evoked APSs became uniformly small. During the

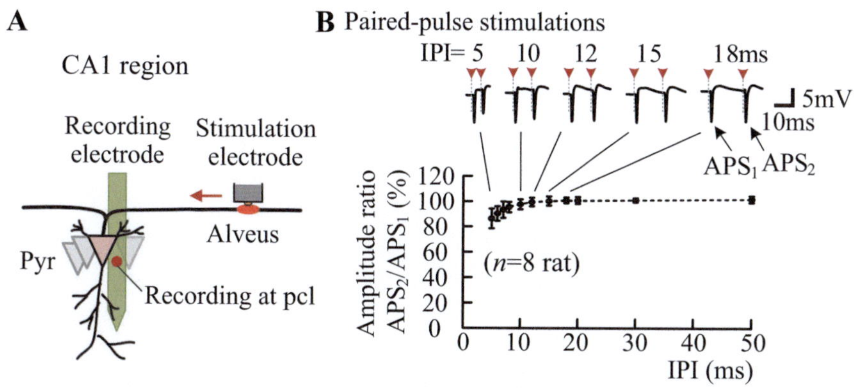

Fig. 6.5 Baseline responses of CA1 neuronal populations to paired-pulse antidromic stimulations with various IPIs. **A** Schematic diagram showing the electrode placements. **B** APS amplitude ratio versus IPI, along with representative APS waveforms. Modified from Wang et al. (2022)

steady A-HFS period, the mean APS amplitudes were similar in groups both with and without bifurcation.

When a 1 min bi-IPI A-HFS was applied with alternating IPI of 12 and 18 ms (ΔIPI = 6 ms), the mean pulse frequency remained at 66.7 Hz. Large APSs appeared at onset with similar amplitudes despite the alternating IPIs (Fig. 6.6C), similar to the APSs induced by paired-pulses (IPI = 12 or 18 ms) at baseline (Fig. 6.5B). However, after 1 to 2 s, as APS amplitude decreased, a bifurcation appeared and persisted until the end of A-HFS (Fig. 6.6D). During the steady A-HFS period, the mean APS amplitude following the shorter IPI = 12 ms (denoted as APS_S) was only 14.8 ±

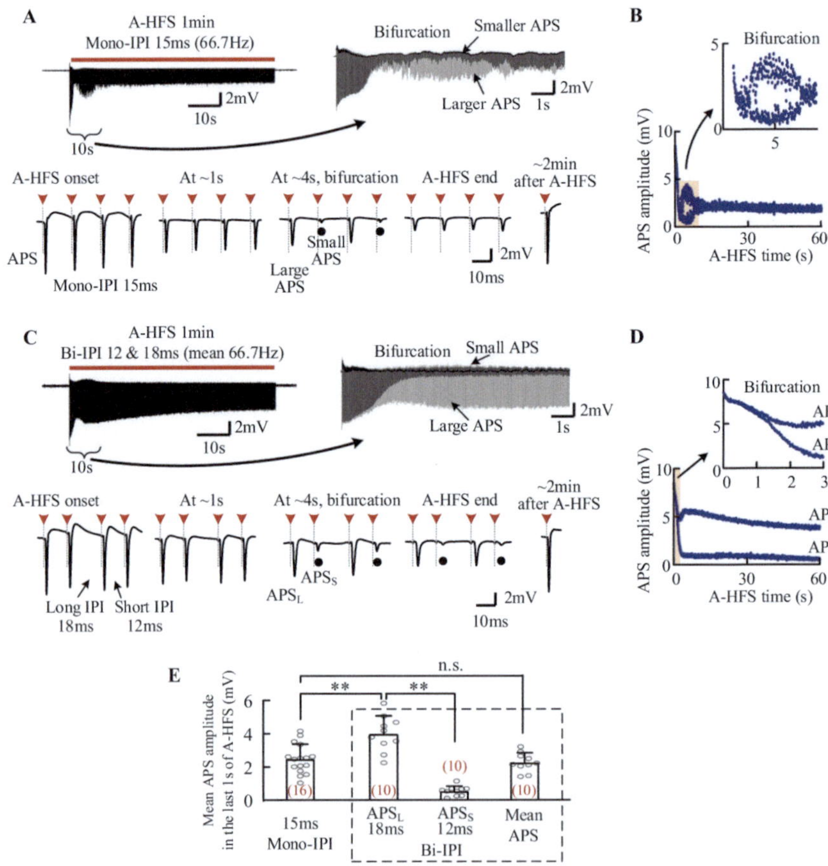

Fig. 6.6 Bifurcations in the APS amplitudes produced by sustained A-HFS. **A** Typical APS signal during 1 min A-HFS with mono-IPI of 15 ms (66.7 Hz) at 0.3 mA intensity. **B** Scatter plot of APS amplitudes from the A-HFS shown in (**A**). **C** Typical APS signal during 1 min A-HFS with bi-IPI (12 and 18 ms) at mean pulse frequency of 66.7 Hz. **D** Scatter plot of APS amplitudes from the A-HFS shown in (**C**). **E** Comparisons of mean APS amplitudes in the final A-HFS second between mono-IPI and bi-IPI groups. $^{**}P < 0.001$, one-way ANOVA with Post hoc Bonferroni tests. The experimental rat numbers are shown in parentheses. Modified from Wang et al. (2022)

9.1% of that following the longer IPI = 18 ms (APS_L). The APS_S/APS_L ratio was far smaller than the IPI ratio of 12/18 = 67%. Additionally, the APS_L amplitude was significantly greater than the APS amplitude induced by mono-IPI A-HFS at 66.7 Hz (Fig. 6.6E). However, the average APS amplitude in bi-IPI A-HFS was similar to that of mono-IPI A-HFS, indicating that both types of A-HFS produced similar levels of neuronal firing given the same number of applied pulses.

These results indicated that even with A-HFS at a constant IPI, a brief period of APS bifurcation can occur during the APS decay phase before stabilization—characterized by alternating larger and smaller amplitudes. Such brief bifurcation in constant A-HFS often appeared at pulse frequencies between 50 and 70 Hz, but not at higher frequencies (e.g., 100 or 200 Hz) or lower frequencies (e.g., 20 or 10 Hz). During bi-IPI A-HFS, the bifurcation persisted until the end of A-HFS, suggesting a gradual transfer of firing from following the shorter IPI to the longer IPI. The APS attenuation during A-HFS results from intermittent axonal block, with the degree of attenuation depending on pulse frequency (see Sect. 5.2). Therefore, we next examined how neuronal populations respond to bi-IPI A-HFS across various mean pulse frequencies.

In the 1 min bi-IPI A-HFS, we fixed the ratio of the shorter IPI (IPI_S) to longer IPI (IPI_L) at 2/3, corresponding to a ΔIPI ratio (ΔIPI/mean IPI) of 40%. We varied the mean A-HFS frequency in a range of 16–200 Hz, corresponding to a ΔIPI range of 25–2 ms. Here, we tentatively use "HFS" to refer to all these stimulations, although neuromodulation stimulations at 50 Hz and above are commonly referred to as HFS (Durand and Bikson 2001). As shown in Fig. 6.7A, during the onsets of bi-IPI A-HFS with various ΔIPI, each pulse consistently induced a large APS with similar amplitude despite the alternating IPIs. The APS amplitudes then decreased as A-HFS progressed, with higher average pulse frequencies leading to more rapid decreases. The mean APS amplitudes at the end of A-HFS were always significantly smaller than their initial values (Fig. 6.7B).

At a lower mean frequency of 16 Hz, the APS amplitudes decreased slightly and slowly, maintaining more than 50% of the initial amplitude until the end of stimulation. No bifurcation appeared during the entire 16 Hz A-HFS, even with a ΔIPI as long as 25 ms. When the mean pulse frequency increased to 50 Hz and above (including 66.7, 100, 133 and 200 Hz), the APS amplitudes decreased more and quickly. The decrease speed was frequency-dependent. The half-amplitude time—the duration until APS amplitude decreased to half its initial value—dropped significantly from $\sim$ 20 s at 50 Hz to $\sim$ 0.2 s at 200 Hz (Fig. 6.7C). A bifurcation appeared along with the APS attenuation and persisted to the end of A-HFS (Fig. 6.7A, *right*). Higher pulse frequencies triggered earlier bifurcation, with initial times ranging from $\sim$ 6 s at 50 Hz to $\sim$ 0.1 s at 200 Hz (Fig. 6.7D). As the mean pulse frequency increased, a greater proportion of APS_S shifted to APS_L. At 133 Hz and above, the APS_S almost disappeared. Only the pulses following IPI_L induced APS, despite the IPI_S/IPI_L ratio remaining at 2/3 (Fig. 6.7A and E).

These results indicate that when the mean pulse frequency is sufficiently high, even a tiny ΔIPI in bi-IPI A-HFS can induce a bifurcation in neuronal responses.

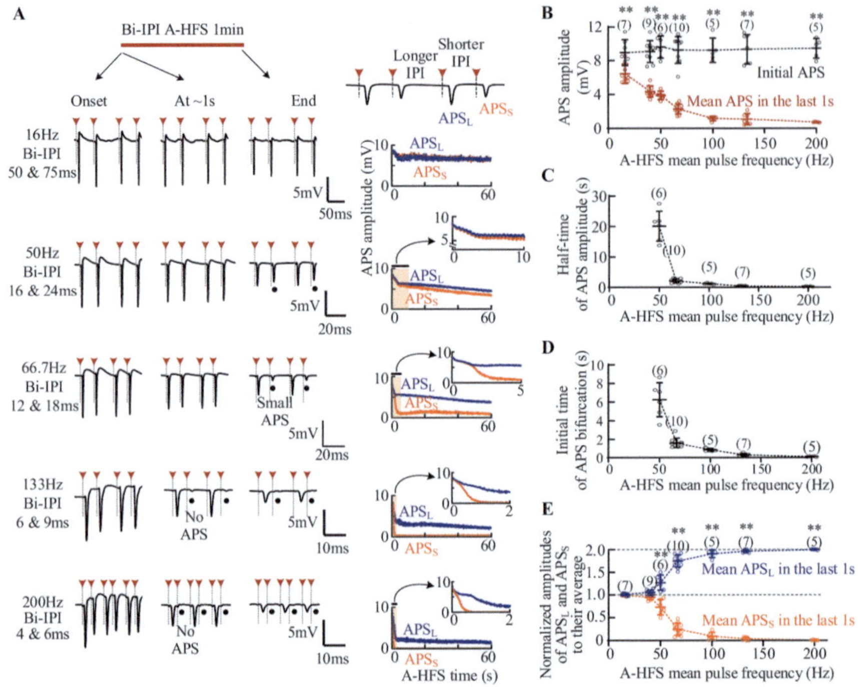

Fig. 6.7 Bifurcations in APS amplitudes produced by bi-IPI A-HFS with a fixed IPI ratio of 2/3. **A** Examples of APS signals during 1 min bi-IPI A-HFS trains with mean pulse frequencies of 16, 50, 66.7, 133 and 200 Hz, respectively. *Left*: APSs at onset, ~ 1 s and end of A-HFS. *Right*: Scatter plots of APS amplitudes from the corresponding A-HFS shown on the left. Blue and orange colors denote the APS_L and APS_S following the longer and shorter IPIs, respectively. **B** Comparisons of mean amplitudes between the initial APS and the mean APS at the final second of the A-HFS with seven different mean pulse frequencies. **C** Half-amplitude time of APS amplitudes versus mean pulse frequency. **D** Initial time of APS amplitude bifurcation versus mean pulse frequency. **E** Comparisons of normalized APS_L and APS_S amplitudes to their average at the final second of bi-IPI A-HFS at seven mean pulse frequencies. In B and E, $^{**}P < 0.001$, paired *t*-test. Modified from Wang et al. (2022)

Since clinical DBS typically uses pulse frequencies between 100 and 200 Hz—particularly around 130 Hz—we set the mean pulse frequency to 133 Hz (with a mean IPI of 7.5 ms) to determine the minimum ΔIPI needed to induce a bifurcation. Figure 6.8A shows examples of APS signals in different periods of bi-IPI A-HFS with ΔIPI of 0, 0.2, 1 and 3 ms, respectively. The scatter plots of APS amplitudes are shown on the right side of the figure. At ΔIPI = 0 ms (mono-IPI A-HFS as control), no bifurcate appeared in the APS amplitudes (Fig. 6.8A1). However, with the other three ΔIPIs, a bifurcation began shortly after A-HFS onset and persisted until the end of A-HFS (Fig. 6.8A2–A4). In the final second of 1 min A-HFS, the amplitude of APS_L following the longer IPI was significantly greater than the amplitude of APS_S following the shorter IPI (Fig. 6.8B). Especially, when the two IPIs were 7.4 and

7.6 ms with a ΔIPI of only 0.2 ms and a ratio ΔIPI/mean IPI = 2.7%, about 20% of neuronal firing transferred from APS_S to APS_L. When ΔIPI increased to 3 ms or above (ΔIPI/mean IPI ≥ 40%), APS_S almost disappeared (Fig. 6.8A and B).

With similar APS amplitudes ($APS_S/APS_L \approx 1.0$) at the onset of A-HFS, the APS_S/APS_L ratios fell significantly below 1 in the final second of the A-HFS with the non-zero ΔIPI values (0.2, 1, 3 and 5 ms). The APS_S/APS_L ratios decreased exponentially as ΔIPI increased, with a determination coefficient $R^2 = 0.99$ (Fig. 6.8C). Additionally, in the final second of A-HFS with different ΔIPI, the mean APS amplitudes (sum of APS amplitudes/pulse number) showed no significant differences (Fig. 6.8D). This indicated that at a same mean frequency of 133 Hz, these A-HFS trains generated similar levels of neuronal firing in the steady period. However,

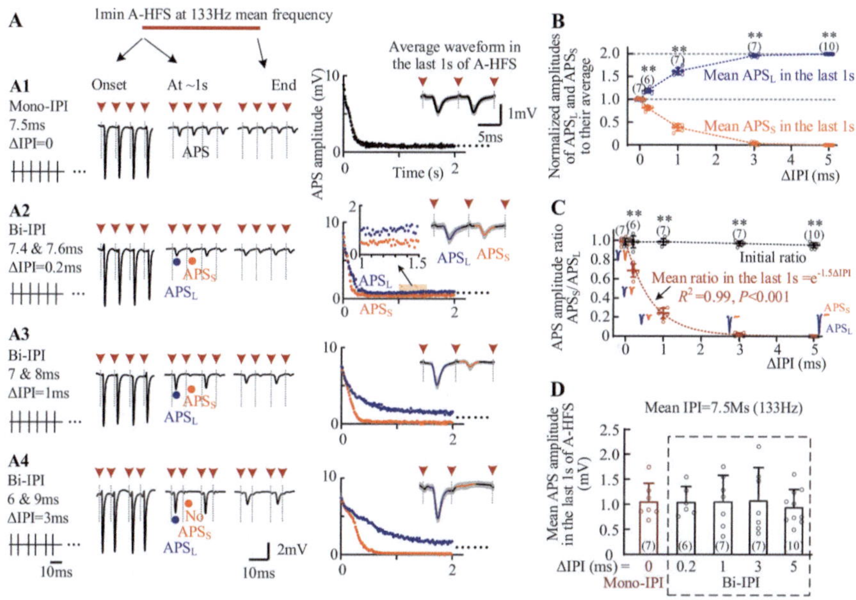

Fig. 6.8 Bifurcations in APS amplitudes produced by bi-IPI A-HFS with different ΔIPI at a fixed 133 Hz mean pulse frequency. **A** Examples of APS signals during 1 min A-HFS trains with mono-IPI (ΔIPI = 0) and bi-IPI (ΔIPI = 0.2, 1 and 3 ms). Left: APSs at onset, ~ 1 s and the end of A-HFS. Right: Scatter plots of APS amplitudes in the initial 2 s A-HFS period. The insets in the upper right show superposed APS waveforms in the final second of A-HFS. Blue and orange colors denote APS_L and APS_S following the longer and shorter IPIs, respectively. **B** Comparisons of the normalized APS_L and APS_S to their average in the final second of 1 min A-HFS with ΔIPI = 0, 0.2, 1, 3 and 5 ms, respectively. $^{**}P < 0.001$, paired *t*-test. **C** Comparisons of amplitude ratio APS_S/APS_L between the initial value and the mean value in the final second of A-HFS under different ΔIPIs. $^{**}P < 0.001$, paired *t*-test. The exponential fitting curve of the mean values in the final second is shown with the fitting equation and the determination coefficient (R^2). **D** Comparison of mean APS amplitudes in the final A-HFS second among mono-IPI and different bi-IPI values (ΔIPI = 0.2, 1, 3 and 5 ms). From Wang et al. (2022)

the temporal structure of the pulse trains (the different IPIs) altered the timing of neuronal responses by redistributing firing between IPI pairs.

Furthermore, when the order of the IPI pairs in an A-HFS train switches from shorter-longer to longer-shorter, the APS amplitudes can quickly adapt to the change. Figure 6.9 shows an example of this change in a bi-IPI A-HFS with a 133 Hz mean frequency at 0.3 mA. Initially, the A-HFS ran with IPI pairs of 7.4 ms followed by 7.6 ms for 30 s. During the steady period, the APS amplitude after the longer IPI (7.6 ms) was approximately double that following the shorter IPI (7.4 ms). At 30 s, the order of IPI pairs switched to 7.6 ms first then 7.4 ms, resulting in two consecutive 7.6 ms IPIs at the switching point. Following the second 7.6 ms IPI (marked by the red arrow in the figure), the induced APS was smaller, indicating a delay in neuronal response to the IPI change. Within several IPIs, the neuronal responses adjusted to produce a larger APS following the longer IPI. The scatter plot shows that the APS amplitudes following odd pulses (blue dots) and even pulses (orange dots) exhibit an "X" pattern around the IPI switch, as one APS group decreasing while the other increasing.

These results indicate that at a sufficiently high mean frequency, even a sub-millisecond difference in IPI can lead to a bifurcation in the neuronal population responses to bi-IPI A-HFS. Besides IPI, we hypothesized that small changes in other

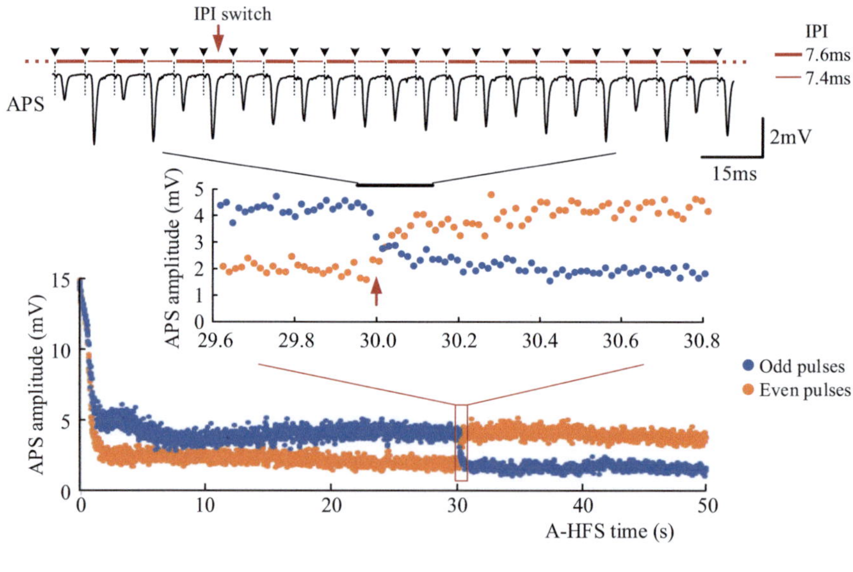

Fig. 6.9 Changes in APS signal after switching between longer and shorter IPI during the steady period of A-HFS with bi-IPI. *Top*: APS signal around the switching point indicated by the red arrow. Dashed lines with black arrows denote removed pulse artifacts. Thick and thin red bars between pulses indicate longer and shorter IPIs respectively, with only a 0.2 ms difference. *Middle*: Enlarged scatter plot of APS amplitudes around the switching point. *Bottom*: Scatter plot of APS amplitudes during 50 s A-HFS, where blue and orange dots represent the APS induced by odd and even pulses, respectively

A-HFS parameters might also generate similar bifurcations in neuronal responses. Therefore, we investigated the neuronal responses to bi-intensity A-HFS.

6.2.2 Bifurcation Responses of Neuronal Population to A-HFS with Bi-Intensity

We set three bi-intensity pairs with small differences: 0.10 and 0.11 mA, 0.30 and 0.31 mA and 0.60 and 0.62 mA. In baseline single-pulse stimulations at these intensities, the APS amplitude increased with pulse intensity (Fig. 6.10A). In separated paired-pulse stimulations (IPI = 7.5 ms) at these bi-intensity pairs (Fig. 6.10B), the weaker intensity pair (0.10 and 0.11 mA) produced an APS_2 amplitude that was slightly greater than the APS_1 amplitude, resulting in an APS_2/APS_1 ratio above one (Fig. 6.10C). With the other two stronger bi-intensity pairs, the APS_2/APS_1 ratios were both below one, indicating a slight inhibitory effect of APS_1 on APS_2 at the 7.5 ms IPI.

We created 1 min A-HFS trains respectively using the three bi-intensity sets to form stimulations with alternating intensities at a constant 133 Hz (IPI = 7.5 ms). As shown in Fig. 6.10D, the APS amplitudes decreased rapidly during initial A-HFS periods, with stronger bi-intensity causing faster decreases. The half-amplitude times of APS attenuation differed significantly among stimulations with different bi-intensity sets (Fig. 6.10E). As APS amplitude decreased, a bifurcation appeared during all the bi-intensity A-HFS trains—smaller APSs followed weaker pulses while larger APSs followed stronger pulses (Fig. 6.10D). This bifurcation response contrasted with both the baseline situation and the initial response to the first pulse pair at A-HFS onset, where the weaker pulse induced a larger APS at the stronger bi-intensities of 0.30 and 0.31 mA and 0.6 and 0.62 mA. When we swapped the order of two intensities in bi-intensity A-HFS, the APS bifurcation still appeared, maintaining larger APS following stronger pulses. The APS_2/APS_1 amplitude ratio increased from ~ 1 at onset to ~ 3 during the final second of A-HFS (Fig. 6.10F), reaching nearly triple the intensity ratios of 1.1 (0.11/0.10) and 1.03 (0.31/0.30 and 0.62/0.60). This showed that a mere 3–10% difference in pulse intensities produced a $\sim 200\%$ difference in APS amplitudes, indicating nonlinear neuronal responses to the intensity changes.

The results demonstrate that while small differences in bi-intensity A-HFS cannot produce substantial changes in baseline neuronal responses, they can generate substantial bifurcation in neuronal responses during sustained stimulations, similar to bi-IPI A-HFS.

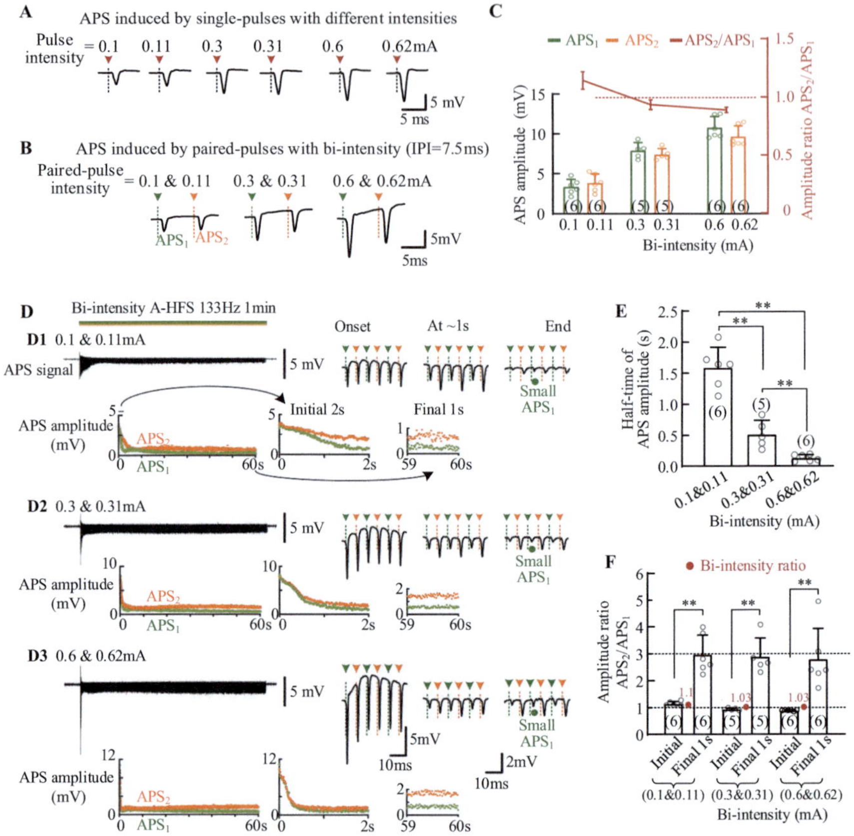

Fig. 6.10 Bifurcation produced by bi-intensity A-HFS at 133 Hz. **A** APS waveforms induced by single-pulses at distinct intensities of 0.10, 0.11, 0.30, 0.31, 0.60 and 0.62 mA. **B** APS waveforms induced by paired-pulses (IPI = 7.5 ms) with three bi-intensity sets of 0.10 and 0.11 mA, 0.30 and 0.31 mA, and 0.60 and 0.62 mA. **C** Comparison of paired-pulse-induced APS_1 and APS_2 amplitudes and their ratio APS_2/APS_1. **D** Typical APS signals produced by 1 min bi-intensity A-HFS with three different paired-intensities. In **D1**, **D2** and **D3**, *Top*: APS signals with enlarged views at three different A-HFS periods. *Bottom*: scatter plots of APS amplitudes during the entire 1 min A-HFS with expanded insets showing the initial 2 s and final 1 s periods. Green and orange colors denote the APS_1 following weaker pulses and the APS_2 following stronger pulses, respectively. **E** Comparisons of the half-amplitude time of APS attenuation during 1 min A-HFS with the three bi-intensity sets. $^{**}P < 0.001$, one-way ANOVA with Post hoc Bonferroni tests. **F** Comparisons of amplitude ratios (APS_2/APS_1) between the initial and final A-HFS seconds at the three bi-intensity sets. $^{**}P < 0.001$, paired *t*-test. Red dots represent the bi-intensity ratios (1.1, 1.03, and 1.03). From Wang et al. (2022)

6.2.3 Different Neuronal States Underlying Similar Evoked APS Sequences

As described above, during sustained A-HFS with either bi-IPI or bi-intensity (referred to as bi-pulse A-HFS), APS bifurcations caused neurons to fire mostly following every other pulse. This produced an APS sequence similar to the pattern induced by mono-IPI or mono-intensity A-HFS (referred to as mono-pulse A-HFS) at half frequency. It raised a question: what were the effects of those pulses that rarely produced APS in bi-pulse A-HFS? To address this question, we next compared neuronal excitability between bi-pulse A-HFS and mono-pulse A-HFS at half frequency by inserting test pulses into the A-HFS trains.

As shown in Fig. 6.11A–C, we compared the A-HFS of bi-IPI (5 and 10 ms) with the A-HFS of mono-IPI (15 ms). During the steady period of the 133 Hz bi-IPI A-HFS, only pulses following the longer IPI (10 ms) induced an APS, while pulses following the shorter IPI (5 ms) induced no obvious APS (Fig. 6.11A *upper*). The mean APS amplitude following the longer IPI was similar to that induced by mono-IPI A-HFS pulses at 66.7 Hz (Fig. 6.11A and B). The test pulse inserted in the middle of the longer IPI in bi-IPI A-HFS induced an APS that was much smaller than the corresponding APS induced in mono-IPI A-HFS (Fig. 6.11A). Although the preceding APSs were similar ($A_{\text{Bi}}/A_{\text{Mono}} \approx 100\%$), the mean APS amplitude induced by the test pulses during bi-IPI A-HFS ($A_{\text{Bi-test}}$) was only $\sim$50% of those during mono-IPI A-HFS ($A_{\text{Mono-test}}$). Across all six insertions spaced 9 s apart during 1 min A-HFS, the $A_{\text{Bi-test}}/A_{\text{Mono-test}}$ ratios were significantly smaller than the $A_{\text{Bi}}/A_{\text{Mono}}$ ratios (Fig. 6.11C). This indicates that neurons had lower excitability during pulse intervals of higher-frequency bi-IPI A-HFS.

Similarly, we compared bi-intensity 133 Hz A-HFS (at an average pulse intensity of 0.3 mA) with mono-intensity 66.7 Hz A-HFS at 0.3 mA (Fig. 6.11D–F) using test pulses. Both A-HFS modes had mono-IPI of 7.5 and 15 ms respectively, with a test pulse inserted every 9 s. In the bi-intensity (0.29 and 0.31 mA) A-HFS, the test pulse was inserted at 11.5 ms after the stronger pulse (i.e., at the middle between 0.29 and 0.31 mA pulses). In the mono-intensity A-HFS, the test pulse was inserted at the same timing of 11.5 ms after an A-HFS pulse (Fig. 6.11D). The mean APS amplitude induced by weaker pulses in bi-intensity A-HFS was significantly smaller than those induced both by stronger pulses and in mono-intensity A-HFS (Fig. 6.11E). Test pulses in bi-intensity A-HFS hardly induced APS ($A_{\text{Bi-test}}$). Across all six insertions during 1 min A-HFS, the APS ratios of $A_{\text{Bi-test}}/A_{\text{Mono-test}} \approx 10\%$ were significantly smaller than the preceding APS ratios of $A_{\text{Bi}}/A_{\text{Mono}} \approx 80\%$ (Fig. 6.11F). Again, this result confirms that neuronal excitability during the pulse interval was lower in higher-frequency A-HFS.

The experiments shown in Fig. 6.11 indicate that the neuronal excitability in the pulse interval can weaken during bi-pulse A-HFS trains with twice the pulse number, even though their APS sequences are similar to mono-pulse A-HFS trains. This means that the additional pulses in the higher-frequency A-HFS create a deeper axonal block, despite similar firing levels. This finding is consistent with the results

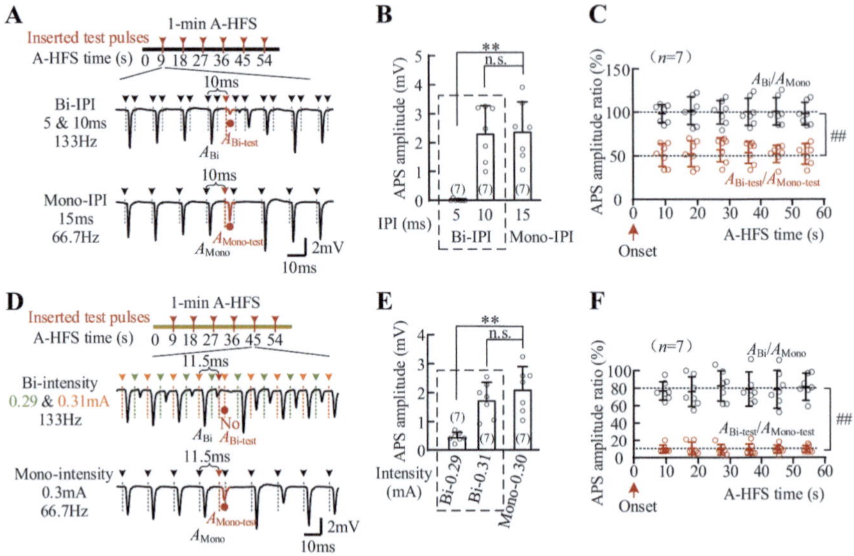

Fig. 6.11 Comparisons of neuronal responses to mono-pulse A-HFS at 66.7 Hz and bi-pulse A-HFS with a mean frequency of 133 Hz. **A** Typical APS signal during the steady periods of bi-IPI and mono-IPI A-HFS with additional test pulses inserted every 9 s (denoted by red arrows). Red dots denote the APSs induced by the test pulses. **B** Comparisons of mean APS amplitudes in the final 1 s of bi-IPI and mono-IPI A-HFS trains. **C** APS amplitude ratios between bi-IPI and mono-IPI A-HFS immediately before (in black) and following (in red) the six inserted test pulses. **D** Typical APS signal during the steady period of bi-intensity and mono-intensity A-HFS with additional test pulses inserted every 9 s. Green and orange arrows indicate the weaker pulses (0.29 mA) and the stronger pulses (0.31 mA), respectively. **E** Comparisons of mean APS amplitudes in the final 1 s of bi-intensity and mono-intensity A-HFS. **F** APS amplitude ratios between bi-intensity and mono-intensity A-HFS immediately before and following the inserted test pulses. In **B** and **E**, $^{**}P < 0.001$, "n.s." $P > 0.05$, one-way ANOVA with Post hoc Bonferroni tests. In **C** and **F**, $^{\#\#}P < 0.001$, paired *t*-test. From Wang et al. (2022)

showing similar neuronal firing levels during steady periods of A-HFS at different frequencies ranging in 50–200 Hz, as described in Sect. 5.2.2 (see Fig. 5.7G).

6.2.4 *Possible Mechanisms of Neuronal Bifurcation Responses to A-HFS*

Bifurcation refers to a radical change in the rate and/or timing pattern of neuronal firing caused by small alterations in external inputs when nonlinear neurons are unstable near a bifurcation point. While previous studies focused mainly on bifurcations in individual neurons, our study explored this phenomenon in neuronal populations. We showed that small changes in both the IPI and intensity of bi-pulse A-HFS can lead to bifurcation in population-level neuronal responses. Additionally, our study was conducted in the intact brain in-vivo, making the results more clinically relevant than in-vitro studies. The mechanism of intermittent axonal block produced by A-HFS suggests that the stimulation-induced depolarization may bring the axonal membrane near a bifurcation point, where sodium and potassium channels can change more rapidly and sensitively than at rest (Hodgkin and Huxley 1952; Kocsis et al. 1983; Zheng et al. 2020). Consequently, tiny changes in stimulation excitatory inputs from varying parameters—such as IPI and pulse intensity—can result in neuronal firing bifurcation, whereas such small changes cannot make substantial difference under baseline conditions.

In addition, a single pulse applied extracellularly at the alveus of hippocampal CA1 region can activate a bundle of axons simultaneously. The coupling between adjacent axons may also contribute to firing bifurcation in neuronal populations. Previous studies have shown that electrical coupling between axons in the hippocampal region can lead to extremely rapid synchronous activity in neuronal populations (Blankenship and Feller 2010; Molchanova et al. 2016; Schmitz et al. 2001). During sustained A-HFS, this coupling mechanism may cause axonal firing to increasingly follow longer IPI or stronger pulses due to their facilitation effects. This firing convergence can amplify the firing difference between pulses with slightly different parameters, resulting in bifurcation.

In summary, with the combined effects of intermittent depolarization block and axonal coupling, even small changes can alter the firing patterns of neuronal populations during sustained bi-pulse A-HFS. Additionally, although the pulses with weaker parameters in bi-pulse A-HFS cannot directly produce substantial neuronal firing, they can maintain the depolarization block on neuronal membrane, preventing neurons from responding to additional activations between pulses (see Fig. 6.11).

In the experiments with inserted test pulses shown in Fig. 6.11A and D, we observed that a pulse insertion produced a series of firing changes in the subsequent A-HFS period that differed from the pre-insertion firing. This implied that a single pulse insertion could generate substantial changes in neuronal responses well beyond the immediate pulses. During high-frequency stimulations, axons may behave like taut strings—a strike can produce a period of vibrations. This led us to investigate the effects of deleting and inserting pulses during HFS with constant parameters as shown below.

6.3 Effects of Pulse Deletion and Insertion During Constant-IPI HFS

6.3.1 *Changes in Evoked APS by Pulse Deletion and Insertion During A-HFS*

During the steady period of 100 Hz A-HFS, we deleted a pulse to extend a 10 ms IPI to a 20 ms gap (Fig. 6.12A1). We labeled the APS evoked by the first A-HFS pulse as APS_{init}, the APS evoked by the pulse preceding the gap (P_0) as APS_0, and the APS that would have been evoked by the deleted pulse (P_1) as APS_1. For the pulses following the gap, we labeled their evoked APSs sequentially as APS_2, APS_3, …, APS_k, with their corresponding pulses labeled as P_k. The APS_1 amplitude was zero due to the absence of P_1 (Fig. 6.12A1 and A2). The 20 ms gap caused the APS_2 amplitude to increase significantly—reaching approximately 3 times the APS_0 amplitude and half of the APS_{init} amplitude (Fig. 6.12A3). However, the following APS_3 was small again. Subsequently, the evoked APSs alternated between large and small amplitudes for a period (Fig. 6.12A2, *top*).

The increase in APS_2 amplitude resulted from more neuronal recovery in the prolonged gap interval (refer to Sect. 5.2.2). By this logic, shortening the interval by inserting a pulse should decrease the subsequent APS. Indeed, inserting a pulse in the middle of 10 ms IPI hardly produced an APS (APS_1) following the subsequent pulse P_1 (Fig. 6.12B1 and B2). The inserted pulse itself neither produced an APS. However, the APS_2—produced by the second pulse P_2 after the insertion—increased significantly to ~ 2 times the APS_0 amplitude and$\sim$1/3 of APS_{init} amplitude (Fig. 6.12B3). Subsequently, an oscillation between larger and smaller APSs appeared (Fig. 6.12B2, *top*), resulting in an APS pattern similar to that produced by the opposite manipulation of deleting a pulse—an unexpected result.

After normalization by APS_0 amplitude, the mean amplitudes of APS_1 through APS_8 following the deleted or inserted pulse showed clear changes: significant decreases in APS_1, APS_3, APS_5 and APS_7, alongside significant increases in APS_2, APS_4, APS_6 and APS_8 (Fig. 6.12C1 and C2). Additionally, the mean normalized APS_2 amplitude was significantly greater following the deleted pulse (3.0 ± 0.38, $n = 7$) than following the inserted pulse (2.0 ± 0.22, $n = 7$; *t*-test, $P < 0.001$). The APS latencies showed no substantial changes following both the deleted and inserted pulse (Fig. 6.12D). These results showed that during the steady period of 100 Hz A-HFS, both shortening the IPI by inserting an pulse and extending it by deleting a pulse can produce similar oscillations in the subsequent APSs.

To investigate how the timing of inserted pulse affects responses, we inserted pulses at nine different time points during the steady period of late 60 s in 2 min A-HFS. The pulses were inserted at 1, 2, … and 9 ms (denoted as $\Delta t = 1, 2, \ldots,$ 9 ms) within the 10 ms IPI. We performed these nine insertions in each A-HFS in random order, spacing them 5 s apart (Fig. 6.13A). At $\Delta t \leq 7$ ms, the inserted pulse hardly produced APS, while the APS_1 evoked by the subsequent pulse (P_1) gradually

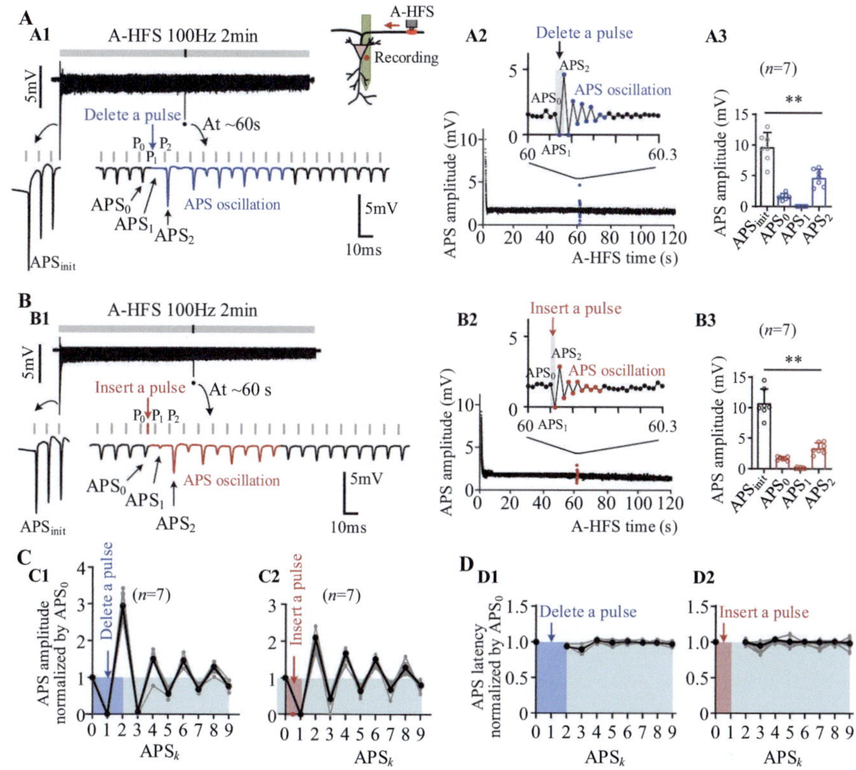

Fig. 6.12 APS changes produced by deleting and inserting a pulse in the steady period of A-HFS. **A A1**: Typical recording of 100 Hz A-HFS with a pulse deletion. The deletion and subsequent APS oscillation are highlighted in blue in the enlarged view. **A2**: Scatter plot of APS amplitudes during the A-HFS shown in **A1**, with an enlarged view around the deletion. **A3**: Amplitude comparison among four APS waveforms: APS_{init}, APS_0, APS_1 and APS_2. **B B1–B3**: APS changes produced by a pulse insertion at the midpoint of IPI in the steady period of 100 Hz A-HFS. The insertion and subsequent APS oscillation are highlighted in red. **C** Changes in APS_k amplitudes (normalized by APS_0) following deletion (**C1**) and insertion (**C2**). Black dots represent the mean of individual experiments in grey dots ($n = 7$). **D** Changes in APS_k latencies (normalized by APS_0) following deletion (**D1**) and insertion (**D2**). APS_1 latency was absent because no APS_1 occurred. In **A3** and **B3**, $^{**}P < 0.001$, one-way repeated measures ANOVA with post hoc Bonferroni tests. Modified from Hu et al. (2023)

decreased to a disappearance as Δt increased. At $\Delta t = 8$ and 9 ms, the inserted pulse produced an APS (also denoted as APS_1), but P_1 produced no APS.

When the inserted pulse occurred around the middle of 10 ms IPI (e.g. $\Delta t =$ 6 ms), the oscillations of subsequent APSs were greater. As shown in the bottom of Fig. 6.13B, two indices were used to evaluate the oscillation in the normalized APS amplitudes. The first was oscillation magnitude, defined as the difference between the maximum and minimum APS following the insertion, i.e., the difference between the normalized amplitude of APS_2 and APS_1. The second was oscillation duration,

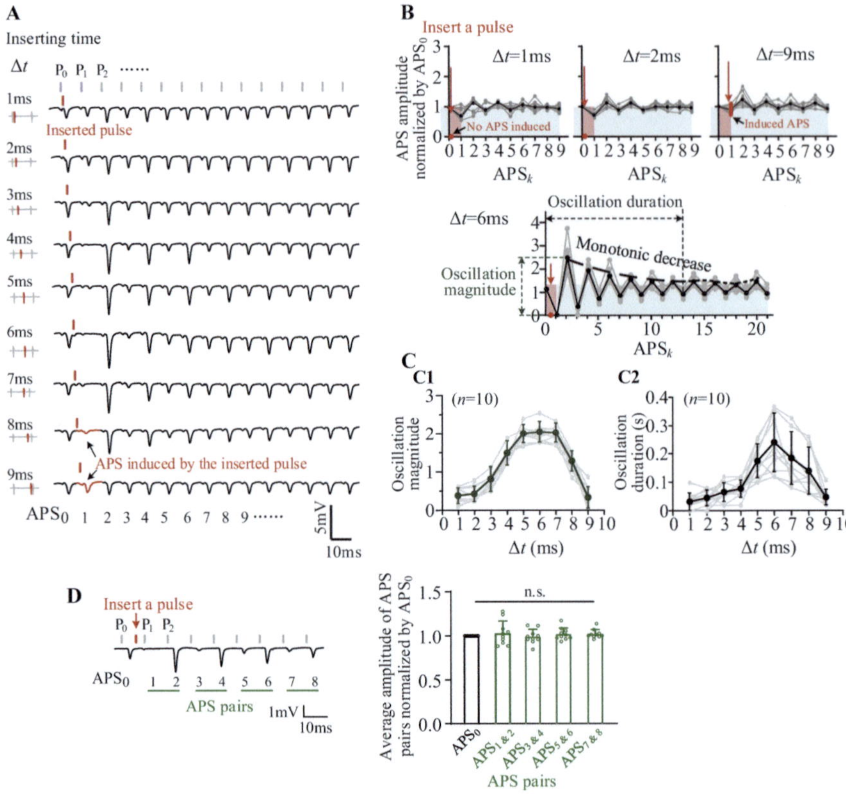

Fig. 6.13 Changes in APS amplitudes induced by pulse insertion at various timing points in IPI. **A** Typical recordings of evoked APS around pulse insertions at timings of $\Delta t = 1, 2 \ldots 9$ ms during the steady period of 100 Hz A-HFS. Red short bars denote the inserted pulses. **B** APS amplitudes normalized by APS_0 following example insertions at $\Delta t = 1$, 2, 9 and 6 ms. Black dots represent mean values of grey dots ($n = 10$). The definitions of magnitude and duration of APS oscillations are shown in the inset of $\Delta t = 6$ ms. **C** The magnitudes (**C1**) and durations (**C2**) of APS oscillations change with the insertion timing Δt. Thick dots denote the mean values of grey dots ($n = 10$). **D** Average amplitudes of adjacent APS pairs following pulse insertion at $\Delta t = 6$ ms were similar to the APS_0 amplitude ($n = 10$). "n.s.", $P > 0.05$, one-way repeated measures ANOVA. Modified from Hu et al. (2023)

defined as the period during which the larger APSs (those with even subscripts) monotonically decreased. Both indices changed non-monotonically with insertion time Δt, showing a pattern of slower increase followed by rapider decrease (Fig. 6.13C). The indices peaked around $\Delta t = 6$ ms. At $\Delta t = 1$ and 9 ms, both the mean oscillation magnitudes and durations were minimal. However, at $\Delta t = 1$ ms, the APS_1 was produced by the A-HFS pulse, whereas at $\Delta t = 9$ ms, it was produced by the inserted pulse.

The APS oscillation produced by the inserted pulse suggested a redistribution of neuronal firing. For example, when inserting a pulse at $\Delta t = 6$ ms, the APS_k alternated between smaller and larger amplitudes, yet the average amplitudes of adjacent APS pairs remained similar to the APS_0 amplitude (Fig. 6.13D). This pattern indicated that the insertion did not significantly change the total neuronal firing but shifted the firing from APS_1 to APS_2, thereby creating a damped oscillation in the subsequent APSs.

To investigate whether more neuronal firing could be postponed through repeated insertions, we added four pulses respectively in four successive IPIs at $\Delta t = 6$ ms, and applied this four-pulse insertion at 60, 80 and 100 s of A-HFS repeatedly (Fig. 6.14A, *top*). The inserted pulses were labeled as p_1, p_2, p_3 and p_4. Neither p_1 nor its immediately subsequent A-HFS pulse produced an APS, which was consistent with the results of single-pulse insertion. The inserted pulses p2, p3 and p4 produced small APSs (shown as red dots in Fig. 6.14A *bottom*), while none of the four A-HFS pulses following the inserted pulses produced any obvious APS (shown as green dots). After the insertion completed, the APS sequence ($APS_{1,\ 2\ \ldots\ldots\ k}$) showed temporary amplitude oscillations, similar to the pattern observed with single-pulse insertion (Fig. 6.14B). Both the mean magnitude and duration of the APS oscillations after the four-pulse insertion showed no significant difference from those following single-pulse insertion (Fig. 6.14C).

The mean duration from the beginning of the four-pulse insertion to the end of APS oscillation was approximately 0.24 s, including a 0.04 s insertion period and a 0.20 ± 0.07 s oscillation period (Fig. 6.14C2). The corresponding duration for the single-pulse insertion was approximately 0.25 s, including a 0.01 s insertion period and a 0.24 ± 0.10 s oscillation period. To compare neuronal firing levels, we calculated the sums of APS amplitudes (normalized by APS_0) over a 0.25 s period for both single- and four-pulse insertions, as well as for the steady period of A-HFS immediately before the insertions. The comparison showed no significant difference in APS amplitude sums among these three periods (Fig. 6.14D).

These results showed that pulse insertion only temporarily redistributed evoked neuronal firing without changing the total firing amount during the transient phases. In addition, multiple pulse insertions in successive IPIs neither postponed more neuronal firing nor produced greater APS oscillation than single-pulse insertion.

6.3.2 *Changes in Evoked OPS by Pulse Insertion During O-HFS*

The above results of antidromically evoked APS were generated by A-HFS applied directly to the axons of recorded CA1 pyramidal neurons without involving synaptic transmissions. We also investigated the effects of the same pulse insertions in the O-HFS at the Schaffer collaterals to orthodromically activate the post-synaptic CA1 pyramidal neurons through monosynaptic transmissions (Fig. 6.15A).

We inserted single pulses at 60, 80 and 100 s during the steady period of 100 Hz O-HFS at $\Delta t = 6$ ms in IPIs by using the identical pulse sequence as A-HFS

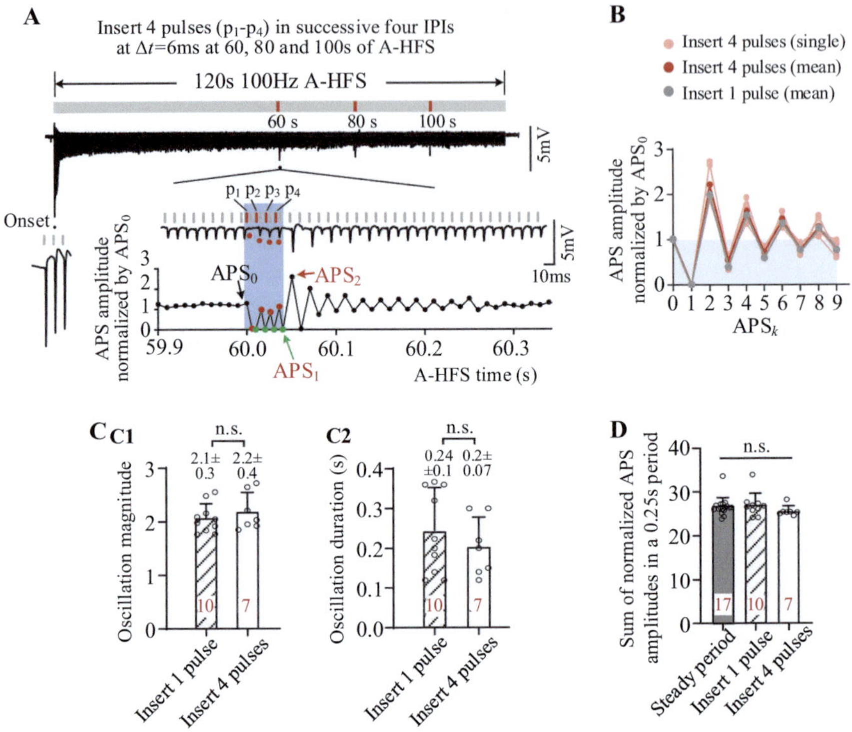

Fig. 6.14 Comparisons of evoked APSs following single-pulse insertion versus four-pulse insertion. **A** Top: Typical recording of 100 Hz A-HFS with three four-pulse insertions at 60, 80 and 100 s. The enlarged insets show APS waveforms at onset and around the inserted pulses (p_1 to p_4) at 60 s, along with normalized APS amplitudes. **B** Comparison of normalized APS amplitudes following single-pulse versus four-pulse insertions. **C** Comparison of APS oscillation magnitude (**C1**) and duration (**C2**) between single-pulse and four-pulse insertions. "n.s.", $P > 0.05$, *t*-test. **D** Comparison of the sums of normalized APS amplitudes in a 0.25 s period among three conditions: steady A-HFS period immediately before insertion, single-pulse insertion period, and four-pulse insertion period. "n.s.", $P > 0.05$, one-way ANOVA. From Hu et al. (2023)

(Fig. 6.15A). The O-HFS pulses immediately following the insertion pulse were sequentially labeled as P_1, P_2, P_3, etc. The pulse insertion produced a large OPS, denoted as OPS_2 because it followed the second pulse (P_2). No more OPS appeared after OPS_2. The OPS_2 waveforms from the three insertions at 60, 80 and 100 s of O-HFS were superimposed and averaged. The mean amplitude of the averaged OPS_2 was significantly smaller than that of OPS_{init} induced by the first O-HFS pulse (Fig. 6.15B1). Similarly, in the responses of A-HFS with identical pulse insertion, the mean amplitude of APS_2 was significantly smaller than that of APS_{init} (Fig. 6.15B2). However, the amplitude ratio $OPS_2/OPS_{init} \approx 2/3$ was significantly greater than the corresponding ratio $APS_2/APS_{init} \approx 1/3$ (Fig. 6.15B3).

A notable observation concerned the timing of OPS_2 occurrence. The OPS_2 appeared following the second pulse (P_2) after the insertion, with a mean latency

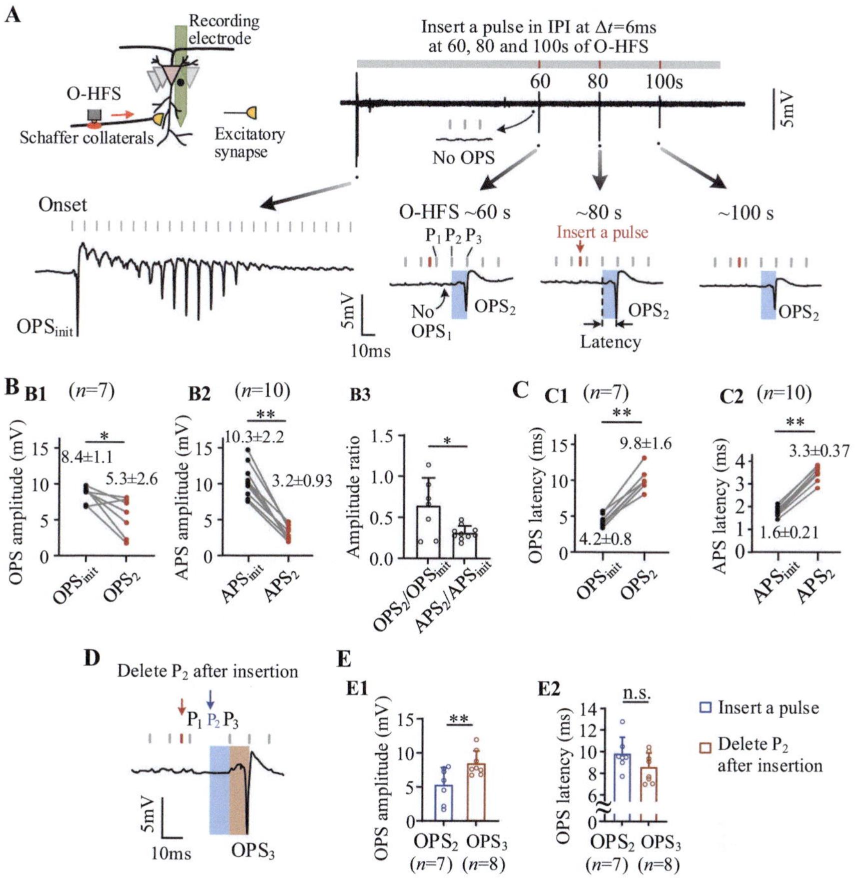

Fig. 6.15 Neuronal responses to single-pulse and four-pulse insertions during the steady period of 100 Hz O-HFS. **A** Typical recording of 2 min O-HFS with single-pulse insertions at 60, 80 and 100 s. The enlarged insets show the OPSs appeared in the initial period of O-HFS and the OPS_2 induced by the inserted pulses (denoted by the red bars). The blue shades highlight the IPIs where OPS_2 appeared. A schematic diagram of O-HFS experiment is illustrated in the upper left corner. **B** Comparisons between the amplitudes of OPS_{init} and OPS_2 (**B1**), between the amplitudes of APS_{init} and APS_2 (**B2**), as well as between OPS_2/OPS_{init} and APS_2/APS_{init} (**B3**). **C** Comparisons between the latencies of OPS_{init} and OPS_2 (**C1**) as well as between the latencies of APS_{init} and APS_2 (**C2**). **D** Typical neuronal response to the removal of P_2 after the single-pulse insertion. **E** Comparisons of mean amplitudes (**E1**) and latencies (**E2**) between the OPS_2 induced by merely a pulse insertion and the OPS_3 induced by a pulse insertion followed by the removal of P_2. n.s. $P > 0.05$, $^{*}P < 0.05$, and $^{**}P < 0.01$, *t*-test or paired *t*-test. Modified from Hu et al. (2023)

of 9.8 ± 1.6 ms measured from P_2, indicating that P_2, not P_1, produced the OPS_2. Additionally, the OPS_2 latency was significantly longer than the OPS_{init} latency of 4.2 ± 0.83 ms (Fig. 6.15C1), which was consistent with the prolonged latency of APSs in A-HFS with single-pulse insertions (Fig. 6.15C2).

To further verify that P_2 produced the OPS_2, we deleted P_2 from the O-HFS train. This resulted in no OPS appearing in the virtual interval between P_2 and P_3 (see the blue shade in Fig. 6.15D). However, a new OPS (labeled as OPS_3) appeared following P_3, with a mean amplitude significantly greater than the OPS_2 induced in the O-HFS with P_2 (Fig. 6.15E1). The mean OPS_3 latency was slightly shorter than that of OPS_2 (Fig. 6.15E2). This experiment confirmed that P_2—not the preceding pulses—produced OPS_2. These results were consistent with the A-HFS results, indicating that the inserted pulse delayed the P_1-induced activation and merged it into the P_2-induced activation. This increased activation generated OPS_2. When P_2 was deleted, the longer IPI allowed more recovery of axonal excitability, thereby resulting in a larger OPS_3.

To investigate the effects of multiple-pulse insertions, the same sequence used in A-HFS with four-pulse insertions was applied as O-HFS (Fig. 6.16A). No OPS occurred during the 40 ms insertion period (dashed boxes in Fig. 6.16A), while the second pulse (P_2) after the insertions induced a large OPS_2. The mean amplitude of OPS_2 was not significantly smaller than that of OPS_{init} (Fig. 6.16B1), while the mean latency of OPS_2 was significantly longer than that of OPS_{init} (Fig. 6.16B2).

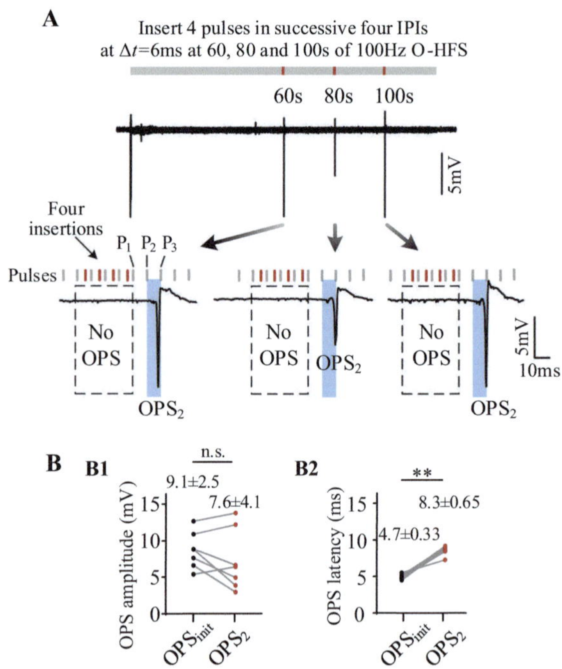

Fig. 6.16 Neuronal responses to four-pulse insertions in four successive IPIs during the steady period of 100 Hz O-HFS. **A** Typical recording of 120 s O-HFS with four-pulse insertions at 60, 80 and 100 s. **B** Comparisons of the amplitudes (**B1**) and latencies (**B2**) between the OPS_{init} induced by the first O-HFS pulse and the OPS_2 induced by the P_2 following four-pulse insertions. n.s. $P > 0.05$, $^{**}P < 0.001$, paired *t*-test, $n = 7$. From Hu et al. (2023)

These results showed that both single-pulse and four-pulse insertions produced large OPS during steady O-HFS period originally without obvious OPS. The OPS was induced by the second pulse (P_2) after the end of both types of insertions, with an extended latency.

6.3.3 *Possible Mechanisms of Neuronal Responses to Pulse Insertion and Deletion During HFS*

The experiment results of pulse deletion and insertion indicate that during steady HFS periods with constant IPI, altering a single pulse can change the neuronal response in a subsequent period, leading to larger population potentials. When axonal fibers undergo sustained HFS, they experience intermittent depolarization block with extended refractory periods. Under this situation, each axon can only fire action potentials in response to some HFS pulses (refer to Sect. 5.2). Through the antidromic pathway without involving synaptic transmissions, the intermittent activation can produce soma APSs with reduced amplitudes. Deleting a pulse to extend an IPI can provide additional time to allow more axons to recover from depolarization block, resulting in increased APS amplitude (i.e., the APS_2 in Fig. 6.12A).

Interestingly, inserting a pulse to shorten an IPI can produce a similar APS sequence including larger APSs (Fig. 6.12B, C). This result can also be explained by the mechanism of intermittent depolarization block. The activation of the inserted pulse can deepen the depolarization block in the membranes, causing the axon firing that would have followed pulse P_1 to be postponed until after pulse P_2, thereby generating an increased APS_2 (Fig. 6.12B). In addition, the increase of APS_2 correlated with the timing of pulse insertion (Fig. 6.13), with substantial increased APS appearing only when the insertion occurred around the middle of IPI. Presumably, when the additional pulse is inserted closer to the preceding A-HFS pulse (P_0 in Fig. 6.13A), it cannot affect the neuronal firing already triggered by P_0. When the pulse is inserted late near the next A-HFS pulse (P_1 in Fig. 6.13A), it can replace P_1 to trigger neuronal firing itself. Therefore, only a pulse inserted in the middle of IPI can optimally delay the recovery process of ongoing axonal block, postponing the neuronal firing from following P_1 to P_2. This firing redistribution can result in a damped oscillation of smaller and larger APSs (Fig. 6.13B). The unchanged total evoked firing, shown by the sum of APS amplitudes, supports the redistribution hypothesis for both single- and four-pulse insertions (Fig. 6.14D). In addition, the four-pulse insertion did not accumulate more firing to make a larger APS_2 than the single-pulse insertion, likely due to the released neuronal firing during the multi-pulse insertion period (see the small APSs induced by the inserted pulses p_2, p_3 and p4 in Fig. 6.14A).

Similarly, the redistribution of axonal firing through pulse insertions can occur during O-HFS at the Schaffer collaterals, which orthodromically activates the CA1 pyramidal neurons through monosynaptic transmissions. A large OPS_2 was induced by the second pulse (P_2) rather than the first pulse (P_1) after both single- and four-pulse insertions (Figs. 6.15A and 6.16A). This timing matched the appearance of APS_2 during A-HFS (Figs. 6.13 and 6.14). The parallel changes in both amplitudes and latencies of the OPS_2 and APS_2 further supported the hypothesis (Fig. 6.15B, C).

However, the activation pathway of O-HFS includes both axonal conduction and synaptic transmission (Fig. 6.15A). It may be argued that HFS-induced synaptic failures, such as transmitter depletion, could explain our observations, as suggested in previous studies (Anderson et. al. 2006; Iremonger et. al. 2006; Rosenbaum et. al. 2014). This would suggest that synaptic failures, rather than axonal blocks, can cause the absence of OPS events during sustained O-HFS. Thus, the large OPS_2 after the inserted pulse might result from enhanced synaptic transmission due to the additional pulse. However, if this were true, the large OPS should have appeared immediately after the inserted pulse or P_1, not after P_2. The temporal distance (> 20 ms) from the OPS_2 to the inserted pulse was too long to benefit from synaptic transmissions generated by the inserted pulse, because the baseline OPS latency (OPS_{init} latency, partly caused by the delay of monosynaptic transmission) was only about 5 ms (Figs. 6.15C1 and 6.16B2). In addition, in the experiments with P_2 deleted, the appearance of OPS_3 after pulse P_3 cannot be explained by synaptic failure correction by the inserted pulse, since the temporal distance (>30 ms) from OPS_3 to the inserted pulse was even longer (Fig. 6.15D). Instead, the deletion of P_2 can provide an additional recovery period for the HFS-induced axonal block. Meantime, the inserted pulse can postpone firing. Together, the effects of pulse insertion and P_2 deletion can result in an OPS_3 greater than OPS_2 (Fig. 6.15E1).

While the neuronal responses to pulse insertions during both A-HFS and O-HFS can be explained by intermittent axonal block, the synaptic transmission in the O-HFS pathway does create notable differences. First, the insertion-induced OPS_2 was greater than the APS_2 (Fig. 6.15B3). However, during the steady O-HFS period before insertion, there was no OPS, whereas small APS continued to follow each pulse during the steady A-HFS period. These differences can stem from the threshold effect of synaptic transmission (Spruston 2008), as shown by the S-shape amplitude curve of single-pulse induced OPS versus stimulus intensity (Fig. 2.7 in Sect. 2.3). After insertion, the axonal activation induced by P_2 in O-HFS should have reached a maximum (corresponding to APS_2 with the greatest increased amplitude during the insertion-induced APS oscillation in A-HFS). This P_2 activation exceeded the thresholds of many synaptic transmissions, thus inducing a large OPS_2 from absence. This threshold effect led to the more pronounced change in neuronal responses during O-HFS than during A-HFS. Additionally, the orthodromic activation from each afferent axon of Schaffer collaterals can innervate many post-synaptic neurons through its axonal terminations, while the antidromic activation from each efferent axon can activate only one neuronal soma. This difference also contributed to the larger amplitude of insertion-induced OPS_2 than APS_2.

Second, the oscillation in APSs (Figs. 6.12, 6.13 and 6.14) was absent in OPSs (Figs. 6.15 and 6.16). This absence can be attributed to both feedforward and feedback local inhibitory circuits (refer to Sect. 2.3). The O-HFS pulses and the pyramidal neuron firing can activate the inhibitory effects of these local circuits, thereby preventing subsequent OPSs after the large OPS_2. In contrast, local inhibitions hardly affect APS generation, which does not involve synaptic integrations.

Taken together, based on the mechanism of HFS-induced intermittent axonal block, inserting pulses can redistribute axonal firing, substantially changing neuronal responses during the steady period in both A-HFS and O-HFS. This redistribution can result in increased synchronized firing of neuronal populations. In addition, in the activation pathway of O-HFS, synaptic transmissions and synaptic integrations in local neuronal circuits can nonlinearly transform activations from the afferent axons through a threshold effect. As a result, the post-synaptic neuron firing in O-HFS differs from the firing directly induced in the same neurons by antidromic activations of A-HFS. Our study provides a potential way to design HFS patterns with varying IPIs by inserting pulses into conventional HFS with a constant frequency. Clinical trials and animal experiments have shown that pulse sequences with real-time varying IPIs can produce diverse stimulation effects, offering more options for neuromodulation therapies (Brocker et al. 2013; Akbar et al. 2016; Karamintziou et al. 2016; Grill 2018; Santos-Valencia et al. 2019; Okun et al. 2022). The next section will show our investigations into the stimulation effects of randomly varying IPIs.

6.4 Neuronal Responses to HFS with Random Pulse Intervals

The above pulse insertion experiments suggest that for instantaneous frequency changes (i.e., IPI changes), HFS-induced neuronal firing by each pulse is not necessarily proportional to its preceding IPI. Our experiments with randomly varying frequency verified this hypothesis in both A-HFS and O-HFS, as described below.

6.4.1 Responses of Neuronal Population to A-HFS with Randomly Varying IPIs

1. **A-HFS with Randomly Varying IPIs in a Larger Range**

As shown in Fig. 6.17A, we compared varying and constant frequency stimulations within an A-HFS train, which was consisted of three periods: a 50 s period of constant 100 Hz stimulation to allow neuronal response to reach steady state, a 10 s period of varying frequencies ranging from 20 to 600 Hz, and a final 20 s period returning to constant 100 Hz. The total A-HFS duration was 80 s. During the varying-frequency period, the mean pulse frequency was maintained at 100 Hz, with IPI ranging from 1.67 to 50 ms.

In the typical APS recording shown in Fig. 6.17A, during the initial 50 s period of 100 Hz constant A-HFS, the APS change was similar to those described previously. At onset, each pulse induced a large APS. Within seconds, the APS amplitude rapidly decreased and stabilized at about 20% of its initial amplitude by the end of the 50 s constant stimulation (Fig. 6.17B). During the subsequent 10 s period with varying frequency, the IPI randomly varied with a Poisson distribution ranging from 1.67 to 50 ms (Fig. 6.17C). The APS amplitude became unstable, fluctuating in a range of

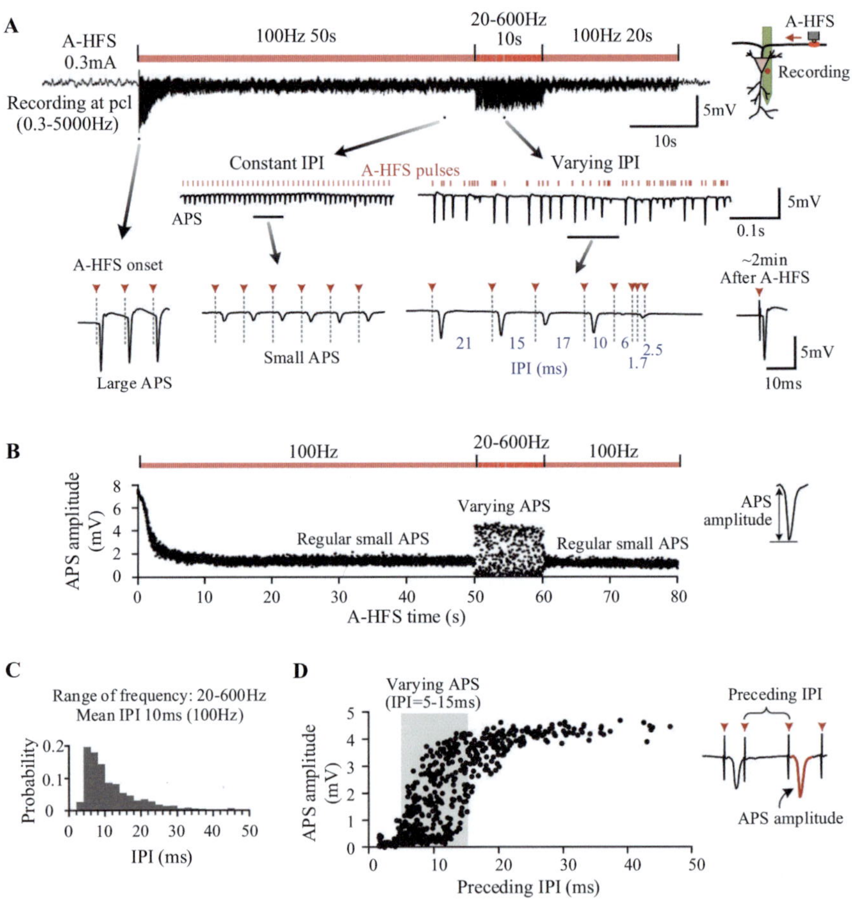

Fig. 6.17 Neuronal responses to A-HFS with randomly varying IPIs. **A** Typical APS signal during 80 s A-HFS with constant IPI (100 Hz, during 0–50 s and 60–80 s) and random IPI (20–600 Hz, during 50–60 s). **B** Scatter plot of the APS amplitudes during the entire A-HFS shown in (**A**). **C** Poisson distribution of the random IPIs (1.67–50 ms) with a mean IPI of 10 ms. **D** Scatter plot of APS amplitudes versus the immediately preceding IPI during the 10 s period with varying IPIs. Modified from Feng et al. (2019)

0–4.5 mV (Fig. 6.17B and D). Once the stimulation returned to constant frequency, the APS reverted to a stable small level. Approximately 2 min after the end of A-HFS, the neuronal responses returned to baseline level, as indicated by the recovered APS (shown in the lower right corner of Fig. 6.17A).

During the 10 s A-HFS period with random IPIs, despite the mean pulse frequency remaining at 100 Hz, the APS amplitudes varied markedly and showed an overall positive correlation with the preceding IPI (Fig. 6.17D). Larger APSs (> 4 mV) followed longer preceding IPIs (25–50 ms), while smaller APSs (< 1 mV) followed shorter preceding IPIs (1.7–5 ms). This pattern reflected that longer intervals facilitated larger APS generation. However, in the 5–15 ms IPI range, the APS amplitude varied in a range of 0–4.5 mV with no clear correlation to IPI (shadow area in Fig. 6.17D). A larger APS could follow a shorter IPI, or vice versa. This result suggested that even when all IPIs were short, slight variations in the IPIs could result in irregular firing in neuronal populations with some larger APSs. To test this hypothesis, we narrowed the IPI range to 5–10 ms (equivalent to 100–200 Hz). This produced markedly different results from the consistently small APSs seen in A-HFS with a constant IPI in the same range.

2. **A-HFS with Randomly Varying IPIs in a Smaller Range**

As shown in Fig. 6.18A, after a 50 s 100 Hz control stimulation period, when the APSs stabilized at a small amplitude, the A-HFS switched to a 10 s period with random IPIs distributed uniformly within 5–10 ms, resulting in a mean pulse frequency of 133 Hz. Notably, no IPI in the varying period exceeded the 10 ms IPI of the constant period. However, during this varying period, many APS amplitudes exceeded those observed during the constant period. This result differed from the larger APSs induced by longer IPIs during the steady A-HFS period with constant frequency (Fig. 5.7F). Additionally, many pulses during the varying period hardly induced APSs (Fig. 6.18A, B).

We compared the neuronal responses to A-HFS between constant and varying IPIs by analyzing APS amplitude distributions during two adjacent 10 s periods: the 40–50 s period with constant 10 ms IPI (Fig. 6.18C) and the 50–60 s period with randomly varying IPIs (Fig. 6.18D). With constant IPI, APS amplitudes showed an approximately normal distribution with a narrower range of 0.83–1.90 mV and a mean of 1.32 mV. In contrast, with varying IPIs, APS amplitudes followed a declining distribution with a wider range of 0–3.22 mV and a mean of 1.03 mV. Additionally, this declining distribution differed from the uniform distribution of varying IPIs (shown in the upper right of Fig. 6.18D), indicating a nonlinear relationship between neuronal responses and IPIs.

The mean frequency during the varying period was 133 Hz, higher than the constant 100 Hz in the control period. Increasing the control frequency to 200 Hz while keeping other parameters unchanged, we observed similar results (Fig. 6.18E–H). The higher control frequency resulted in a decrease in the steady APS amplitude with a mean of 0.64 mV and a narrower range of 0.36–1.15 mV during the 40–50 s constant period (Fig. 6.18G). However, during the 50–60 s period with randomly

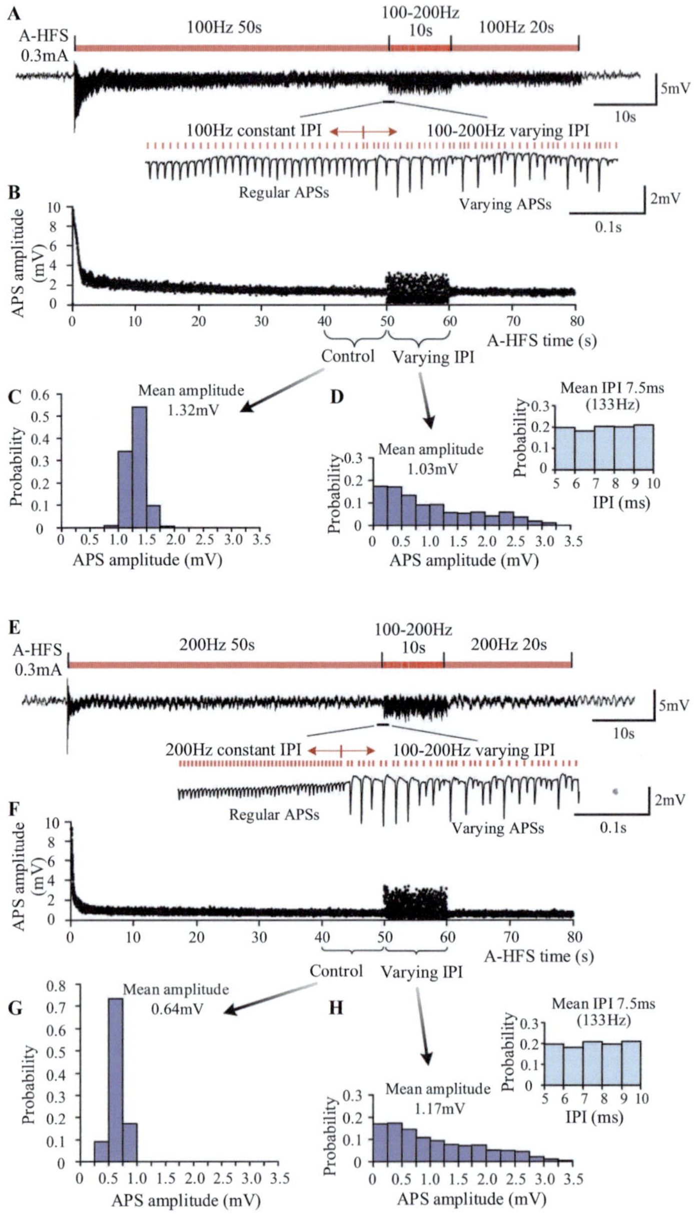

Fig. 6.18 Neuronal responses to A-HFS with randomly varying IPIs of 5–10 ms (100–200 Hz). **A** Typical APS signal during 80 s A-HFS with constant IPI (100 Hz, during 0–50 s and 60–80 s) and random IPI (100–200 Hz, during 50–60 s). **B** Scatter plot of the APS amplitudes during the entire A-HFS shown in (**A**). **C** and **D** APS amplitude distributions during the 10-s control period before varying IPIs (**C**) and the 10-s period with varying IPIs (**D**), with the plot of IPI distribution (upper right). **E**–**H** Corresponding plots as in **A**–**D** for A-HFS with identical stimulation paradigms shown in (**A**), but with increased frequency of 200 Hz in the constant IPI periods. From Feng et al. (2019)

varying IPIs (still ranging in 5–10 ms), the APS amplitudes increased to a wider range of 0–3.38 mV with a mean of 1.17 mV (Fig. 6.18H).

Statistical data showed that while the initial APS amplitudes induced by the first pulse of A-HFS were similar (Fig. 6.19A), the mean steady APS amplitude of 200 Hz A-HFS was significantly smaller than that of 100 Hz A-HFS during the control periods of both 40–50 s and 60–70 s (the periods immediately before and after the 10 s varying period). This difference occurred because higher-frequency constant A-HFS can suppress APS more by causing deeper axonal block. However, during the 10 s varying period inserted in the middle of 200 Hz control A-HFS, both the mean and the interquartile range of APS amplitudes were not significantly different from those with 100 Hz control A-HFS (Fig. 6.19B). This result indicated that neuronal responses to the varying IPIs were independent of the preceding APS suppression level.

Interestingly, despite the narrow range of varying IPIs in 5–10 ms (equivalent to 200–100 Hz), the amplitude ranges of APSs produced by these IPIs were far beyond the steady APS amplitudes produced by constant IPI at 100 or 200 Hz. When IPI varied randomly, the maximum APS amplitude was significantly greater than those during the preceding control periods of both 100 and 200 Hz A-HFS (Fig. 6.19C). Additionally, some pulses with varying IPIs failed to induce any APS (amplitude = 0), whereas pulses with constant IPI consistently induced APS (Fig. 6.18). As shown in Fig. 6.20A, the frequency of the constant periods was set to 133 Hz to match the mean frequency of the varying period. Using the APS area, a more accurate index reflecting the neuronal firing level (Theoret et al. 1984), the sum of APS areas per second during varying A-HFS was not significantly different from that during constant A-HFS (Fig. 6.20B). However, the interquartile range of APS area distribution was significantly larger during varying A-HFS than during constant A-HFS (Fig. 6.20C). These results showed that small random variations in IPIs generated APS amplitude changes that significantly exceeded the range of steady APS amplitudes caused by constant A-HFS. Moreover, the varying IPIs affected only the timing of neuronal firing without significantly altering the total firing level.

Note: unless otherwise specified in this book, the pulse intensity was set to induce APS or OPS at about 3/4 of the maximum amplitude, usually around 0.3 mA. At this unsaturated intensity, the amplitude of APS or OPS correlates linearly with its area (Theoret et al. 1984). Therefore, we often used the more intuitive measure of amplitude to compare the relative number of neurons involved in firing (i.e., firing level).

3. **Small IPI Variations in A-HFS Adjust Neuronal Firing Timing**

The timing of neuronal firing during varying A-HFS depended on both the preceding IPI and the preceding APS. As illustrated in Fig. 6.21A, we defined the following terms: current APS, preceding APS, current IPI, and preceding IPI. Using these definitions, we analyzed the relationship between APS amplitude and IPI length during varying A-HFS through scatter plots.

During the 10-s A-HFS period with randomly varying IPIs ranging in 5–10 ms, larger APSs (> 1.8 mV) only followed longer IPIs (7.5–10 ms) rather than shorter IPIs

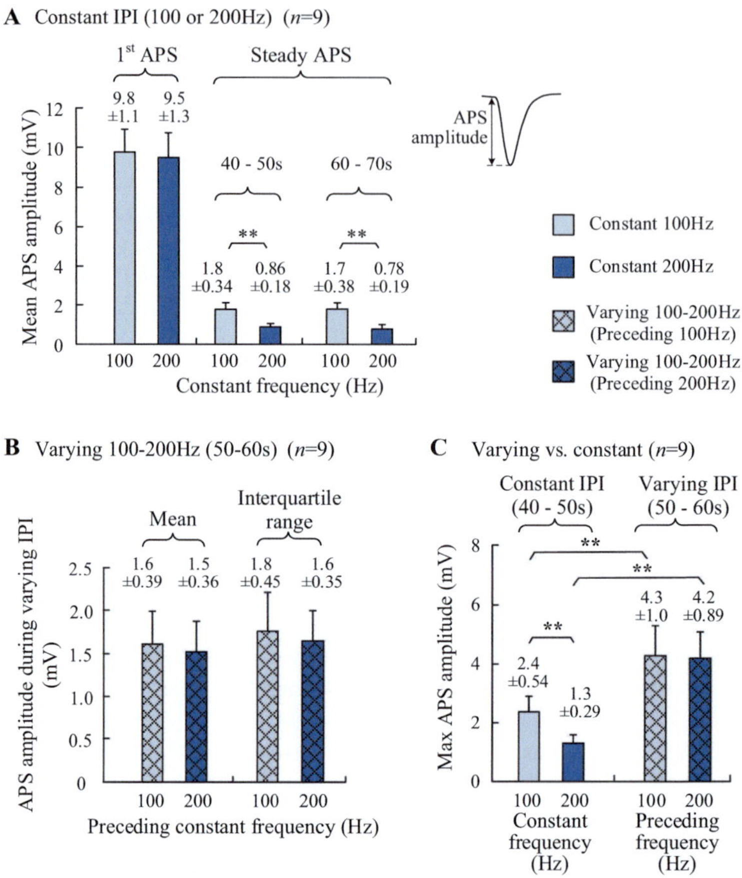

Fig. 6.19 Comparisons of APS amplitudes during different periods of A-HFS. **A** Amplitudes of the initial APS (induced by the first pulse at A-HFS onset) and the steady APS (mean amplitudes during 40–50 s and 60–70 s periods) between constant 100 and 200 Hz A-HFS. **B** Mean and the interquartile range of APS amplitudes during the 10 s periods (50–60 s) with randomly varying IPIs (100-200 Hz), preceded by constant periods at 100 and 200 Hz. **C** Maximum APS amplitudes during steady constant periods at 100 and 200 Hz, and during varying IPI periods preceded by constant periods at 100 and 200 Hz. $^{**}P < 0.01$, $n = 9$, t-test. Modified from Feng et al. (2019)

(5–7.5 ms, Fig. 6.21B). Although smaller APSs sometimes occurred after longer IPIs, they were often preceded by larger APSs (Fig. 6.21C and D). Even when two longer IPIs occurred consecutively in the uniform distribution of varying IPIs (Fig. 6.21E), two larger APSs never appeared consecutively (the shade area in Fig. 6.21C). This indicated that a larger preceding APS prevented a second larger APS from being immediately induced by the next pulse. In other words, the neuronal firing was history-dependent. A longer IPI was necessary but not sufficient for producing a larger APS, as some longer IPIs did not induce larger APSs. This occurred because the longest IPI was only 10 ms, not long enough. As shown in Fig. 6.17D, when the

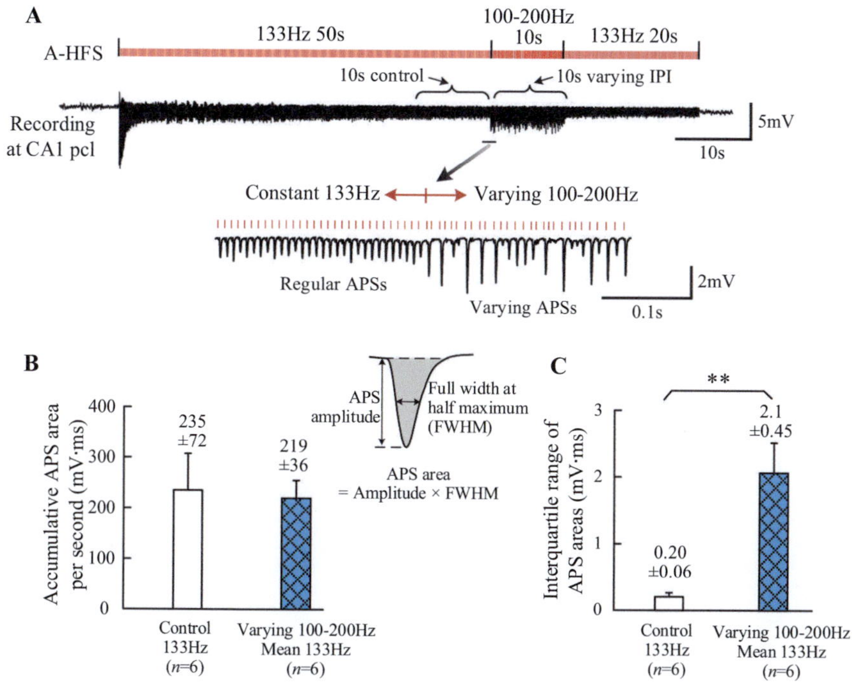

Fig. 6.20 Comparisons of neuronal firing between constant IPI period (control) and varying IPI period with a same mean frequency of 133 Hz. **A** Typical APS signal during 80 s A-HFS, consisting of constant IPI periods (133 Hz, during 0-50 s and 60–80 s) and varying IPI period (100–200 Hz, mean 133 Hz, during 50–60 s). **B** Comparison of cumulative APS areas between the control periods and the varying IPI periods ($P = 0.64$, $n = 6$, t-test). **C** Comparison of the interquartile ranges of APS areas between the two periods ($^{**}P < 0.01$, $n = 6$, t-test). From Feng et al. (2019)

IPI was sufficiently long (e.g., > 20 ms), the pulse following the long IPI consistently induced a larger APS.

Note: The range of varying IPIs was set at 5–10 ms, which exceeded the latencies of APSs induced during A-HFS (see Fig. 5.2). Therefore, each APS was produced by its immediately preceding pulse. Additionally, the APS amplitude was independent of the current IPI (Fig. 6.21F), since the APS was induced by the preceding pulse rather than the succeeding one.

These results indicate that small variations in IPI can significantly alter the firing timing of neuronal somata and increase firing randomness during sustained antidromic activation from axonal A-HFS.

4. **Different Neuronal Responses to an Identical Set of Varying IPIs with Different Orders**

The experimental data in Fig. 6.21 suggest that changing the order of an identical set of varying IPIs can affect the neuronal responses during A-HFS. For example, we compared APS signals produced by A-HFS trains with a same set of IPIs arranged

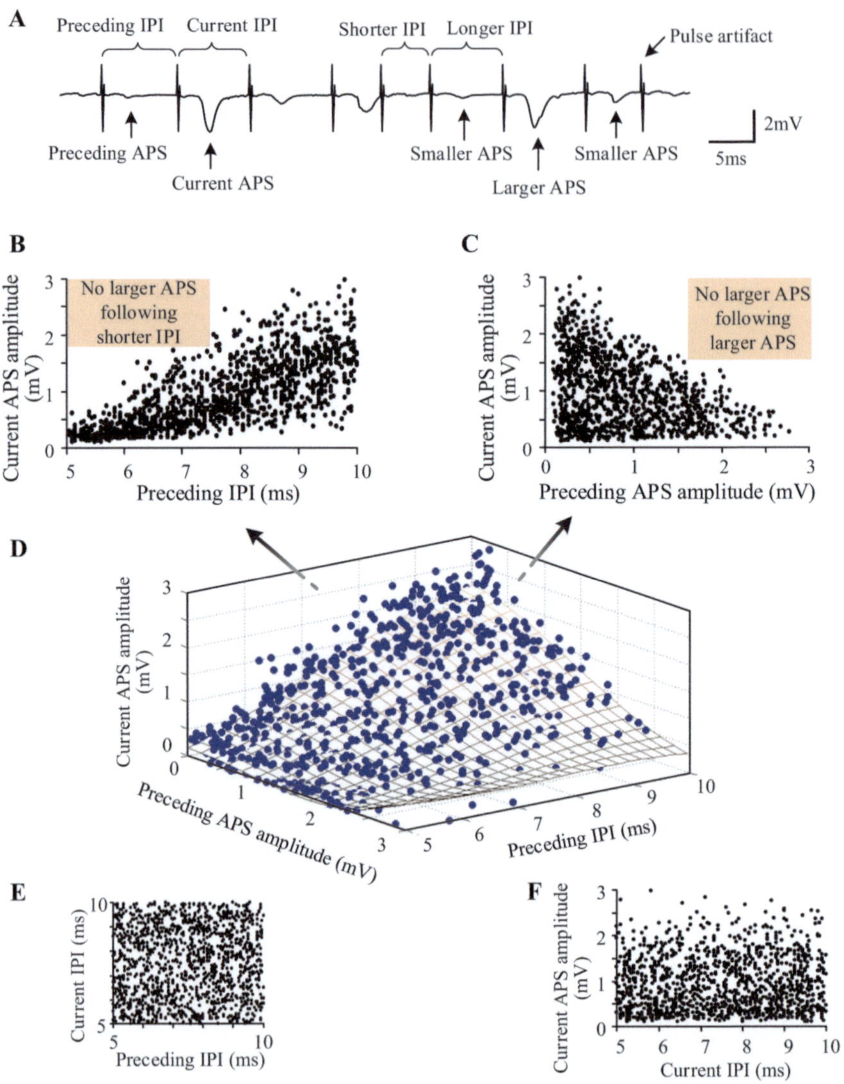

Fig. 6.21 Relationships among the neighboring APS amplitudes and IPI lengths during A-HFS with randomly varying IPIs. **A** A segment of APS signal illustrating the definitions of various APSs and IPIs. **B** Scatter plot of current APS amplitudes versus preceding IPI lengths. **C** Scatter plot of current APS amplitudes versus preceding APS amplitudes. **D** Three-dimensional scatter plot of current APS amplitudes versus both preceding APS amplitudes and preceding IPI lengths, with a grid fitting surface highlighting the distribution trend. **E** Plot of current IPI lengths versus preceding IPI lengths. **F** Plot of current APS amplitudes versus current IPI lengths. From Feng et al. (2019)

in two different orders: from longest to shortest, and random. The two IPI sequences were set to a 10-s duration, with IPIs uniformly ranging in 5-10 ms. Given the 20 kHz sampling rate used in our recordings, we set the pulse time resolution to 0.05 ms. At this resolution, the 5–10 ms range yielded 101 distinct IPI lengths. To achieve a uniform distribution, we repeated each IPI length 13 times and added one extra IPI for every five IPI lengths. This yielded 1334 IPIs ($101 \times 13 + 21$) with a total length of $10.005 \approx 10$ s. This set of IPIs was then arranged in two different ways to produce two sequences: one decreased from longest to shortest, while the other was randomized using MATLAB "randperm" function. Both sequences contained the identical set of IPIs, differing only in their order.

As shown in Fig. 6.22A1, the 10 s decreasing IPI sequence was applied following an initial 20 s pulse sequence at constant 100 Hz to form a total 30 s A-HFS. In the 10–20 s constant period, the APS amplitude stabilized at about 20% of its initial level. Based on the mean APS amplitude in this steady period, the normalized APS amplitude showed a decrease from ~ 1 to ~ 0.3 during the final 10 s period as IPI decreased from 10 to 5 ms (Fig. 6.22A2). Due to the uniform distribution of the varying IPIs, the IPI decrease over time was quasi-linear rather than exactly linear (Fig. 6.22A3). The APS amplitude showed a strong correlation with its first preceding IPI (termed as IPI_1) with a correlation coefficient of 0.94 (Fig. 6.22A4). When this stimulation was followed by a gradually increasing IPI sequence from 5 to 10 ms, the APSs exhibited a reverse change.

When the 10 s decreasing IPI sequence was replaced by the randomly varying IPI sequence (Fig. 6.22B1), the normalized APS amplitudes fluctuated "randomly" between 0 and 2.4 (Fig. 6.22B2). Although these random IPIs came from the identical IPI set with the same range of 5–10 ms and uniform distribution (Fig. 6.22B3), the APS amplitude showed no linear correlation with IPI_1, as both larger and smaller APSs occurred following longer IPI_1 intervals (Fig. 6.22B4 upper). Nevertheless, the APS amplitude showed a positive correlation with the difference between IPI_1 and IPI_2 ($\Delta IPI = IPI_1\text{-}IPI_2$) (Fig. 6.22B4 bottom). This indicated that a longer IPI_1 accompanied by a shorter IPI_2 (the second preceding IPI) can produce a larger APS. For detailed statistical analyses of these experiments, please refer to our paper (Zheng et al. 2020).

The HFS sequences with time-varying IPIs can generate diverse stimulation effects, thereby offering promising options for clinical applications. While the design of IPI sequences currently lacks theoretical guidance, our experimental results indicate that neuronal firing can be mainly influenced by the preceding intervals IPI_1 and IPI_2. Based on this finding, we can design pulse sequences with IPIs varying only within a small range (such as 5–10 ms) to generate various APS sequences for different effects of neural modulations. We have tested this approach, as detailed in our reports (Zheng et al. 2021; Zheng et al. 2022).

5. **APS Latencies Remain Stable During Random-IPI A-HFS**

During the 10 s period of randomly varying IPIs (5–10 ms) inserted in the steady period of A-HFS (Fig. 6.23A), while the APS amplitudes changed substantialy (Fig. 6.23B1), the APS latencies remained stable, matching the constant periods

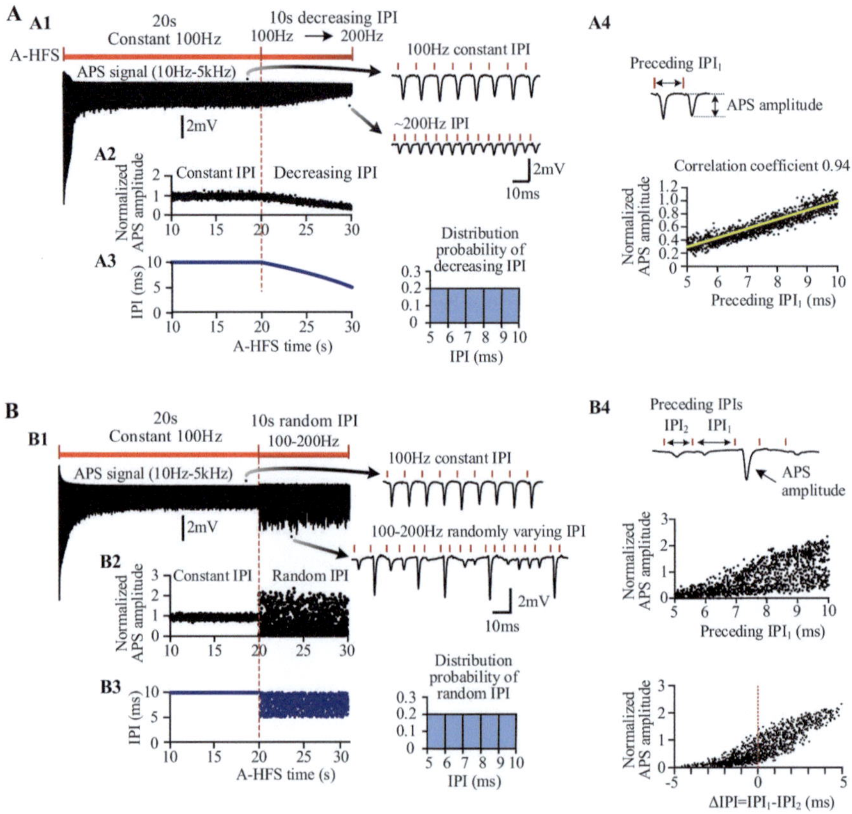

Fig. 6.22 Neuronal responses to A-HFS with an identical set of IPIs arranged in different orders. **A** APSs evoked by pulses with gradually decreasing IPIs. **A1**: Typical APS signal during a 30 s A-HFS consisting of an initial 20 s constant IPI period (10 ms, 100 Hz) followed by a 10 s gradually decreasing IPI period (10–5 ms, 100–200 Hz). **A2**: Scatter plot of normalized APS amplitudes during the 10 to 30 s A-HFS period. **A3**: Scatter plot of IPIs (left) and the IPI distribution (right). **A4**: Correlation between normalized APS amplitude and preceding IPI_1 during the 10 s period with decreasing IPI. **B** APSs evoked by pulses with random IPIs. **B1–B4**: Corresponding to **A1-A4**. **B4** *bottom*: Scatter plot of normalized APS amplitudes versus the differences between IPI_1 and IPI_2. From Zheng et al. (2020)

before and after the varying period (Fig. 6.23B2). Similar to those described in Sect. 5.2, Fig. 6.23B shows a substantial increase in APS latency occurred in the initial 4.5 s period of A-HFS as APS amplitude decreased. The scatter plot of APS amplitude versus latency in this period showed a significant negative correlation, with a determination coefficient of $R^2 = 0.97$ (Fig. 6.23C1). However, during the 50–60 s period of A-HFS with varying IPIs, the determination coefficient fell to $R^2 = 0.27$ (Fig. 6.23C2), indicating no correlation. As described in Sect. 5.2.3, when an extended gap was inserted during the steady period of constant A-HFS, a "fast recovery" occurred in APS amplitude but not in APS latency (see Fig. 5.11). Here the stability in APS latencies with varying IPIs was consistent with the neuronal responses to the inserted gap described in Sect. 5.2.3.

6. A-HFS with Randomly Varying IPIs Throughout

In the A-HFS with varying IPIs described above, we initiated the stimulation with a constant IPI period to allow the induced APS to reach a steady state before switching to varying IPIs. This paradigm served two purposes. First, it allowed direct comparison between varying and constant A-HFS, with the constant period serving as a

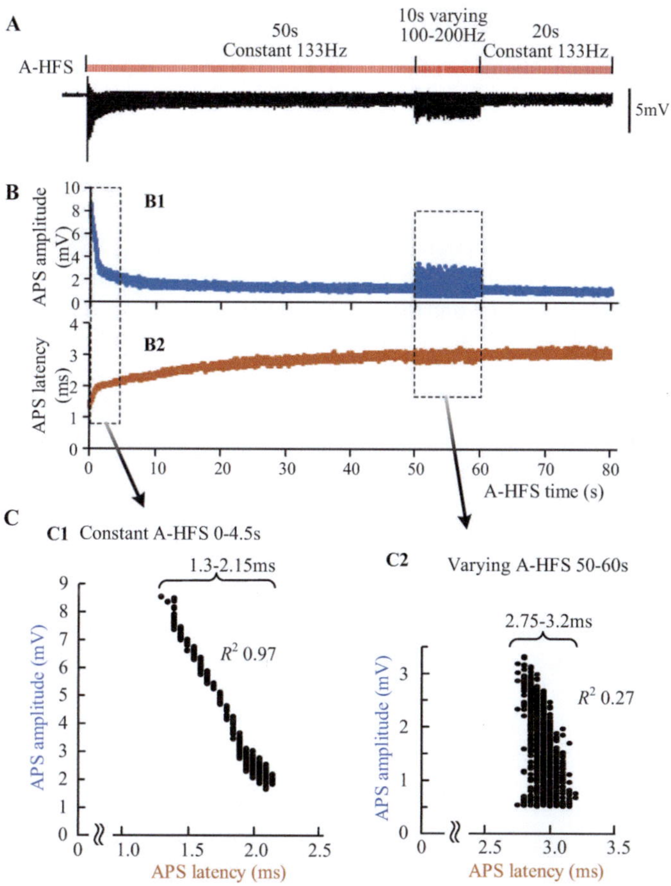

Fig. 6.23 Stable APS latencies during A-HFS period with randomly varying IPIs. **A** Typical APS signal during 80 s A-HFS comprising three periods: initial 50 s and final 20 s at constant 133 Hz, and a middle 10 s period (50–60 s) with randomly varying IPIs at a mean 133 Hz ranging in 100–200 Hz. **B** Scatter plots showing APS amplitudes (**B1**) and latencies (**B2**) throughout the A-HFS period. **C** Relationship between APS amplitude and latency during the initial constant period of 0–4.5 s (**C1**) and during the varying period of 50–60 s (**C2**). Note that only APSs with amplitudes greater than 0.5 mV were included in **C2** to ensure accurate latency measurements

control in the same A-HFS to minimize interference. Second, it brought neuronal responses to a steady state before switching to varying A-HFS, enabling comparison of steady responses. However, one may argue that the neuronal responses to varying A-HFS could simply represent a transient response caused by the transition from constant to varying periods, rather than a steady-state response. To address this concern, we next examined neuronal responses to A-HFS with varying IPIs throughout the entire stimulation period.

Figure 6.24A1 shows a typical APS signal during A-HFS with randomly varying IPIs ranging in 100–200 Hz (a mean of 133 Hz). The corresponding IPI distribution is shown in Fig. 6.22B3. As a control, Fig. 6.24A2 shows an A-HFS with a constant 133 Hz. At the onset of both A-HFS trains, each pulse induced a large APS with a similar amplitude. After ~ 4 s of A-HFS, the APS amplitude decreased to a steady state. At this period, the APSs induced by varying A-HFS showed marked variation, while those induced by constant A-HFS were uniformly small. This difference persisted until the end of A-HFS, as shown in the scatter plots of APS amplitudes (grey dots in Fig. 6.24B1 and B2). Nevertheless, their average APS amplitudes per 0.1 s followed bi-exponential fittings (brown dots and red curves in Fig. 6.24B1 and B2) and showed similar APS trends during both A-HFS trains. These average APS amplitudes decreased rapidly in the initial period, with a half-life time shorter than 1 s (0.31 s for varying A-HFS and 0.50 s for constant A-HFS). However, the half-life time of the maximum APS amplitudes per 0.1 s differed significantly between the two A-HFS trains (blue dots and curves in the Fig. 6.24B1 and B2): 3.4 s for varying A-HFS and 0.52 s for constant A-HFS. The statistical data are detailed in our report (Hu et al. 2021).

Like constant A-HFS, varying A-HFS showed an initial transient phase before stabilizing at a steady phase. However, small IPI variations of 5–10 ms in varying A-HFS substantially delayed the decay of maximum APS. Although both A-HFS paradigms showed similar average neuronal firing levels, the larger APS occurrences during varying A-HFS indicated stronger synchronized firing with potential stronger effects in neuronal networks. Additionally, the random synchronous firing (APSs) generated by varying stimulations may eliminate pathological rhythmic activity in brain disorders.

6.4.2 Neuronal Responses to O-HFS with Randomly Varying IPIs

As previously described, APSs generated by axonal A-HFS occur without synaptic transmissions. To investigate how downstream post-synaptic neurons respond to axonal O-HFS with time-varying IPIs, we applied identical pulse sequences at the Schaffer collaterals while maintaining the CA1 recording site. The orthodromic activation pathway of O-HFS involves axonal conduction, synaptic transmission, and post-synaptic integration.

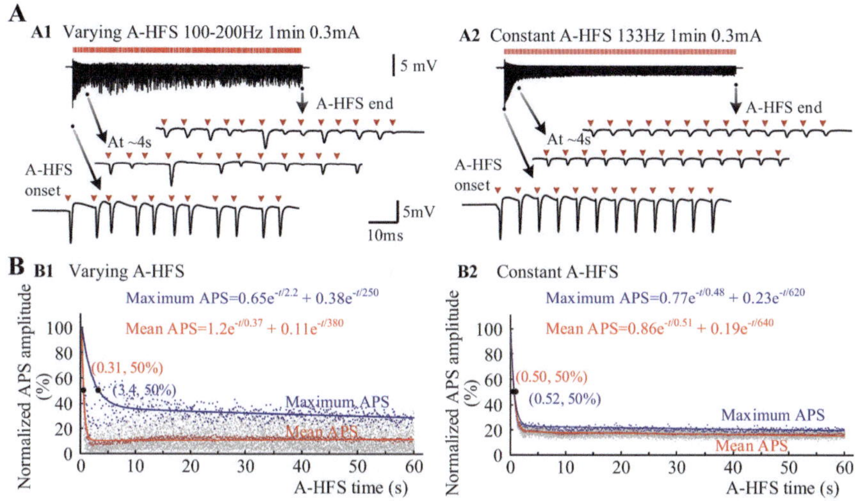

Fig. 6.24 Comparison of the neuronal responses to constant and varying A-HFS trains. **A** Typical APS signal during varying A-HFS (**A1**) and constant A-HFS (**A2**). **B** Normalized APS amplitude (grey dots) during varying A-HFS (**B1**) and constant A-HFS (**B2**), with fitting curves showing the mean and the maximum APS amplitudes per 0.1 s (brown and blue dots). Larger black dots on the falling phases of fitting curves denote half-value time points. Modified from Hu et al. (2021)

1. Neuronal Responses to O-HFS with Inserted Periods of Varying IPIs

Similar to the A-HFS described above, the O-HFS train combined both constant and varying IPI periods. To verify the differences in neuronal responses, we switched between the two stimulation modes twice. The O-HFS train began with a 50 s constant 100 Hz period, followed by a 10 s period with randomly varying IPIs. It then returned to the constant period and repeated this 60 s pulse sequence once again. Finally, the stimulation concluded with a 20 s constant period, resulting in a total O-HFS duration of 140 s (Fig. 6.25A). The IPIs in the two 10 s varying periods were uniformly distributed within the range of 5–10 ms (100–200 Hz), identical to that shown in Fig. 6.22B3.

The neuronal responses to the initial 50 s constant O-HFS was similar to that described in Sect. 5.3. After the first large OPS at the onset and subsequent OPS events, the neuronal responses in the late period of 50-s constant O-HFS entered a steady state without obvious OPS (Fig. 6.25A). However, when the constant stimulation switched to varying stimulation, OPS reappeared and persisted throughout the 10-s varying period. Upon returning to 100 Hz constant stimulation, OPS disappeared again. The neuronal responses maintained consistent patterns during the second round of varying stimulation. As shown in Fig. 6.25B, when we increased the pulse frequency in constant stimulations from 100 to 200 Hz while keeping other parameters unchanged, OPS events also occurred during the two 10-s periods of varying stimulation. No obvious OPS appeared during the steady period of 200 Hz constant stimulations. The mean pulse frequency of varying stimulation was 133 Hz. Setting

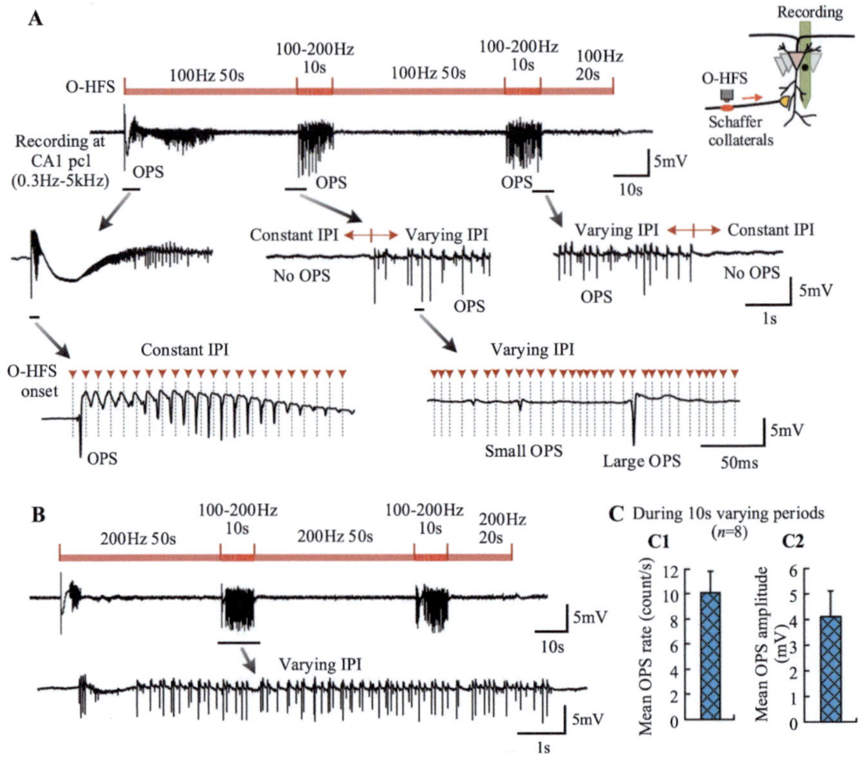

Fig. 6.25 Neuronal responses to O-HFS trains with randomly varying IPIs. **A** Typical recording of neuronal responses to a 140 s O-HFS train, consisting of constant IPI periods (100 Hz during 0–50 s, 60–110 s, and 120–140 s) and two periods of randomly varying IPIs (100–200 Hz during 50–60 s and 110–120 s). **B** Typical recording using a similar stimulation sequence as in (**A**), but with the constant IPI at 200 Hz. **C** The mean OPS rate (**C1**) and the mean OPS amplitude (**C2**) during the two 10-s periods with varying IPIs. Modified from Feng et al. (2019)

the constant stimulations to 133 Hz did not change the responses during varying stimulation, as detailed in our report (Hu et al. 2019). Additionally, during the varying periods, not every pulse induced an OPS—the OPS rate was much lower than the mean pulse frequency (Fig. 6.25C1), indicating that the synchronous neuronal firing (OPS) required a cumulative effect of multiple pulses. Moreover, the OPS amplitude showed considerable variation, with an average amplitude much smaller than the initial OPS induced at O-HFS onset.

To investigate the OPS pattern during varying periods, we analyzed the relationship between OPS amplitudes and preceding IPIs. As shown in Fig. 6.26A, during a 10 s varying period, OPSs appeared either separately or in bursts. We labeled the four preceding pulses before an OPS as P_1, P_2, P_3 and P_4, with their corresponding preceding IPIs as IPI_1, IPI_2, IPI_3 and IPI_4. The random IPIs uniformly distributed within the range of 5–10 ms. The distributions of OPS amplitude and occurrence probability versus the length of IPI_1 and IPI_4 were approximately uniform (Fig. 6.26B1 and B4), indicating no significant correlation between the OPSs and these IPIs. However, both the OPS amplitude and occurrence probability showed correlation with IPI_2 and IPI_3 to some extent. When IPI_2 was longer or IPI_3 was shorter, the OPS amplitude was greater and the probability of OPS occurrence was higher (Fig. 6.26B2 and B3). The three-dimensional scatter plot in Fig. 6.26C illustrates these relationships.

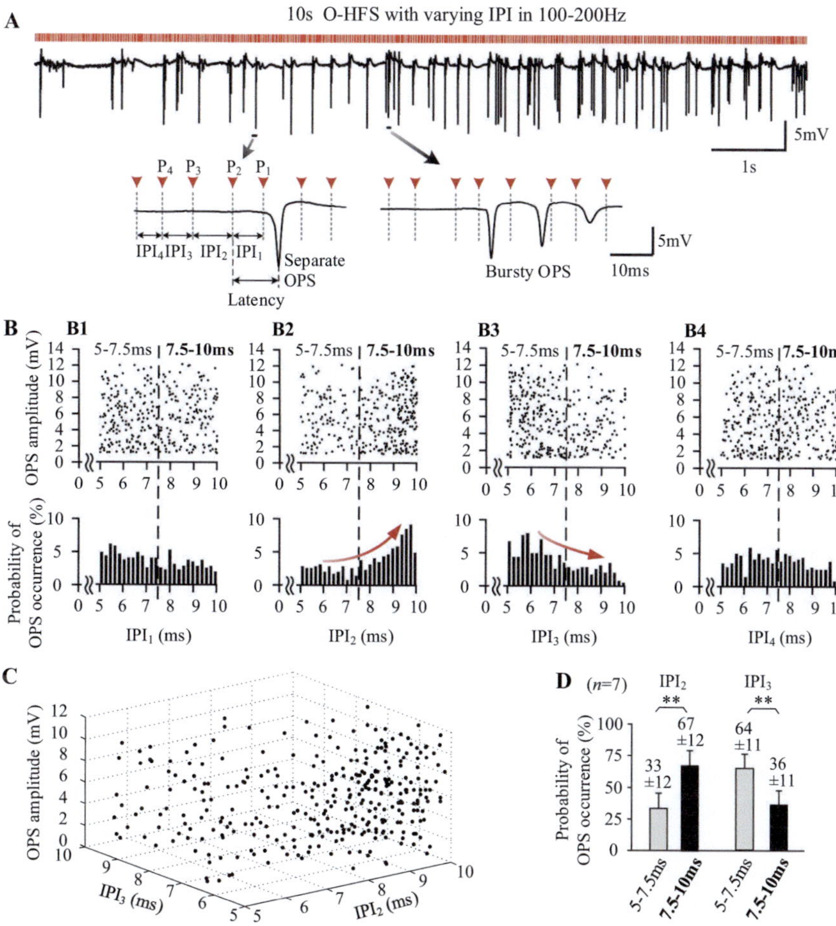

Fig. 6.26 Relationship between OPS occurrence and IPI length during O-HFS with randomly varying IPIs. **A** Example of 10-s recording during O-HFS with varying IPIs ranging in 5–10 ms. **B** Scatter plots showing OPS amplitude (upper row) and OPS occurrence probability (bottom row) in relation to the lengths of four preceding IPIs (**B1**-**B4**) during varying O-HFS. The vertical dashed lines divide the varying IPIs (5–10 ms) into two equal sub-ranges. **C** Three-dimensional scatter plot of OPS amplitudes versus the lengths of IPI_2 and IPI_3. **D** Comparisons of OPS occurrence probability between two IPI sub-ranges (5–7.5 ms and 7.5–10 ms) for IPI_2 and IPI_3. $^{**}P < 0.01$, paired t-test, $n = 7$ rats. Modified from Hu et al. (2019)

We divided the IPI range equally into two sub-ranges: 5–7.5 ms (shorter) and 7.5–10 ms (longer). Statistical data showed that OPS appeared approximately twice as frequently with the longer IPI_2 compared to shorter IPI_2 (Fig. 6.26D left), while OPS appeared more frequently with shorter than longer IPI_3 (Fig. 6.26D right). This correlations between OPS occurrence and the two IPIs showed that pulses P_2 and P_3 were more likely to induce OPS than P_1 and P_4. This occurred because the synaptic transmission in the orthodromic activation pathway delayed downstream neuronal responses to stimulation activation. Under baseline situations, a single pulse typically induced an OPS with a latency of about 5 ms. Sustained axonal stimulation could extend this latency. During the varying O-HFS period, the OPS latency could exceed most of the IPIs ranging in 5–10 ms. Therefore, P_1 was unlikely to induce OPS, while P_2 or P_3 were more likely triggers. However, if P_3 induced an OPS, its latency would exceed the combined duration of IPI_1 and IPI_2 (> 10 ms). Our another study showed that during sustained axonal O-HFS, the latency of most downstream neuronal firing was shorter than 10 ms (Wang et al. 2018), making P_3 an unlikely trigger. Thus, P_2 had the highest probability of inducing OPS (Fig. 6.26A, lower left). The experimental data confirmed that a longer IPI_2 combined with a shorter IPI_3 facilitated P_2 to induce a larger OPS (Fig. 6.26).

2. **O-HFS with Randomly Varying IPIs Throughout**

The above experiments investigated OPS events induced during a 10-s varying period temporarily inserted into the steady period of constant O-HFS. We also explored neuronal responses to 3-min O-HFS entirely with randomly varying IPIs of 5–10 ms (Fig. 6.27A). OPSs appeared continuously during the early O-HFS period before ceasing at about 65 s. The mean OPS occurrence rate within this 65-s period was 16 count/s (Fig. 6.27B). This OPS duration was much longer than that observed during the initial period of constant O-HFS at both 100 and 200 Hz (refer to Sect. 5.3.1). During the late period of 3-min O-HFS when obvious OPSs were absent, unit spikes (MUA) exhibited higher density than the baseline level before O-HFS. With a shorter O-HFS duration (e.g., 1 min), population spikes (PS) could even continue for a while after O-HFS ended, similar to the after-discharge described in Sect. 8.1. Strong epileptiform discharges could even lead to spreading suppression (SD).

These results indicate that small random variations in IPIs can substantially enhance the activation effect of O-HFS on post-synaptic neurons, leading to persistent synchronous discharges in downstream neurons. These discharges resemble the OPSs that appeared before axonal block during the initial O-HFS period. This means that this time-varying stimulation paradigm can overcome the activation attenuation caused by sustained constant HFS. The resulting persistent excitation may offer therapeutic benefits for certain diseases.

3. **Reproducibility of Neuronal Responses to A-HFS and O-HFS with Randomly Varying IPIs**

The activation pathways of A-HFS and O-HFS differ essentially in whether or not involving synaptic transmissions. An identical pulse sequence with time-varying IPIs

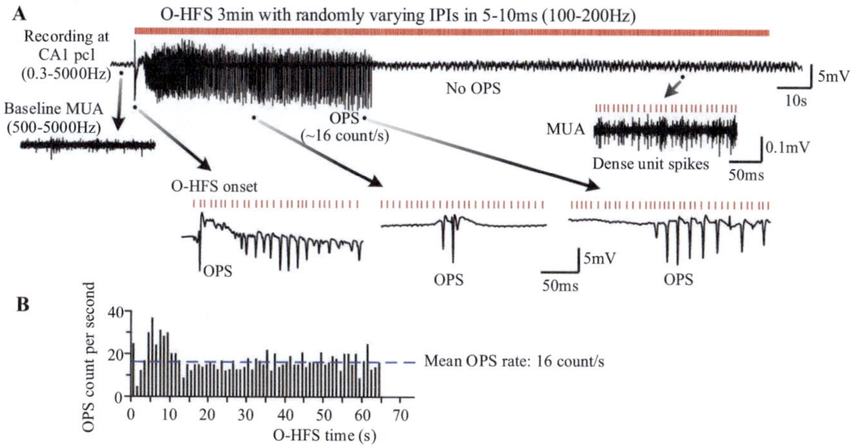

Fig. 6.27 Neuronal responses to 3-min O-HFS with randomly varying IPIs of 5–10 ms throughout. **A** Example of a 3-min O-HFS with randomly varying IPIs throughout. Red small bars in the enlarged insets indicate the removed pulse artifacts. **B** OPS counts per second during the initial 70-s period of the O-HFS shown in (**A**). The blue dashed line represents the average OPS rate during this period

can induce different neuronal responses through these distinct activation pathways with different repeatability.

As shown in Fig. 6.28A, we applied an identical 1-min pulse sequence with randomly varying IPIs (5–10 ms) as both A-HFS and O-HFS. Both stimulations continuously induced PSs (APSs or OPSs) throughout the entire 1-min period. In the six rat experiments in which we applied both stimulations once, the first pulse induced similar APS or OPS. Subsequently, the APS patterns during A-HFS were consistent across experiments, though the APS amplitudes triggered by each pulse varied markedly. The resulting APS sequence showed good repeatability (Fig. 6.28B and C, left). However, during O-HFS with the identical pulse sequence, both the timings and amplitudes of induced OPSs lacked consistency and repeatability across experiments, except for a few pulses in the initial period (Fig. 6.28B and C, right). These results indicate that time-varying stimulations can produce definite firing patterns in directly stimulated neurons. This allows for designing various pulse sequences to meet different requirements of neuronal firing patterns, as detailed in our previous report (Zheng et al. 2021). However, when synaptic transmissions are involved, post-synaptic neurons cannot precisely replicate a definite firing pattern.

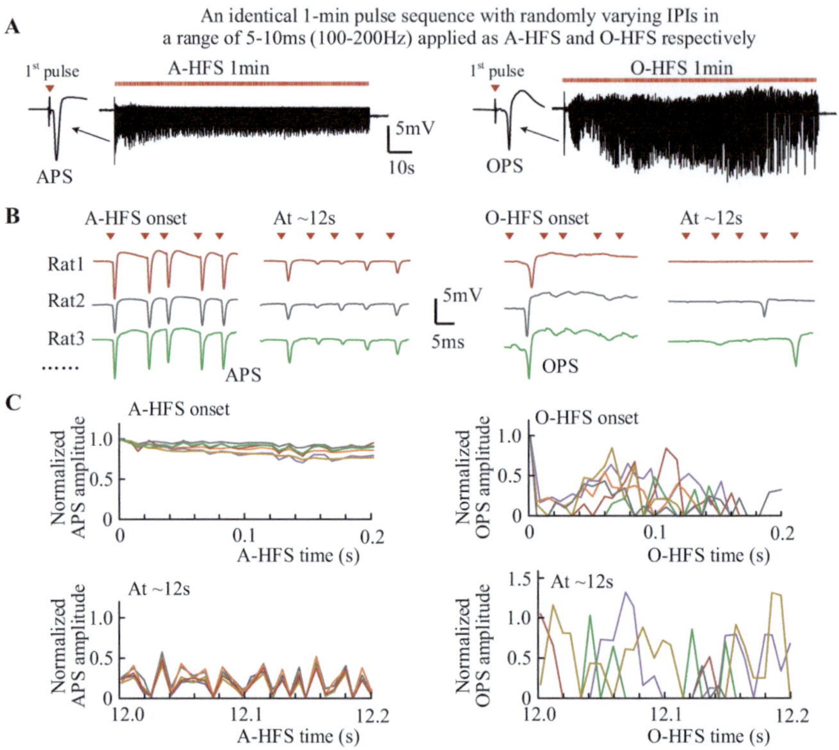

Fig. 6.28 Comparison of neuronal response repeatability between A-HFS and O-HFS with random IPI variations. **A** Examples of 1-min A-HFS and O-HFS recordings using an identical pulse sequence. **B** Examples of induced APSs and OPSs during the initial and around 12 s periods of A-HFS and O-HFS from three rat experiments. **C** Overlaid plots showing normalized APS and OPS amplitudes from six rat experiments at onset (top) and ~ 12 s (bottom) of A-HFS and O-HFS

6.5 Summary

This chapter presents our findings on HFS paradigms with time-varying parameters, including gradient frequency and intensity, alternating bi-parameters, pulse insertion and deletion in regular stimulations, and randomly varying IPIs.

During the initial period of HFS, gradual changes in pulse intensity and frequency can progressively expand the stimulated area, preventing the onset response—a phenomenon caused by synchronous firing of numerous neurons. During sustained A-HFS activation without synaptic transmission involvement, neuronal responses to gradually varying stimulations can change smoothly without abrupt transitions. Alternatively, some stimulation paradigms with small variations can generate abrupt neuronal responses, due to the nonlinear dynamic properties of neuronal membranes under HFS-induced intermittent depolarization block. For example, during steady A-HFS with slight variations in bi-IPI or bi-intensity, the amplitude of induced

APSs can undergo nonlinear bifurcation. This demonstrates a nonlinear magnification of the small differences in stimulation parameters into substantial changes in neuronal responses. Additionally, adding or removing a pulse in A-HFS or O-HFS with constant parameters can significantly alter subsequent neuronal responses. Furthermore, when pulse frequency (or IPI) varies randomly, neuronal responses no longer correlate simply with the preceding IPI, unlike in situations with constant and gradient parameter stimulations. Small changes in randomly varying IPIs can substantially alter the firing patterns of neuronal populations.

To investigate the mechanisms underlying these experimental phenomena, we used a computational model to simulate neuronal responses to various stimulations. The model incorporated potassium ion accumulation in the peri-axonal spaces, which generated HFS-induced depolarization block at the axonal membrane. Our simulation results were consistent with the experimental findings and showed the effects of nonlinear dynamics in membrane ion channels, as detailed in our report (Zheng et al. 2020). In addition to the nonlinear dynamics of ion channels described in Sect. 1.3, random characteristics at various structural levels—known as "noise" (Mino and Grill 2002; Faisal et al. 2008)—may also contribute to the changes in neuronal responses observed in our experiments.

During intermittent axonal block induced by HFS, the stimulation paradigms with time-varying parameters can generate more synchronized neuronal firing than those with constant parameters, resulting in stronger modulation of neuronal populations in the projection region. This provides a way to adjust the "dose" of stimulation, potentially offering more treatment options for various neurological disorders. For example, stimulations with random IPI variations may help improve minimal consciousness states (Quinkert and Pfaff 2012), representing one of the new DBS paradigms under development (Hess et al. 2013; Gunduz et al. 2017; Grill 2018).

In summary, our findings in this chapter show that under the intermittent depolarization block produced by continuous HFS, axons can behave like taut strings—small changes in the applied stimulation can produce distinct "melodies". These varied neuronal responses to time-varying high-frequency pulse stimulations provide new clues for developing diverse neuromodulation approaches.

References

Akbar U, Raike RS, Hack N et al (2016) Randomized, blinded pilot testing of nonconventional stimulation patterns and shapes in Parkinson's disease and essential tremor: evidence for further evaluating narrow and biphasic pulses. Neuromodulation 19(4):343–356

Anderson TR, Hu B, Iremonger K et al (2006) Selective attenuation of afferent synaptic transmission as a mechanism of thalamic deep brain stimulation-induced tremor arrest. J Neurosci 26(3):841–850

Ansó J, Benjaber M, Parks B et al (2022) Concurrent stimulation and sensing in bi-directional brain interfaces: a multi-site translational experience. J Neural Eng 19(2):026025

Atherton LA, Prince LY, Tsaneva-Atanasova K (2016) Bifurcation analysis of a two-compartment hippocampal pyramidal cell model. J Comput Neurosci 41(1):91–106

Benabid AL, Pollak P, Louveau A et al (1987) Combined (thalamotomy and stimulation) stereotactic surgery of the VIM thalamic nucleus for bilateral Parkinson disease. Appl Neurophysiol 50(1–6):344–346

Bhadra N, Foldes EL, Ackermann DM et al (2009) Reduction of the onset response in high frequency nerve block with amplitude ramps from non-zero amplitudes. Annu Int Conf IEEE Eng Med Biol Soc 2009:650–653

Bhadra N, Kilgore KL (2005) High-frequency electrical conduction block of mammalian peripheral motor nerve. Muscle Nerve 32(6):782–790

Blankenship AG, Feller MB (2010) Mechanisms underlying spontaneous patterned activity in developing neural circuits. Nat Rev Neurosci 11(1):18–29

Brocker DT, Swan BD, Turner DA et al (2013) Improved efficacy of temporally non-regular deep brain stimulation in Parkinson's disease. Exp Neurol 239:60–67

Cai Z, Feng Z, Guo Z et al (2017) Novel stimulation paradigms with temporally-varying parameters to reduce synchronous activity at the onset of high frequency stimulation in rat hippocampus. Front Neurosci 11:563

Cai Z, Feng Z, Hu H et al (2018) Design of a novel stimulation system with time-varying paradigms for investigating new modes of high frequency stimulation in brain. Biomed Eng Online 17(1):90

De Maesschalck P, Wechselberger M (2015) Neural excitability and singular bifurcations. J Math Neurosci 5(1):29

Deco G, Jirsa VK, Robinson PA et al (2008) The dynamic brain: from spiking neurons to neural masses and cortical fields. PLoS Comput Biol 4(8):e1000092

Faisal AA, Selen LP, Wolpert DM (2008) Noise in the nervous system. Nat Rev Neurosci 9(4):292–303

Fan Y, Holden AV (1993) Bifurcations, burstings, chaos and crises in the rose-hindmarsh model for neuronal-activity. Chaos, Solitons Fractals 3(4):439–449

Feng Z, Ma W, Wang Z et al (2019) Small changes in inter-pulse-intervals can cause synchronized neuronal firing during high-frequency stimulations in rat hippocampus. Front Neurosci 13:36

Gerges M, Foldes EL, Ackermann DM et al (2010) Frequency- and amplitude-transitioned waveforms mitigate the onset response in high-frequency nerve block. J Neural Eng 7(6):066003

Gilbert Z, Mason X, Sebastian R et al (2023) A review of neurophysiological effects and efficiency of waveform parameters in deep brain stimulation. Clin Neurophysiol 152:93–111

Grill WM (2018) Temporal pattern of electrical stimulation is a new dimension of therapeutic innovation. Curr Opin Biomed Eng 8:1–6

Gu H, Chen S (2014) Potassium-induced bifurcations and chaos of firing patterns observed from biological experiment on a neural pacemaker. Sci China (Technol Sci) 57(5):864–871

Gu H, Zhao Z, Jia B et al (2015) Dynamics of on-off neural firing patterns and stochastic effects near a sub-critical hopf bifurcation. PLoS ONE 10(4):e0121028

Gunduz A, Foote KD, Okun MS (2017) Reengineering deep brain stimulation for movement disorders: emerging technologies. Curr Opin Biomed Eng 4:97–105

Hahn PJ, Durand DM (2001) Bistability dynamics in simulations of neural activity in high-extracellular-potassium conditions. J Comput Neurosci 11(1):5–18

Hess CW, Vaillancourt DE, Okun MS (2013) The temporal pattern of stimulation may be important to the mechanism of deep brain stimulation. Exp Neurol 247:296–302

Hodgkin AL, Huxley AF (1952) A quantitative description of membrane current and its application to conduction and excitation in nerve. J Physiol 117(4):500–544

Hosain MK, Kouzani A, Tye S (2014) Closed loop deep brain stimulation: an evolving technology. Australas Phys Eng Sci Med 37(4):619–634

Hu H, Feng Z, Wang Z et al (2019) The altering effect of high frequency stimulation on brain neurons with small changes in the lengths of inter-pulse-intervals. Prog Biochem Biophys 46(8):804–811 (in Chinese)

Hu Y, Feng Z, Wang Z et al (2021) Transient neuronal responses to high-frequency pulse stimulation in rat hippocampus. Prog Biochem Biophys 48(7):827–835 (in Chinese)

Hu Y, Feng Z, Zheng L et al (2023) Adding a single pulse into high-frequency pulse stimulations can substantially alter the following episode of neuronal firing in rat hippocampus. J Neural Eng 20(1):016004

Iremonger KJ, Anderson TR, Hu B et al (2006) Cellular mechanisms preventing sustained activation of cortex during subcortical high-frequency stimulation. J Neurophysiol 96(2):613–621

Jia B, Gu H, Xue L (2017) A basic bifurcation structure from bursting to spiking of injured nerve fibers in a two-dimensional parameter space. Cogn Neurodyn 11(2):189–200

Karamintziou SD, Deligiannis NG, Piallat B et al (2016) Dominant efficiency of nonregular patterns of subthalamic nucleus deep brain stimulation for Parkinson's disease and obsessive-compulsive disorder in a data-driven computational model. J Neural Eng 13(1):016013

Khandhar SM, Heath SL, Ostrem SL et al (2005) Efficacy and tolerability of rapid cyclical stimulation of the subthalamic nucleus in Parkinson's disease. In: Ninth international congress of Parkinson's disease and movement disorders (Hoboken, NJ), vol 20, p S162

Kocsis JD, Malenka RC, Waxman SG (1983) Effects of extracellular potassium concentration on the excitability of the parallel fibres of the rat cerebellum. J Physiol 334(1):225–244

Kuncel AM, Birdno MJ, Swan BD et al (2012) Tremor reduction and modeled neural activity during cycling thalamic deep brain stimulation. Clin Neurophysiol 123(5):1044–1052

Kuncel AM, Cooper SE, Wolgamuth BR et al (2006) Clinical response to varying the stimulus parameters in deep brain stimulation for essential tremor. Mov Disord 21(11):1920–1928

Ma J, Tang J (2017) A review for dynamics in neuron and neuronal network. Nonlinear Dyn 89(3):1569–1578

Miles JD, Kilgore KL, Bhadra N et al (2007) Effects of ramped amplitude waveforms on the onset response of high-frequency mammalian nerve block. J Neural Eng 4(4):390–398

Mino H, Grill WM Jr (2002) Effects of stochastic sodium channels on extracellular excitation of myelinated nerve fibers. IEEE Trans Biomed Eng 49(6):527–532

Molchanova SM, Huupponen J, Lauri SE et al (2016) Gap junctions between CA3 pyramidal cells contribute to network synchronization in neonatal hippocampus. Neuropharmacology 107:9–17

Okun MS, Hickey PT, Machado AG et al (2022) Temporally optimized patterned stimulation (TOPS®) as a therapy to personalize deep brain stimulation treatment of Parkinson's disease. Front Hum Neurosci 16:929509

Paydarfar D, Forger DB, Clay JR (2006) Noisy inputs and the induction of on-off switching behavior in a neuronal pacemaker. J Neurophysiol 96(6):3338–3348

Quinkert AW, Pfaff DW (2012) Temporal patterns of deep brain stimulation generated with a true random number generator and the logistic equation: effects on CNS arousal in mice. Behav Brain Res 229(2):349–358

Rabinovich MI, Varona P, Selverston AI et al (2006) Dynamical principles in neuroscience. Rev Mod Phys 78(4):1213–1265

Rosenbaum R, Zimnik A, Zheng F et al (2014) Axonal and synaptic failure suppress the transfer of firing rate oscillations, synchrony and information during high frequency deep brain stimulation. Neurobiol Dis 62:86–99

Santos-Valencia F, Almazán-Alvarado S, Rubio-Luviano A et al (2019) Temporally irregular electrical stimulation to the epileptogenic focus delays epileptogenesis in rats. Brain Stimul 12(6):1429–1438

Schmitz D, Schuchmann S, Fisahn A et al (2001) Axo-axonal coupling: a novel mechanism for ultrafast neuronal communication. Neuron 31(5):831–840

Spruston N (2008) Pyramidal neurons: dendritic structure and synaptic integration. Nat Rev Neurosci 9(3):206–221

Theoret Y, Brown A, Fleming SP et al (1984) Hippocampal field potential: a microcomputer aided comparison of amplitude and integral. Brain Res Bull 12(5):589–595

Wang Z, Feng Z, Yuan Y et al (2022) Bifurcations in the firing of neuronal population caused by a small difference in pulse parameters during sustained stimulations in rat hippocampus in vivo. IEEE Trans Biomed Eng 69(9):2893–2904

Yang J, Duan Y, Xing J et al (2006) Responsiveness of a neural pacemaker near the bifurcation point. Neurosci Lett 392(1–2):105–109

Zhang X, Gu H, Guan L (2019) Stochastic dynamics of conduction failure of action potential along nerve fiber with Hopf bifurcation. Sci China-Technol Sci 62(9):1502–1511

Zhang X, Gu H, Ma K (2020) Dynamical mechanism for conduction failure behavior of action potentials related to pain information transmission. Neurocomputing 387:293–308

Zheng L, Feng Z, Hu H et al (2020) The appearance order of varying intervals introduces extra modulation effects on neuronal firing through non-linear dynamics of sodium channels during high-frequency stimulations. Front Neurosci 14:397

Zheng L, Feng Z, Hu Y et al (2021) Adjust neuronal reactions to pulses of high-frequency stimulation with designed inter-pulse-intervals in rat hippocampus in vivo. Brain Sci 11(4):509

Chapter 7
Neuronal Responses to Both Negative and Positive Pulses During HFS

Abstract This chapter presents our findings that challenge the conventional view—that negative pulses are more effective than positive pulses in extracellular stimulations. During sustained high-frequency stimulations (HFS) with alternating monophasic pulses of opposite polarities, the firing induced by positive pulses became comparable to that induced by negative pulses. Furthermore, the two types of pulses activated different subpopulations of neurons. For neurons that responded to both types of pulses at baseline, positive pulses actually surpassed negative pulses during HFS. Our computational simulations revealed complex interactions between the two pulse types in their activation mechanisms on axonal membrane.

Keywords Pulse polarity · Monophasic stimulation · Biphasic stimulation · Positive pulse · Negative pulse · Neuron subpopulation

In extracellular electrical stimulations, negative pulses can activate neuronal membranes more effectively than positive pulses. A negative pulse depolarizes the membrane near the stimulation site while hyperpolarizing it at a distance by returning current. Conversely, a positive pulse hyperpolarizes the membrane nearby while depolarizing it at a distance. As the electrical field from the stimulation site diminishes rapidly with distance, the depolarization produced by a positive pulse at a distance is much weaker than that produced by a negative pulse near the stimulation site. This makes negative pulses more efficient for activation. In clinical neuromodulation applications, single-phase pulses are rarely used because they may cause tissue damage and electrode corrosion (refer to Sect. 1.4.4). Instead, charge-balanced biphasic pulses—consisting of a preceding negative phase followed by a positive phase—are commonly used. The negative phase provides activation while the positive phase balances the charges.

The positive phase in biphasic pulses can potentially weaken the effect of the preceding negative phase. If the depolarization caused by the negative phase has not fully initiated an action potential before the positive phase begins, the resulting hyperpolarization may interrupt the action potential generation (Cappaert et al. 2013). To eliminate this impact, asymmetric charge-balance waveforms have been developed,

Z. Feng, *Electrical Brain Stimulation: Mechanisms and Modulation of Neuronal Activity*, https://doi.org/10.1007/978-981-95-4145-4_7

including smaller and wider square or non-square waveforms for positive phases (Montgomery 2017; Foutz and McIntyre 2010; Haji Ghaffari et al. 2020). Additionally, adding a short interphase gap (IPG) between the negative and positive phases can provide sufficient time for the negative phase-induced depolarization to initiate an action potential, thereby reducing the impact of positive phase (Gorman and Mortimer 1983; Weitz et al. 2014; Carlyon et al. 2005; Deprez et al. 2018). However, previous studies have shown that for myelinated nerve fibers in the brain, the "vulnerable window" of hyperpolarization interference is only about 100 μs. Furthermore, under the super-threshold activation from a negative pulse with sufficient intensity and width, this vulnerable window can virtually disappear. Therefore, the immediately following positive phase likely has no significant effect on the activation of preceding negative phase (van den Honert and Mortimer 1979). Our experimental results from stimulations on axonal fibers in the rat hippocampus CA1 region support this point. Using a pulse width of 100 μs per phase with sufficient intensities, the activation effect of the biphasic pulses was similar to that of single-phase negative pulses (Hu et al. 2015). In some cases, the following positive phase may even enhance the activation effect of the preceding negative phase (Zheng et al. 2022). Additionally, the IPG effect can depend on whether unipolar or bipolar stimulation is used. In our bipolar stimulation experiments (a method commonly used in clinic), adding an IPG did not improve the efficiency of preceding negative phase and sometimes even weakened it, consistent with other report (Eickhoff and Jarvis 2021). Therefore, the positive phase in biphasic pulses does not necessarily reduce stimulation efficiency.

Nevertheless, the addition of balance phase can double electric energy consumption, regardless of its possible adverse effect. Could the positive phase be beneficial in specific stimulation situations? Particularly, during sustained HFS such as in neuromodulation therapies, the efficiencies of positive and negative phases could change. On slender neuronal axons, a positive pulse can hyperpolarize the axonal membrane near the stimulation electrode while depolarizing the membranes on both flanks. These depolarizations can produce action potentials that propagate bidirectionally—a phenomenon known as "virtual cathode" activation (Ranck 1975; Rattay 1989; Basser and Roth 2000). Although positive pulses typically show much lower activation efficiency than negative pulses, our experiments with axonal HFS on CA1 pyramidal neurons revealed surprising results: During intermittent axonal block induced by sustained HFS, the activation effect of positive pulses can match or even exceed that of negative pulses (Feng et al. 2022; Hu et al. 2023), as presented below.

7.1 Neuronal Responses to A-HFS with Alternating Opposite-Polarity Pulses

7.1.1 *Effect of Positive Pulses During the Steady Period of A-HFS with Alternating Pulses*

To investigate the effects of monophasic pulses with opposite polarities during HFS, we separated the negative and positive phases of biphasic pulses by adding an IPG between them to create a stimulation sequence with alternating negative and positive pulses. We applied this sequence as A-HFS at alveus, using a bipolar concentric electrode (Mode CBCSG75, refer to Sect. 3.4) with the diameters of inner and outer poles of 75 and 250 μm respectively, and both poles having a height of 100 μm with a separation distance of 100 μm (Fig. 7.1A). The 16-channel recording array (Mode A1 × 16-Poly2-5 mm–50 s-177) positioned at ~ 1.3 mm upstream of the stimulation site. We still analyzed the APS signal recorded in the CA1 pcl to evaluate neuronal responses. During baseline tests, the $\mathrm{APS}_{\top}$ induced by a single negative pulse (denoted as $\top$) was much larger than the $\mathrm{APS}_{\perp}$ induced by a positive pulse (denoted as $\perp$) (Fig. 7.1A, *left*), indicating a higher activation efficiency of the $\top$ pulse.

1. **Activation Effect of Positive Pulses**

In a sequence of 100 Hz monophasic A-HFS with alternating negative and positive pulses (denoted as $\mathrm{A\text{-}HFS}_{\top\perp}$), the occurrence frequencies of both $\top$ and $\perp$ pulses were 50 Hz. At the onset of 2-min $\mathrm{A\text{-}HFS}_{\top\perp}$, the two types of pulses alternately produced larger $\mathrm{APS}_{\top}$ and smaller $\mathrm{APS}_{\perp}$, respectively (Fig. 7.1B), similar to the single-pulse stimulations at baseline. As the $\mathrm{A\text{-}HFS}_{\top\perp}$ continued, the APS amplitudes decreased rapidly. Interestingly, the $\mathrm{APS}_{\perp}$ first decreased to almost disappearance before returning to a steady level, while the $\mathrm{APS}_{\top}$ decreased monotonically to a steady level. After approximately 40 s stimulation, the amplitudes of APSs produced by both types of pulses became similar (Fig. 7.1B, C1). In the experiments with 100 Hz $\mathrm{A\text{-}HFS}_{\top\perp}$, the amplitudes between the initial $\mathrm{APS}_{\top}$ and $\mathrm{APS}_{\perp}$ ($A_{\top\mathrm{init}}$ and $A_{\perp\mathrm{init}}$) differed significantly (Fig. 7.1C2), with the mean $A_{\perp\mathrm{init}}$ being only about half of the mean $A_{\top\mathrm{init}}$. However, the mean amplitudes of steady APSs ($A_{\top\mathrm{end}}$ and $A_{\perp\mathrm{end}}$) were similar (Fig. 7.1C3). The $A_{\top\mathrm{end}}$ and $A_{\perp\mathrm{end}}$ were ~ 20% and ~ 40% of $A_{\top\mathrm{init}}$ and $A_{\perp\mathrm{init}}$, respectively (Fig. 7.1C2, C3), indicating different suppression ratios of APSs generated by the two types of pulses.

In addition, during $\mathrm{A\text{-}HFS}_{\top\perp}$, the latencies of all APSs gradually increased, with the mean $\mathrm{APS}_{\perp}$ latency being shorter than the mean $\mathrm{APS}_{\top}$ latency (Fig. 7.1C4), which persisted throughout the entire $\mathrm{A\text{-}HFS}_{\top\perp}$ period ($L_{\perp\mathrm{init}}$ vs. $L_{\top\mathrm{init}}$ in Fig. 7.1C5 and $L_{\perp\mathrm{end}}$ vs. $L_{\top\mathrm{end}}$ in Fig. 7.1C6). The latencies of both APS types nearly doubled from their initial to steady levels. After the end of $\mathrm{A\text{-}HFS}_{\top\perp}$, the APSs induced by single-pulse test stimulations returned to baseline level within about 2 min (Fig. 7.1B, upper right).

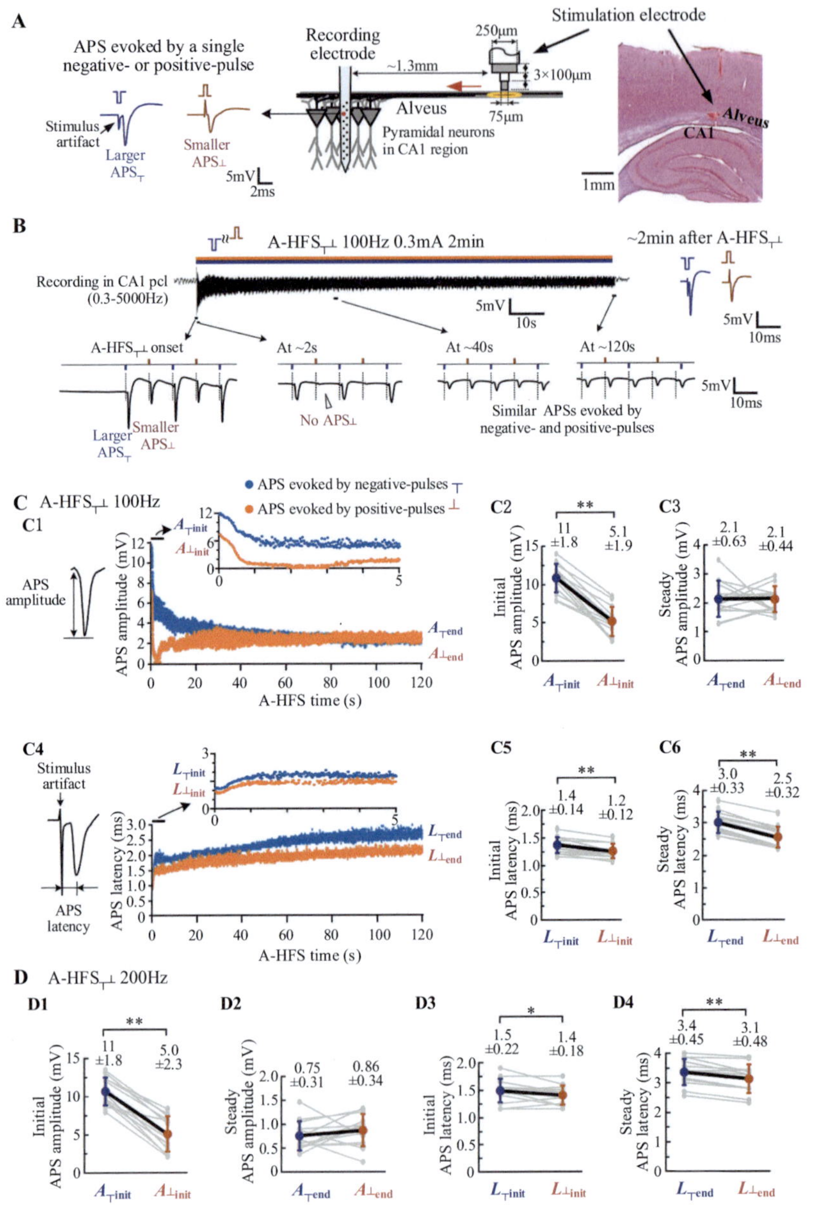

A
APS evoked by a single negative- or positive-pulse
Stimulus artifact
Larger APS┬
Smaller APS⊥
5mV
2ms
Recording electrode
~1.3mm
250μm
3×100μm
75μm
Stimulation electrode
Alveus
Pyramidal neurons in CA1 region
CA1
1mm
B
A-HFS┬⊥ 100Hz 0.3mA 2min
~2min after A-HFS┬⊥
Recording in CA1 pcl (0.3-5000Hz)
5mV
10s
10ms
A-HFS┬⊥ onset
At ~2s
At ~40s
At ~120s
No APS⊥
Similar APSs evoked by negative- and positive-pulses
C A-HFS┬⊥ 100Hz
C1
APS evoked by negative-pulses ┬
APS evoked by positive-pulses ⊥
APS amplitude
APS amplitude (mV)
A-HFS time (s)
C2
C3
Initial APS amplitude (mV)
Steady APS amplitude (mV)
11 ±1.8
5.1 ±1.9
2.1 ±0.63
2.1 ±0.44
C4
Stimulus artifact
APS latency
APS latency (ms)
C5
C6
Initial APS latency (ms)
Steady APS latency (ms)
1.4 ±0.14
1.2 ±0.12
3.0 ±0.33
2.5 ±0.32
D A-HFS┬⊥ 200Hz
D1
D2
D3
D4
11 ±1.8
5.0 ±2.3
0.75 ±0.31
0.86 ±0.34
1.5 ±0.22
1.4 ±0.18
3.4 ±0.45
3.1 ±0.48

◂**Fig. 7.1** Neuronal responses to A-HFS$_{\top\bot}$ with alternating negative and positive pulses at 100 and 200 Hz in the rat hippocampal CA1 region. **A** Schematic diagram showing the electrode positions in the CA1 region, along with APS waveforms induced by single $\top$ and $\bot$ pulses (left). A histological picture of sagittal slice shows the SE trace to the alveus, indicated by a bloodstain trail through cortex to the dorsal hippocampus (right). **B** Typical APS signal during 2-min 100 Hz A-HFS with alternating $\top$ and $\bot$ pulses. **C** For 100 Hz A-HFS ($n = 15$ rats): typical scatter plot of APS amplitudes induced by each pulse (**C1**), comparisons of initial APS amplitudes ($A_{\top init}$ vs. $A_{\bot init}$, **C2**) and steady APS amplitudes ($A_{\top end}$ vs. $A_{\bot end}$, **C3**), as well as typical scatter plot of APS latencies (**C4**), and comparisons of initial APS latencies ($L_{\top init}$ vs. $L_{\bot init}$, **C5**) and steady APS latencies ($L_{\top end}$ vs. $L_{\bot end}$, **C6**). The $A_{\top end}$, $A_{\bot end}$, $L_{\top end}$ and $L_{\bot end}$ were averages from the final second of A-HFS. **D** For 200 Hz A-HFS ($n = 13$ rats): comparisons of initial (**D1**) and steady (**D2**) APS amplitudes, as well as initial (**D3**) and steady (**D4**) APS latencies. $^{*}P < 0.05$, $^{**}P < 0.01$, paired t-test, $n = 15$ or 13. From Feng et al. (2022)

When the pulse frequency of A-HFS$_{\top\bot}$ was increased from 100 to 200 Hz, the change of APS amplitudes showed similar patterns (Fig. 7.1D1 and D2), with the steady values decreasing to below ~ 10% of the initial value for APS$_{\top}$s and below ~ 20% for APS$_{\bot}$s, approximately half of the ratios observed at 100 Hz A-HFS$_{\top\bot}$. The mean APS$_{\bot}$ latency was significantly shorter than that of APS$_{\top}$s (Fig. 7.1D3, D4).

These results indicated that despite the significant differences in their initial activations, the $\bot$ pulse activation was not weaker than that of $\top$ pulses during sustained A-HFS$_{\top\bot}$ when the induced APSs were suppressed. We then compared APS changes between monophasic-pulse A-HFS$_{\top\bot}$ and biphasic-pulse A-HFS (denoted as A-HFS$_{++}$), using identical pulse frequency and intensity but different consumptions of electric energy.

2. **Comparison of APS signals during monophasic-pulse A-HFS$_{\top\bot}$ and biphasic-pulse A-HFS$_{++}$**

The neuronal responses to A-HFS$_{++}$ have been described in details in Chap. 5, simplified as A-HFS. For comparison, an example of 100 Hz A-HFS$_{++}$ is shown in Fig. 7.2A again. Each pulse of the A-HFS$_{++}$ produced an APS, denoted as APS$_{+}$. During the steady period of A-HFS$_{++}$, the APS$_{+}$ amplitude decreased to a level similar to that of A-HFS$_{\top\bot}$ (Fig. 7.2B1). Meanwhile, the APS$_{+}$ latency gradually increased to a steady level, falling between the latencies of APS$_{\bot}$ and APS$_{\top}$ during A-HFS$_{\top\bot}$ (Fig. 7.2B2).

Statistical data showed that at the onset of A-HFSs, the mean amplitudes of A_{+init} and $A_{\top init}$ showed no significant difference, and both were significantly greater than $A_{\bot init}$ (Fig. 7.2C1). This difference disappeared during the steady period (Fig. 7.2C2). Additionally, the mean latencies of the three types of APSs showed significant differences in both initial and steady periods of A-HFSs. Biphasic-pulses produced APS latencies (L_{+init} and L_{+end}) significantly longer than those by positive-pulses ($L_{\bot init}$ and $L_{\bot end}$), but shorter than those by negative-pulses ($L_{\top init}$ and $L_{\top end}$) (Fig. 7.2C3, C4). When the pulse frequency was increased from 100 to 200 Hz, similar results were observed, except that the mean latency L_{+end} was significantly shorter than both $L_{\top end}$ and $L_{\bot end}$ (Fig. 7.2D).

To compare the mean firing levels produced by A-HFS$_{\top\bot}$ and A-HFS$_{++}$, the amplitudes of APS$_{\top}$ and APS$_{\bot}$ induced by A-HFS$_{\top\bot}$ were averaged (shown by the rightmost

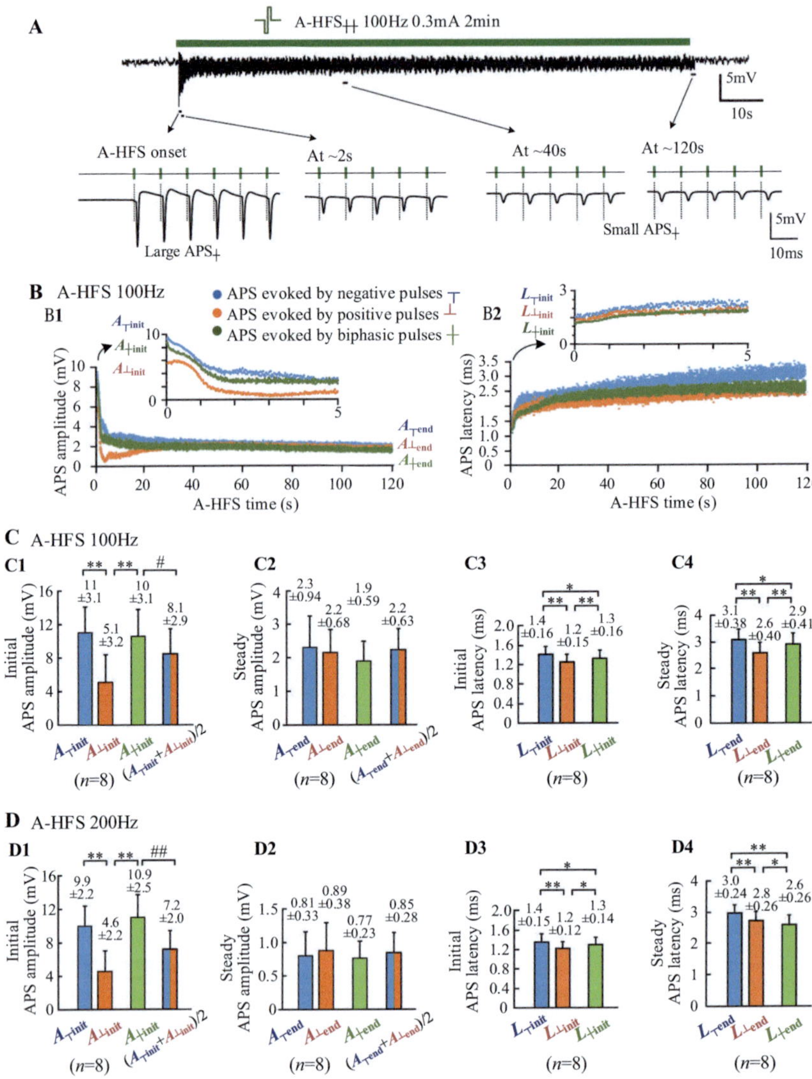

Fig. 7.2 Comparisons of APS amplitudes and latencies between A-HFS$_{\top\perp}$ and A-HFS$_{++}$. **A** Typical APS$_{+}$ signal during 2-min 100 Hz A-HFS$_{++}$. **B** Scatter plots showing APS amplitude (left) and latency (right) induced by each biphasic-pulse (green) during the A-HFS$_{++}$ shown in (**A**), along with corresponding data from APS$_{\top}$ and APS$_{\perp}$ (blue and orange, respectively) during A-HFS$_{\top\perp}$ in the same rat experiment. **C** Comparisons of initial and steady values of APS amplitudes (**C1** and **C2**) and latencies (**C3** and **C4**) among APS$_{\top}$, APS$_{\perp}$ and APS$_{+}$ induced during 100 Hz A-HFSs. (**C1**) and (**C2**) also include comparisons between the APS$_{+}$ values and the average values of APS$_{\top}$ and APS$_{\perp}$ (rightmost two-colored bar). **D** Corresponding comparisons for 200 Hz A-HFS. $^{*}P < 0.05$, $^{**}P < 0.01$, two-way ANOVA with post-hoc Bonferroni tests, $n = 8$. $^{\#}P < 0.05$, $^{\#\#}P < 0.01$, paired t-test, $n = 8$. From Feng et al. (2022)

two-colored bar in Fig. 7.2C1, C2, D1, D2). For both 100 and 200 Hz A-HFSs, at onset, the average amplitudes $(A_{\top init} + A_{\perp init})/2$ were significantly smaller than the amplitude $A_{\dagger init}$ (Fig. 7.2C1, D1). During the steady period, the $(A_{\top end} + A_{\perp end})/2$ became similar to the $A_{\dagger end}$ (Fig. 7.2C2, D2), indicating similar neuronal firing levels for both types of A-HFS. However, A-HFS$_{\dagger\dagger}$ consumed twice the electric energy of A-HFS$_{\top\perp}$.

3. **Safety of Monophasic-pulse A-HFS$_{\top\perp}$**

During A-HFS$_{\top\perp}$, a gap existed between the negative and positive pulses—10 ms at 100 Hz and 5 ms at 200 Hz. Could this gap disrupt the charge balance between positive and negative pulses and cause neuronal damage? To address this concern, we compared the recovery of neuronal activity after A-HFS$_{\top\perp}$ versus A-HFS$_{\dagger\dagger}$. Single test pulses with parameters identical to A-HFS pulses were applied at ~ 30 s intervals after A-HFS ended. Both types of A-HFS showed over 90% recovery in APS amplitudes and latencies within ~ 2 min after A-HFS, indicating no obvious neuronal damages caused by either A-HFS (Fig. 7.3).

7.1.2 Opposite-Polarity Pulses Activate Different Subpopulations of Neurons During A-HFS$_{\top\perp}$

The above results showed that despite significant differences in the initial APSs, the steady-state APSs during both A-HFS$_{\top\perp}$ and AHFS$_{\dagger\dagger}$ were suppressed to a similar level. This suppression indicated that each pulse activated only a small portion of the neuronal population that was originally responsive during baseline and at A-HFS onset. Since the $\top$ and $\perp$ pulses activate different sites along an axon (Brocker and Grill 2013; Ranck 1975; Rattay 1999), we hypothesized that the suppressed APSs induced by the two types of pulses during A-HFS$_{\top\perp}$ could be generated by distinct subpopulations of neurons.

To test this hypothesis, we utilized the refractory period following APSs to determine whether the two opposite pulses activated the same neuronal population. As shown in Fig. 7.4A, in baseline conditions, pairs of biphasic-pulses were applied with progressively shortened IPIs. The first pulse was designated as the control pulse and the second as the test pulse. At an IPI of 1.5 ms, the test pulse induced a smaller APS (denoted as $A_{\dagger test}$). When the IPI decreased to 0.8 ms, no $A_{\dagger test}$ was induced, indicating a baseline refractory period of ~ 1 ms. However, during sustained 100 Hz A-HFS$_{\dagger\dagger}$ with identical parameters, an additional pulse inserted 5 ms after an A-HFS pulse induced no APS (denoted as $A_{\dagger ins}$ in Fig. 7.4B), indicating that A-HFS prolonged the refractory period to longer than 5 ms (refer to Sect. 5.2.2).

Pairs of monophasic-pulses were applied with a leading $\top$ pulse followed by a $\perp$ pulse in baseline conditions (Fig. 7.4C). At an IPI of 1.5 ms, the $\perp$ pulse induced only a smaller APS (denoted as $A_{\perp test}$). When the IPI decreased to 0.8 ms, no $A_{\perp test}$ was induced due to the refractory period. This suggested that the neuronal population

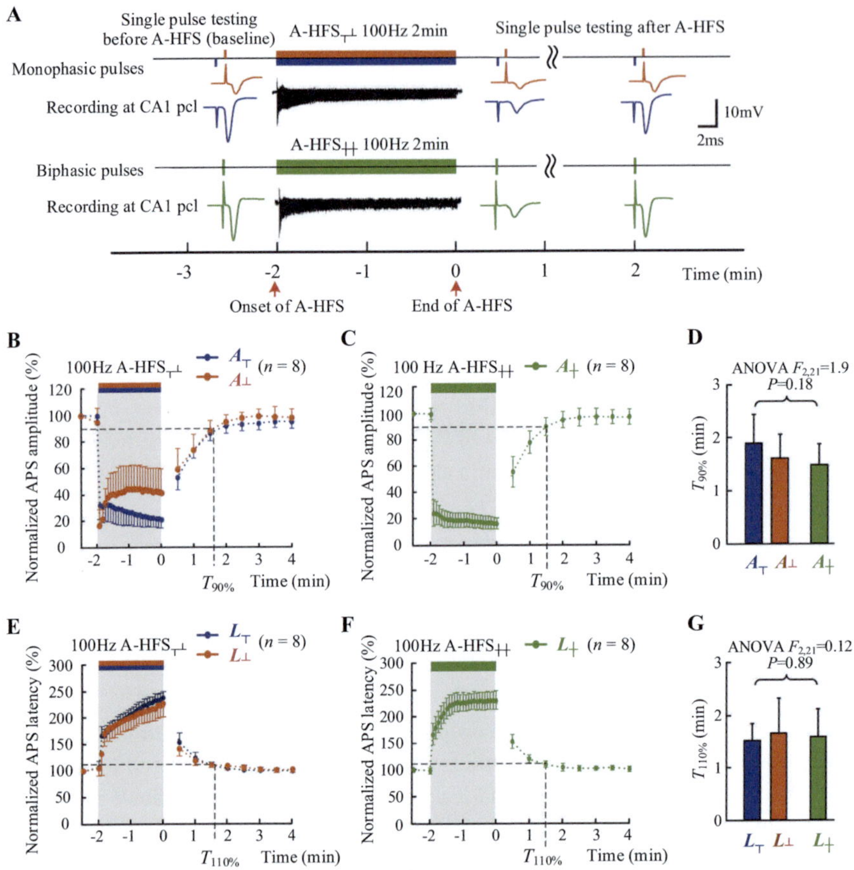

Fig. 7.3 Recoveries of neuronal responses after A-HFS with alternating monophasic-pulses and A-HFS with biphasic-pulses. **A** Typical APSs to single-pulse tests before and after both types of A-HFS. **B** and **C** Changes in the mean APS amplitudes normalized to baseline values for both A-HFS types. The shadow denotes the A-HFS period during which the first data point was from the single APS induced by the first A-HFS pulse, while each subsequent data point represented an average APS value from the last second of each 5-s interval. Post-A-HFS data points were from single-pulse tests at 30-s intervals. **D** Comparison of mean recovery times for APS amplitudes to reach 90% of baseline value ($T_{90\%}$) calculated by linear interpolation as illustrated in (**B**) and (**C**). **E–G** APS latency data corresponding to the APS amplitude data in (**B–D**). The recovery time of APS latency was measured when the increased latency returned to 110% of baseline value ($T_{110\%}$). From Feng et al. (2022)

activated by the ⊥ pulse was fully included within the population activated by the ⊤ pulse. With an IPI shorter than the refractory period, the subsequent ⊥ pulse could not reactivate the neurons that had just fired in response to the leading ⊤ pulse.

During sustained 100 Hz A-HFS$_{\top\perp}$, an additional pulse of ⊤ or ⊥ was inserted in the middle of the 10 ms IPI—5 ms after an A-HFS$_{\top\perp}$ pulse (Fig. 7.4D). These insertions generated different APSs. When the polarity of the inserted pulse was

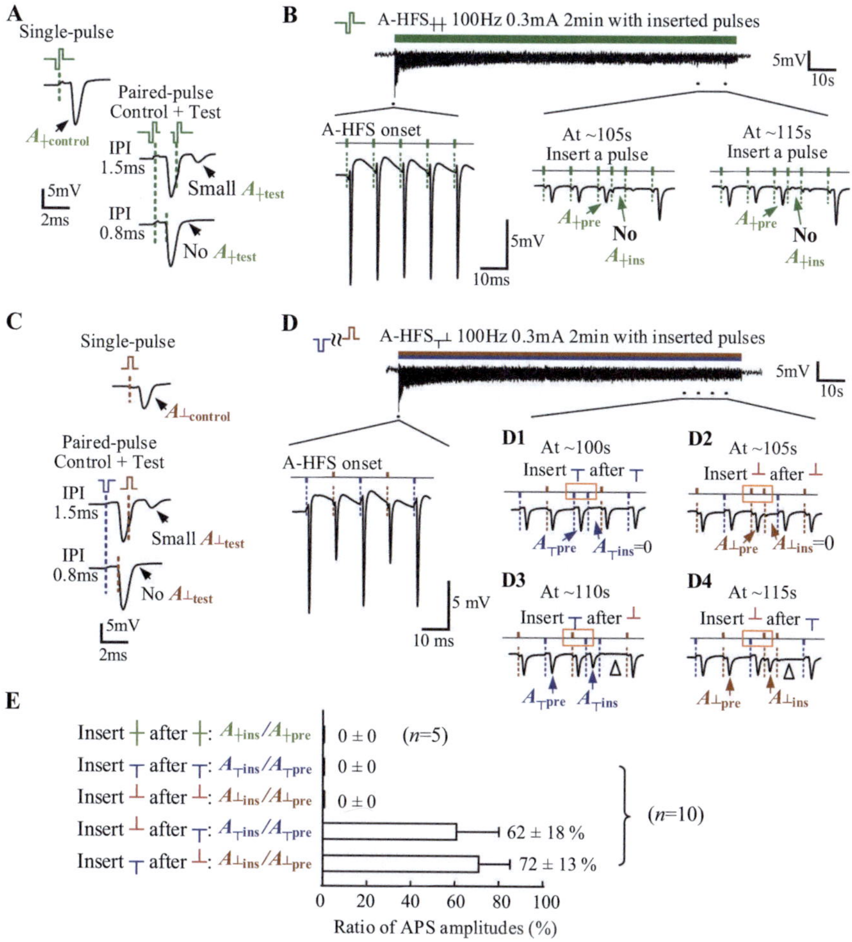

Fig. 7.4 Different activation patterns during sustained A-HFS$_{++}$ and A-HFS$_{\top\perp}$. **A** Examples of the control APS ($A_{+\text{control}}$) induced by a single biphasic-pulse (left) and the APSs induced by paired biphasic-pulses with IPI of 1.5 and 0.8 ms (right) under baseline conditions. **B** During 2 min 100 Hz A-HFS$_{++}$, test pulses inserted at ~ 105 and ~ 115 s in the middle of 10 ms IPIs produced no APS ($A_{+\text{ins}} = 0$). **C** Examples of the control APS ($A_{\perp\text{control}}$) induced by a single ⊥ pulse and the APSs induced by paired ⊤ and ⊥ pulses with IPI of 1.5 and 0.8 ms in baseline conditions. **D** During 2 min 100 Hz A-HFS$_{\top\perp}$, test-pulses inserted at ~ 100, ~ 105, ~ 110 and ~ 115 s in the middle of 10 ms IPI formed four types of tests (orange boxes in **D1**–**D4**): ⊤ after ⊤, ⊥ after ⊥, ⊤ after ⊥ and ⊥ after ⊤. **E** Amplitude ratios between the APSs induced by the inserted test pulses and those induced by the preceding A-HFS$_{\top\perp}$ pulses of the same polarity. From Feng et al. (2022)

same as the preceding A-HFS$_{\top\perp}$ pulse, it induced no APS (Fig. 7.4D1, D2), similar to the results during A-HFS$_{++}$ (Fig. 7.4B). However, when the inserted pulse had opposite polarity to the preceding pulse, it induced a substantial APS (Fig. 7.4D3, D4) with a mean amplitude of 60%–70% of the APS induced by the last previous pulse with the same polarity (Fig. 7.4E). Meantime, the inserted pulse prevented any APS following the subsequent A-HFS$_{\top\perp}$ pulse of the same polarity (denoted by the hollow triangles in Fig. 7.4D3, D4), which further demonstrating the effect of extended refractory period.

These results showed that monophasic-pulses of opposite polarities controlled distinct neuronal subpopulations during sustained A-HFS$_{\top\perp}$, although initially a $\perp$ pulse only controlled a subset of neurons responding to a $\top$ pulse. The following additional experiment further verified this finding. We set an A-HFS$_{\top\perp}$ train by repeating a paired-pulse (IPI = 2.5 ms) at 50 Hz for 2 min (Fig. 7.5A). This resulted in the IPIs of A-HFS$_{\top\perp}$ alternating between 2.5 and 17.5 ms—with each $\perp$ pulse following the shorter 2.5 ms IPI and each $\top$ pulse following the longer 17.5 ms IPI. At the A-HFS$_{\top\perp}$ onset, due to the refractory period, the induced APS$_{\perp}$ following the shorter IPI was significantly smaller than both its baseline control (APS$_{\perp control}$) and the APS$_{\top}$ following the longer IPI (Fig. 7.5B, *left*). The APS$_{\perp}$ then almost disappeared in the period of 0.6–17 s before reappearing with an increasing amplitude until it reached the APS$_{\top}$ level. During the steady period of A-HFS$_{\top\perp}$, the amplitude ratio of APS$_{\perp steady}$/APS$_{\perp control}$ was significantly greater than that of APS$_{\top steady}$/APS$_{\top control}$ (Fig. 7.5B, *right*). The APS$_{\perp}$ showed a smaller suppression rate (~ 65%) than that of APS$_{\top}$ (~ 81%). In contrast, when the same alternating IPI sequence (2.5 and 17.5 ms) was applied to A-HFS$_{++}$, the APS following shorter IPIs disappeared completely (refer to Sect. 6.2.1). The persistent APS$_{\perp steady}$ following APS$_{\top steady}$ with the short 2.5 ms IPI demonstrated that APS$_{\perp steady}$ was unaffected by the refractory period of the preceding APS$_{\top steady}$. The result again confirmed that A-HFS$_{\top\perp}$ pulses of opposite polarities activated different subpopulation neurons. Then, which subset of neurons was activated by each type of pulses?

As described in Sect. 5.2.1, during the steady period of 50 Hz biphasic-pulse A-HFS$_{++}$, the suppression rate of APS$_{+}$ was approximately 60%. Similarly, in the 100 Hz A-HFS$_{\top\perp}$, where the $\perp$ pulses also had a frequency of 50 Hz, the suppression rate of APS$_{\perp}$ during the steady period was approximately 60%, too (Fig. 7.5B). Based on this, we inferred that the neuronal population originally activated by both $\top$ and $\perp$ pulses could become responsive only to the $\perp$ pulses—not the $\top$ pulses—during the steady period of A-HFS$_{\top\perp}$. At baseline, the activation area of $\top$ pulse was larger than and encompassed that of $\perp$ pulse. Could sustained A-HFS$_{\top\perp}$ have made the activation capability of $\perp$ pulses to surpass that of $\top$ pulses? To test this hypothesis, we designed the following experiment to make identical activation areas for both pulse types at baseline, then compared the neuronal responses to them during the steady period of A-HFS$_{\top\perp}$.

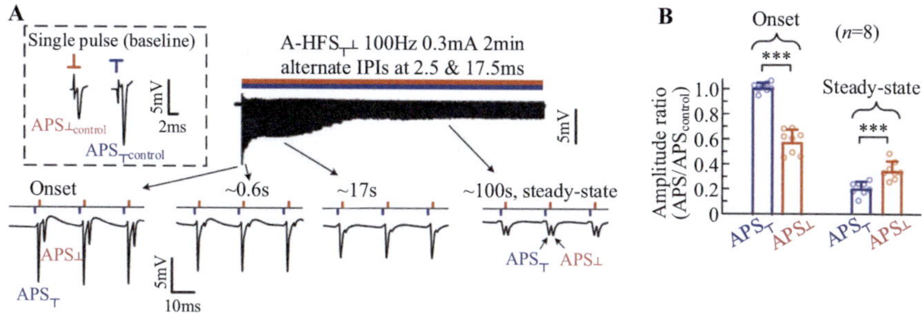

Fig. 7.5 Neuronal responses to A-HFS$_{\top\perp}$ with alternating IPIs of 2.5 and 17.5 ms. **A** Typical APS signal during 2 min 100 Hz A-HFS$_{\top\perp}$ with alternating IPIs. APS$_{\top control}$ and APS$_{\perp control}$, induced by single $\top$ and $\perp$ pulses in baseline, are shown in the upper left. **B** Comparison of the normalized amplitudes of APS$_{\top}$ and APS$_{\perp}$ at the onset and steady period of A-HFS$_{\top\perp}$. $^{***}P < 0.001$, paired *t*-test, $n = 8$ rats. Modified from Hu et al. (2023)

7.1.3 Only Positive Pulses Can Activate Neurons Initially Responsive to Both Pulse Types

At an identical intensity, the activation area of a $\top$ pulse was larger than that of a $\perp$ pulse. To equalize the activation areas, we reduced the $\top$ pulse intensity. However, using different intensities while keeping the same 100 μs pulse width would disrupt the charge balance during A-HFS$_{\top\perp}$. To preserve this balance, we added an opposite phase with 1/40 intensity and 40 × width immediately following each $\top$ and $\perp$ pulse, creating quasi-monophasic-pulses (denoted as $\top_Q$ and $\perp_Q$). Their corresponding APSs were denoted as APS$_{\top Q}$ and APS$_{\perp Q}$ (Fig. 7.6A1). Under baseline conditions, these pulses produced activation effects similar to those without a balance phase, as verified by comparable APS amplitudes (Fig. 7.6A2). This indicated that the balance phase did not substantially affect the action of the leading phase.

During baseline tests, we kept the intensity of $\perp_Q$ while gradually reduced the intensity of $\top_Q$ to match the APS$_{\top Q}$ amplitude with the APS$_{\perp Q}$ amplitude (Fig. 7.6B1). The mean APS amplitude evoked by $\top_Q$ at ~ 0.1 mA was similar to that evoked by $\perp_Q$ at ~ 0.3 mA (Fig. 7.6B2). Additionally, in paired-pulse stimulations of both "$\top_Q\perp_Q$" and "$\perp_Q\top_Q$" with IPIs ranging from 1 to 2.5 ms (Fig. 7.6C), the first APS showed consistent amplitude whether evoked by $\top_Q$ or $\perp_Q$. The second APS was fully suppressed at 1 ms IPI and emerged progressively as IPI increased. These results indicated that $\top_Q$ and $\perp_Q$ shared a similar activation area.

These pulses were then applied alternately to form the A-HFS$_{\top Q\perp Q}$ with a constant 10 ms IPI (Fig. 7.6D1). Initially, the amplitudes of APS$_{\top Q}$ and APS$_{\perp Q}$ were similar, though both gradually decreased. During the steady period of the A-HFS$_{\top Q\perp Q}$, the APS$_{\top Q}$ almost disappeared while $\perp_Q$ pulses continued to produce clear APS$_{\perp Q}$ events. The mean amplitudes of the two APS types differed significantly (Fig. 7.6D2). Additionally, when we removed all $\top_Q$ pulses from the 100 Hz A-HFS$_{\top Q\perp Q}$ to create an A-HFS$_{\perp Q}$ with pure $\perp_Q$ pulses at 50Hz (Fig. 7.6E1), the ratio of APS$_{\perp Q_steady}$/

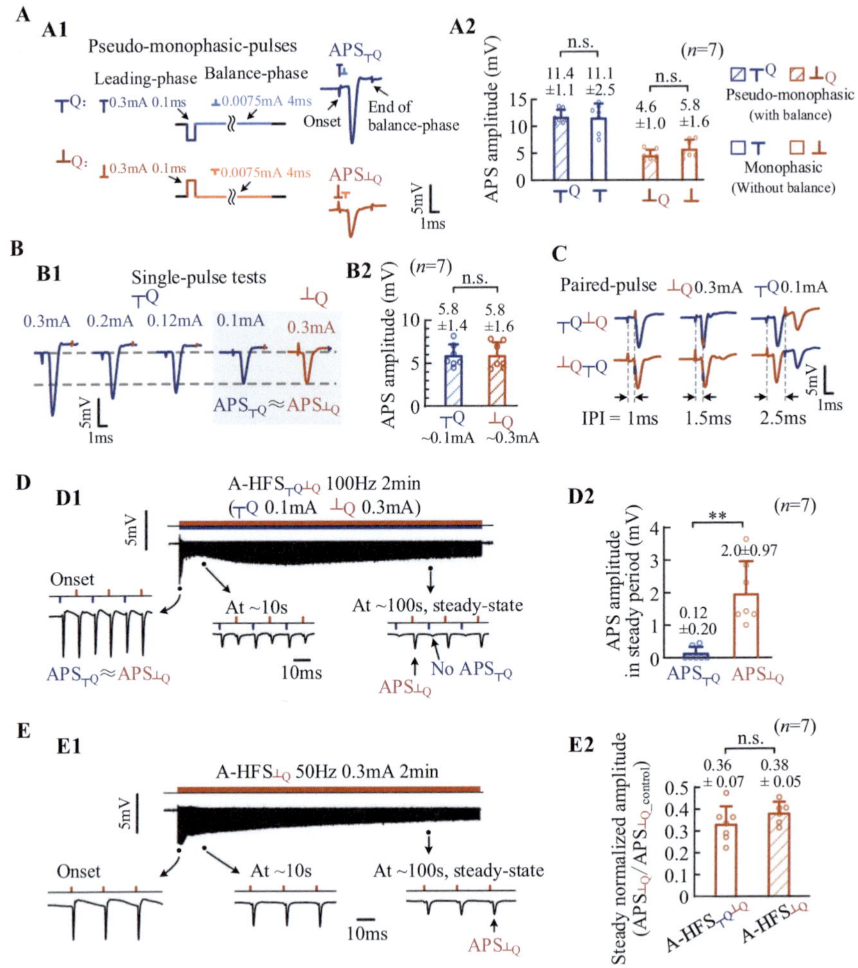

Fig. 7.6 Responses of neurons within the shared area activated by quasi-monophasic negative- and positive-pulses. **A A1**: Schematic diagram showing the creations of quasi-monophasic-pulses with a balance phase, denoted as $\top_Q$ and $\perp_Q$ (left), along with typical APSs induced by these pulses (right). **A2**: Comparison of APS amplitudes induced by negative- and positive-pulses with and without a balance phase. **B B1**: APSs induced by single-pulses of 0.3–0.1 mA $\top_Q$ and 0.3 mA $\perp_Q$. **B2**: Comparison of the APS amplitudes induced by ~ 0.1 mA $\top_Q$ and ~ 0.3 mA $\perp_Q$. **C** APSs produced by paired-pulse stimulations of "$\top_Q\perp_Q$" and "$\perp_Q\top_Q$" (0.1 mA $\top_Q$ and 0.3 mA$\perp_Q$) with short IPIs of 1, 1.5 and 2.5 ms. **D D1**: Typical APS signal during 100 Hz A-HFS$_{\top Q\perp Q}$ alternating between 0.1 mA $\top_Q$ and 0.3 mA $\perp_Q$. **D2**: Comparison of APS$_{\top Q}$ and APS$_{\perp Q}$ amplitudes in the steady period of A-HFS$_{\top Q\perp Q}$. **E E1**: Typical APS signal during 50 Hz A-HFS$_{\perp Q}$ with mere 0.3 mA $\perp_Q$ pulses. **E2**: Comparison of mean amplitude ratios APS$_{\perp Q_steady}$/APS$_{\perp Q_control}$ between A-HFS$_{\top Q\perp Q}$ (shown in **D1**) and A-HFS$_{\perp Q}$ (shown in **E1**). In **A2**, **B2**, and **D2**: n.s. $P > 0.05$, $^{**}P < 0.01$, paired t-test. In **E2**: n.s. $P > 0.05$, t-test, $n = 7$. From Hu et al. (2023)

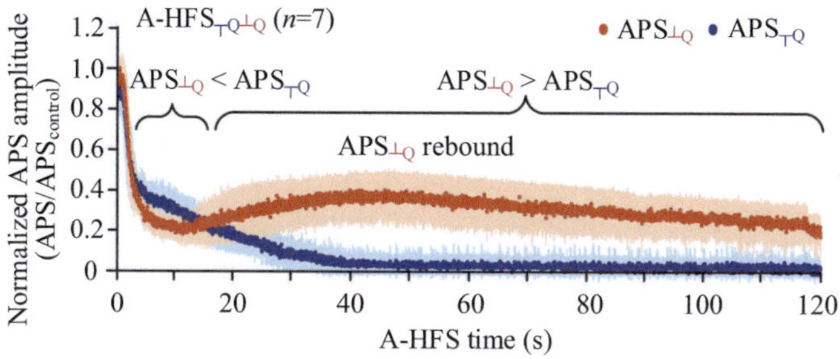

Fig. 7.7 Scatter plot showing mean normalized amplitudes of APS$_{\top Q}$ and APS$_{\perp Q}$ produced by each pulse of A-HFS$_{\top Q \perp Q}$ alternating between ~ 0.1 mA $\top$Q and ~ 0.3 mA $\perp$Q. The light-colored ranges represent ± standard deviation. From Hu et al. (2023)

APS$_{\perp Q_control}$ in the A-HFS$_{\perp Q}$ was comparable to that in the A-HFS$_{\top Q \perp Q}$ (Fig. 7.6E2). These results showed that during the steady period of A-HFS$_{\top Q \perp Q}$, only $\perp_Q$ pulses were able to produce APSs in neurons within the shared activation area of both polarity pulses, similar to the action of $\perp_Q$ pulses in A-HFS$_{\perp Q}$ without $\top_Q$ pulses.

Furthermore, statistical data showed that during the ~ 2 to ~ 15 s A-HFS$_{\top Q \perp Q}$ period, the mean amplitude of APS$_{\top Q}$ exceeded that of APS$_{\perp Q}$ before eventually decreasing to below APS$_{\perp Q}$ in the late period (Fig. 7.7). That is, during the transition period of declining APSs, the activation of $\perp_Q$ initially dropped below that of $\top_Q$ but later rebounded to surpass it. Since current experimental techniques cannot effectively track pulse-induced action potentials in individual neuronal axons in the brain, we employed a computational axon model to simulate neuronal responses to both $\top$ and $\perp$ pulses during A-HFS$_{\top\perp}$ to investigate the possible underlying mechanisms.

7.2 Simulation Study of Axonal Activity During HFS with Alternating Opposite-Polarity Pulses

In addition to our findings shown in Chap. 5, the experimental results in Sect. 7.1 also demonstrate that HFS-induced suppression of neuronal activations must have originated in the axons rather than the somata. Otherwise, if the suppression had originated in the somata, the action potentials (AP) initially generated on axons by the two types of pulses during A-HFS$_{\top\perp}$ would not have produced distinct APS$_{\top}$ and APS$_{\perp}$ suppressions, let alone the mere APS$_{\perp Q}$ responses during sustained A-HFS$_{\top Q \perp Q}$. Therefore, to focus on the axonal mechanisms, we simulated the neuronal responses to A-HFS$_{\top\perp}$ using the model of myelinated axons in the NEURON simulation environment (Guo et al. 2018; Hu et al. 2023).

As shown in Fig. 7.8A, the axon model consisted of 21 Ranvier nodes (denoted as Node$_0$ to Node$_{20}$) and 20 myelinated internode segments. Please refer to the literature for detailed model parameters, including morphologic parameters, passive and active electrical properties of axon membrane, as well as the accumulation and clearance mechanisms of K^+ in the narrow peri-axonal space (Guo et al. 2018; Bellinger et al. 2008; Beurrier et al. 2001). The structure and materials of the stimulation electrode (SE) in the model were consistent with those used in the rat experiments (refer to Sect. 3.4.2). The electrical field potential of A-HFS$_{\top\perp}$, delivered by the SE, was simulated in COMSOL Multiphysics 5.3 (COMSOL Inc. Sweden) and then loaded into the NEURON to mimic the bipolar stimulation in the rat experiments (Zheng et. al. 2020). The SE tip center was positioned above the central node (Node$_{10}$) of the axon. To analyze the mechanisms of axonal responses to the alternating $\top$ and $\perp$ pulses during A-HFS$_{\top\perp}$, the following model variables at each axonal node were recorded and analyzed: the membrane potential (V_m), the inactivation variable of Na^+ channels (h_Na), and the K^+ concentration in the narrow peri-axonal space ($[K^+]_o$).

The h_Na variable represents the inactivation state of Na^+ channels in the HH model of neuronal membrane (refer to Sect. 1.3). When the membrane is depolarized, voltage-gated Na^+ channels open, triggering an action potential. These Na^+ channels then rapidly enter an inactive state, becoming unresponsive to external stimuli during their refractory period. Once the membrane potential repolarizes, the inactive Na^+ channels enter a closed state, after which they can be reactivated. However, if the membrane remains depolarized, the Na^+ channels stay inactive. The h_Na value ranges from 0 to 1, where smaller values indicate more severe Na^+ channel inactivation, making the axonal membrane harder to trigger action potentials. Therefore, analyzing h_Na values can reveal the change of neuronal excitability caused by stimulations.

As shown in Fig. 7.8B, since the central node (Node$_{10}$) of the axon is located directly under the SE, a $\top$ pulse can depolarize and initiate an AP at Node$_{10}$ while simultaneously hyperpolarizing the nodes in the two flanking regions (e.g. Node$_8$, denoted by the blue ◀ in the left column of Fig. 7.8B). In contrast, a $\perp$ pulse can depolarize and initiate APs at the flanking nodes (e.g. Node8 and Node12) while simultaneously hyperpolarizing Node$_{10}$ (Fig. 7.8B, right column). With single-pulse stimulations, the AP initiated by either a $\top$ or $\perp$ pulse can successfully conduct to the end of axon (e.g. Node$_0$), provided that the $\top$ pulse-induced AP at Node$_{10}$ is strong enough to overcome the hyperpolarization at flanking nodes (Kiernan and Bostock 2000; van de Steene et al. 2020; Zheng et. al. 2022). We define an activation that can successfully conduct to Node$_0$ as a successful activation; otherwise, it is a failed activation.

During A-HFS$_{\top\perp}$, membrane potentials at the Node$_0$ (V_m_Node$_0$, Fig. 7.8C) show three types of firing (denoted as Firing-A, -B and -C) on axons at varying distances from the SE (Fig. 7.8A). Firing-A and -B occur on axons within the activated area of both $\top$ and $\perp$ pulses, while Firing-C appears only on axons activated by $\top$ pulses. The axons showing Firing-A are closest to the SE, while those showing Firing-C are farthest away. As shown in Fig. 7.8C, in the onset period, the Firing-A axon follows each $\top$ and $\perp$ pulse to fire an AP and successfully conducts it. However, during intermediate and steady periods, this axon fails at $\top$ pulses while succeeding only at $\perp$ pulses. The Firing-B axon initially behaves like Firing-A but during the intermediate period, it succeeds only at $\top$ pulses. Later, it switches to succeeding

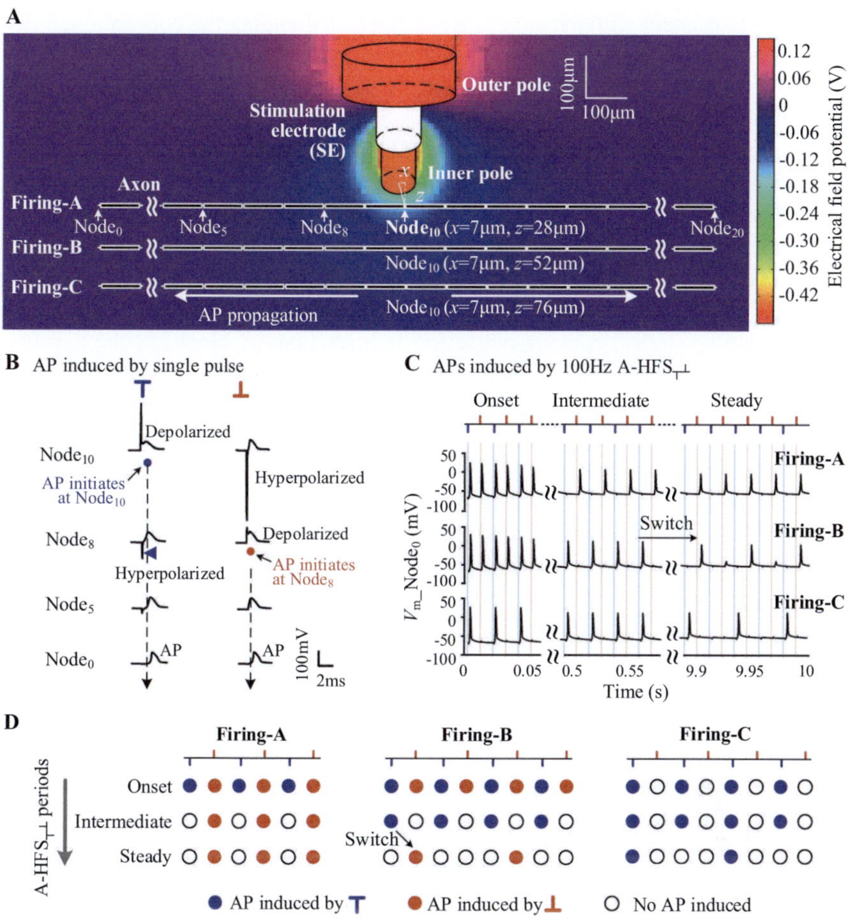

Fig. 7.8 Computational model of axons positioned at different distances from the stimulation electrode (SE) and their responses to A-HFS$_{\top\perp}$. **A** Picture of the simulated electrical field potentials generated by a $\top$ pulse. **B** Depolarization, hyperpolarization, and AP generated by single $\top$ (left) and $\perp$ (right) pulses at different nodes along an axon. **C** Membrane potentials at Node$_0$ (V_{m}_Node$_0$) generated in the three types of axons shown in (**A**) during the onset, intermediate and steady periods of 100 Hz A-HFS$_{\top\perp}$. **D** Schematic summary of the three firing types shown in (**C**). Blue and orange dots represent successful activations by $\top$ and $\perp$ pulses respectively, while black circles represent failed activations. From Hu et al. (2023)

at $\perp$ pulses through the steady period. The Firing-C axon responds exclusively to $\top$ pulses throughout the entire A-HFS$_{\top\perp}$.

The schematic summary of the three firing types (Fig. 7.8D) shows how their integration reproduces the observations from rat experiments (Fig. 7.7). The axons initially activated by both $\top$ and $\perp$ pulses respond only to $\perp$ pulses during the steady period of A-HFS$_{\top\perp}$. Even when $\top$ pulses temporarily dominate in the intermediate period, successful activations eventually switch to $\perp$ pulses alone. These simulation results are consistent with the experimental data, challenging the conventional view that ⊤ pulses are more effective than ⊥ pulses at activation. To reveal the underlying mechanisms, we next examine the dynamic changes in the axonal membrane from our simulation results.

The simulation shows that increased $[K^+]_o$ in the peri-axonal space plays a critical role in generating axon activation failures during A-HFS$_{\top\perp}$ (Fig. 7.9). At Node$_{10}$, where ⊤ pulses initiate AP, sustained A-HFS$_{\top\perp}$ increases $[K^+]_o$_Node$_{10}$ (Fig. 7.9A), leading to partial inactivation of Na^+ channels. The inactivation variable of Na^+ channels at Node$_{10}$ (*h*_Na_Node$_{10}$) decreases (Fig. 7.9B), resulting in reduced amplitude of the initial AP with ⊤ pulses (Fig. 7.9C). These decreased APs at Node$_{10}$ fail to propagate to Node$_0$, causing activation failures (axonal block) at the ⊤ pulses (Fig. 7.9D). A similar axonal block should occur in the activations of ⊥ pulses that initiate APs at Node$_8$. However, ⊥ pulses can still generate successful AP in some axons (Firing-A and Firing-B in Fig. 7.8D). The following simulation results show how interactions between ⊤ and ⊥ pulses affect the generations of Firing-A and Firing-B, which succeed only at ⊥ pulses during the steady period of A-HFS$_{\top\perp}$.

Figure 7.10 illustrates an axon with Firing-A, showing the simulation data from Node$_{10}$ to Node$_0$ (top to bottom). At the onset of A-HFS$_{\top\perp}$ (Fig. 7.10A), the axon

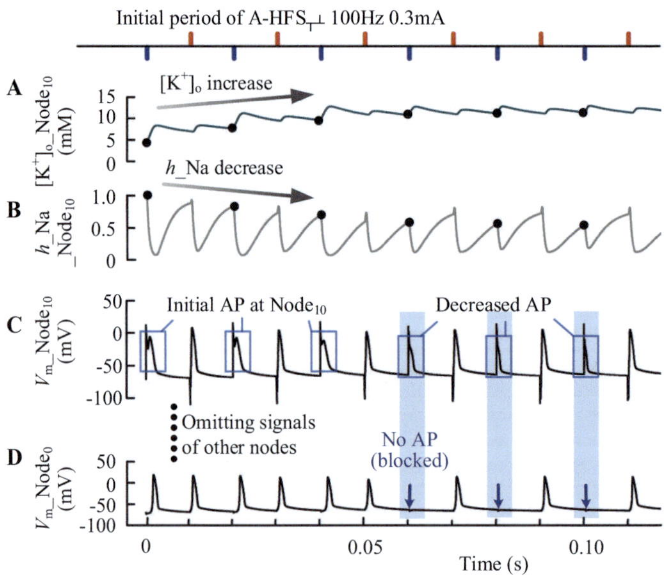

Fig. 7.9 Increased $[K^+]_o$ generating axonal block during A-HFS$_{\top\perp}$. **A** $[K^+]_o$ in the peri-axonal space near Node$_{10}$ ($[K^+]_o$_Node$_{10}$). **B** Inactivation variable of Na^+ channels at Node$_{10}$ (*h*_Na_Node$_{10}$). In **A** and **B**, black dots on the curves mark the values when ⊤ pulses arrive. **C** Membrane potential at Node$_{10}$ (V_m_Node$_{10}$). The APs initiated by ⊤ pulses at Node$_{10}$ are indicated by blue boxes. **D** Membrane potential at Node$_0$ (V_m_Node$_0$). Blue shadows denote the failures of AP conduction to Node$_0$, representing axonal block. From Hu et al. (2023)

responds to each ⊤ and ⊥ pulses to initiate an AP at $Node_{10}$ and $Node_8$, respectively. These APs can successfully conduct to $Node_0$. During the intermediate period (Fig. 7.10B), as *h*_Na_$Node_{10}$ gradually decreases, the amplitude of APs initiated by ⊤ pulses at $Node_{10}$ also decrease. This reduces the AP conduction ability, eventually leading to conduction failure due to the strong hyperpolarization at $Node_8$ (highlighted by the blue ◀ within the blue shadow in Fig. 7.10B). However, the attenuation of AP conduction to $Node_8$ allows *h*_Na at $Node_8$ to recover, resulting in increased *h*_Na_$Node_8$ when the next ⊥ pulse arrives (denoted by the orange arrow line on *h*_Na_$Node_8$). The ⊥ pulse then initiates an increased AP at Node8, which successfully conducts to $Node_0$ (shown in green shadow in Fig. 7.10B). Simultaneously, this $Node_8$-AP propagates in the opposite direction to $Node_{10}$, inducing an AP (denoted by the red ▲ on V_m_$Node_{10}$) larger than the AP induced by the previous ⊤ pulse. This $Node_{10}$-AP further decreases *h*_Na_$Node_{10}$ before the next ⊤ pulse. Repeatedly, the ⊥-induced AP on $Node_{10}$ causes the *h*_Na_$Node_{10}$ to decrease progressively by the arrival of each subsequent ⊤ pulse until the ⊤ pulses can no longer initiate APs at $Node_{10}$, resulting in generation failure of AP (Fig. 7.10C). In this process, once a ⊤ pulse fails to make a successful AP, subsequent ⊤ pulses cannot induce any propagable APs again.

During the steady period of A-$HFS_{⊤⊥}$ (Fig. 7.10C), the decreased *h*_Na at each axon node is steady. When a ⊤ pulse arrives, *h*_Na_$Node_{10}$ is too small to initiate an AP. However, a ⊥ pulse can still initiate a propagable AP at $Node_8$. Although the amplitude of $Node_8$-AP decreases obviously due to the decreased *h*_Na_$Node_8$, the AP is still able to propagate successfully since no hyperpolarization blocks its conduction pathway.

For an axon with Firing-B (Fig. 7.11), at the onset of A-$HFS_{⊤⊥}$ (Fig. 7.11A), its firing resembles Firing-A, with successful AP reaching $Node_0$ following each pulse. However, during the intermediate period (Fig. 7.11B), as *h*_Na decreases, the first conduction failure occurs with the AP induced by a ⊥ pulse (shown in the orange shadow in the left of Fig. 7.11B), not by a ⊤ pulse. This occurs because this axon locates farther from the SE than the axon with Firing-A, resulting in a weaker activation of ⊥ pulses at $Node_8$. Similarly, a ⊤ pulse produces weaker hyperpolarization at the flanking regions of this axon. The hyperpolarization is insufficient to block the conduction of AP initiated at $Node_{10}$, despite the initial AP having decreased due to decreased *h*_Na_$Node_{10}$. Consequently, the ⊥ pulse, being less effective at depolarizing the neuronal membrane, causes AP conduction failure before the ⊤ pulse does.

Nevertheless, after about 1.2 s of stimulation, as *h*_Na_$Node_{10}$ continues to decrease, the further decreased AP—initiated by a ⊤ pulse at $Node_{10}$—fails to overcome the hyperpolarization in its conduction pathway, resulting in a conduction failure (shown in the blue shadow in the right of Fig. 7.11B). Once a ⊤ pulse fails to generate a propagable AP, the failure facilitates *h*_Na_$Node_8$ to recover, allowing a ⊥ pulse to induce a successful AP (indicated by the orange arrow line on *h*_Na_$Node_8$ and the green shadow in the right of Fig. 7.11B). This successful AP results in a further decrease in *h*_Na_$Node_{10}$ when a ⊤ pulse arrives, preventing the pulse from inducing a successful AP. Consequently, the neuronal response switches from a

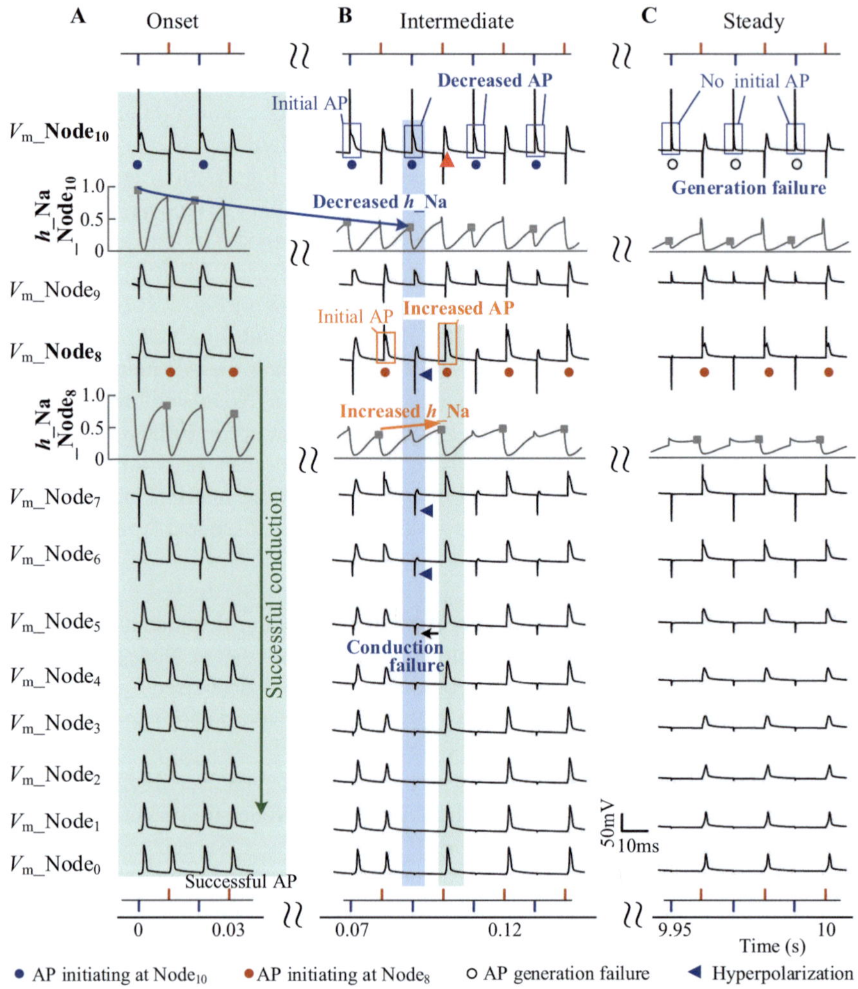

Fig. 7.10 Changes in membrane potential (V_m) at each axon node during onset (**A**), intermediate (**B**) and steady (**C**) periods of A-HFS$_{\top\perp}$ in a Firing-A neuron. The h_Na curves at Node$_{10}$ (h_Na_Node$_{10}$) and at Node$_8$ (h_Na_Node$_8$) are shown below their corresponding V_m curves. Small squares on h_Na_Node$_{10}$ and h_Na_Node$_8$ respectively highlight the values when the $\top$ and $\perp$ pulses arrive. From Hu et al. (2023)

transition period with only $\top$-induced APs to one with only $\perp$-induced APs (indicated by the red line at V_m_Node$_0$ in the bottom of Fig. 7.11B). The subsequent process is similar to that of Firing-A, but with periodic conduction failures of APs initiated by $\perp$ pulses (indicated by the orange shadows in Fig. 7.11C). $\top$ pulses no longer generate propagable APs due to failure at Node$_{10}$.

In summary, the simulation results show that axonal responses to A-HFS$_{\top\perp}$ decline due to two failures: AP conduction failure and AP generation failure. Both failures

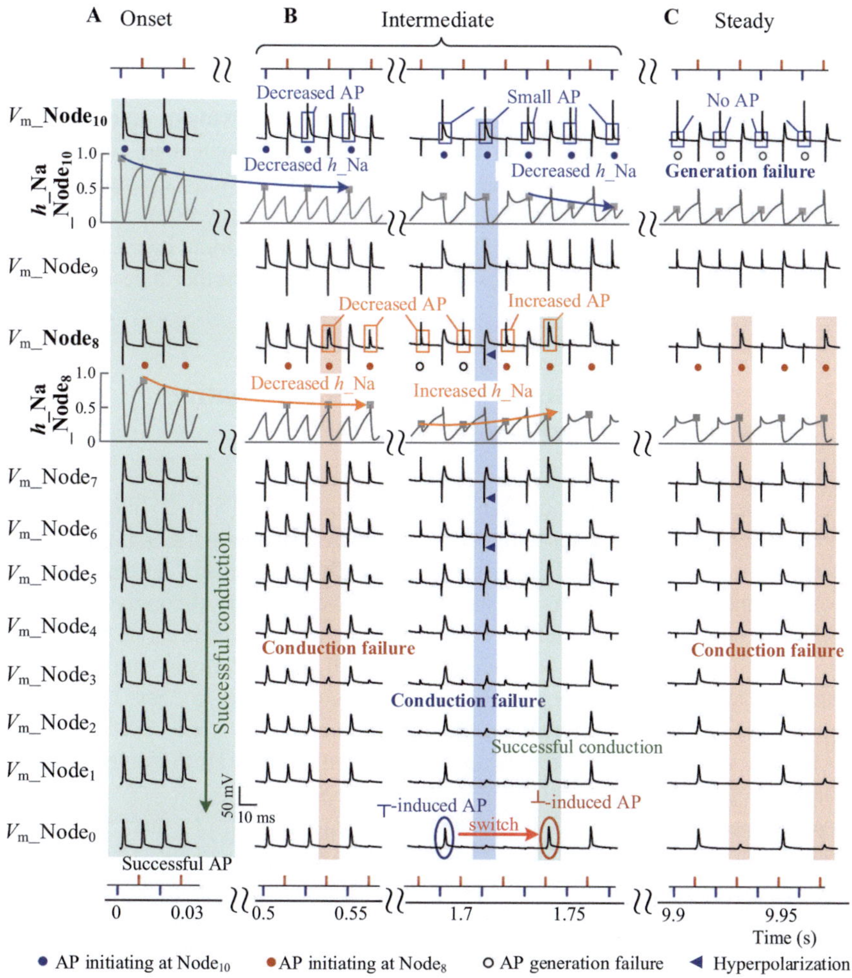

Fig. 7.11 Changes in membrane potential (V_m) at each axon node during onset (**A**), intermediate (**B**) and steady (**C**) periods of A-HFS$_{\top\perp}$ in a Firing-B neuron. The h_Na_Node$_{10}$ and h_Na_Node$_8$ curves are shown below their corresponding V_m curves. Small squares on h_Na_Node$_{10}$ and h_Na_Node$_8$ respectively highlight the values when the $\top$ and $\perp$ pulses arrive. From Hu et al. (2023)

originate from the HFS-induced increase of $[K^+]_o$ in the narrow peri-axonal space, which elevates the axonal membrane potential and then inactivates Na^+ channels partially. Due to hyperpolarization in the conduction pathway, an AP conduction failure eventually occurs upon a $\top$ pulse. Once this occurs, subsequent $\top$ pulses cannot succeed again due to the effect of $\perp$ pulse-induced APs. Additionally, AP conduction failure can first occur with a $\perp$ pulse-induced AP, resulting in the transition period where $APS_{\top} > APS_{\perp}$, as showed in Fig. 7.7. The transition period cannot prevent the eventual conduction failure of $\top$ pulse-induced AP. Ultimately, no $\top$ pulse but $\perp$

pulses can elicit AP successfully in the shared activation area of both polarity pulses. Therefore, the interactions among the neuronal responses to the opposite pulses result in the advantage of $\perp$ pulses.

Nevertheless, this simulation model only includes axons without other neuronal structures. While this simplification helps highlight axonal mechanisms in the stimulations, it may introduce bias by neglecting the influence of other structures. Additionally, besides the depolarization block at stimulated axons caused by increased $[K^+]_o$, the neuronal responses to A-HFS$_{\top\perp}$ could also involve more mechanisms or a combination of multiple mechanisms—all of which require further investigations.

7.3 Neuronal Responses to O-HFS with Alternating Opposite-Polarity Pulses

The above results indicate that opposite-polarity pulses can activate distinct subpopulations of neuronal axons when applied alternately in sustained A-HFS. If these pulse sequences were used as O-HFS to orthodromically activate downstream neurons through synaptic transmissions, the alternately activated axon subpopulations could produce strong excitatory inputs to postsynaptic neurons through numerous axonal terminals. Since these axonal activations occurred within a 10 ms or shorter IPI, the spatial and temporal integration of these synaptic inputs (refer to Sect. 2.2.3) could amplify these activations, enhancing the responses of postsynaptic neurons. To test this hypothesis, we applied the pulse sequences used in the A-HFS$_{\top\perp}$ to the Schaffer collaterals of rat hippocampal CA1 region as O-HFS$_{\top\perp}$.

We conducted the O-HFS$_{\top\perp}$ at three pulse frequencies (100, 133, and 200 Hz) and compared them with biphasic-pulse O-HFS$_{++}$ at the same frequencies. Figure 7.12A shows an example of 2 min O-HFS$_{\top\perp}$ at 133 Hz. In the early stimulation period, OPSs continuously occurred. The synchronous firing of neuronal populations resembled epileptiform discharge and lasted about 43 s. We termed this the "OPS period". In the later O-HFS$_{\top\perp}$ period, no obvious OPSs appeared (though occasional OPSs reappeared in some rats). Figure 7.12B shows O-HFS$_{++}$ applied to the same rat for comparison. The OPS$_1$ induced by the first pulse showed similar amplitude to that in O-HFS$_{\top\perp}$, but the OPS period lasted less than 10 s. No obvious OPSs appeared in the later O-HFS$_{++}$ period. At all three frequencies, O-HFS$_{\top\perp}$ consistently produced longer mean OPS periods than O-HFS$_{++}$ (Fig. 7.12C). The durations of OPS periods for both types of O-HFS decreased as pulse frequency increased. This result was consistent with the changes in the durations of initial OPS periods during O-HFS$_{++}$ at 50, 100 and 200 Hz (refer to Sect. 5.3.1).

During O-HFS$_{\top\perp}$, the frequency of each pulse type was half the stimulation frequency. Therefore, we next compared 200 Hz O-HFS$_{\top\perp}$ with 100 Hz O-HFS$_{++}$ (Fig. 7.13). Although the first $\top$ pulse of O-HFS$_{\top\perp}$ and $+$ pulse of O-HFS$_{++}$ produced similar OPS$_1$ amplitudes at onset, during the initial 0.2 s period, O-HFS$_{\top\perp}$ produced smaller OPSs than O-HFS$_{++}$. However, as stimulation continued, the subsequent

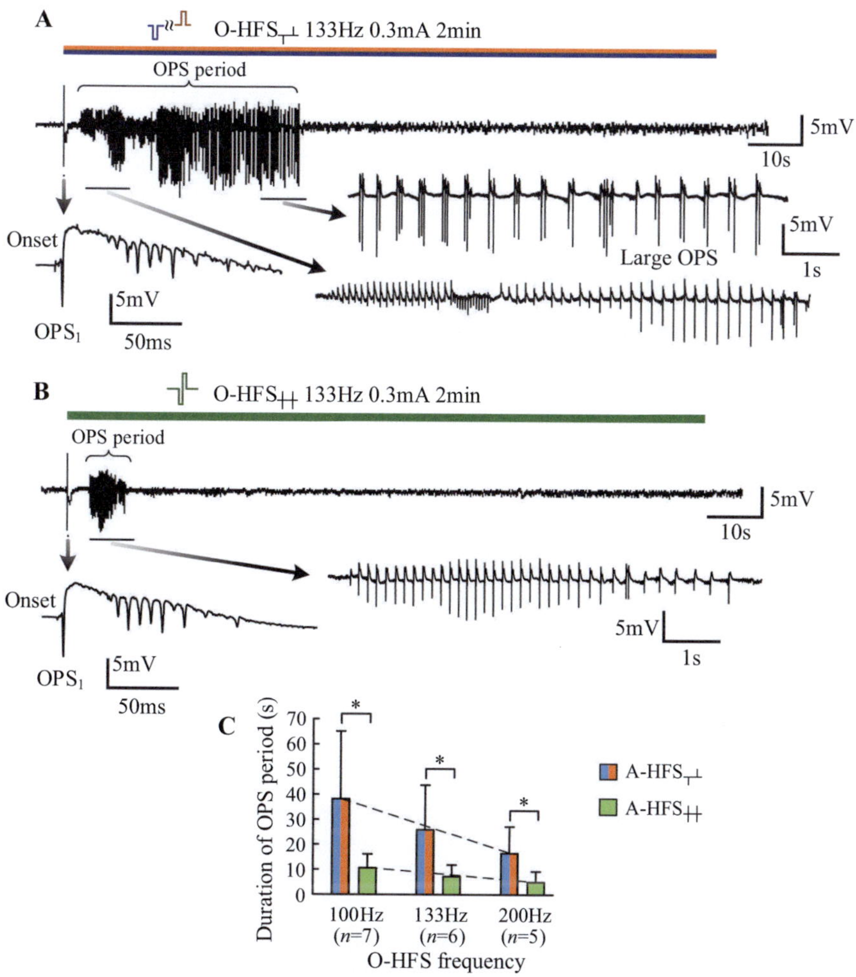

Fig. 7.12 Comparison of neuronal responses to O-HFS$_{\top\perp}$ and O-HFS$_{++}$ in the Schaffer collaterals of rat hippocampal CA1 region. **A**, **B** Typical recordings of O-HFS$_{\top\perp}$ and O-HFS$_{++}$ at 133 Hz. **C** Comparisons of the mean durations of OPS periods produced by O-HFS$_{\top\perp}$ and O-HFS$_{++}$ at three frequencies (100, 133, and 200 Hz). $^{*}P < 0.05$, paired *t*-test

OPSs in O-HFS$_{\top\perp}$ became larger than those in O-HFS$_{++}$ (Fig. 7.13A, B). Additionally, the mean OPS period duration in 200 Hz O-HFS$_{\top\perp}$ was longer than that in 100 Hz O-HFS$_{++}$ (Fig. 7.13C). This indicated that when the pulse frequency of each polarity in O-HFS$_{\top\perp}$ is same as that of O-HFS$_{++}$, O-HFS$_{\top\perp}$ can produce stronger activation.

These results indicate that the activations of distinct subpopulation neurons by the opposite-polarity pulses during O-HFS$_{\top\perp}$ can enhance the excitatory effect on downstream neurons in the projection regions through synaptic integrations, thereby reducing the activation attenuation caused by axonal block.

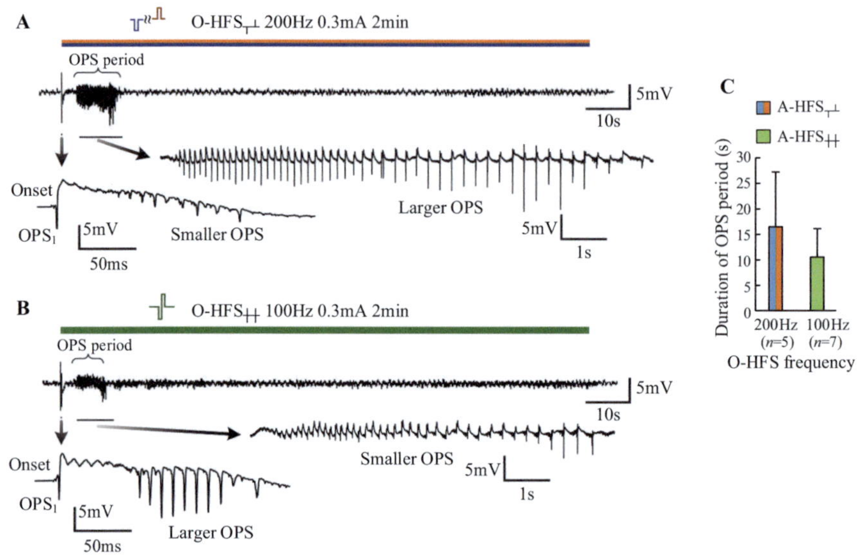

Fig. 7.13 Comparison between neuronal responses to $\text{O-HFS}_{\top\perp}$ at 200 Hz and O-HFS_{++} at 100 Hz. **A, B** Typical recordings of $\text{O-HFS}_{\top\perp}$ at 200 Hz and O-HFS_{++} at 100 Hz, respectively. **C** Comparison of the durations of OPS periods produced by the two types of O-HFS

7.4 Summary

The findings presented in this chapter challenge the conventional view of negative pulses being more effective than positive pulses. Usually, in extracellular stimulations, depolarization caused by a positive pulse is much weaker than that by a negative pulse at the same intensity, because a positive pulse depolarizes the axonal membranes on flanks of the stimulation site through a "virtual cathode" mechanism. Therefore, in baseline conditions, the APS induced by a positive pulse was much smaller than that induced by a negative pulse (Fig. 7.1C2, D1). However, during sustained HFS, the firing level produced by positive pulses became similar to that of negative pulses in the directly simulated neurons. Additionally, the two types of pulses activated distinct subpopulations of neurons. For neurons responsive to both types of pulses at baseline, positive pulses became more effective than negative pulses during steady HFS periods. Furthermore, alternating HFS with opposite monophasic-pulses is equivalent to inserting a short gap between the negative and positive phases of biphasic pulses. It can achieve safety similar to biphasic-pulse stimulation (refer to Fig. 7.3).

The simulation results of our computational axon model show the possible mechanisms behind neuronal responses to alternating monophasic-pulses of opposite polarity. The actions of the two types of pulses do not cancel each other out, but rather exhibit complex interactions. During HFS, the axonal membrane initially generates continuous action potentials (AP), which increases extracellular potassium concentration ($[K^+]_o$). The elevated $[K^+]_o$ raises the membrane potential and partially inactivates Na^+ channels. As the stimulation sustains, both types of pulses eventually begin to fail to induce and conduct APs. However, in the shared activation area of both types of pulses, the activation effect of positive pulses surpasses that of negative pulses.

These findings provide clues for designing new stimulation patterns for neuromodulation. During sustained HFS with alternating negative- and positive-pulses, the opposite pulses activate two subpopulations of neurons, whose discharges are not constrained by the refractory period from each other's firing (Figs. 7.4 and 7.5), resulting in enhanced activations on projection neurons (Figs. 7.12 and 7.13). The stimulation electrodes used in our experiments were coaxial bipolar electrodes with an inner polar surface area of only 0.028mm^2 (see Fig. 3.13 in Sect. 3.4.2)—about one-third of the outer polar surface area. This small inner polar can create higher current density and stronger electric field near the electrode tip, making it the active electrode in the bipolar stimulation. Despite the small active region, our experimental results indicate that alternating negative- and positive-pulses can generate substantial synchronous discharge in downstream postsynaptic neurons by amplifying the activations induced in the directly stimulated axons.

This type of alternating stimulations with opposite-polarity pulses may improve the efficacy of certain neuromodulation therapies. For example, previous studies on the thresholds and loudness of cochlear implants (CI) in patients have shown that HFS with alternating monophasic-pulses of opposite polarity at specific intervals is more effective than biphasic-pulses (van Wieringen et al. 2005, 2006). The results in this chapter suggest that this improvement may be due to the enhanced HFS activations. Other studies have shown that one of the important roles of DBS therapy is to generate new neuronal activity to replace and mask pathological activity (Chiken and Nambu 2016; Lee et al. 2019). The enhanced excitatory effect from opposite pulses could therefore improve DBS efficiency. Conventional HFS used in current DBS can cause axonal depolarization block and synaptic transmitter depletion—which help block pathological activity but reduce the excitatory effect of stimulation, limiting its ability to produce new neuronal activity. Our simulation results in Sect. 7.2 show that HFS with alternating negative- and positive-pulses can maintain the axonal block effect. Rat experiments with A-HFS confirm that this stimulation paradigm can achieve comparable axonal block to biphasic-pulse HFS (Fig. 7.2). Therefore, the alternating stimulations may both enhance the excitatory effect of stimulation and prevent the spread of pathological neuronal activity. Additionally, this method consumes only half the electric energy of biphasic-pulse HFS at the same stimulation frequency, making it more energy-efficient.

Some neural prostheses that use electrical stimulations could potentially utilize this characteristic of monophasic pulses with opposite polarities to activate different neuron subpopulations. Some implanted neural prostheses use multi-contact microelectrodes or arrays to deliver stimulation pulses for neural coding, such as auditory midbrain implants (AMI) (Lim and Lenarz 2015), visual cortical prostheses (VCP) (Fernández et al. 2021), and tactile implants in the somatosensory cortex

(Hughes et al. 2021). Delivering opposite-polarity pulses alternately through a stimulation contact could add a new dimension to neural coding by activating different subgroups of neurons. These neuronal responses could be fine-tuned by varying pulse parameters, including intervals between opposite polarity pulses, pulse intensity ratios, and frequency ratios, to encode neural information. However, the practical implementation of such approach requires more research.

References

Basser PJ, Roth BJ (2000) New currents in electrical stimulation of excitable tissues. Annu Rev Biomed Eng 2:377–397

Bellinger SC, Miyazawa G, Steinmetz PN (2008) Submyelin potassium accumulation may functionally block subsets of local axons during deep brain stimulation: a modeling study. J Neural Eng 5(3):263–274

Beurrier C, Bioulac B, Audin J et al (2001) High-frequency stimulation produces a transient blockade of voltage-gated currents in subthalamic neurons. J Neurophysiol 85(4):1351–1356

Brocker DT, Grill WM (2013) Principles of electrical stimulation of neural tissue. In: Handbook of clinical neurology, vol 116. Elsevier, pp 3–18

Cappaert NL, Ramekers D, Martens HC et al (2013) Efficacy of a new charge-balanced biphasic electrical stimulus in the isolated sciatic nerve and the hippocampal slice. Int J Neural Syst 23(1):1250031

Carlyon RP, van Wieringen A, Deeks JM et al (2005) Effect of inter-phase gap on the sensitivity of cochlear implant users to electrical stimulation. Hear Res 205(1–2):210–224

Chiken S, Nambu A (2016) Mechanism of deep brain stimulation: Inhibition, excitation, or disruption? Neuroscientist 22(3):313–322

Deprez M, Luyck K, Luyten L et al (2018) An evaluation of the effect of pulse-shape on grey and white matter stimulation in the rat brain. Sci Rep 8(1):752

Eickhoff S, Jarvis JC (2021) Pulse shaping strategies for electroceuticals: a comprehensive survey of the use of interphase gaps in miniature stimulation systems. IEEE Trans Biomed Eng 68(5):1658–1667

Feng Z, Zheng L, Yuan Y et al (2022) Cathodic- and anodic-pulses can alternately activate different sub-populations of neurons during sustained high-frequency stimulation of axons in rat hippocampus. J Neural Eng 19(1):016030

Fernández E, Alfaro A, Soto-Sánchez C et al (2021) Visual percepts evoked with an intracortical 96-channel microelectrode array inserted in human occipital cortex. J Clin Invest 131(23):e151331

Foutz TJ, McIntyre CC (2010) Evaluation of novel stimulus waveforms for deep brain stimulation. J Neural Eng 7(6):066008

Gorman PH, Mortimer JT (1983) The effect of stimulus parameters on the recruitment characteristics of direct nerve-stimulation. IEEE Trans Biomed Eng 30(7):407–414

Guo Z, Feng Z, Wang Y et al (2018) Simulation study of intermittent axonal block and desynchronization effect induced by high-frequency stimulation of electrical pulses. Front Neurosci 12:858

Haji Ghaffari D, Finn KE, Jeganathan VSE et al (2020) The effect of waveform asymmetry on perception with epiretinal prostheses. J Neural Eng 17(4):045009

Hu N, Feng Z, Guo Z et al (2015) Different stimulation effects between monophasic pulses and biphasic pulses in deep brain stimulation. Chin J Biomed Eng 34(5):548–557 (in Chinese)

Hu Y, Feng Z, Zheng L et al (2023) Interactions between cathodic- and anodic-pulses during high-frequency stimulations with the monophasic-pulses alternating in polarity at axons—experiment and simulation studies. J Neural Eng 20(5):056021

Hughes CL, Flesher SN, Weiss JM et al (2021) Perception of microstimulation frequency in human somatosensory cortex. Elife 10:e65128

Kiernan MC, Bostock H (2000) Effects of membrane polarization and ischaemia on the excitability properties of human motor axons. Brain 123(Pt 12):2542–2551

Lee DJ, Lozano CS, Dallapiazza RF et al (2019) Current and future directions of deep brain stimulation for neurological and psychiatric disorders. J Neurosurg 131(2):333–342

Lim HH, Lenarz T (2015) Auditory midbrain implant: Research and development towards a second clinical trial. Hear Res 322:212–223

Montgomery EB (2017) Deep brain stimulation programming: mechanisms, principles and practice (2nd ed). Oxford University Press

Ranck JB (1975) Which elements are excited in electrical stimulation of mammalian central nervous system: a review. Brain Res 98(3):417–440

Rattay F (1989) Analysis of models for extracellular fiber stimulation. IEEE Trans Biomed Eng 36(7):676–682

Rattay F (1999) The basic mechanism for the electrical stimulation of the nervous system. Neuroscience 89(2):335–346

van de Steene T, Tanghe E, Tarnaud T et al (2020) Sensitivity study of neuronal excitation and cathodal blocking thresholds of myelinated axons for percutaneous auricular vagus nerve stimulation. IEEE Trans Biomed Eng 67(12):3276–3287

van den Honert C, Mortimer JT (1979) The response of the myelinated nerve fiber to short duration biphasic stimulating currents. Ann Biomed Eng 7(2):117–125

van Wieringen A, Carlyon RP, Laneau J et al (2005) Effects of waveform shape on human sensitivity to electrical stimulation of the inner ear. Hear Res 200(1–2):73–86

van Wieringen A, Carlyon RP, Macherey O et al (2006) Effects of pulse rate on thresholds and loudness of biphasic and alternating monophasic pulse trains in electrical hearing. Hear Res 220(1–2):49–60

Weitz AC, Behrend MR, Ahuja AK et al (2014) Interphase gap as a means to reduce electrical stimulation thresholds for epiretinal prostheses. J Neural Eng 11(1):016007

Zheng L, Feng Z, Hu H et al (2020) The appearance order of varying intervals introduces extra modulation effects on neuronal firing through non-linear dynamics of sodium channels during high-frequency stimulations. Front Neurosci 14:397

Zheng L, Feng Z, Xu Y et al (2022) An anodic phase can facilitate rather than weaken a cathodic phase to activate neurons in biphasic-pulse axonal stimulations. Front Neurosci 16:823423

Chapter 8
Effects of High-Frequency Stimulations on Epileptiform Discharges

Abstract This chapter presents our studies on the inhibitory effects of electrical stimulations on the epileptiform discharges generated by three methods: brief high-frequency stimulation and two epileptogenic agents (4-aminopyridine and picrotoxin). Results showed that both pulse and sinusoidal high-frequency stimulations (HFS) at the Schaffer collaterals (the hippocampal CA1 afferent fibers), can suppress epileptiform spikes induced by these methods. The stimulation patterns and action mechanisms can vary with the epileptogenic methods, involving processes such as desynchronization and depolarization block. The findings suggest customizing stimulation parameters (e.g., duration) and modes (e.g., open-loop and closed-loop) is essential for effectively suppressing various types of epileptiform discharges.

Keywords Epileptiform discharge · 4-Aminopyridine · Picrotoxin · Sustained stimulation · Brief stimulation · Population spike · Discharge suppression

Epilepsy is a common brain disorder affecting approximately 1% of the global population. Currently, about 30% of patients have drug-resistant and refractory epilepsy (Davis and Gaitanis 2020; Smolarz et al. 2021). For these patients, neuromodulation provides a potential therapy through methods like vagal nerve electrical stimulation (VNS), trigeminal nerve stimulation (TNS), and deep brain stimulation (DBS) that targets brain regions such as the thalamus and hippocampus (Boon et al. 2007; Nune et al. 2015; Davis and Gaitanis 2020; Geller et al. 2017; Jobst et al. 2017). Adaptive DBS, also known as closed-loop DBS—has shown clear efficacy and received clinical approval in several countries (Vetkas et al. 2022; Simpson et al. 2022; Sun and Morrell 2014; Cukiert and Lehtimäki 2017; Li and Cook 2018). However, the mechanisms through which DBS treats epilepsy remain incompletely understood.

The main cause of epileptiform discharges is an imbalance between inhibition and excitation in neural networks, which results in excessive neuronal excitation and abnormal synchronous discharges. In field potentials and electroencephalograms, these discharges manifest as epileptiform spikes, indicating increased and synchronized neuronal firing. Therefore, reducing neuronal excitability or/and weakening firing synchronicity can suppress epileptiform discharges.

Z. Feng, *Electrical Brain Stimulation: Mechanisms and Modulation of Neuronal Activity*, https://doi.org/10.1007/978-981-95-4145-4_8

How does DBS control epileptiform discharges—by reducing neuronal firing, weakening firing synchronicity, or both? The narrow pulses delivered by DBS can potentially depolarize neuronal membranes and enhance synchronous firing in directly stimulated neurons (Lowet et al. 2022). However, through the complex neural networks of the brain, the ultimate effect of DBS may differ from its direct action. For instance, when the stimulation activates inhibitory interneurons or their axons and terminals, the resulting inhibition can suppress the firing of principal neurons thereby reducing excitatory inputs to projection areas (Lafreniere-Roula et al. 2010; Prescott et al. 2013; Chiken and Nambu 2013; Birdno et al. 2014). In early DBS application, using HFS to treat Parkinson's disease presented reversible functional impairment, suggesting that HFS can suppress neuronal firing. The observations of decreased firing and silent periods in neurons near the stimulation site supported this view (Dostrovsky et al. 2000; Filali et al. 2004; Lafreniere-Roula et al. 2010). Paradoxically, studies have shown that even when DBS inhibited neuronal somata locally and reduced their firing, the narrow pulses could still activate these neurons' axons, leading to increased neuronal firing in downstream projection areas (McIntyre et al. 2004; Birdno et al. 2014).

Moreover, excessive depolarization produced by stimulation can inactivate membrane Na^+ channels, resulting in depolarization block and suppressed neuronal firing. For example, excessive depolarization in the soma membrane can interrupt epileptiform discharge (Bragin et al. 1997b; Durand and Bikson 2001). In axons, such depolarization can block action potential conduction (refer to the simulation results shown in Sect. 7.2). At synapses, overactivation of the presynaptic membrane can deplete neurotransmitters and disrupt synaptic transmission (Neher and Sakaba 2008; Anderson et al. 2006; Iremonger et al. 2006; Rosenbaum et al. 2014). Under normal conditions, when HFS is applied to the afferent pathway of Schaffer collaterals in the hippocampal CA1 region, it can initially produce strong activation in downstream neurons and induce large population spikes. However, sustained HFS can produce random neuronal firing with reduced synchronicity (refer to Sect. 5.3 for details). This raises a question: can asynchronous activations from axonal HFS suppress epileptiform activity in neurons through over increased excitation? We investigated this hypothesis in the rat hippocampal CA1 region using two epileptiform models: post-stimulation discharges triggered by brief HFS trains and discharges induced by local injections of epileptogenic agents.

8.1 Effect of Sustained Axonal HFS on Neuronal Population Discharges

We applied O-HFS trains with biphasic-pulses at 100 Hz to the Schaffer collaterals in two durations (5 s and 1 min) respectively and recorded the CA1 neuronal responses downstream of the stimulation site (Fig. 8.1A). The 5-s O-HFS produced an after-discharge (AD) with large population spikes (PS) that lasted about 10 s after the

O-HFS ended (Fig. 8.1A1, upper). However, during the 1-min O-HFS, following the same initial 5 s stimulation, only small PSs occurred, and no AD appeared after the O-HFS ended (Fig. 8.1A1, bottom). We termed this period of small PSs during HFS as the "HFS-discharge". The neuronal responses were similar during the initial 5 s periods of both O-HFS durations (enlarged views of ① and ② in Fig. 8.1A2), but the subsequent responses differed (enlarged view of ③ in Fig. 8.1A2). The mean duration of the ADs following 5 s O-HFS showed no significant difference from that of the HFS-discharges within 1 min O-HFS (Fig. 8.1B1). However, the mean PS amplitude in the ADs was significantly larger than that in HFS-discharges (Fig. 8.1B2), indicating that the HFS beyond 5 s suppressed PSs. Although the AD-PSs showed larger amplitudes, they had smaller mean full width at half maximum (FWHM, Fig. 8.1B3) and lower mean rate (Fig. 8.1B4). This resulted in similar sums of PS areas per second during both ADs and HFS-discharges (Fig. 8.1B5). Therefore, while the overall firing levels were similar, the HFS-discharge showed lower synchronicity.

To further investigate the effect of continuous O-HFS on PS discharges, two 5-s trains of O-HFS were applied with a 5-s interval (Fig. 8.2). The first O-HFS produced an AD with large PSs. The second O-HFS significantly suppressed PS amplitudes, but large PSs reappeared after it ended. This effect was consistent at both 100 Hz and 200 Hz O-HFS frequencies.

These results reveal an interesting phenomenon that persisting the excitatory inputs by extending O-HFS can suppress rather than enhance PS discharges. Previous studies have shown that AD stems from impaired inhibitions of GABAergic synapses on pyramidal neurons (Fujiwara-Tsukamoto et al. 2007; Ye and Kaszuba 2017). In the initial period of O-HFS, the strong excitatory inputs—from both O-HFS and produced pyramidal neuron discharges—can activate interneurons in the CA1 local inhibitory circuits through feedforward and feedback connections (Fig. 8.1A, refer to Sect. 2.3 for details). These activations can trigger GABAergic synapses on post-synaptic pyramidal neurons. When these inhibitory synapses are over activated by HFS, excessive chloride ion inflow and bicarbonate ion outflow can result in rapid changes in ion concentrations across post-synaptic membranes (Isomura et al. 2003). This can reverse the normal hyperpolarized synaptic potential into depolarization, converting GABAergic synapses on post-synaptic pyramidal neurons from inhibitory to excitatory (Staley et al. 1995; McCarren and Alger 1985; Pearce et al. 1995). Additionally, brief O-HFS trains on the Schaffer collateral axons can enhance synaptic transmission efficiency through synaptic plasticity (Zucker and Regehr 2002; Rotman et al. 2011; Hennig 2013). These factors can disrupt the balance between inhibition and excitation on pyramidal neurons, leading to AD occurrence. Reports have shown that brief HFS trains can produce ADs in various brain regions of both animals and humans (Jobst et al. 2010; Shigeto et al. 2013; Hannan et al. 2020; Lesser et al. 1999).

The decreased PS amplitude during long O-HFS indicated a suppression from sustained O-HFS. An amplitude decrease in PSs can stem from two factors: fewer neurons participating in the discharge or/and reduced firing synchronicity among neurons. The PSs in HFS-discharge exhibited a larger mean FWHM in waveform (Fig. 8.1B3) and a higher firing rate (Fig. 8.1B4), showing more frequent but less synchronized in the discharges. Additionally, the mean durations of HFS-discharges

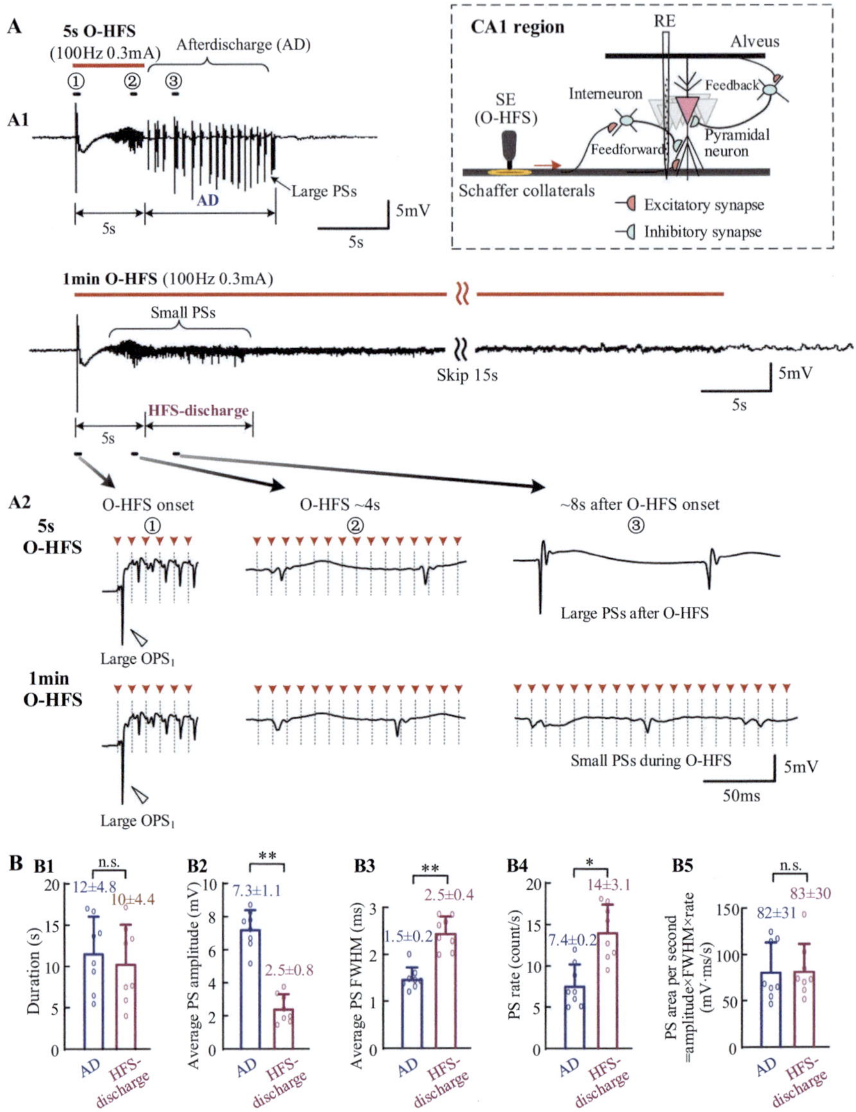

Fig. 8.1 Suppression of neuronal population discharges by continuous axonal O-HFS. **A A1**: Typical recordings in the CA1 pyramidal cell layer showing the neuronal discharges produced by a 5-s brief O-HFS (upper) and a 1-min long O-HFS (bottom) in a same rat experiment. The schematic diagram in the upper right shows the CA1 neural circuits and electrode placements. **A2**: Enlarged views during different O-HFS periods. **B** Comparisons between ADs and HFS-discharges for duration (**B1**), average PS amplitude (**B2**), average PS FWHM (**B3**), PS rate (**B4**), and sum of PS areas per second (**B5**). $^{*}P < 0.05$, $^{**}P < 0.01$, paired t-tests, $n = 8$ rats

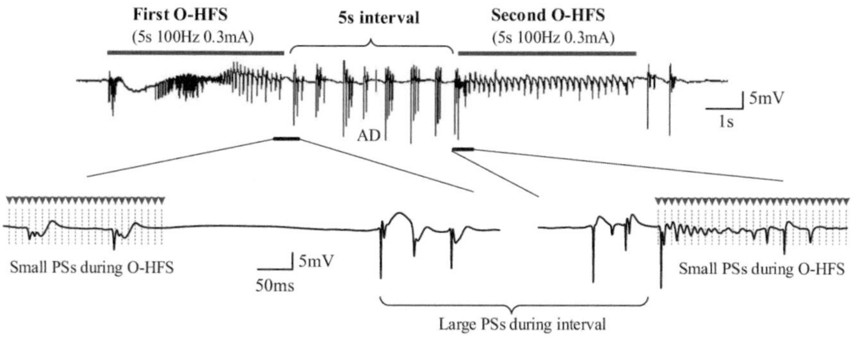

Fig. 8.2 Example of neuronal responses to two 5-s brief O-HFS trains separated by a 5-s interval

and ADs were similar (Fig. 8.1B1) with comparable sum areas of PSs (Fig. 8.1B5), suggesting that the overall firing levels during the two discharge types were similar. Therefore, decreased firing synchronicity can be a possible mechanism of the PS suppression during HFS-discharges (Wang et al. 2021). The unit spike analysis in Sect. 5.3 also supports this point, showing that sustained O-HFS has an excitatory effect on downstream postsynaptic neurons, but with reduced synchronicity.

The above experiments showed that prolonged O-HFS can suppress PSs; however, the AD epileptiform activity used as a control was itself induced by the initial O-HFS. To validate the suppressive effect of O-HFS on epileptiform discharges, we tested it further using other epileptiform models as below.

8.2 Effect of HFS on Epileptiform Activity Induced by 4-Aminopyridine

We used 4-aminopyridine (4-AP), an epileptogenic agent, to create an epileptiform model. 4-AP is a potassium channel blocker that can slow the repolarization of action potentials, prolong the depolarization state and increase the excitability of neuronal membrane (Bean 2007; Storm 1987). Additionally, the prolonged depolarization caused by 4-AP can increase the release of excitatory neurotransmitters like glutamates from presynaptic membranes (Tibbs et al. 1989; Thomsen and Wilson 1983). It can also increase the release of inhibitory neurotransmitters like GABA (Peña and Tapia 1999, 2000). However, overactivation of GABAergic synapses can weaken and reverse their inhibitory effect. Collectively, these effects enable 4-AP to generate epileptiform activity (Perreault and Avoli 1992; Lévesque et al. 2013; Chiang et al. 2013).

To apply 4-AP locally in the rat hippocampal CA1 region, we made a drug delivery device by using a segment of model 7 puncture needle as an injection tubing, then connected its tail to a microinjector through silicone tubing (Fig. 8.3). The needle tip was implanted alongside the recording electrode at the same posterior bregma

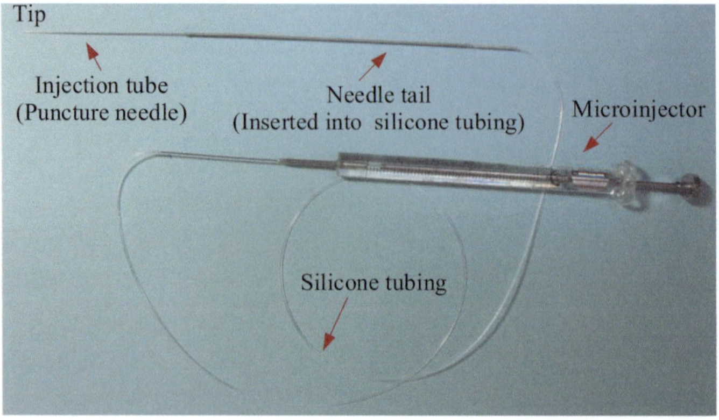

Fig. 8.3 Custom-built injection device for local drug delivery to the rat hippocampal CA1 region

coordinate, but closer to the midline—about 1 mm from the recording electrode (Fig. 8.4A). To induce epileptiform discharges, 4-AP solution (40 mM, 0.5–1 μL) was injected into the hippocampal CA1 region near the recording site.

In baseline recording before 4-AP administration, no PS occurred (Fig. 8.4B). After applying 4-AP, epileptiform discharges gradually developed (Fig. 8.4C), producing waveforms consistent with previous studies (Chiang et al. 2013). Two types of epileptiform waveforms appeared: large-amplitude PSs and small positive wavelets (SPW) at a frequent around 100 Hz. Some PSs were immediately followed by SPW (see the enlarged insets in Fig. 8.4C). During the epileptiform period, a 2-min O-HFS was applied on the Schaffer collaterals. While the first O-HFS pulse produced a large PS, only small PSs and SPWs occurred during subsequent O-HFS (Fig. 8.4D). Statistical data (Fig. 8.4E) showed that both 100 Hz and 200 Hz O-HFS significantly reduced PS rates and amplitudes (Fig. 8.4E1, E2). Consequently, the sum of PS amplitudes per second decreased significantly compared to pre-O-HFS control values (Fig. 8.4E3). After the end of O-HFS, all these measurements returned to their control levels (Fig. 8.4E1–E3).

Figure 8.5 further shows the suppressive effect of O-HFS on PSs by intermittent 5-s O-HFS trains applied at 5-s intervals. During the total 2-min period of 24 trains at 200 Hz O-HFS, PSs rarely occurred except at O-HFS onsets. However, large PSs occurred during all of the intervals (Fig. 8.5A1, A2). The PSs at the O-HFS onsets showed significant variations (Fig. 8.5A3–A5). When O-HFS started during SPW, it produced continuous SPW without obvious PS (Fig. 8.5A1, A3). When O-HFS started immediately after a large PS, the first O-HFS pulse produced either a small PS (Fig. 8.5A2, A4) or a large PS (Fig. 8.5A5), with no more obvious PSs following. Among the 24 O-HFS trains shown in Fig. 8.5A, the first pulses produced large PSs in 15 trains (highlighted in orange in the aligned plots of 24 onset signals in Fig. 8.5B).

To analyze the PS changes during 4-AP-induced epileptiform discharges, we used a bandpass filter (10–3000 Hz) to remove frequency components outside the PS waveform range. A threshold of − 2 mV was then set to detect PSs (Fig. 8.5A).

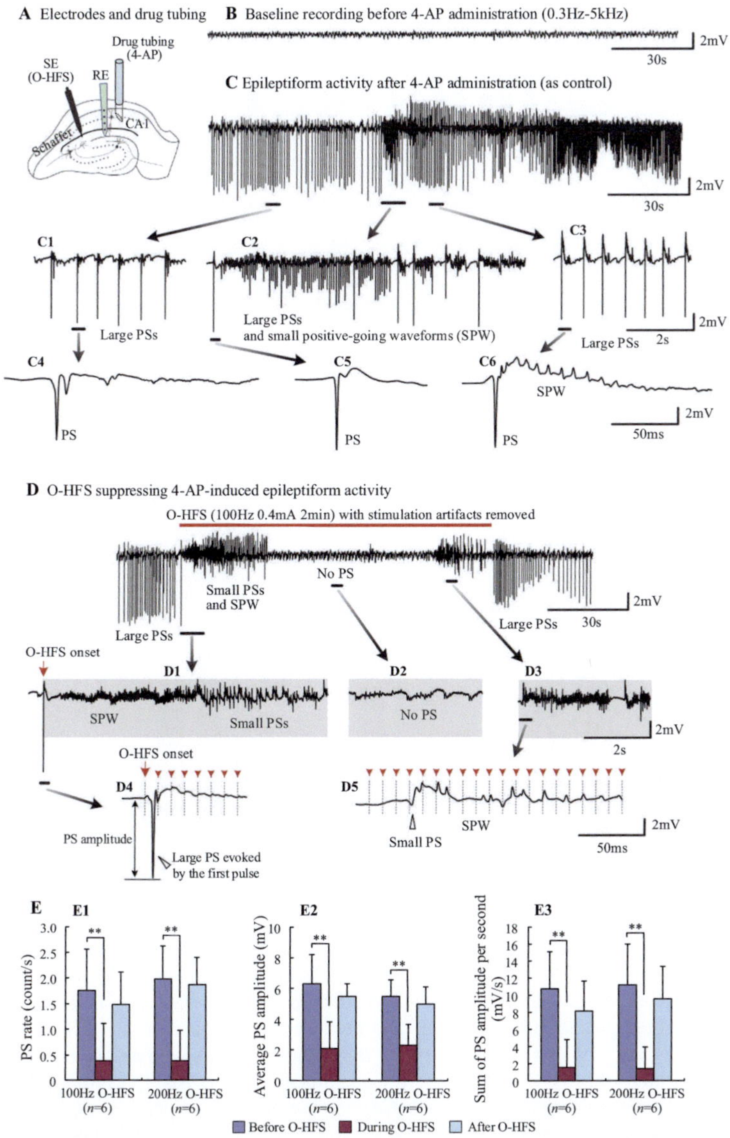

Fig. 8.4 Effect of O-HFS on 4-AP-induced epileptiform discharges in the rat hippocampal CA1 region. **A** Schematic diagram showing the placements of two electrodes and 4-AP drug tubing. **B** Baseline recording in the CA1 pyramidal cell layer. **C** 4-AP-induced epileptiform discharges with enlarged insets. **D** Application of O-HFS (2 min, 100 Hz) during 4-AP-induced epileptiform discharges. Red horizontal bar and grey shadows indicate stimulation period, and red arrows indicate the removed pulse artifacts. **E** Comparisons of PS rate (**E1**), average PS amplitude (**E2**), and sum of PS amplitudes per second (**E3**) before, during, and after 100 Hz and 200 Hz O-HFS. $^{**}P < 0.001$, ANOVA and post hoc Bonferroni multiple comparison tests, $n = 6$

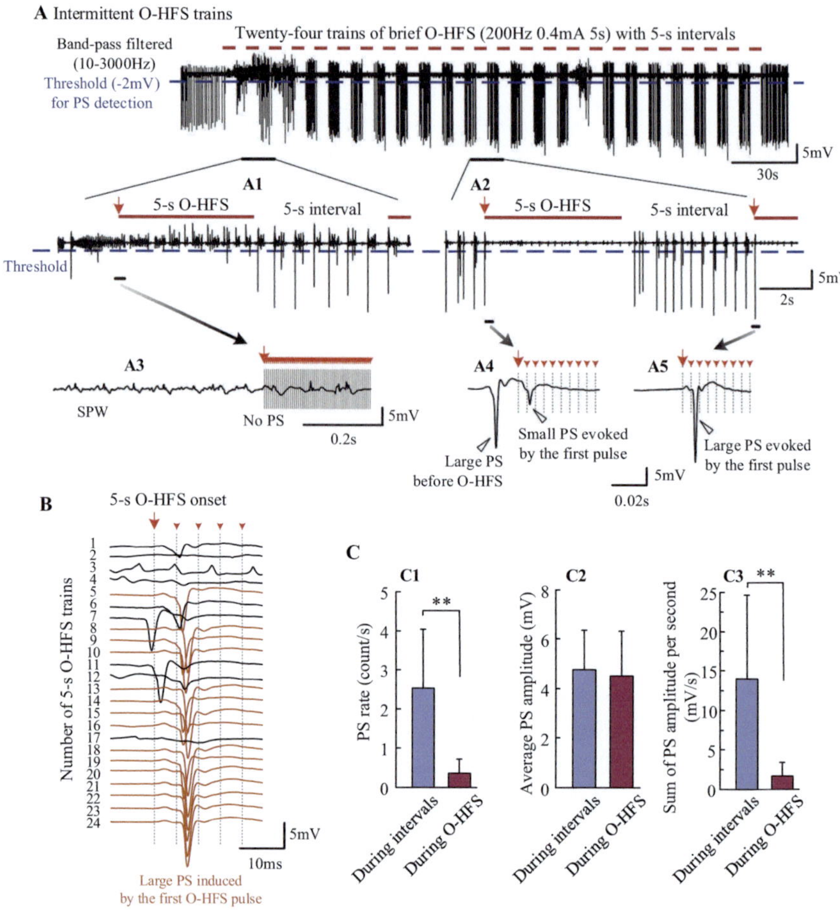

Fig. 8.5 Effect of intermittent O-HFS trains on 4-AP-induced epileptiform PSs. **A** Typical neuronal responses to 24 brief O-HFS trains (200 Hz, 5 s) at 5-s intervals during the epileptiform discharges induced by 4-AP, recorded in the CA1 pyramidal cell layer. To detect PSs using a threshold method, low-frequency field potentials below 10 Hz and high-frequency noise above 3 kHz were filtered out. **B** Aligned plots of the 24 onsets of O-HFS trains shown in (**A**), with orange highlighting the large PSs induced by the first O-HFS pulses. **C** Comparisons of PS rate (**C1**), average PS amplitude (**C2**), and sum of PS amplitudes per second (**C3**) between the O-HFS periods and intervals, from the pooled data of five 100 Hz and four 200 Hz O-HFS trials in five rats. $^{**}P < 0.001$, paired t-test, $n = 9$

Consistent with the results from the 2-min continuous O-HFSs shown in Fig. 8.4E, the mean PS rate was significantly lower during the intermittent 5-s O-HFS periods than 5-s interval periods (Fig. 8.5C1). While PS amplitudes showed no significant difference between O-HFS and interval periods—due to single large PSs at O-HFS onsets (Fig. 8.5C2), the sum of PS amplitudes per second was significantly smaller during O-HFS periods than interval periods (Fig. 8.5C3).

These results showed that O-HFS at afferent fiber axons can rapidly and persistently suppress large PSs in 4-AP-induced epileptiform discharges. This suppression could occur due to a depolarization block of downstream CA1 pyramidal neurons, caused by overexcitation from O-HFS excitatory inputs. To test this hypothesis, we applied paired-pulse antidromic test stimulations (ATS) to the alveus to determine whether the CA1 pyramidal neurons had lost their firing ability during O-HFS.

As shown in Fig. 8.6A, during baseline recording, paired-pulse ATS with a 5-ms interval produced two large APSs (APS_1 and APS_2) with similar amplitudes. After 4-AP administration, during an interval of epileptiform discharges without PSs, an ATS also produced two large APSs followed by SPW (ATS #1 in Fig. 8.6B1). ATS paired-pulses were then applied every 5 s (marked by orange dots). The second ATS, applied during SPW period, produced no obvious APS (ATS #2 in Fig. 8.6B1). The onset of O-HFS during this SPW period also failed to produce large PS (Fig. 8.6B2), consistent with the pattern shown in Fig. 8.5A3. During the 2-min continuous O-HFS period, the ATS pulses produced only small APSs at most (Fig. 8.6B3). Throughout the O-HFS period, the APSs induced by ATSs were significantly suppressed (Fig. 8.6B, bottom). Moreover, this APS suppression was particularly pronounced during SPW periods. Unlike the baseline situation shown in Fig. 8.6A, APS_2 induced by the second pulse of ATS was significantly smaller than APS_1. After the end of O-HFS, ATSs produced large APSs again (Fig. 8.6B4).

The effects of O-HFS at 100 and 200 Hz were similar. The pooled data from both frequencies showed that during the 4-AP-induced epileptiform period, the amplitudes of ATS-induced APS_1 and APS_2 were significantly smaller during O-HFS than before O-HFS (Fig. 8.6C). Additionally, the amplitude ratio of APS_2/APS_1 decreased progressively from baseline control (before 4-AP administration), to the 4-AP-induced epileptiform period, and finally during the O-HFS application in the epileptiform period (Fig. 8.6D). The decreased APS_2 indicated an extension of the APS_1 refractory period. These substantial APS suppressions indicated that synchronous firing was prevented by depolarization block due to excessive neuronal excitation. Therefore, O-HFS can suppress 4-AP-induced epileptiform PSs by driving downstream neurons into a depolarization block state.

8.3 Effect of Sinusoidal Electrical Stimulation on 4-AP-Induced Epileptiform Activity

In addition to the biphasic-pulse O-HFS stimulations used in the above experiments, sinusoidal stimulations at the Schaffer collaterals can also modulate the firing in downstream neurons (Wang et al. 2020). We investigated whether sinusoidal stimulations could suppress 4-AP-induced epileptiform discharges (Guo et al. 2016). We first determined the appropriate sinusoidal intensity (peak-to-peak current amplitude). As shown in Fig. 8.7A, the baseline recording showed spontaneous unit spikes of neurons (marked by the arrows). When 50 Hz sinusoidal stimulation was applied

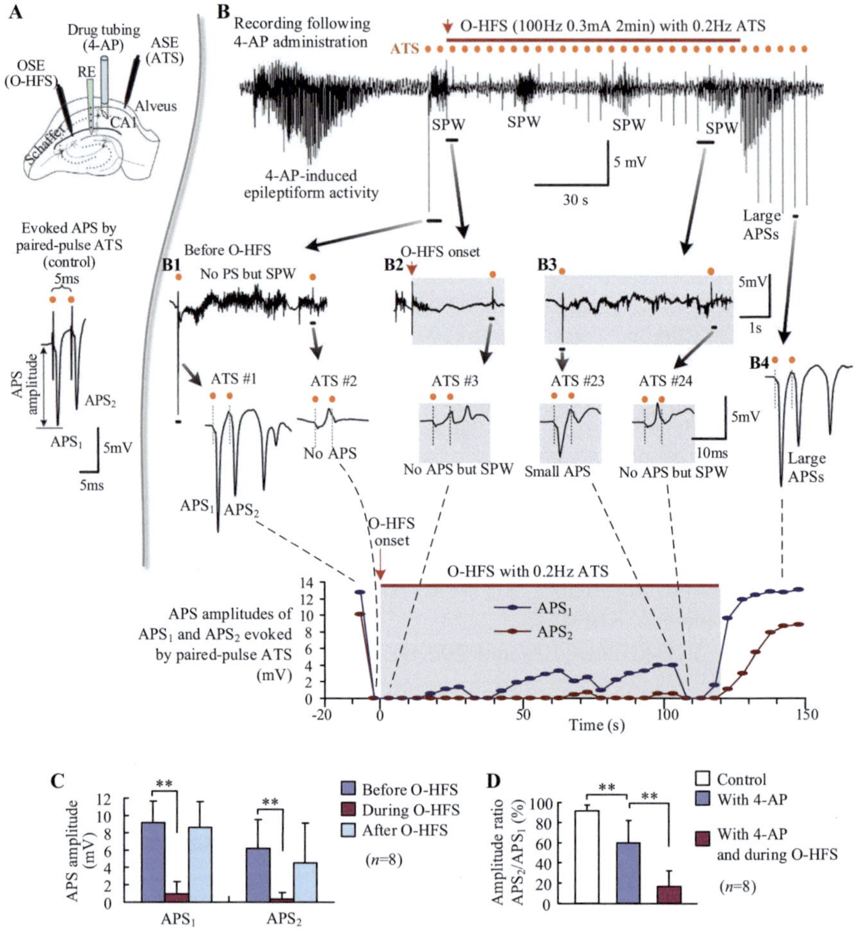

Fig. 8.6 Evaluating the excitability of CA1 pyramidal neurons using paired-pulse antidromic test stimulation (ATS). **A** Schematic diagram illustrating the placements of three electrodes and a 4-AP drug tubing in the hippocampal CA1 region (upper), as well as the large APS_1 and APS_2 induced by an ATS at baseline (bottom). **B** During the 4-AP-induced epileptiform period, ATS pulses were applied every 5 s (orange dots) and a 2-min 100 Hz O-HFS was applied simultaneously (red bar). The ATSs are numbered and the O-HFS period is shaded in grey in the enlarged insets. The bottom plot shows the changes of APS_1 and APS_2 amplitudes over time, with the O-HFS period starting at time zero. **C** Comparison of APS_1 and APS_2 amplitudes before, during, and after O-HFS (five 100 Hz and three 200 Hz trains). **D** Comparison of amplitude ratios (APS_2/APS_1) at baseline, during the 4-AP-induced epileptiform period, and during additional O-HFS application. In figures **C** and **D**, $^{**}P < 0.01$, ANOVA and post hoc Bonferroni multiple comparison tests, $n = 8$

with increasing intensities, its modulation on neuronal firing gradually increased (Fig. 8.7B). At 15 μA intensity (Fig. 8.7B1), unit spikes appeared randomly across different phases of sinusoidal waves, indicating minimal effect on neuronal firing. At 30μA intensity (Fig. 8.7B2), the firing showed a clear phase-locked pattern, with most spikes occurring at the beginning of sinusoidal descending phases. When the intensity increased to 50 μA, the PSs formed by synchronous firing of neuronal populations appeared (marked by orange dots in Fig. 8.7B3).

We set the sinusoidal stimulation at an intensity that was able to modulate neuronal firing but without triggering large PS discharges. As shown in Fig. 8.8A, we applied a 40 μA 50 Hz sinusoidal stimulation for 2 min during the 4-AP-induced epileptiform period. Under the stimulation, 4-AP-induced large PSs disappeared while SPW remained. Statistical data showed that during the 2-min stimulation at a mean intensity of 32 ± 9.7 μA ($n = 5$ rats), the PS rate, PS amplitude, and sum of PS amplitudes were significantly smaller than their pre-stimulation levels (Fig. 8.8B). The results indicate that sinusoidal stimulation can suppress 4-AP-induced epileptiform discharges.

The experiment results from both pulse and sinusoidal stimulations show that orthodromic high-frequency stimulations can suppress the synchronous discharges

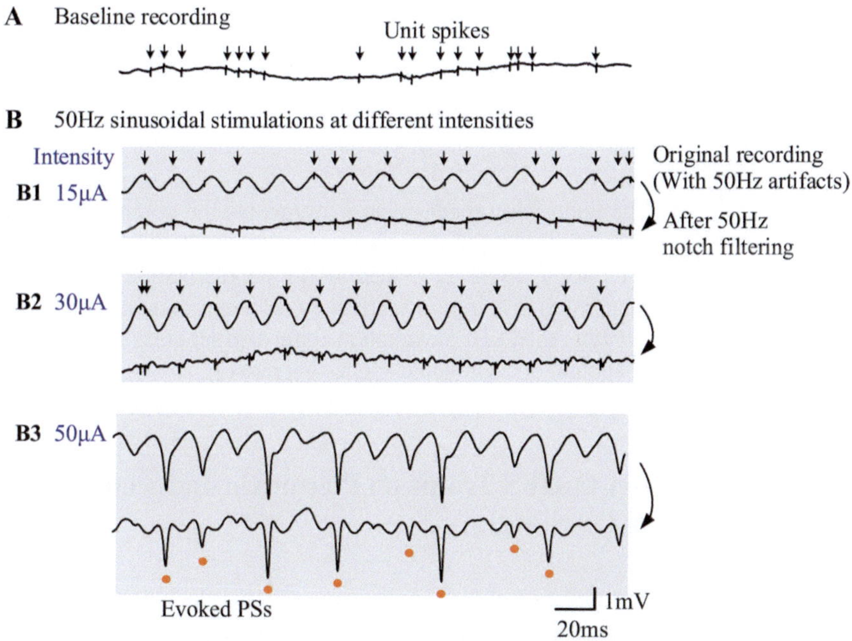

Fig. 8.7 Modulating downstream neurons by 50 Hz sinusoidal stimulations at the Schaffer collaterals. **A** Typical baseline recording in the CA1 pyramidal cell layer, where small arrows indicate unit spikes from individual neurons. **B** Typical recordings during 50 Hz sinusoidal stimulations at three different intensities. Grey shadings indicate stimulation periods, with small arrows indicating unit spikes and orange dots indicating PSs. Modified from Guo et al. (2016)

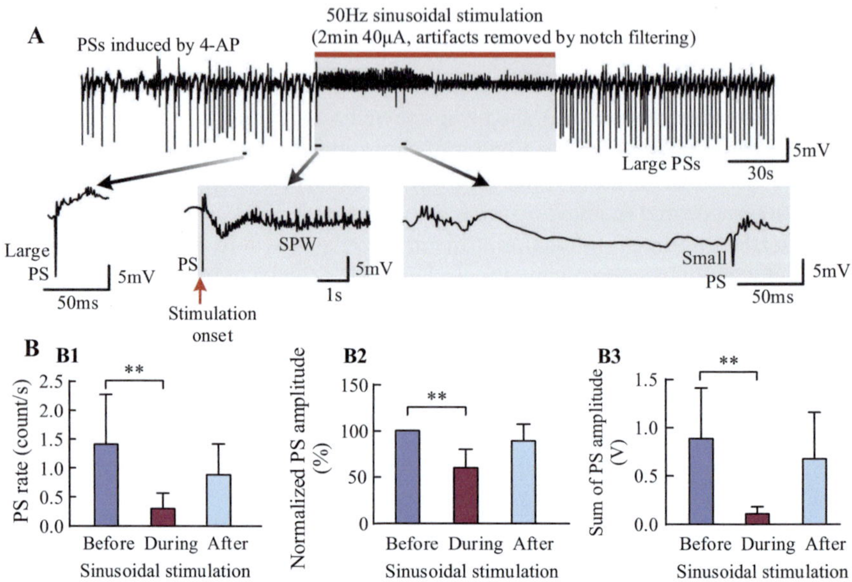

Fig. 8.8 Suppressing 4-AP-induced epileptiform PSs by orthodromic sinusoidal stimulation in the hippocampal CA1 region. **A** Typical recording during 4-AP-induced epileptiform discharges with a 2-min period of 50 Hz sinusoidal stimulation. The red bar and grey shadings indicate stimulation period. **B** Comparisons of PS rate (**B1**), normalized PS amplitude (**B2**), and sum of PS amplitudes within 2 min (**B3**) before, during, and after sinusoidal stimulations. $^{**}P < 0.01$, ANOVA and post hoc Bonferroni multiple comparison tests, $n = 5$ rats. Modified from Guo et al. (2016)

generated by 4-AP. The underlying mechanism may involve the stimulation excitatory inputs further elevating the depolarization of downstream CA1 pyramidal neurons under 4-AP-induced discharges, leading to depolarization block and preventing neuronal synchronous firing, as verified by ATS experiments (Fig. 8.6). We then investigated whether the O-HFS excitation could also suppress epileptiform discharges induced by another epileptogenic agent—picrotoxin.

8.4 Effect of Brief O-HFS Trains on Picrotoxin-Induced Epileptiform Discharges

Picrotoxin (PTX) is an antagonist of the $GABA_A$ receptor. By blocking GABAergic inhibitory synapses in local circuits, it can destroy the balance between neuronal excitation and inhibition, leading to epileptiform discharges.

In baseline recording, no PSs occurred in normal LFP signal (Fig. 8.9A). After injecting PTX (4 mM, 2–4 μL) through the drug tubing into the CA1 region near recording site, epileptiform discharges gradually appeared in the cell layer recording. The discharges were characterized by continuous bursts, each consisting of multiple

PSs (Fig. 8.9B). During the PTX-induced epileptiform period, we applied 100 Hz O-HFS at the Schaffer collaterals for 1 min, which significantly increased the mean PS rate (Fig. 8.9C) from 0.68 ± 0.27 counts/s before O-HFS to 5.7 ± 1.0 counts/s during the O-HFS periods ($P < 0.01$, paired t-test, $n = 6$ rats). However, mean PS amplitude before stimulation (7.1 ± 1.2 mV) was not significantly different from that during O-HFS (7.3 ± 1.1 mV, $P = 0.58$, paired t-test, $n = 6$).

This result suggests that sustained O-HFS can enhance rather than suppress PTX-induced epileptiform PSs. However, during the early stage of O-HFS, there was always a PS suppression period (Fig. 8.9C), indicating that short O-HFS might suppress PSs. Clinical studies have shown that adaptive neural stimulation (RNS), using sub-second brief stimulations, can suppress epileptiform seizures (Geller et al. 2017; Nune et al. 2015). Based on this clue, we designed a closed-loop stimulation system (see Sect. 3.5.6 for details) and applied 1-s brief O-HFS trains immediately following the first PSs of PTX-induced bursts. As shown in Fig. 8.10A, these brief O-HFS trains significantly suppressed the subsequent PSs in bursts, though SPW still occurred. The sum of PS amplitudes in each burst was significantly smaller with O-HFS than without O-HFS (Fig. 8.10B), with the first PS in bursts before O-HFS

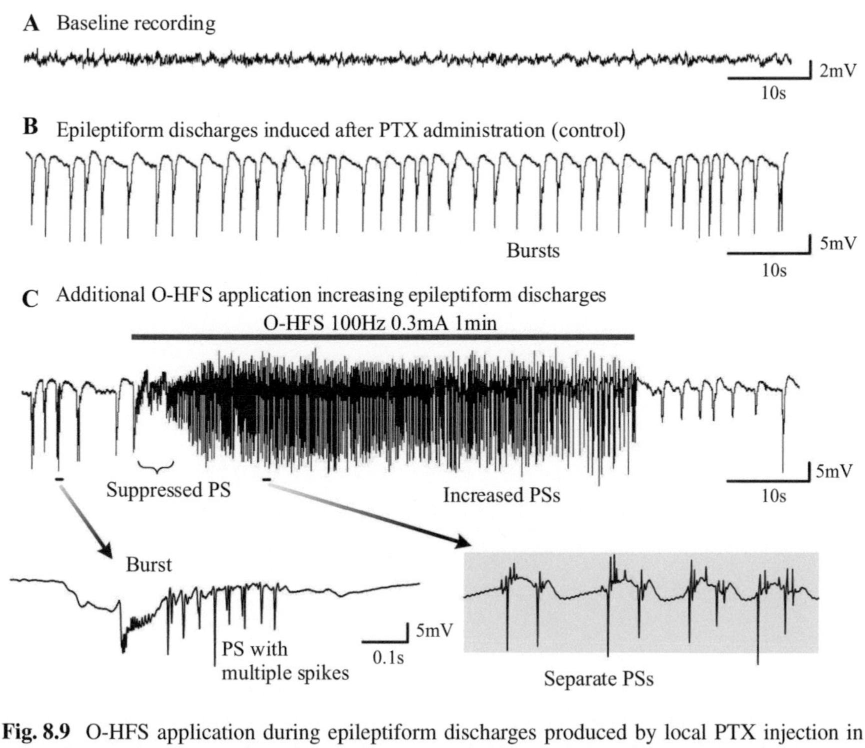

Fig. 8.9 O-HFS application during epileptiform discharges produced by local PTX injection in the rat hippocampal CA1 region. **A** Baseline recording from the pyramidal cell layer before PTX administration. **B** Epileptiform discharges produced by PTX injection. **C** O-HFS application during epileptiform discharges (denoted by red bar and grey shading)

initiation being the main contributor to the PS amplitudes during O-HFS. Similarly, applying a 0.3-s brief O-HFS train triggered by the first PS of burst consistently suppressed subsequent PSs in each burst (Fig. 8.10C).

Similar to the method described in Sect. 8.2, we used ATS at the axons of CA1 pyramidal neurons to test their ability to generate synchronous firing (see Fig. 8.6A), and used single-pulse ATSs here. As shown in Fig. 8.11A, a PTX-induced burst began with a large initial PS of multiple spikes, followed by subsequent PSs and SPW waves. During such a burst, the application of ATS pulses at 10 Hz persistently induced large APSs and SPW waves (Fig. 8.11B, orange dots indicate ATSs). However, when we simultaneously applied a 1-s O-HFS, ATSs failed to induce large APSs during the O-HFS period, and only SPW waves appeared (Fig. 8.11C). The mean APS amplitude decreased significantly during O-HFS compared to the control period before O-HFS. After O-HFS ended, ATSs again induced large APSs comparable to those in the control period (Fig. 8.11D). The result indicates that O-HFS application during bursts can impair the synchronous firing ability of pyramidal neurons.

These results show that brief O-HFS trains can suppress large PSs within PTX-induced bursts. Based on the changes in ATS-induced APSs, we can infer the underlying mechanisms: When PTX blocks the inhibitory $GABA_A$ synapses and

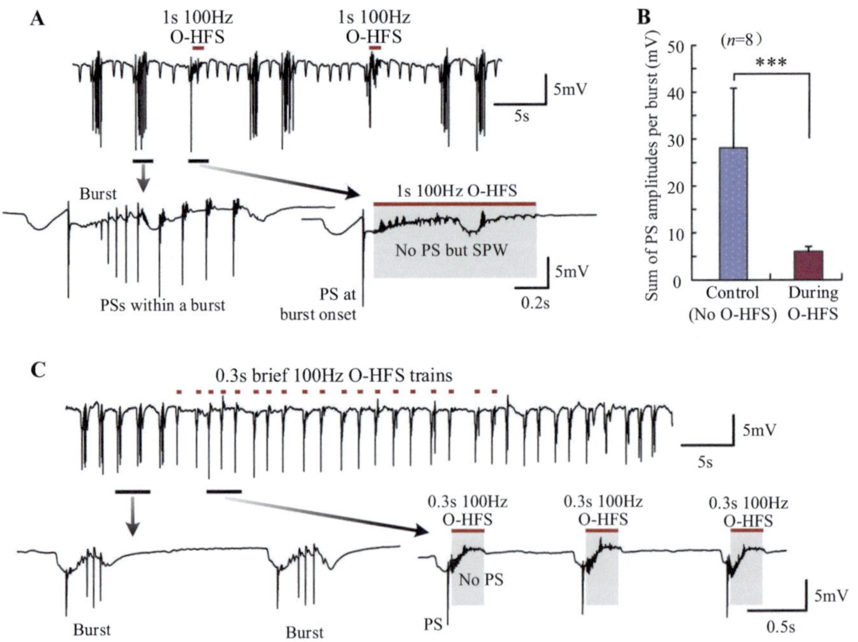

Fig. 8.10 Effects of brief O-HFS trains on PTX-induced epileptiform PSs in bursts. **A** Suppressive effects of 1-s brief O-HFS trains on PTX-induced PSs. **B** Comparison of the sum of PS amplitudes in bursts between with and without 1-s O-HFS stimulation. $^{***}P < 0.001$, paired t-test, $n = 8$ rats. **C** Suppressive effects of 0.3-s brief O-HFS trains on PSs in PTX-induced burst discharges

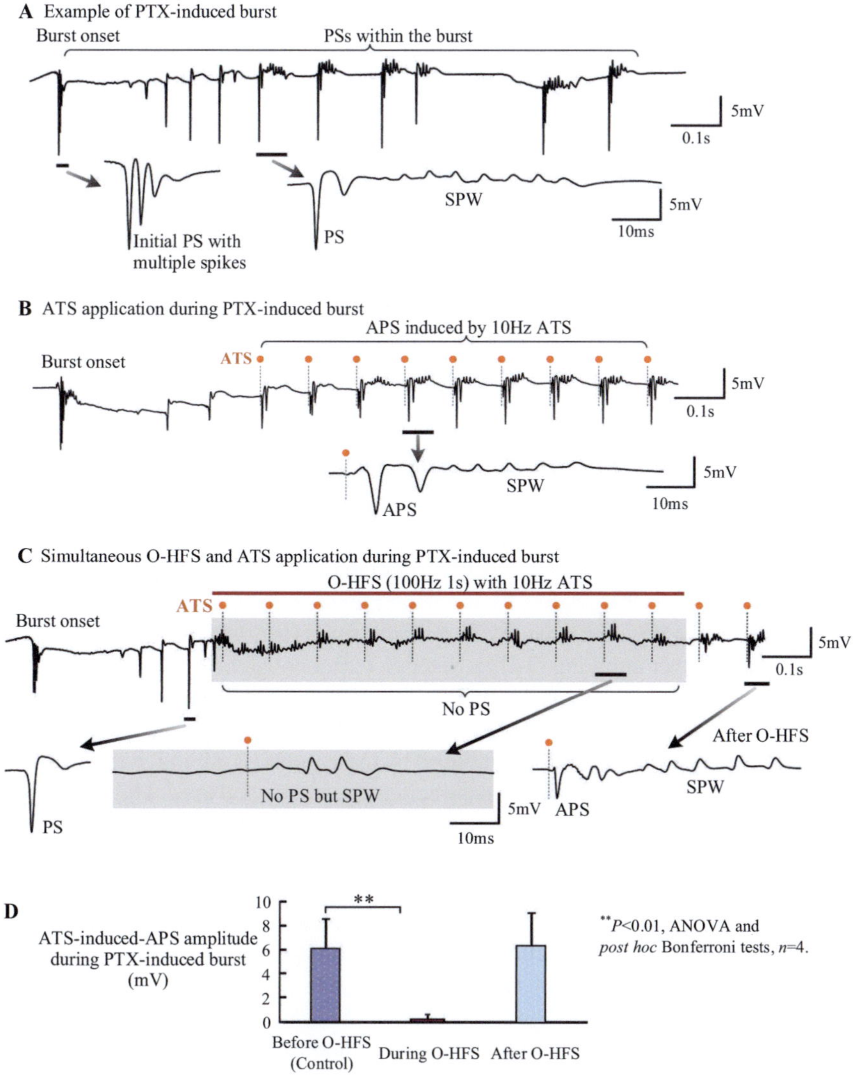

Fig. 8.11 Changes in the firing ability of CA1 pyramidal neurons during brief O-HFS application within PTX-induced bursts. **A** Example of PTX-induced burst recorded in the CA1 pyramidal cell layer. **B** Typical neuronal responses to 10-Hz ATS pulses (orange dots) during a PTX-induced burst. **C** Simultaneous applications of O-HFS (red bar and grey shading) and 10-Hz ATS pulses during a PTX-induced burst. **D** Comparison of mean APS amplitudes induced by ATS pulses across three periods. Modified from Cao et al. (2016)

increases the excitability of pyramidal neurons, epileptiform bursts occur and ATSs can induce large APSs. The addition of brief O-HFS application can further elevate the neuronal membrane depolarization, resulting in a depolarization block in the post-synaptic pyramidal neurons that prevents their firing. The decrease or absence of ATS-induced APSs confirmed the present of this block (Fig. 8.11C). In contrast, when long rather than brief O-HFS is applied, the sustained stimulation can produce intermittent depolarization block in the directly stimulated axons. This can attenuate the O-HFS excitation to downstream pyramidal neurons—not enough to cause depolarization block but sufficient to enhance synchronous discharges in these neurons, as shown in Fig. 8.9C.

8.5 Possible Mechanisms Underlying Neuronal Responses to O-HFS Applications in Different Epileptiform Models

This chapter shows the effects of electrical stimulations on epileptiform discharges. We induced these discharges locally in the rat hippocampal CA1 region using three methods: short O-HFS trains and two epileptogenic agents (4-AP and PTX). We investigated whether applying O-HFS to the Schaffer collaterals—the CA1 afferent fiber—could suppress the epileptiform discharges of pyramidal neurons downstream from the stimulation site. Our results demonstrate that both pulse and sinusoidal O-HFS can suppress epileptiform PSs induced by these methods, although the effective stimulation patterns vary according to the epileptogenic method used. The underlying mechanisms can involve desynchronization and depolarization block, as analyzed below.

Under normal physiological conditions, a single-pulse stimulation with sufficient intensity at the Schaffer collaterals can strongly excite downstream neuron populations and induce their synchronous firing to form a large population spike (PS) (see Sect. 2.3). A brief train of HFS pulses can also induce epileptiform firing—after-discharge (AD) consisting of PSs (Leung 1987; Bragin et al. 1997a, 1997b). Although sustained HFS can lead to axonal block and weaken the HFS excitation on downstream neurons, this excitation can still increase the firing of these neurons (see Sect. 5.3 for details). This excitatory effect can explain why animal and clinical trials have shown that HFS can enhance epileptiform activity (Gloor et al. 1982; Lado 2006; Feddersen et al. 2007). However, experiments on in-vitro brain slices have shown that HFS using both sinusoidal and pulse waveforms can suppress epileptiform activity in the hippocampus. Additionally, these stimulations increased the extracellular potassium concentration ($[K^+]_o$), suggesting that the suppress may result from desynchronized activation and/or depolarization block of neuronal membranes (Bikson et al. 2001; Durand and Bikson 2001). These two mechanisms can also explain the results from the three types of epileptiform models shown in this chapter.

(1) **Suppressive effect of sustained O-HFS on PSs in after-discharge (AD)**

The smaller and more frequent PSs during "HFS-discharge" than during AD indicated a desynchronization effect of HFS. As shown in Sect. 5.3.1, large APSs induced by ATSs during sustained O-HFS suggested that this desynchronization originated from the pre-synaptic asynchronous excitations of the Schaffer collaterals, not from post-synaptic neurons. The asynchronous inputs resulted from O-HFS-induced intermittent depolarization block at stimulated axons. Additionally, sustained O-HFS can eliminate synaptic potentiation caused by initial brief stimulation (Zhou et al. 2017). During sustained O-HFS, the induced axonal block can also avoid excessive excitations to downstream interneurons, allowing GABAergic synapses to restore their inhibitory effect and prevent large PS events, as shown in Sect. 5.3. Therefore, unlike brief O-HFS which has an epileptogenic effect, sustained O-HFS can prevent the formation of large epileptiform PSs under certain conditions.

(2) **Suppressive effect of O-HFS on 4-AP-induced discharges**

Epileptogenic agent 4-AP can increase neuronal excitability by blocking potassium ion channels, resulting in epileptiform activity. When injected locally into the hippocampus, 4-AP produced two types of epileptiform waveforms—large PSs and small positive waves (SPW) in our experiments, similar to those reported in previous studies (Perreault and Avoli 1992; Chiang et al. 2013). Brain slice experiments with both intracellular and extracellular recordings have shown that PS results from synchronized action potentials of neuronal populations, while SPW represents incomplete action potentials appearing as fluctuations on extended depolarization potentials (Avoli et al. 1993). This indicates that during SPW, neuronal membranes stay in a depolarized state. During 4-AP-induced epileptiform discharges, both pulse and sinusoidal O-HFS maintained SPW activity but did not induce large PSs, likely because the stimulation further excited the downstream neurons to a level sufficient to reach depolarization block (refer to Figs. 8.4, 8.5, 8.6 and 8.8). The significant decrease in ATS-induced APSs confirmed the reduced synchronous firing ability in downstream neurons. The extension of refractory period shown by paired-pulse ATS tests further supported that the post-synaptic downstream neurons were in a prolonged depolarized state. Therefore, the suppression of O-HFS on 4-AP-induced large PSs can be caused by excessive depolarization of downstream neurons to an inactive state. Even when directly stimulated axons experience intermittent blocking, sustained O-HFS can still deliver sufficient excitation to create a depolarization block in post-synaptic pyramidal neurons during 4-AP-induced discharges.

(3) **Suppressive effect of brief O-HFS on PTX-induced epileptiform bursts**

Previous literature has reported that HFS pulses can significantly shorten epileptiform discharge duration and reduce discharge intensity (Schiller and Bankirer 2007; McIntyre et al. 2004). Several possible mechanisms have been proposed to explain the results. First, HFS may enhance the activity of inhibitory neurons, particularly GABAergic interneurons, thereby enhancing synaptic inhibitions on principal neurons and suppressing their epileptiform discharges (Schiller and Bankirer 2007; Dostrovsky et al. 2000; Chiken and Nambu 2013). Second, sustained HFS may lead to axonal conduction block and depletion of synaptic neurotransmitters, preventing the spreading of epileptiform activity (Jensen and Durand 2009). Third, HFS may cause neurons to remain in a depolarized state, leading to ion channel inactivation (depolarization block) and preventing neuronal membranes from generating action potentials (Bikson et al. 2001; Durand and Bikson 2001). However, the first two mechanisms cannot explain the suppressive effect of brief O-HFS trains on epileptiform bursts shown in Sect. 8.4. This is because our epileptiform model was induced by the $GABA_A$ antagonist PTX, and $GABA_A$ is the main neurotransmitter of inhibitory synapses in the hippocampal region (Andersen et al. 2007). Since these inhibitory synapses were blocked by PTX, O-HFS could not have acted through them. Moreover, our experiments in Sect. 8.4 show that brief O-HFS trains lasting only 1 s or less suppressed PSs, while sustained O-HFS lasting for 1 min—sufficient time to cause axonal block—produced strong epileptiform discharges. Therefore, the suppression likely did not occur through axonal conduction block or synaptic neurotransmitter depletion. Instead, the combination of inhibitory synapse block by PTX and additional excitation from O-HFS can lead to excessive excitation, causing depolarization block in neurons that can prevent the generation of large epileptiform spikes.

The schematic diagrams in Fig. 8.12 summarize the distinct excitation levels of neurons and their local neural circuits in the CA1 region under different conditions. Pink and blue colors represent excitatory and inhibitory synapses, respectively. The symbol "−" represents inhibitory effects, while different " + " symbols represent various excitatory inputs and enhanced excitations. The sum of "−" and " + " symbols within the cell body of pyramidal neuron (depicted as a triangle) determines its excitation level. As shown in Fig. 8.12A, under normal physiological conditions (baseline), pyramidal neurons maintain a balance between excitatory and inhibitory inputs and exhibit "random" spontaneous firing without synchronous epileptiform discharges. Figure 8.12B illustrates the situation during brief O-HFS-induced AD when excitatory synapses at the terminals of the Schaffer collaterals are enhanced and inhibitory synapses of interneurons are transformed into excitatory ones. These combined changes increase the excitability of pyramidal neurons, resulting in epileptiform discharges. Figure 8.12C shows the "HFS-discharge" period during sustained O-HFS when inhibitory synapses tend to recovery and O-HFS continues to enhance excitatory synapses (though with a weaker strength than during initial O-HFS). These changes result in pyramidal neurons firing with less synchronized small PSs and increased unit spikes. Figure 8.12D shows the situation of sustained O-HFS application during 4-AP-induced epileptiform discharges. Both 4-AP and O-HFS enhance excitatory synapses. Additionally, 4-AP transforms inhibitory synapses into excitatory ones and directly prolongs neuronal depolarization. These combined effects greatly enhance the depolarization of pyramidal neurons, leading to depolarization block. Figure 8.12E illustrates the application of brief O-HFS during PTX-induced bursts. PTX blocks inhibitory synapses. When combined with strong excitatory inputs from brief O-HFS, the changes cause depolarization block in pyramidal neurons. In each subgraph of Fig. 8.12, the net count of "+" symbols (after subtracting "−" symbols) in the cell body of pyramidal neuron can indicate its overexcitation level, from lowest to highest: (C) < (B) < (E) < (D). Figure 8.12C and B respectively

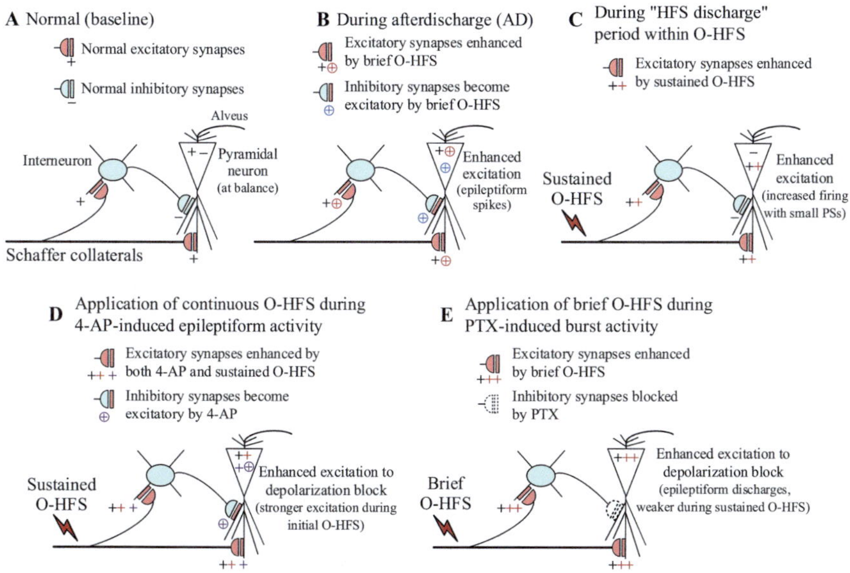

Fig. 8.12 Schematic diagrams showing excitation states of CA1 pyramidal neurons and local neural circuits under various situations. Pink and blue respectively represent excitatory and inhibitory synapses. The symbol "–" denotes inhibitory input, while "+" symbols in different colours and forms (with or without circles) denote various types of excitatory enhancement

show neuron states with increased firing and epileptiform discharge. Figure 8.12E, D show overexcitation states that lead to depolarization block. These analyses contain inferences that require further validations.

8.6 Summary

The experiment results presented in this chapter show that high-frequency stimulations can suppress epileptiform discharges through the mechanisms of desynchronization and depolarization block. In our experiments, we mainly used electrical stimulations of narrow pulses. The direct effect of these pulses on neuronal membranes is depolarization (Lowet et al. 2022)—an excitatory rather than inhibitory effect. Especially, depolarization block—which results from further increase of excitation beyond already abnormal levels—can suppress epileptiform discharges. Such suppressive effect through depolarization block acts like "fighting fire with fire" (Gwinn and Spencer 2004). Similar phenomena have been observed in clinical studies. For instance, brief HFS can induce intense after-discharges in human brains (Lesser et al. 1999; Motamedi et al. 2002), which is typically epileptogenic. However, this same stimulation pattern can also suppress epileptiform activity in the human brain (Lesser et al. 1999; Sun and Morrell 2014).

Our findings demonstrate that suppressing epileptiform discharges requires stimulation patterns matching the underlying pathological mechanisms. Different epileptiform situations require different stimulation parameters (e.g., short versus long duration) and approaches (e.g., open-loop versus closed-loop stimulation). In addition, epileptiform PSs—which represent synchronized action potentials from neuronal populations—are characteristic waveforms in epileptiform activity. Continuous PS discharges indicate severe seizures that can cause symptoms such as brain dysfunction and myoclonus. Thus, suppressing PS discharges is crucial for effective epilepsy treatments. Our findings provide new insights for developing DBS treatments for epilepsy.

References

Andersen P, Morris R, Amaral D et al (2007) The hippocampus book. Oxford University Press

Anderson TR, Hu B, Iremonger K et al (2006) Selective attenuation of afferent synaptic transmission as a mechanism of thalamic deep brain stimulation-induced tremor arrest. J Neurosci 26(3):841–850

Avoli M, Psarropoulou C, Tancredi V et al (1993) On the synchronous activity induced by 4-aminopyridine in the CA3 subfield of juvenile rat hippocampus. J Neurophysiol 70(3):1018–1029

Bean BP (2007) The action potential in mammalian central neurons. Nat Rev Neurosci 8(6):451–465

Bikson M, Lian J, Hahn PJ et al (2001) Suppression of epileptiform activity by high frequency sinusoidal fields in rat hippocampal slices. J Physiol 531(Pt 1):181–191

Birdno MJ, Tang W, Dostrovsky JO et al (2014) Response of human thalamic neurons to high-frequency stimulation. PLoS ONE 9(5):e96026

Boon P, Vonck K, De Herdt V et al (2007) Deep brain stimulation in patients with refractory temporal lobe epilepsy. Epilepsia 48(8):1551–1560

Bragin A, Csicsvári J, Penttonen M et al (1997a) Epileptic afterdischarge in the hippocampal–entorhinal system: current source density and unit studies. Neuroscience 76(4):1187–1203

Bragin A, Penttonen M, Buzsáki G (1997b) Termination of epileptic afterdischarge in the hippocampus. J Neurosci 17(7):2567–2579

Cao J, Feng Z, Guo Z et al (2016) Mechanism of suppression of epileptiform burst by closed-loop electrical stimulation. Chin J Biomed Eng 35(1):79–87 (in Chinese)

Chiang C, Lin C, Ju M et al (2013) High frequency stimulation can suppress globally seizures induced by 4-AP in the rat hippocampus: an acute in vivo study. Brain Stimul 6(2):180–189

Chiken S, Nambu A (2013) High-frequency pallidal stimulation disrupts information flow through the pallidum by GABAergic inhibition. J Neurosci 33(6):2268–2280

Cukiert A, Lehtimäki K (2017) Deep brain stimulation targeting in refractory epilepsy. Epilepsia 58(Suppl 1):80–84

Davis P, Gaitanis J (2020) Neuromodulation for the treatment of epilepsy: a review of current approaches and future directions. Clin Ther 42(7):1140–1154

Dostrovsky JO, Levy R, Wu J et al (2000) Microstimulation-induced inhibition of neuronal firing in human globus pallidus. J Neurophysiol 84(1):570–574

Durand DM, Bikson M (2001) Suppression and control of epileptiform activity by electrical stimulation: a review. Proc IEEE 89(7):1065–1082

Feddersen B, Vercueil L, Noachtar S et al (2007) Controlling seizures is not controlling epilepsy: a parametric study of deep brain stimulation for epilepsy. Neurobiol Dis 27(3):292–300

Filali M, Hutchison WD, Palter VN et al (2004) Stimulation-induced inhibition of neuronal firing in human subthalamic nucleus. Exp Brain Res 156(3):274–281

Fujiwara-Tsukamoto Y, Isomura Y, Imanishi M et al (2007) Distinct types of ionic modulation of GABA actions in pyramidal cells and interneurons during electrical induction of hippocampal seizure-like network activity. Eur J Neurosci 25(9):2713–2725

Geller EB, Skarpaas TL, Gross RE et al (2017) Brain-responsive neurostimulation in patients with medically intractable mesial temporal lobe epilepsy. Epilepsia 58(6):994–1004

Gloor P, Olivier A, Quesney LF et al (1982) The role of the limbic system in experiential phenomena of temporal lobe epilepsy. Ann Neurol 12(2):129–144

Guo Z, Feng Z, Yu Y et al (2016) Sinusoidal stimulation trains suppress epileptiform spikes induced by 4-ap in the rat hippocampal CA1 region in-vivo. Annu Int Conf IEEE Eng Med Biol Soc. 2016:5817–5820

Gwinn RP, Spencer DD (2004) Fighting fire with fire: brain stimulation for the treatment of epilepsy. Clin Neurosci Res 4(1–2):95–105

Hannan S, Faulkner M, Aristovich K et al (2020) Optimised induction of on-demand focal hippocampal and neocortical seizures by electrical stimulation. J Neurosci Methods 346:108911

Hennig MH (2013) Theoretical models of synaptic short term plasticity. Front Comput Neurosci 7:45

Iremonger KJ, Anderson TR, Hu B et al (2006) Cellular mechanisms preventing sustained activation of cortex during subcortical high-frequency stimulation. J Neurophysiol 96(2):613–621

Isomura Y, Sugimoto M, Fujiwara-Tsukamoto Y et al (2003) Synaptically activated Cl– accumulation responsible for depolarizing GABAergic responses in mature hippocampal neurons. J Neurophysiol 90(4):2752–2756

Jensen AL, Durand DM (2009) High frequency stimulation can block axonal conduction. Exp Neurol 220(1):57–70

Jobst BC, Darcey TM, Thadani VM et al (2010) Brain stimulation for the treatment of epilepsy. Epilepsia 51(s3):88–92

Jobst BC, Kapur R, Barkley GL et al (2017) Brain-responsive neurostimulation in patients with medically intractable seizures arising from eloquent and other neocortical areas. Epilepsia 58(6):1005–1014

Lado FA (2006) Chronic bilateral stimulation of the anterior thalamus of kainate-treated rats increases seizure frequency. Epilepsia 47(1):27–32

Lafreniere-Roula M, Kim E, Hutchison WD et al (2010) High-frequency microstimulation in human globus pallidus and substantia nigra. Exp Brain Res 205(2):251–261

Lesser RP, Kim SH, Beyderman L et al (1999) Brief bursts of pulse stimulation terminate afterdischarges caused by cortical stimulation. Neurology 53(9):2073–2081

Leung LW (1987) Hippocampal electrical activity following local tetanization. I. Afterdischarges. Brain Res 419(1–2):173–187

Lévesque M, Salami P, Behr C et al (2013) Temporal lobe epileptiform activity following systemic administration of 4-aminopyridine in rats. Epilepsia 54(4):596–604

Li MCH, Cook MJ (2018) Deep brain stimulation for drug-resistant epilepsy. Epilepsia 59(2):273–290

Lowet E, Kondabolu K, Zhou S et al (2022) Deep brain stimulation creates informational lesion through membrane depolarization in mouse hippocampus. Nat Commun 13(1):7709

McCarren M, Alger BE (1985) Use-dependent depression of ipsps in rat hippocampal pyramidal cells in vitro. J Neurophysiol 53(2):557–571

McIntyre CC, Grill WM, Sherman DL et al (2004) Cellular effects of deep brain stimulation: model-based analysis of activation and inhibition. J Neurophysiol 91(4):1457–1469

Motamedi GK, Lesser RP, Miglioretti DL et al (2002) Optimizing parameters for terminating cortical afterdischarges with pulse stimulation. Epilepsia 43(8):836–846

Neher E, Sakaba T (2008) Multiple roles of calcium ions in the regulation of neurotransmitter release. Neuron 59(6):861–872

Nune G, DeGiorgio C, Heck C (2015) Neuromodulation in the treatment of epilepsy. Curr Treat Options Neurol 17(10):43

Pearce RA, Grunder SD, Faucher LD (1995) Different mechanisms for use-dependent depression of two GABAA-mediated IPSCs in rat hippocampus. J Physiol 484(2):425–435
Peña F, Tapia R (1999) Relationships among seizures, extracellular amino acid changes, and neurodegeneration induced by 4-aminopyridine in rat hippocampus: a microdialysis and electroencephalographic study. J Neurochem 72(5):2006–2014
Peña F, Tapia R (2000) Seizures and neurodegeneration induced by 4-aminopyridine in rat hippocampus in vivo: role of glutamate- and GABA-mediated neurotransmission and of ion channels. Neuroscience 101(3):547–561
Perreault P, Avoli M (1992) 4-aminopyridine-induced epileptiform activity and a GABA-mediated long-lasting depolarization in the rat hippocampus. J Neurosci 12(1):104–115
Prescott IA, Dostrovsky JO, Moro E et al (2013) Reduced paired pulse depression in the basal ganglia of dystonia patients. Neurobiol Dis 51:214–221
Rosenbaum R, Zimnik A, Zheng F et al (2014) Axonal and synaptic failure suppress the transfer of firing rate oscillations, synchrony and information during high frequency deep brain stimulation. Neurobiol Dis 62:86–99
Rotman Z, Deng PY, Klyachko VA (2011) Short-term plasticity optimizes synaptic information transmission. J Neurosci 31(41):14800–14809
Schiller Y, Bankirer Y (2007) Cellular mechanisms underlying antiepileptic effects of low- and high-frequency electrical stimulation in acute epilepsy in neocortical brain slices in vitro. J Neurophysiol 97(3):1887–1902
Shigeto H, Boongird A, Baker K et al (2013) Systematic study of the effects of stimulus parameters and stimulus location on afterdischarges elicited by electrical stimulation in the rat. Epilepsy Res 104(1–2):17–25
Simpson HD, Schulze-Bonhage A, Cascino GD et al (2022) Practical considerations in epilepsy neurostimulation. Epilepsia 63(10):2445–2460
Smolarz B, Makowska M, Romanowicz H (2021) Pharmacogenetics of drug-resistant epilepsy (review of literature). Int J Mol Sci 22(21):11696
Staley KJ, Soldo BL, Proctor WR (1995) Ionic mechanisms of neuronal excitation by inhibitory GABAA receptors. Science 269(5226):977–981
Storm JF (1987) Action potential repolarization and a fast after-hyperpolarization in rat hippocampal pyramidal cells. J Physiol 385:733–759
Sun FT, Morrell MJ (2014) Closed-loop neurostimulation: the clinical experience. Neurotherapeutics 11(3):553–563
Thomsen RH, Wilson DF (1983) Effects of 4-aminopyridine and 3,4-diaminopyridine on transmitter release at the neuromuscular junction. J Pharmacol Exp Ther 227(1):260–265
Tibbs GR, Barrie AP, Van Mieghem FJ et al (1989) Repetitive action potentials in isolated nerve terminals in the presence of 4-aminopyridine: effects on cytosolic free Ca^{2+} and glutamate release. J Neurochem 53(6):1693–1699
Vetkas A, Fomenko A, Germann J et al (2022) Deep brain stimulation targets in epilepsy: systematic review and meta-analysis of anterior and centromedian thalamic nuclei and hippocampus. Epilepsia 63(3):513–524
Wang Z, Feng Z, Yuan Y et al (2021) Suppressing synchronous firing of epileptiform activity by high-frequency stimulation of afferent fibers in rat hippocampus. CNS Neurosci Ther 27(3):352–362
Wang Z, Feng Z, Zheng L et al (2020) Sinusoidal stimulation on afferent fibers modulates the firing pattern of downstream neurons in rat hippocampus. J Integr Neurosci 19(3):413–420
Ye H, Kaszuba S (2017) Inhibitory or excitatory? Optogenetic interrogation of the functional roles of GABAergic interneurons in epileptogenesis. J Biomed Sci 24(1):93
Zhou W, Feng Z, Qiu C et al (2017) Sustained high-frequency stimulations change the effects on neural networks induced by short stimulations. Prog Biochem Biophys 44(9):769–775 (in Chinese)
Zucker RS, Regehr WG (2002) Short-term synaptic plasticity. Annu Rev Physiol 64:355–405

Chapter 9
Neuromodulation Through Axonal Stimulations in Brain

Abstract This chapter summarizes our research on axonal electrical stimulations described in Part II. It presents the critical role of axonal responses in brain stimulations, discusses the high-frequency stimulation (HFS)-induced axonal block and its resulting neuronal responses, analyzes the effects of different time-varying stimulations under the situation of axonal block, and explores the potential post-stimulation effects of DBS, applications of closed-loop stimulations, and the relations between DBS and brain-computer interfaces.

Keywords Axonal role · Intermittent axonal block · Deep brain stimulation · Post-stimulation effect

9.1 Role of Axonal Responses in Deep Brain Stimulation

Since the late 1980s, DBS has been used in clinic to treat some neurological disorders, such as Parkinson's disease, primary tremor, and refractory epilepsy (Benabid et al. 1987; Krack et al. 2003; Krauss et al. 2021; Li and Cook 2018). It has also shown promise in treating other neurological and psychiatric disorders, including obsessive–compulsive disorder, depression, obesity, anorexia, drug addiction, Alzheimer's disease, and intellectual disabilities. Additionally, it has even shown potential for improving memory and awakening patients from vegetative states (Sullivan et al. 2021; Alagapan et al. 2023; Shivacharan et al. 2022; Picton et al. 2024; Dang et al. 2023; Rissardo et al. 2023). Notably, its clinical applications have preceded understanding of its mechanisms (Miocinovic et al. 2013). While practical experience has driven its development, the precise mechanisms of DBS remain inconclusive. A deeper understanding of the mechanisms can guide the development of new stimulation paradigms to treat more brain disorders.

In the early years, as an alternative to neurosurgical ablation, the main DBS effect was thought to inhibit neuronal firing, which may occur through two possible mechanisms. First, HFS of DBS can generate dense excitatory inputs, causing sustained

Z. Feng, *Electrical Brain Stimulation: Mechanisms and Modulation of Neuronal Activity*, https://doi.org/10.1007/978-981-95-4145-4_9

depolarization on neuronal membranes. This depolarization can inactivate voltage-gated sodium channels and create a depolarization block, preventing neurons from producing action potentials (Beurrier et al. 2001; Burbaud et al. 1994). Second, HFS may activate presynaptic axons (or axonal terminals) of inhibitory synapses, increasing the release of inhibitory neurotransmitters and enhancing inhibitions on postsynaptic neurons (Boraud et al. 1996; Dostrovsky et al. 2000). However, studies on monkeys with Parkinson's disease have shown a more complex phenomenon: even when neuronal firing decreased near HFS site, the stimulation actually increased neuronal firing in downstream projection areas. Meantime, the stimulation effectively alleviated symptoms of motor disorders in this situation (Anderson et al. 2003; Vitek 2002; Hashimoto et al. 2003). These results suggest that HFS can act through mechanisms beyond simply inhibiting neuronal firing (Deniau et al. 2010).

Moreover, even when the soma of a neuron is inhibited and fails to generate action potentials, HFS can still excite its axon to generate and propagate action potentials. This phenomenon—known as "local inhibition and distant excitation" (Florence et al. 2016; Herrington et al. 2016)—occurs because axons have shorter chronaxies, making them more responsive to the HFS narrow pulses used in DBS than other neuronal structures (Ranck 1975). The pulses can independently induce action potentials in axons and transmit them bidirectionally outward. The axonal firing propagating antidromically to somata can disrupt the soma original firing. In the opposite direction, the pulse-induced axonal firing propagating orthodromically to axonal terminals can activate synapses that either excite or inhibit downstream postsynaptic neurons. Animal studies have indeed confirmed that HFS can simultaneously inhibit the somata while exciting axons to generate action potentials (Nowak and Bullier 1998). Mathematical models have also shown that during HFS, the soma firing may differ from the axon firing (McIntyre et al. 2004). When inhibitory synapses prevent the soma from responding to excitatory inputs, the axon can still generate action potentials in response to stimulation pulses, replacing the soma's original firing.

In neural networks, stimulation-induced axon firing can propagate through single and multiple synaptic connections to reach numerous neurons across brain regions. The firing transmissions can trigger the releases of neurotransmitters and other chemical substances to produce wide effects. Therefore, HFS-induced axonal activity can play a crucial role in DBS therapies (Herrington et al. 2016; Feng et al. 2018). Experimental studies have shown that these axonal activations can excite neurons in downstream projection regions (Reese et al. 2011; Cleary et al. 2013), consistent with the experimental results presented in Sect. 5.3 of this book. Notably, axonal fibers occupy a larger volume in brain tissue than the other neuronal structures like somata and dendrites (Buzsáki 2006). The white matter, formed by axons, makes up approximately half of the human brain volume (Fields 2008), with the remaining half volume occupied by other neuronal structures, ventricles, and blood vessels. In the directly stimulated area near stimulation electrode, there are three types of axons: afferent fibers from upstream neurons, efferent fibers from local neurons, and passing fibers. All these axons readily respond to HFS narrow pulses. Therefore, axonal activity has become increasingly recognized as fundamental to neural electrical stimulation therapies (Kent et al. 2015; Girgis and Miller 2016; Montgomery

2017; Howell et al. 2019; Jakobs et al. 2019). The discovery of HFS-induced intermittent axonal block has provided a new direction for revealing brain stimulation mechanisms and developing stimulation paradigms.

9.2 Role of Intermittent Axonal Block Induced by HFS

9.2.1 HFS-Induced Axonal Block

Narrow pulses can powerfully depolarize neuronal membranes (Lowet et al. 2022). Under normal physiological conditions, this depolarization can trigger action potentials in neurons. The typical refractory period of neurons is approximately 1 ms. DBS pulses at frequencies within 50–200 Hz have intervals exceeding the refractory period. Thus, most neurons—especially fast-responding axons for neural signal conductions—should be able to fire in response to every pulse in this frequency range. Given that peripheral nerve block requires kilohertz stimulations, debate has continued about whether DBS pulse at around 100 Hz can block axon firing and conduction. Although sustained axonal HFS can indeed decrease its excitatory effect on downstream neurons compared to initial period, the question is where this decrease stems from, the presynaptic axons or the synapses? Some researchers have attributed it to synaptic failures, considering that intense activations may deplete neurotransmitters through continuous releases, causing synaptic fatigues and failures (Anderson et al. 2006; Iremonger et al. 2006; Rosenbaum et al. 2014; Farokhniaee and McIntyre 2019). However, experiments on in-vitro brain slices from the thalamus and basal ganglia have shown that axons fail to produce action potentials following each pulse at frequencies above 50 Hz, resulting in axonal block (Shen and Johnson 2008; Zheng et al. 2011; Rosenbaum et al. 2014). Similar axonal block has been observed in brain slice experiments from other regions, including the cerebral cortex and hippocampus (Chomiak and Hu 2007; Jensen and Durand 2009). Our in-vivo experiments in intact rat brains further confirmed the HFS-induced axonal block. Nevertheless, the structural differences in axons across brain regions and peripheral nerves—such as thickness and myelin sheath presence—can influence the extent and progression of HFS-induced axonal block.

As presented in Sect. 5.2, we studied axonal HFS by utilizing the clear layered structures in the hippocampal region—especially the alveus axon fibers covering the dorsal surface of hippocampus beneath ventricle. The alveus, formed by the axons of CA1 pyramidal neurons, allowed us to use small bipolar stimulation electrodes to focus the stimulated area. We verified that HFS at around 100 Hz can produce intermittent axonal block by combining both antidromic and orthodromic stimulations in experiments. When the antidromically-evoked APSs significantly decreased during steady A-HFS period—indicating that the stimulation of pyramidal neurons' axons failed to generate soma action potentials—the orthodromic test pulses (OTS) were still able to activate these somata. This result clearly showed that the failures occurred in the axons rather than the somata, providing strong evidence of axonal block. Furthermore, our series HFS experiments have provided additional supports. For instance, as shown in Chap. 7, the distinct soma responses to the opposite-polarity pulses during A-HFS$_{T\perp}$ also indicated the response attenuation originated in the axons. Otherwise, if the axons had reliably generated action potentials following each pulse, the somata would have received identical antidromic signals, making it impossible to distinguish between and respond differently to pulse polarities. Although these experiments have confirmed the axonal block, they cannot exclude other mechanisms, such as synaptic neurotransmitter depletion in the orthodromic pathway. Presumably, both mechanisms may coexist or dominate at different stages during HFS. Neurotransmitter depletion may precede axonal block in early HFS period. Once axonal block occurs, the resulting decrease in the excitatory inputs on synapses can then allow neurotransmitter to replenish.

The mechanism of HFS-induced axonal block may involve increased extracellular potassium ion concentration ($[K^+]_o$) (Bellinger et al. 2008; Liu et al. 2009). The HFS-induced intense firing at onset can rapidly elevate $[K^+]_o$, preventing the axonal membrane from repolarizing promptly after generating an action potential. The membrane can thus maintain a prolonged depolarized state and become unresponsive to continuous HFS pulses (Guo et al. 2018; Zheng et al. 2020)—as shown in the simulation results of axon model in Sect. 7.2. Although $[K^+]_o$ can play a key role, other ionic changes across the axonal membrane may also contribute to axonal depolarization block, including the accumulation of sodium ions ($[Na^+]_i$) inside the membrane (Zang and Marder 2021). Additionally, structural factors like axonal branching and HFS-induced changes in axonal morphology and biophysical properties can promote the development of axon block (Chéreau et al. 2017; Chomiak and Hu 2007; Rama et al. 2018). Therefore, axonal block may result from multiple interacting mechanisms.

HFS has be used to block signal conduction in the spinal cord and peripheral nerves, achieving therapeutic efficacies such as pain relief (Mekhail et al. 2020). However, a complete block in axonal fibers requires HFS above kilohertz frequencies (Arle et al. 2016). Typical DBS using pulse frequencies below 200 Hz can only produce a "partial" or "intermittent" block. This means that the HFS can still trigger action potentials in axons but at a frequency lower than the stimulation pulses, thereby exciting post-synaptic neurons and establishing new firing activity to replace original activity in the axon projection areas. This effect resembles "ablation" or "information loss" (Lowet et al. 2022). The newly produced firing has some randomization (see Sect. 5.3), which can eliminate synchronous firing between neurons, modify their original burst firing patterns, and suppress rhythmic activity in neural networks (Feng et al. 2017; Yu et al. 2016; Barow et al. 2014; Lee et al. 2011). In addition, during this partial block situation, individual axons may fire at different rates. Some axons in the stimulated fiber may even completely stop firing until HFS ends.

Our experimental results also show that during HFS-induced axonal block, although pulses can intermittently trigger axons to fire, signals from upstream neurons should have been completely blocked. During O-HFS, the stimulation increased the firing of downstream neurons. However, once the O-HFS ended, these neurons immediately stopped firing and remained silent for seconds before gradually resuming firing (see Fig. 5.19). This indicates that during O-HFS, the downstream neurons were driven by the stimulation, not by the upstream neurons—had upstream control remained, the downstream neurons should have maintained at least partial firing after the stimulation ceased. The mechanism can involve O-HFS-induced intermittent firing propagating antidromically along axons to somata, suppressing the original activity at somata and altering their excitability (refer to Sects. 5.2.3 and 5.2.4 for details). The silent period of neuronal firing that occurred immediately after the end of A-HFS supports this mechanism (see Fig. 5.13). Consequently, when the O-HFS control ceased, the upstream somata needed time to recover their original activity before resuming control over downstream neurons. This produced the "silent period" in downstream neurons after O-HFS ended (Fig. 5.19).

These results verify that the intermittent depolarization block caused by axonal HFS can interrupt signal transmissions between upstream and downstream neurons, thereby disrupting pathological neural activity, as proposed by previous reports (Chiken and Nambu 2014). Besides signal blocking, the axonal HFS can bilaterally control both upstream somata and downstream post-synaptic neurons. Its action on downstream neurons and their networks is crucial for brain stimulation therapy. Our experiment results show that the intermittent block (or intermittent recovery) of stimulated axons can create randomness in the HFS-induced firing, weakening the phase-locking between induced firing and stimulation pulses, thereby reducing firing synchronization between individual axons. This leads to asynchronous activity in downstream neuronal populations (see Sect. 5.3). In addition, HFS with constant pulse intervals can produce non-uniform "cluster" firing (Sect. 5.2.5). Such HFS-induced activity may eliminate pathological synchronous activity—for example, suppressing excessive synchronous activity in epilepsy (Chap. 8)—thereby achieving therapeutic efficacies.

9.2.2 *Diverse Stimulation Effects Under Intermittent Axonal Block*

During intermittent depolarization block, small changes in axonal membrane potential can substantially change the state of membrane ion channels (refer to the potential-dependent changes in HH model parameters shown in Fig. 1.11). Besides the nonlinear dynamics of membrane ion channels, the random characteristics ("noise") in various structural levels of the nervous system (Mino and Grill 2002; Faisal et al. 2008) may promote the changes in HFS-induced neuronal firing. As a result, even slight variations in HFS parameters can trigger diverse neuronal responses. Chapter 6 presents our findings with time-varying stimulation patterns, including HFS with gradually varying frequencies and intensities, alternating dual-parameters, single-pulse insertion and deletion, and randomly varying frequencies (IPIs).

Among these findings, the neuronal responses to randomly varying frequencies were surprising. Our results of constant HFS at 50–200 Hz showed that at lower frequencies (such as 50 Hz), neuronal firing strongly phase-locked with the stimulation pulses. As the HFS frequency increased, this phase-locking weakened and neuronal firing tended to random. Following these results, I originally expected that with IPIs varying randomly within a small range, the HFS would further randomize neuronal firing and reduce firing synchronization among neurons. However, our experiments of HFS trains with 5–10 ms randomly varying IPIs showed an opposite effect—the randomized pulse timings actually produced large PSs occasionally (see Sect. 6.4). This means that during HFS-induces axonal block, the stimulation can still activate downstream neuron populations strongly since PS events represent strong synchronized firing. Additionally, within a same small varying IPI range of 5–10 ms, different neuronal responses can be produced by designing various IPI sequences or adjusting IPI orders (Zheng et al. 2021). Notably, stimulation patterns with time-varying parameters offer broad options for neural modulation therapies beyond constant patterns. Like pharmaceutical treatments, these time-varying stimulation patterns can provide various "dosages" and "efficacies" for clinical applications, helping to expand the potential uses of DBS technology.

In addition, under intermittent block situation, axonal responses to opposite-polarity pulses can differ significantly from those under normal conditions. During the stimulations with alternating opposite-polarity pulses, the two pulse types can respectively activate two sub-groups of neurons. For axons that can response to both types of pulses at baseline, they only response to positive pulses rather than negative pulses during HFS of alternating pulses. This finding also provides valuable clues for developing new stimulation paradigms (Chap. 7).

Our studies presented in Chap. 8 show that excitatory axonal stimulations—delivered through both open-loop sustained HFS or closed-loop brief HFS—can suppress epileptiform spikes in downstream neurons. These suppressions may involve mechanisms of desynchronization and depolarization block. The effects of HFS patterns with time-varying parameters on abnormal neural activity require further study. Under intermittent depolarization block, for example, randomly-varying frequency stimulation can excite downstream neurons more strongly than constant frequency stimulation. Such stimulations might suppress certain types of epileptiform activity more effectively through a "fight fire with fire" effect (Gwinn and Spencer 2004)—though this hypothesis requires experimental verification.

9.2.3 *Experimental Methods for Brain Stimulation Investigations Used in This Book*

Conventional DBS primarily targets the thalamus and globus pallidus to treat movement disorders. Theoretically, external electrical stimulations can potentially treat various neurological diseases through modulating neural activity in specific brain regions. Therefore, new DBS targets for different conditions are continuously being explored. For instance, studies have shown that cerebellar stimulations can enhance cognitive recovery in rodents with prefrontal traumatic injury, offering potentials for rehabilitating patients with trauma-induced cognitive impairments (Chan et al. 2022).

The hippocampus is a common pathogenic region for brain diseases such as epilepsy and Alzheimer's disease. It has served as a potential target for DBS treatment of refractory epilepsy in clinical practice (Vetkas et al. 2022; Geller et al. 2017; Wu and Sharan 2013). The hippocampal region has distinct structural layers—respectively containing axon bundles, somata, and dendrites. This feature enables precise positioning of stimulation and recording electrodes in different neuronal structures. Additionally, hippocampal neurons are densely arranged, facilitating the detection and measurement of biomarkers as quantitative indicators. For instance, the dense cell bodies in the soma layer enables the formation of large population spikes (PS) in the extracellular space during synchronous firing. The PS amplitude and area can indicate the amount of firing neurons and serve as feedback indicators for closed-loop control, as shown in Sect. 8.4. Although a PS cannot determine the absolute number of firing neurons, it can be used to compare relative quantities. Other extracellular signals can also be detected, including action potentials of individual neurons (unit spike), field excitatory postsynaptic potentials (fEPSP), and local field potentials (LFP). Recordings of these electrical signals from population and individual neurons simultaneously can both provide feedback indicators for DBS controls and reveal valuable insights into DBS mechanisms.

Beyond brain stimulations, neuromodulation technology encompasses various types of nerve stimulations widely used in clinical treatment. These include vagus nerve stimulation (VNS) for refractory epilepsy and migraines, spinal cord stimulation (SCS) for chronic pain, and sacral nerve stimulation (SNS) for bladder voiding dysfunction and fecal incontinence (Gurbani et al. 2016; Silberstein et al. 2020; Lam et al. 2023; Kaaki and Gupta 2020; Hull et al. 2013). Additional applications include cochlear implants (CI) for deafness, hypoglossal nerve stimulation (HNS) for sleep apnea, and functional electrical stimulation (FES) for restoring disabled limb function (Gay et al. 2022; Mashaqi et al. 2021; Pellot-Cestero et al. 2023). These techniques all work by applying electrical stimulations to nerve fibers—neuronal axons—either for blocking signal transmission or for producing axonal activity to modulate neuronal or muscle activity. Although our experiments in this book focused on the hippocampal region, our findings on axonal HFS (particularly from A-HFS experiments without involving synaptic transmissions) have general implications, providing valuable insights for stimulation applications in other brain regions and peripheral nerves.

In addition, we utilized electrophysiological methods in our studies, using microelectrode array recordings to directly measure neuronal activity with high spatiotemporal resolution. Research on DBS mechanisms has primarily relied on neuronal signal recordings from both animal and human being studies. These recordings fall into two categories: near-field and far-field. This book focuses on near-field recordings, which includes invasive recording of extracellular action potentials from surrounding neurons, as well as local field potentials. In contrast, far-field recordings include electroencephalogram (EEG) from the scalp and electrocorticogram (ECoG) from beneath the skull or dura mater. They have lower spatiotemporal resolution than near-field recordings due to signal attenuations by brain tissue, dura mater, skull, and skin. They mainly capture low-frequency synaptic potential signals. Although normal neuronal action potentials cannot be detected through far-field recording, integrated action potentials can become visible when large groups of neurons fire together abnormally—for instance, the spike waveforms appeared in the EEG signals of epilepsy patients. Only near-field recordings can detect action potentials from individual neurons, though EEG offers the advantage of being non-invasive. Beyond these direct electrical measurements, imaging techniques such as functional magnetic resonance imaging (fMRI) and positron emission tomography (PET) can indirectly track neural activity through blood flow changes. In addition, biochemical methods can examine neurotransmitter changes to provide more insights into the DBS mechanisms (Udupa and Chen 2015).

9.3 Applications and Prospects of Brain Electrical Stimulations

9.3.1 Post-stimulation Effects and Closed-Loop Stimulations

Conventional clinical applications of neuromodulation technologies, including DBS, primarily rely on real-time stimulation actions to control disease symptoms—once stimulation stops, symptoms return (Temperli et al. 2003). Our research on brain stimulations discussed in Part II also mainly focuses on neuronal responses during HFS. However, studies have shown that neurons can experience a period of inhibition after the end of HFS at a frequency around 100 Hz—commonly used by DBS. For example, significant reductions in excitatory postsynaptic currents were observed in rat primary motor cortex and subthalamic nucleus (STN) after HFS ended (Iremonger et al. 2006; Shen and Johnson 2008). Similarly, HFS in the globus pallidus internal segment (GPi) produced reduced post-stimulation neuronal firing in its thalamic projection area (Muralidharan et al. 2017). In both human GPi and our rat hippocampal experiments (Chap. 5), neuronal firing ceased immediately for several seconds after HFS ended (Lafreniere-Roula et al. 2010; Feng et al. 2017;

Wang et al. 2018). These findings suggest that post-stimulation neuronal inhibition can occur in many stimulated brain regions. Importantly, clinical DBS studies have shown that Parkinson's patients experienced better outcome when they had longer periods of neuronal firing silence after 100 Hz STN stimulations (Milosevic et al. 2018), highlighting the clinical relevance of post-stimulation inhibitions.

Multiple mechanisms may cause reduced or absent neuronal firing after the end of HFS. In downstream projection neurons, increased inhibitions from GABAergic synapses may suppress neuronal firing when STN and GPi stimulations end (Chiken and Nambu 2013; Milosevic et al. 2018). For directly stimulated neurons, decreased excitability and elevated firing threshold can result in a silent period after HFS (Beurrier et al. 2001)—a finding supported by our experimental results described in Chap. 5. The antidromic conduction of HFS-induced axonal activity can affect the upstream somata (Sects. 5.2.3 and 5.2.4). This, combined with reduced excitatory inputs due to axonal block and synaptic failures, can produce an inactivity period in downstream neurons after the end of stimulation (Figs. 5.13 and 5.19). Furthermore, our experiments with both A-HFS and O-HFS in the rat hippocampal region showed clear differences in neuronal firing during and after HFS. While the stimulation enhanced neuronal firing during HFS, the firing decreased or completely stopped immediately after the HFS ended. This reflects a transition period of neuronal firing back to spontaneous activity after the removal of HFS modulation. Nevertheless, these post-stimulation inhibitory periods are very brief compared to the stimulation periods.

Conventional DBS, using continuous HFS lasting minutes to days or longer, can safely treat brain diseases. However, brief HFS lasting seconds or less, despite using identical pulses, can create animal epilepsy models through a "kindling" effect (Musto et al. 2009; Lothman and Williamson 1993). These brief HFS trains can cause irreversible long-term effects, including synaptic changes through plasticity mechanisms (Malenka 1994; Martin et al. 2000). In contrast, prolonged continuous HFS makes DBS reversible but restricts its therapeutic effects to the stimulation period. To achieve lasting effects after stimulation ends, or to have therapeutic benefits maintain between brief stimulations rather than only during them, alterative stimulation patterns are necessary. Animal studies have shown that certain patterns, such as burst stimulations, can produce therapeutic effects lasting several hours after stimulation (Spix et al. 2021). Our findings of different neuronal responses during initial and steady HFS periods have also revealed the distinctions between brief and prolonged HFS. Additionally, our studies on time-varying HFS and closed-loop HFS have provided new insights for developing stimulation patterns (detailed in Chapts. 6–8). Research into post-stimulation effects is one of important directions in DBS development—it can reduce stimulation duration, conserve electric energy, minimize treatment risks, and promote designs of sophisticated closed-loop stimulation paradigms.

Closed-loop DBS (also known as adaptive brain stimulation) has advantages including improved efficiency and efficacy, reduced side effects, and lower energy consumption. Although it has been used to treat epilepsy, open-loop stimulation remains the standard for other conditions like Parkinson's disease. The implementation of closed-loop DBS requires disease-specific biomarkers for feedback signals. Currently, two main types of neuro-electrical signals serve this purpose: rhythmic waves in local field potential (LFP) and action potentials from individual neurons. LFP, which contains lower frequency signals, represents the integrated electrical activity of neuronal populations measured locally. For example, β rhythmic waves (12–30 Hz) and γ rhythmic waves (60–90 Hz) in subthalamic nucleus LFP have been shown to correlate with motor dysfunction and bradykinesia in Parkinson's disease, showing promise as biomarkers for closed-loop DBS treatment (Bouthour et al. 2019; Little and Brown 2020). LFP biomarkers have also shown significant potential for adaptive stimulation in treating psychiatric disorders, such as major depression and obsessive–compulsive disorder, thereby advancing DBS applications in psychiatric treatment (Provenza et al. 2021; Sullivan et al. 2021). In refractory epilepsy treatment with adaptive stimulations, determining reliable feedback biomarkers remains challenging and requires more research to establish accurate epilepsy prediction signals (Ryvlin et al. 2021). The population spike (PS) recorded from the hippocampal soma layer, as introduced in this book, can serve as a biomarker for epileptiform activity to control closed-loop stimulations (refer to Sect. 8.4).

Additionally, when studying axonal stimulation—such as applying HFS to the CA1 alveus formed by pyramidal neuron axons—the soma ensemble can serve as an effective detector for recording APSs to reflect the axonal responses. The somatic membrane has a much larger surface area than thin axons, creating a stronger extracellular electric field during action potentials for accurate measurements from both population and single neurons (see Sect. 5.2). In the stratum oriens (basal dendritic layer) of hippocampal CA1 region, a composite action potential (CAP) can also be measured at the alveus axons near the soma layer to study axonal activity (Jensen and Durand 2009). However, CAPs have only sub-millivolt amplitudes—much smaller than somatic APSs—and can be influenced by somatic signals. Nevertheless, in spinal cord electrical stimulation, real-time recording of CAPs, also known as evoked compound action potential (ECAP), can achieve closed-loop spinal nerve stimulation (Mekhail et al. 2020; 2022). ECAP represents the extracellular potential created by the current leaking from the Ranvier nodes near the recording site when stimulation-evoked action potentials travel along axonal bundles.

In addition to electrophysiological biomarkers, biochemical substances can also serve as biomarkers for closed-loop DBS. When DBS alters neuronal firing and corrects pathological discharges, the resulting neural signals can travel to axon terminals and modify neurotransmitter releases. Real-time measurement of neurotransmitter concentrations, through electrochemical detection methods like voltammetry, can provide feedback signals to regulate stimulators, enabling closed-loop stimulations (Rojas Cabrera et al. 2020).

9.3.2 Relations Between Deep Brain Stimulation and Brain-Computer Interface

DBS uses electrodes implanted in the brain to connect stimulation devices with brain tissue, forming a neural interface that belongs to the broad category of brain-computer interface (BCI; also known as brain-machine interface, BMI). Open-loop DBS is a unidirectional interface, while closed-loop DBS is bidirectional, involving both the delivery of stimulation signals to the brain and the collection of neural feedback signals (Bouthour et al. 2019). DBS and BCI developed for different purposes. Medical experts and neurologists pioneered DBS as a treatment for brain diseases by applying electrical signals to modulate neural activity. Meanwhile, scientists and engineers with expertise in computational algorithms created BCI for robotics and neural rehabilitation, primarily aiming to read brain signals for controlling machines and computers. However, as both fields progress, they increasingly converge.

In its narrow sense, BCI conventionally refers to controlling machines, equipment, prostheses, and robots by decoding neuronal signals collected through intracranial electrodes (Chaudhary et al. 2016; Ajiboye et al. 2017; Willett et al. 2023). Alternatively, non-invasive BCI has also been used by collecting scalp EEG signals (Orban et al. 2022). While BCI technology primarily aims to assist disabled people, it also has a broader goal for advancing human intelligence through brain reading and writing (Yuste et al. 2017). In brain reading, a significant milestone occurred in 2006 when researchers first achieved mind-controlled machines by recording and decoding neural signals from the human brain (Hochberg et al. 2006). For brain writing, DBS represents a prototype by delivering electrical stimulations through implanted electrodes. Recent studies have shown that real-time closed-loop DBS applying to the human prefrontal cortex during sleep—synchronizing with endogenous slow-wave activity—can enhance memory consolidation (Geva-Sagiv et al. 2023). Beyond treating neurological diseases, DBS also aims to enhance brain function and human intelligence, aligning with BCI broader goals. Although BCI initially focused on "brain reading" and DBS on "brain writing", both technologies are now evolving toward bidirectional capabilities and to a same destination through different paths.

References

Ajiboye AB, Willett FR, Young DR et al (2017) Restoration of reaching and grasping movements through brain-controlled muscle stimulation in a person with tetraplegia: a proof-of-concept demonstration. Lancet 389(10081):1821–1830

Alagapan S, Choi KS, Heisig S et al (2023) Cingulate dynamics track depression recovery with deep brain stimulation. Nature 622(7981):130–138

Anderson ME, Postupna N, Ruffo M (2003) Effects of high-frequency stimulation in the internal globus pallidus on the activity of thalamic neurons in the awake monkey. J Neurophysiol 89(2):1150–1160

Anderson TR, Hu B, Iremonger K et al (2006) Selective attenuation of afferent synaptic transmission as a mechanism of thalamic deep brain stimulation-induced tremor arrest. J Neurosci 26(3):841–850

Arle JE, Mei LZ, Carlson KW et al (2016) High-frequency stimulation of dorsal column axons: potential underlying mechanism of paresthesia-free neuropathic pain relief. Neuromodulation 19(4):385–397

Barow E, Neumann WJ, Brücke C et al (2014) Deep brain stimulation suppresses pallidal low frequency activity in patients with phasic dystonic movements. Brain 137(Pt 11):3012–3024

Bellinger SC, Miyazawa G, Steinmetz PN (2008) Submyelin potassium accumulation may functionally block subsets of local axons during deep brain stimulation: a modeling study. J Neural Eng 5(3):263–274

Benabid AL, Pollak P, Louveau A et al (1987) Combined (thalamotomy and stimulation) stereotactic surgery of the VIM thalamic nucleus for bilateral Parkinson disease. Appl Neurophysiol 50(1–6):344–346

Beurrier C, Bioulac B, Audin J et al (2001) High-frequency stimulation produces a transient blockade of voltage-gated currents in subthalamic neurons. J Neurophysiol 85(4):1351–1356

Boraud T, Bezard E, Bioulac B et al (1996) High frequency stimulation of the internal Globus Pallidus (GPi) simultaneously improves Parkinsonian symptoms and reduces the firing frequency of GPi neurons in the MPTP-treated monkey. Neurosci Lett 215(1):17–20

Bouthour W, Megevand P, Donoghue J et al (2019) Biomarkers for closed-loop deep brain stimulation in Parkinson disease and beyond. Nat Rev Neurol 15(6):343–352

Burbaud P, Gross C, Bioulac B (1994) Effect of subthalamic high frequency stimulation on substantia nigra pars reticulata and globus pallidus neurons in normal rats. J Physiol Paris 88(6):359–361

Buzsáki G (2006) Rhythms of the brain. Oxford University Press, Oxford

Chan HH, Hogue O, Mathews ND et al (2022) Deep cerebellar stimulation enhances cognitive recovery after prefrontal traumatic brain injury in rodent. Exp Neurol 355:114136

Chaudhary U, Birbaumer N, Ramos-Murguialday A (2016) Brain-computer interfaces for communication and rehabilitation. Nat Rev Neurol 12(9):513–525

Chéreau R, Saraceno GE, Angibaud J et al (2017) Superresolution imaging reveals activity-dependent plasticity of axon morphology linked to changes in action potential conduction velocity. Proc Natl Acad Sci USA 114(6):1401–1406

Chiken S, Nambu A (2013) High-frequency pallidal stimulation disrupts information flow through the pallidum by GABAergic inhibition. J Neurosci 33(6):2268–2280

Chiken S, Nambu A (2014) Disrupting neuronal transmission: mechanism of DBS? Front Syst Neurosci 8:33

Chomiak T, Hu B (2007) Axonal and somatic filtering of antidromically evoked cortical excitation by simulated deep brain stimulation in rat brain. J Physiol 579(Pt 2):403–412

Cleary DR, Raslan AM, Rubin JE et al (2013) Deep brain stimulation entrains local neuronal firing in human globus pallidus internus. J Neurophysiol 109(4):978–987

Dang Y, Wang Y, Xia X et al (2023) Deep brain stimulation improves electroencephalogram functional connectivity of patients with minimally conscious state. CNS Neurosci Ther 29(1):344–353

Deniau JM, Degos B, Bosch C et al (2010) Deep brain stimulation mechanisms: beyond the concept of local functional inhibition. Eur J Neurosci 32(7):1080–1091

Dostrovsky JO, Levy R, Wu J et al (2000) Microstimulation-induced inhibition of neuronal firing in human globus pallidus. J Neurophysiol 84(1):570–574

Faisal AA, Selen LP, Wolpert DM (2008) Noise in the nervous system. Nat Rev Neurosci 9(4):292–303

Farokhniaee A, McIntyre CC (2019) Theoretical principles of deep brain stimulation induced synaptic suppression. Brain Stimul 12(6):1402–1409

Feng Z, Guo Z, Wang Z (2018) Advances in understanding the mechanisms of deep brain stimulation. Prog Biochem Biophys 45(12):1197–1203 (in Chinese)

Feng Z, Wang Z, Guo Z et al (2017) High frequency stimulation of afferent fibers generates asynchronous firing in the downstream neurons in hippocampus through partial block of axonal conduction. Brain Res 1661:67–78

Fields RD (2008) White matter matters. Sci Am 298(3):42–49

Florence G, Sameshima K, Fonoff ET et al (2016) Deep brain stimulation: more complex than the inhibition of cells and excitation of fibers. Neuroscientist 22(4):332–345

Gay RD, Enke YL, Kirk JR et al (2022) Therapeutics for hearing preservation and improvement of patient outcomes in cochlear implantation—progress and possibilities. Hear Res 426:108637

Geller EB, Skarpaas TL, Gross RE et al (2017) Brain-responsive neurostimulation in patients with medically intractable mesial temporal lobe epilepsy. Epilepsia 58(6):994–1004

Geva-Sagiv M, Mankin EA, Eliashiv D et al (2023) Augmenting hippocampal-prefrontal neuronal synchrony during sleep enhances memory consolidation in humans. Nat Neurosci 26(6):1100–1110

Girgis F, Miller JP (2016) White matter stimulation for the treatment of epilepsy. Seizure 37:28–31

Guo Z, Feng Z, Wang Y et al (2018) Simulation study of intermittent axonal block and desynchronization effect induced by high-frequency stimulation of electrical pulses. Front Neurosci 12:858

Gurbani S, Chayasirisobhon S, Cahan L et al (2016) Neuromodulation therapy with vagus nerve stimulation for intractable epilepsy: a 2-year efficacy analysis study in patients under 12 years of age. Epilepsy Res Treat 2016:9709056

Gwinn RP, Spencer DD (2004) Fighting fire with fire: brain stimulation for the treatment of epilepsy. Clin Neurosci Res 4(1–2):95–105

Hashimoto T, Elder CM, Okun MS et al (2003) Stimulation of the subthalamic nucleus changes the firing pattern of pallidal neurons. J Neurosci 23(5):1916–1923

Herrington TM, Cheng JJ, Eskandar EN (2016) Mechanisms of deep brain stimulation. J Neurophysiol 115(1):19–38

Hochberg LR, Serruya MD, Friehs GM et al (2006) Neuronal ensemble control of prosthetic devices by a human with tetraplegia. Nature 442(7099):164–171

Howell B, Choi KS, Gunalan K et al (2019) Quantifying the axonal pathways directly stimulated in therapeutic subcallosal cingulate deep brain stimulation. Hum Brain Mapp 40(3):889–903

Hull T, Giese C, Wexner SD et al (2013) Long-term durability of sacral nerve stimulation therapy for chronic fecal incontinence. Dis Colon Rectum 56(2):234–245

Iremonger KJ, Anderson TR, Hu B et al (2006) Cellular mechanisms preventing sustained activation of cortex during subcortical high-frequency stimulation. J Neurophysiol 96(2):613–621

Jakobs M, Fomenko A, Lozano AM et al (2019) Cellular, molecular, and clinical mechanisms of action of deep brain stimulation: a systematic review on established indications and outlook on future developments. EMBO Mol Med 11(4):e9575

Jensen AL, Durand DM (2009) High frequency stimulation can block axonal conduction. Exp Neurol 220(1):57–70

Kaaki B, Gupta D (2020) Medium-term outcomes of sacral neuromodulation in patients with refractory overactive bladder: a retrospective single-institution study. PLoS ONE 15(7):e0235961

Kent AR, Swan BD, Brocker DT et al (2015) Measurement of evoked potentials during thalamic deep brain stimulation. Brain Stimul 8(1):42–56

Krack P, Batir A, Van Blercom N et al (2003) Five-year follow-up of bilateral stimulation of the subthalamic nucleus in advanced Parkinson's disease. N Engl J Med 349(20):1925–1934

Krauss JK, Lipsman N, Aziz T et al (2021) Technology of deep brain stimulation: current status and future directions. Nat Rev Neurol 17(2):75–87

Lafreniere-Roula M, Kim E, Hutchison WD et al (2010) High-frequency microstimulation in human globus pallidus and substantia nigra. Exp Brain Res 205(2):251–261

Lam CM, Latif U, Sack A et al (2023) Advances in spinal cord stimulation. Bioengineering 10(2):185

Lee KH, Hitti FL, Chang SY et al (2011) High frequency stimulation abolishes thalamic network oscillations: an electrophysiological and computational analysis. J Neural Eng 8(4):046001

Li MCH, Cook MJ (2018) Deep brain stimulation for drug-resistant epilepsy. Epilepsia 59(2):273–290

Little S, Brown P (2020) Debugging adaptive deep brain stimulation for Parkinson's disease. Mov Disord 35(4):555–561

Liu HL, Roppolo JR, de Groat WC et al (2009) The role of slow potassium current in nerve conduction block induced by high-frequency biphasic electrical current. IEEE Trans Biomed Eng 56(1):137–146

Lothman EW, Williamson JM (1993) Rapid kindling with recurrent hippocampal seizures. Epilepsy Res 14(3):209–220

Lowet E, Kondabolu K, Zhou S et al (2022) Deep brain stimulation creates informational lesion through membrane depolarization in mouse hippocampus. Nat Commun 13(1):7709

Malenka RC (1994) Synaptic plasticity in the hippocampus: LTP and LTD. Cell 78(4):535–538

Martin SJ, Grimwood PD, Morris RG (2000) Synaptic plasticity and memory: an evaluation of the hypothesis. Annu Rev Neurosci 23(1):649–711

Mashaqi S, Patel SI, Combs D et al (2021) The hypoglossal nerve stimulation as a novel therapy for treating obstructive sleep apnea—a literature review. Int J Environ Res Public Health 18(4):1642

McIntyre CC, Grill WM, Sherman DL et al (2004) Cellular effects of deep brain stimulation: model-based analysis of activation and inhibition. J Neurophysiol 91(4):1457–1469

Mekhail N, Levy RM, Deer TR et al (2020) Long-term safety and efficacy of closed-loop spinal cord stimulation to treat chronic back and leg pain (Evoke): a double-blind, randomised, controlled trial. Lancet Neurol 19(2):123–134

Mekhail N, Levy RM, Deer TR et al (2022) Durability of clinical and quality-of-life outcomes of closed-loop spinal cord stimulation for chronic back and leg pain: a secondary analysis of the evoke randomized clinical trial. JAMA Neurol 79(3):251–260

Milosevic L, Kalia SK, Hodaie M et al (2018) Neuronal inhibition and synaptic plasticity of basal ganglia neurons in Parkinson's disease. Brain 141(1):177–190

Mino H, Grill WM Jr (2002) Effects of stochastic sodium channels on extracellular excitation of myelinated nerve fibers. IEEE Trans Biomed Eng 49(6):527–532

Miocinovic S, Somayajula S, Chitnis S et al (2013) History, applications, and mechanisms of deep brain stimulation. JAMA Neurol 70(2):163–171

Montgomery EB (2017) Deep brain stimulation programming: mechanisms, principles and practice (2nd ed). Oxford University Press

Muralidharan A, Zhang J, Ghosh D et al (2017) Modulation of neuronal activity in the motor thalamus during GPi-DBS in the MPTP nonhuman primate model of Parkinson's disease. Brain Stimul 10(1):126–138

Musto AE, Samii MS, Hayes JF (2009) Different phases of afterdischarge during rapid kindling procedure in mice. Epilepsy Res 85(2–3):199–205

Nowak LG, Bullier J (1998) Axons, but not cell bodies, are activated by electrical stimulation in cortical gray matter. Exp Brain Res 118(4):489–500

Orban M, Elsamanty M, Guo K et al (2022) A review of brain activity and EEG-based brain-computer interfaces for rehabilitation application. Bioengineering (Basel) 9(12):768

Pellot-Cestero JE, Herring EZ, Graczyk EL et al (2023) Implanted electrodes for functional electrical stimulation to restore upper and lower extremity function: history and future directions. Neurosurgery 93(5):965–970

Picton B, Wong J, Lopez AM et al (2024) Deep brain stimulation as an emerging therapy for cognitive decline in Alzheimer's disease: systematic review of evidence and current targets. World Neurosurg 184:253-266.e2

Provenza NR, Sheth SA, Dastin-van Rijn EM et al (2021) Long-term ecological assessment of intracranial electrophysiology synchronized to behavioral markers in obsessive-compulsive disorder. Nat Med 27(12):2154–2164

Rama S, Zbili M, Debanne D (2018) Signal propagation along the axon. Curr Opin Neurobiol 51:37–44

Ranck JB (1975) Which elements are excited in electrical stimulation of mammalian central nervous system: a review. Brain Res 98(3):417–440

Reese R, Leblois A, Steigerwald F et al (2011) Subthalamic deep brain stimulation increases pallidal firing rate and regularity. Exp Neurol 229(2):517–521

Rissardo JP, Vora NM, Tariq I et al (2023) Deep brain stimulation for the management of refractory neurological disorders: a comprehensive review. Medicina (Kaunas) 59(11):1991

Rojas Cabrera JM, Price JB, Rusheen AE et al (2020) Advances in neurochemical measurements: a review of biomarkers and devices for the development of closed-loop deep brain stimulation systems. Rev Anal Chem 39(1):188–199

Rosenbaum R, Zimnik A, Zheng F et al (2014) Axonal and synaptic failure suppress the transfer of firing rate oscillations, synchrony and information during high frequency deep brain stimulation. Neurobiol Dis 62:86–99

Ryvlin P, Rheims S, Hirsch LJ et al (2021) Neuromodulation in epilepsy: state-of-the-art approved therapies. Lancet Neurol 20(12):1038–1047

Shen K, Johnson SW (2008) Complex EPSCs evoked in substantia nigra reticulata neurons are disrupted by repetitive stimulation of the subthalamic nucleus. Synapse 62(4):237–242

Shivacharan RS, Rolle CE, Barbosa DAN et al (2022) Pilot study of responsive nucleus accumbens deep brain stimulation for loss-of-control eating. Nat Med 28(9):1791–1796

Silberstein SD, Yuan H, Najib U et al (2020) Non-invasive vagus nerve stimulation for primary headache: a clinical update. Cephalalgia 40(12):1370–1384

Spix TA, Nanivadekar S, Toong N et al (2021) Population-specific neuromodulation prolongs therapeutic benefits of deep brain stimulation. Science 374(6564):201–206

Sullivan CRP, Olsen S, Widge AS (2021) Deep brain stimulation for psychiatric disorders: from focal brain targets to cognitive networks. Neuroimage 225:117515

Temperli P, Ghika J, Villemure JG et al (2003) How do Parkinsonian signs return after discontinuation of subthalamic DBS? Neurology 60(1):78–81

Udupa K, Chen R (2015) The mechanisms of action of deep brain stimulation and ideas for the future development. Prog Neurobiol 133:27–49

Vetkas A, Fomenko A, Germann J et al (2022) Deep brain stimulation targets in epilepsy: systematic review and meta-analysis of anterior and centromedian thalamic nuclei and hippocampus. Epilepsia 63(3):513–524

Vitek JL (2002) Mechanisms of deep brain stimulation: Excitation or inhibition. Mov Disord 17(S3):S69–S72

Wang Z, Feng Z, Wei X (2018) Axonal stimulations with a higher frequency generate more randomness in neuronal firing rather than increase firing rates in rat hippocampus. Front Neurosci 12:783

Willett FR, Kunz EM, Fan C et al (2023) A high-performance speech neuroprosthesis. Nature 620(7976):1031–1036

Wu C, Sharan AD (2013) Neurostimulation for the treatment of epilepsy: a review of current surgical interventions. Neuromodulation 16(1):10–24

Yu Y, Feng Z, Cao J et al (2016) Modulation of local field potentials by high-frequency stimulation of afferent axons in the hippocampal CA1 region. J Integr Neurosci 15(1):1–17

Yuste R, Goering S, Arcas BAY et al (2017) Four ethical priorities for neurotechnologies and AI. Nature 551(7679):159–163

Zang Y, Marder E (2021) Interactions among diameter, myelination, and the Na/K pump affect axonal resilience to high-frequency spiking. Proc Natl Acad Sci USA 118(32):e2105795118

Zheng F, Lammert K, Nixdorf-Bergweiler BE et al (2011) Axonal failure during high frequency stimulation of rat subthalamic nucleus. J Physiol 589(Pt 11):2781–2793

Zheng L, Feng Z, Hu H et al (2020) The appearance order of varying intervals introduces extra modulation effects on neuronal firing through non-linear dynamics of sodium channels during high-frequency stimulations. Front Neurosci 14:397

Zheng L, Feng Z, Hu Y et al (2021) Adjust neuronal reactions to pulses of high-frequency stimulation with designed inter-pulse-intervals in rat hippocampus in vivo. Brain Sci 11(4):509

Appendix
MATLAB Program for the HH Model

This MATLAB program for solving the HH model is modified from the textbook "Application of Numerical Methods in Biomedical Engineering" (Dunn et al. 2009). It consists of four MATLAB files with ".m" extensions: two scripts and two functions. The first script, "constants_variables.m", calculates the rate constants, time constants and gating variables as functions of membrane voltage. The second script, "HH_simulation.m", solves the model equations. To run these scripts, simply type the script name after the MATLAB command prompt "≫". The scripts call two functions: "rate_constants.m" for calculating rate constants and "hodgkin_huxley_equations.m" for the model equations. These four MATLAB files are detailed below, with comment lines denoted by "%". Figure 1.11 displays the output of running "≫ constants_variables", while Fig. 1.12 displays the outputs of running "≫ HH_simulation" with different external current (Iapp) settings. The membrane potential curves in the left column of Fig. 1.12 are produced by adding the outputs of variable "v" to a resting potential of − 70 mV.

1. **Script "constants_variables.m"**

```
clc; clear all;
% Calculate rate constants for membrane potentials ranging from -100 to +100mV
v=[-100:1:100];
[alpha_n,beta_n,alpha_m,beta_m,alpha_h,beta_h]=rate_constants(v);

% Calculate time constants and gating variables at steady-state
tau_n=1./(alpha_n+beta_n);
n_ss=alpha_n.*tau_n;
tau_m=1./(alpha_m+beta_m);
m_ss=alpha_m.*tau_m;
tau_h=1./(alpha_h+beta_h);
h_ss=alpha_h.*tau_h;
```


Z. Feng, *Electrical Brain Stimulation: Mechanisms and Modulation of Neuronal Activity*, https://doi.org/10.1007/978-981-95-4145-4

```
% Plot the curves of rate constants against the membrane potential as showed in
Fig. 1.11A and B
clf; figure(1); plot(v, alpha_n,v, alpha_m,v, alpha_h)
axis([-100 100 0 10])
xlabel('Membrane potential (mV)'); ylabel('(1/ms)')
text(75,1.2,'\alpha _n'); text(35,3,'\alpha _m'); text(-75,4,'\alpha _h');
figure(2); plot(v, beta_n,v, beta_m,v, beta_h)
axis([-100 100 0 10])
xlabel('Membrane potential (mV)'); ylabel('(1/ms)')
text(-75,4,'\beta _n'); text(0,0.8,'\beta _m'); text(15,8,'\beta _h');

% Plot the curves of time constants against the membrane potential as showed in
Fig. 1.11C
figure(3); plot(v,tau_n,v,tau_m,v,tau_h)
axis([-100 100 0 10])
xlabel('Membrane potential (mV)'); ylabel('Time constant (ms)')
text(-75,4,'\tau _n'); text(0,0.8,'\tau _m'); text(15,8,'\tau _h');

% Plot the steady value curves of gating variables as showed in Fig. 1.11D
figure(4); plot(v,n_ss,v,m_ss,v,h_ss)
axis([-100 100 0 1])
xlabel('Membrane potential (mV)'); ylabel('Steady-state value')
text(-35,0.1,'n_\infty'); text(25,0.4,'m_\infty');
text(-20,0.8,'h_\infty');
```

2. **Script "HH_simulation.m"**

```
clc; clear all;
% Calculate the initial values for gating variables—the values at resting potential
v=0; [alpha_n,beta_n,alpha_m,beta_m,alpha_h,beta_h]=rate_constants(v);
tau_n=1./(alpha_n+beta_n);
n_ss=alpha_n.*tau_n;
tau_m=1./(alpha_m+beta_m);
m_ss=alpha_m.*tau_m;
tau_h=1./(alpha_h+beta_h);
h_ss=alpha_h.*tau_h;
fprintf('\n Initial values of gating variables, i.e, the value at rest potential:')
fprintf('\n n_ss= %5.4g \n m_ss= %5.4g \n h_ss= %5.4g ', n_ss,m_ss,h_ss)

% Using MATLAB function "ode45", an ordinary differential equation solver, to
solve the HH equations
```

```
% The array "yzero" contains four variables that define the initial values of membrane potential "v" (measured as difference from resting potential), along with gating variables m, n and h. Setting v=8 means an initial membrane depolarization of 8mV, mimicking an external stimulus. By adjusting v, you can simulate how the neuronal membrane responds to both subthreshold and super-threshold stimuli. The "tspan=[0,20]" sets the simulation duration from 0 to 20ms. To change the duration, simply modify the value "20".
yzero=[8,n_ss,m_ss,h_ss];
tspan=[0,20];
[t,y]=ode45(@hodgkin_huxley_equations,tspan,yzero);

% Calculate the conductance of K+ and Na+ channels. ggK and ggNa represent the maximum conductance of K+ and Na+ channels, respectively.
ggK=36; ggNa=120;
gK=ggK*y(:,2).^4; gNa=ggNa*y(:,3).^3.*y(:,4);

% Plot the simulation results, including the membrane potential (containing the induced action potential waveform), the gating variables and the conductance of K+ and Na+ ion channels.
clf; figure(1); plot(t,y(:,1),'k');
xlabel('Time (ms)'); ylabel('Membrane potential (mV)')
figure(2); plot(t,y(:,2:4));
xlabel('Time (ms)'); ylabel('Gate variables')
text(7,0.6,'\leftarrow n(t)'); text(4.5,0.9,'\leftarrow m(t)');
text(7,0.25,'\leftarrow h(t)')
figure(3); plot(t,gK,t,gNa);
xlabel('Time (ms)'); ylabel('Conductance')
text(7,6,'g _K'); text(3.6,25,'g _{Na}');
```

3. **Function "rate_constants.m"**

```
function [alpha_n,beta_n,alpha_m,beta_m,alpha_h,beta_h]=rate_constants(v)
% Calculate rate constants from the specified membrane potential "v".
alpha_n=0.01*(10-v)./(exp((10-v)/10)-1);
beta_n=0.125*exp(-v/80);
alpha_m=0.1*(25-v)./(exp((25-v)/10)-1);
beta_m=4*exp(-v/18);
alpha_h=0.07*exp(-v/20);
beta_h=1./(exp((30-v)/10)+1);
```

4. **Function "hodgkin_huxley_equations.m"**

```
function dy=hodgkin_huxley_equations(t,y)
```

```
% Set the values of constants, including the maximum conductance, the equilibrium membrane potentials, and the membrane capacitance. Iapp represents Im in Eq (1.55) and functions as an applied stimulus current, which can be configured as a step or square waveform by using an "if" statement. A positive Iapp produces membrane depolarization, while a negative value causes hyperpolarization. The simulation results shown in Fig. 1.12 are produced by adjusting Iapp. When applying Iapp alone, set the initial membrane potential v to 0 (in the array "yzero" in script "HH_simulation.m").
ggK=36; ggNa=120; ggL=0.3;
vK=-12; vNa=115; vL=10.6;
Cm=1;
% Configure Iapp to generate simulation results of a biphasic-pulse stimulation as shown in Fig. 1.12G (with a setting of yzero=[0,n_ss,m_ss,h_ss] in "HH_simulation. m").
Iapp=20;
if t>=0.75 Iapp=-20; end
if t>=1.5 Iapp=0; end

% Calculate rate constants
v=y(1); n=y(2); m=y(3); h=y(4);
[alpha_n,beta_n,alpha_m,beta_m,alpha_h,beta_h]=rate_constants(v);

% HH equations
dy=[(-ggK*n^4*(v-vK)-ggNa*m^3*h*(v-vNa)-ggL*(v-vL)+Iapp)/Cm
alpha_n*(1-n)-beta_n*n
alpha_m*(1-m)-beta_m*m
alpha_h*(1-h)-beta_h*h];
```

MIX
Papier aus verantwortungsvollen Quellen
Paper from responsible sources
FSC® C105338

If you have any concerns about our products, you can contact us on
ProductSafety@springernature.com

In case Publisher is established outside the EU, the EU authorized representative is:
Springer Nature Customer Service Center GmbH
Europaplatz 3, 69115 Heidelberg, Germany

Printed by Libri Plureos GmbH
in Hamburg, Germany